de Gruyter Studies in Mathematics

Herbert Amann

Ordinary Differential Equations

An Introduction to Nonlinear Analysis

Translated from the German by Gerhard Metzen

 Walter de Gruyter
Berlin · New York 1990

Author

Herbert Amann

Mathematisches Institut
Universität Zürich
CH-8001 Zürich, Switzerland

Translator

Gerhard Metzen

Department of Mathematical Sciences
Memphis State University
Memphis, Tennessee 38152, USA

Series Editors

Heinz Bauer
Mathematisches Institut
der Universität
Bismarckstrasse 1 ½
D-8520 Erlangen, FRG

Jerry L. Kazdan
Department of Mathematics
University of Pennsylvania
209 South 33rd Street
Philadelphia, PA 19104-6395, USA

Eduard Zehnder
ETH-Zentrum/Mathematik
Rämistrasse 101
CH-8092 Zürich
Switzerland

Title of the German original edition: *Gewöhnliche Differentialgleichungen.* ISBN 3-11-009573-4
Publisher: Walter de Gruyter & Co., Berlin · New York 1983

1980 Mathematics Subject Classification (1985 Revision): Primary: 34-01. Secondary: 54H20; 93D05.

⊚ Printed on acid-free paper which falls within the guidelines of the ANSI to ensure permanence and durability.

Library of Congress Cataloging-in-Publication Data

Amann, H. (Herbert), 1938—
   [Gewöhnliche Differentialgleichungen. English]
   Ordinary differential equations : an introduction to nonlinear
analysis / Herbert Amann ; translated from the German by
Gerhard Metzen.
    p. cm. — (De Gruyter studies in mathematics ; 13)
   Translation of: Gewöhnliche Differentialgleichungen.
   Includes bibliographical references.
   ISBN 0-89925-552-3 (alk. paper) :
   1. Differential equations. 2. Nonlinear functional analysis.
I. Title. II. Series.
QA372.A43 1990   89-25928
515′.352 — dc20   CIP

Deutsche Bibliothek Cataloging-in-Publication Data

Amann, Herbert:
Ordinary differential equations : an introduction to nonlinear
analysis / Herbert Amann. Transl. from the German by Gerhard
Metzen. — Berlin ; New York : de Gruyter, 1990
   (De Gruyter studies in mathematics ; 13)
   Einheitssacht.: Gewöhnliche Differentialgleichungen < engl. >
   ISBN 3-11-011515-8
NE: GT

Typesetting with TEX: Gerhard Metzen, Memphis, and Danny Lee Lewis, Berlin. Printing: Gerike
GmbH, Berlin. Binding: Dieter Mikolai, Berlin. Cover design: Rudolf Hübler, Berlin.

Preface to the English Edition

The present English translation differs from the German original only in that some proofs have been simplified and mistakes and misprints have been corrected. I am grateful to my students and many colleagues for pointing out these inaccuracies.

I should like to thank Professor G. Metzen for his efforts in translating this book and for his critical comments. My special thanks go to Walter de Gruyter Publishers for making this translation possible. Last but not least I thank my 'comma sniffer' for the painstaking corrections of the galley proofs.

Zürich, March 1990   *Herbert Amann*

> "Among all the disciplines of mathematics, the *theory of differential equations* is the most important one. All areas of physics pose problems which lead to the integration of differential equations. In fact, it is the theory of differential equations which shows the way to understanding all time-dependent natural phenomena. If, on the one hand, the theory of differential equations has extreme *practical* significance, then, on the other hand, it attains a corresponding *theoretical* importance because it leads in a rational way to the study of new functions or classes of functions."
>
> *Sophus Lie (1894)*

Preface

The present book offers an introduction into the theory of ordinary differential equations. It will be attempted to give the reader some insight into the wider connections into which this theory is embedded. I have not shied away from going further back sometimes and treat problems which usually are not found in text books on ordinary differential equations. So, for instance, the foundations of the calculus of variations are examined and an essentially self-contained treatment of the Brouwer degree, as well as a proof of Borsuk's theorem, are presented. In connection with the Poincaré-Bendixson theory, the m-dimensional winding number of a vector field will be introduced and its relation to the Brouwer degree will be demonstrated.

The theory of ordinary differential equations is a central branch of analysis, in fact, of mathematics as a whole. There exist numerous links to other parts of mathematics and to many other scientific disciplines. Large parts of modern linear and nonlinear functional analysis were developed to treat the multitude of questions from the area of differential equations. This is true, in the first place, for the theory of partial differential equations. However, also for ordinary differential equations, it is of great benefit to assume the somewhat more abstract point of view of functional analysis. Through this the whole theory gains precision and transparency.

While writing this book – which originated from lectures given at the universities of Bochum, Kiel and Zürich (partly under different titles) – my view has always been aimed at the theory of evolution equations. For this reason I have developed the theory, in some places, in a slightly more general setting –

for instance, the fundamentals of semiflows are treated in metric spaces – than is necessary and usual for ordinary differential equations. Characteristically, ordinary differential equations can be treated with much less technical effort than can partial differential equations, which clearly brings to light the intuitive, geometric framework. This makes it possible to introduce students, in a simple and natural manner, to general principles which form the basis for large parts of the theory of evolution equations.

By necessity, the selection of the material is, of course, subjective. I have made every attempt, however, to introduce the most important methods and proof techniques in the general theory of (initial value problems for) ordinary differential equations. Two exceptions concern the well-developed and penetrating stability theory of Hamiltonian systems, as well as the general theory of structural stability, that is to say, the more difficult topological dynamics. There exist excellent treatments of both areas by more competent authors.

Boundary value problems are not treated since, in my opinion, they more properly belong to the functional analytic approach of elliptic boundary value problems. Next to giving an introduction into the important dynamic theory of ordinary differential equations, my goal is to prepare the student for the study of evolution equations in infinite dimensional spaces.

The present book also represents an *introduction to nonlinear functional analysis*. Even though the problems throughout are finite dimensional, one is naturally led to the methods of nonlinear functional analysis in connection with questions from the area of ordinary differential equations. For instance, the existence problem of periodic solutions to nonautonomous equations leads to fixed point problems, which can be tackled by the techniques of degree theory. Likewise, studying the behavior of solutions, as a parameter is changed continuously, gives cause for the analysis of bifurcation problems. Their importance extends far beyond the theory of ordinary differential equations.

The functional analytic machinery will first be introduced in simple, finite dimensional situations and will only be put to use where it can be employed with benefit. However, the proofs are constructed – if this is possible without too much additional expenditure – so that they remain valid in the infinite dimensional case. The single exception here is the theory of linear differential equations, for which I have employed the methods of linear algebra. Since, in the case of partial differential equations mainly unbounded operators appear, it seems to me to be sensible to call on the functional analytic spectral theory only when it is actually needed.

This book is addressed to students in their second year of studies. Next to the common background in linear algebra I assume the student to be familiar with calculus of several variables, in particular, with the coordinate-free approach as expounded in most of the newer text books. Occasionally I have made use of alternating differential forms. However, these few sections are not necessary for the understanding of the remaining parts of the book.

Previous experience with differential equations or functional analysis is not required. I tried hard, on the one hand, to keep the prerequisites to a minimum for the purpose of addressing a wider audience, and, on the other, to motivate the reader to become thoroughly familiar with the functional analytic point of view in order to also study more complex problems – as, for example, partial differential equations – with benefit.

The thoroughly logical construction of the theory begins with chapter two. The first chapter plays a special role. The function of this chapter is largely motivational, and it serves to show "where differential equations come from" and what some of the typical questions are that are analyzed in this discipline. Here the discussion consists mainly of plausibility arguments and the prerequisites are varied. At some places it is expected that the reader has a certain mathematical maturity, because some concepts and techniques are used without prior introduction. The subsequent chapters, however, are not affected by this.

The easily understandable examples and model problems, discussed on a heuristic basis in the first chapter – in particular in the first section – will again be taken up in subsequent chapters, explicitly or implicitly, to test the available theory. The presentation of diffusion problems merely serves to furnish the reader with a view beyond the narrow confines of ordinary differential equations.

Special thanks are due to Mrs. S. Brawer for the meticulous typing of the manuscript and the patient implementation of the many changes, from the first draft to the final manuscript. I would also like to thank the publisher for the pleasant cooperation during the completion of this book.

Zürich, June 1983   *Herbert Amann*

Contents

Notation

$\mathbb{N}$ $:= \{0, 1, 2, \ldots\}$ is the set of all natural numbers.

$\mathbb{N}^*$ $:= \mathbb{N} \setminus \{0\}$.

$\mathbb{R}_+$ is the set of all nonnegative real numbers.

$\mathbb{K}$ $:= \mathbb{R}$ or $\mathbb{C}$.

$|\cdot|$ denotes, in general, the Euclidean norm in $\mathbb{K}^m$, but it can also denote the norm in an arbitrary normed vector space (NVS) (depending on the context).

$\mathbb{B}(x, r)$ denotes the open ball with center x and radius r in a metric space, especially in a NVS. In the latter case we also write $x + r\mathbb{B}$.

$\mathbb{B}^m$ is the open unit ball in $\mathbb{R}^m$.

$\mathbb{S}^m$ is the m-dimensional unit sphere, i.e., $\mathbb{S}^m = \partial \mathbb{B}^{m+1}$.

$\mathcal{L}(E, F)$ is the NVS of all continuous linear maps $T : E \to F$, where E and F are NVS's. The norm in $\mathcal{L}(E, F)$ is defined by

$$\|T\| := \sup\{\|Tx\| \mid \|x\| \leq 1\}.$$

It is well-known that $\mathcal{L}(E, F)$ is a Banach space whenever F is a Banach space.

$\mathcal{L}(E)$ $:= \mathcal{L}(E, E)$.

$\mathcal{GL}(E)$ is the group of all invertible operators in $\mathcal{L}(E)$.

A^T denotes the transpose of the matrix A.

$\mathbb{M}^m(\mathbb{K})$ is the ring of all $m \times m$ matrices with elements in $\mathbb{K}$.

$(\cdot \mid \cdot)$ denotes the Euclidean (Hermitian) inner product in $\mathbb{K}^m$ or in an arbitrary inner product space.

$\langle \cdot, \cdot \rangle$ denotes the duality pairing between E', the dual of a NVS E, and E itself, i.e., $\langle x', x \rangle$ represents the value of $x' \in E'$ at the point $x \in E$.

2^X is the power set of the set X, i.e., the set of all subsets of X.

$\mathrm{span}(M)$ is the linear subspace spanned by the subset M of a vector space.

$B[x]^k$ is the abbreviation for $B[x, x, \ldots, x]$, where B denotes a k-linear map.

Chapter I
Introduction

The goal of this chapter is to display some typical problems from the theory of (ordinary) differential equations and to motivate the ensuing detailed investigations. For this reason, we will briefly go into some of the "modern" and very "classical" areas of application, those which are closely connected with the theory of differential equations and have always profoundly influenced its development, and still do, namely ecological questions (population models), the classical calculus of variations, classical mechanics and diffusion processes. We will try to point out deeper connections and, in particular, abstract functional-analytic generalizations which play an increasingly important role in the much more complex theory of partial differential equations.

In many cases, we will substitute strict proofs by plausibility arguments in the first four sections of this chapter. The mathematical basis, which serves to put these plausibility arguments on "solid ground," will, in fact, be worked out in succeeding chapters.

The last section of this chapter deals with some elementary integration techniques. The goal of this analysis is to study the behavior of some typical differential equations by means of explicit calculations. With this we want to point out those phenomena on simple, explicitly solvable differential equations, which will repeatedly occur later in much greater generality. Moreover, it will be shown that in most cases an explicit solution of a differential equation is not particularly useful, but that a powerful theory, one which permits the deduction of qualitative statements about existence and long time behavior, or facilitates general geometric insight, is much more valuable.

1. Ecological Models

(1.1) Population Models. We let $p(t)$ denote the *population* of a given species at time t (e.g., the earth population, the carps in a pond, or the atoms of some radioactive substance). Then $\dot{p}(t)/p(t)$ is the total *growth rate* (= change with respect to time, $\dot{p} = dp/dt$, related to the total population) at time t. The total growth rate will, in general, be a function of time and the population p itself, that is to say,

$$\frac{\dot{p}(t)}{p(t)} = r(t, p). \tag{1}$$

In a *closed system*, i.e., without migration, we have

$$r(t, p) = g(t, p) - s(t, p),$$

where

$$g(t, p) \text{ is the } birth\ rate$$

and

$$s(t, p) \text{ is the } death\ rate.$$

If g and s – or more generally, r – are known functions, the evolution of the population over time is described by $p = p(t)$, where p must satisfy differential equation (1), that is, it must satisfy the *growth equation*

$$\dot{p} = r(t, p)p. \tag{2}$$

We must therefore find a (continuously differentiable) function $p : I \to \mathbb{R}$ on some interval $I \subseteq \mathbb{R}$ – ideally on $\mathbb{R}$ or $[t_0, \infty)$ – which satisfies differential equation (2) and at time $t = t_0$ must take on the prescribed (known) value

$$p(t_0) = p_0. \tag{3}$$

Such a problem is called an *initial value problem* (IVP) and is denoted by

$$\dot{p} = r(t, p)p, \quad p(t_0) = p_0. \tag{4}$$

This means, however, that every *solution* $p : I \to \mathbb{R}$ must satisfy the equation

$$\dot{p}(t) = r(t, p(t))p(t), \quad \forall t \in I,$$

and the *initial condition* (3).

(a) *Constant Growth Rate.* The simplest special case is obtained if we set

$$r(t, p) = \alpha, \quad \forall (t, p) \in \mathbb{R} \times \mathbb{R},$$

where $\alpha \in \mathbb{R}$. If $p \neq 0$, the following holds

$$\dot{p} = \alpha p \iff \frac{\dot{p}}{p} = \alpha \iff \frac{d}{dt} \ln|p(t)| = \alpha$$

$$\iff \ln|p(t)| = \alpha t + c \iff |p(t)| = e^{\alpha t + c} = c_1 e^{\alpha t}$$

for some constant $c \in \mathbb{R}$ and $c_1 := e^c$. Since, by the last equation, p can never change its sign, and since we must have $p(t_0) = p_0$, it finally follows that the solution of the IVP is given by the formula

$$p(t) = p_0 e^{\alpha(t - t_0)}, \quad \forall t \in \mathbb{R}. \tag{5}$$

In particular, we see that – for every $p_0 \in \mathbb{R}$ – the IVP

$$\dot{p} = \alpha p, \quad p(t_0) = p_0,$$

has a unique solution given by the function in (5).

We also read off from (5) that $p(t) \to \infty$ as $t \to \infty$, whenever $\alpha > 0$ ("unlimited growth"), while $\lim_{t \to \infty} p(t) = 0$ if $\alpha < 0$ ("extinction").

(b) *The Logistic Equation.* Assuming a constant growth rate is in most cases very unrealistic. For example, in general one does not observe unlimited growth. We therefore assume now that there exists a "limiting population" $\xi > 0$ such that when p exceeds the value ξ, the growth rate becomes negative, that is, we assume that

$$r(t, p) \leq 0 \quad \text{if} \quad p \geq \xi.$$

A particularly simple situation occurs when r is a linear function of p:

$$r(t, p) = \beta(\xi - p), \quad \forall p \in \mathbb{R},$$

where $\beta, \xi > 0$. With these assumptions the growth equation takes on the special form

$$\dot{p} = \alpha p - \beta p^2 = (\alpha - \beta p)p, \quad \alpha := \beta \xi. \tag{6}$$

This is the so-called *equation of limited growth* or the *logistic equation.*

In order to solve, or "*integrate,*" equation (6), we proceed as in case (a), that is, we *separate the variables* (namely, p and t, where the latter does not appear explicitly). We write (6) in the form

$$\frac{\dot{p}}{(\alpha - \beta p)p} = 1, \tag{7}$$

where we of course exclude the cases $p = 0$ and $p = \alpha/\beta = \xi$. If F denotes an antiderivative of $1/(\alpha - \beta p)p$, i.e., if

$$F'(p) = 1/(\alpha - \beta p)p,$$

then (7) is evidently equivalent to

$$\frac{d}{dt}(F(p(t))) = 1,$$

and hence to

$$F(p(t)) = t + c \tag{8}$$

for some constant $c \in \mathbb{R}$. Here we may choose an arbitrary antiderivative F, since we know that any two such functions differ by a constant, which can of course be absorbed in c.

Using partial fractions, we immediately obtain

$$F(p) = \int \frac{dp}{(\alpha - \beta p)p} = \frac{1}{\alpha} \int \frac{dp}{p} + \frac{\beta}{\alpha} \int \frac{dp}{\alpha - \beta p} = \ln \left| \frac{p}{\alpha - \beta p} \right|^{1/\alpha}.$$

Hence (8) is equivalent to

$$\left| \frac{p}{\alpha - \beta p} \right| = e^{\alpha(t+c)} = c_1 e^{\alpha t}, \tag{9}$$

where $c_1 := e^{\alpha c}$. For $t = t_0$ we must have $p(t) = p_0$, and so if we assume that $p_0 \neq 0$ and $p_0 \neq \xi = \alpha/\beta$, we must have

$$\left| \frac{p_0}{\alpha - \beta p_0} \right| = c_1 e^{\alpha t_0}.$$

This implies that

$$\left| \frac{p(t)}{p_0} \right| = \left| \frac{\alpha - \beta p(t)}{\alpha - \beta p_0} \right| e^{\alpha(t-t_0)}.$$

From (9) we read off that $p(t) \neq 0$ and $\alpha - \beta p(t) \neq 0$ for all $t \in \mathbb{R}$ if this happens to be true at time $t = t_0$. In particular, we have $(\alpha - \beta p(t))(\alpha - \beta p_0)^{-1} > 0$, and so it follows that

$$\frac{p(t)}{p_0} = \frac{\alpha - \beta p(t)}{\alpha - \beta p_0} e^{\alpha(t-t_0)},$$

that is,

$$p(t) = \frac{\alpha p_0}{\beta p_0 + (\alpha - \beta p_0) e^{-\alpha(t-t_0)}}, \qquad \forall t \in \mathbb{R}. \tag{10}$$

The derivation also shows that for $p_0 \neq 0$ and $p_0 \neq \xi = \alpha/\beta$, the function p, given by (10), is the unique solution of the IVP

$$\dot{p} = (\alpha - \beta p)p, \qquad p(t_0) = p_0.$$

From (10) we read off that

$$p(t) \nearrow \xi \quad \text{as} \quad t \to \infty, \quad \text{if} \quad 0 < p_0 < \xi$$

and

$$p(t) \searrow \xi \quad \text{as} \quad t \to \infty, \quad \text{if} \quad p_0 > \xi.$$

By differentiating the differential equation we get

$$\ddot{p} = (\dot{p})^{\cdot} = (\alpha - 2\beta p)(\alpha - \beta p)p.$$

Hence

$$\ddot{p} > 0, \quad \text{if} \quad p \in (0, \xi/2) \cup (\xi, \infty)$$

and

$$\ddot{p} < 0, \quad \text{if} \quad \xi/2 < p < \xi.$$

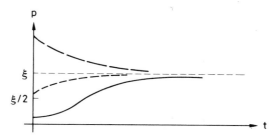

Figure 1: Graphs of solution (10) for distinct initial values

Therefore the graph of f has, for different values of p_0, the qualitative behavior described in Fig. 1.

In particular, if $0 < p_0 < \xi/2$, then increased growth occurs initially ($\ddot{p} > 0$) until the population reaches the value $\xi/2$.
From this time onward the growth slows down ($\ddot{p} < 0$) and, in any case, the population tends toward the limit ξ.

This behavior can also be derived from a direct analysis of the differential equation

$$\dot{p} = \alpha p - \beta p^2.$$

For small positive values of p, the term αp dominates and the equation closely approximates the differential equation with constant growth (case (a)). Hence for small values of p the solution is convex and increasing. For large values of p the term $-\beta p^2$ prevails and "curbs" the growth, which amounts to a slackening in rise.

The term βp^2 can be interpreted as a number which is proportional to the average number of "encounters" of p individuals. It is therefore possible to interpret $-\beta p^2$ as a "*social friction term*" which slows down the growth.

For more complicated situations it is of great importance that one can analyze all *qualitative features* of the equation $\dot{p} = \alpha p - \beta p^2$ without explicitly solving it. To do this, one interprets $p(t)$ as a point in the "*phase space*" $\mathbb{R}$, whose motion ("*flow*") is described by the differential equation $\dot{p} = \alpha p - \beta p^2$. Thus, if $t \mapsto p(t)$ is a solution of the differential equation, then $\dot{p}(t)$ represents the velocity at which the point is moving. If at every point $p \in \mathbb{R}$ we plot the value $f(p) := \alpha p - \beta p^2$ as a vector, we obtain the *direction field* of the differential equation $\dot{p} = f(p)$. This direction field indicates at which velocity and in which direction the point p moves. Stated differently: If $t \mapsto p(t)$ is a solution of the differential equation $\dot{p} = f(p)$, then $f(p)$ is the tangent vector to the path $t \mapsto p(t)$ at $p = p(t)$ in $\mathbb{R}$.

In our case, the points $p = 0$ and $p = \xi$ are *stationary points*, that is, if $p_0 = 0$ or $p_0 = \xi$, then $p(t) = p_0$ for all time. The direction field has the following qualitative features:

Figure 2: The direction field for $\dot{p} = \alpha p - \beta p^2$

From this we deduce (if one accepts that the equation $\dot{p} = f(p)$ in fact has solutions which are defined for all time) that every point $p_0 \in (0, \infty)$ approaches asymptotically (i.e., as $t \to \infty$) the stationary point ξ. The length of the arrows determines the "speed," while the "acceleration" $\ddot{p}$ is obtained – as above – by differentiating the differential equation.

(1.2) Predator-Prey Models. We now consider a slightly more complicated two-species model, where

$$x(t) \text{ denotes the prey population at time } t$$

and

$$y(t) \text{ denotes the predator population at time } t$$

(e.g., redear – bass). Now growth equation (2) applies to each population, where, of course, the growth rate of one population influences that of the other, that is,

$$\dot{x} = r_1(t, x, y)x, \quad \dot{y} = r_2(t, x, y)y. \tag{11}$$

We now have a *system of* two coupled *differential equations* (of "first order and in explicit form," i.e., only first order derivatives occur and they are expressed in terms of the other variables), the general *growth equation for two-species models*.

In our models we now assume that the prey is the only source of subsistence for the predator species, and that the prey has an unlimited food supply. As in the one-population model above, we will consider two cases; namely, the case of unlimited and the case of limited growth.

(a) *Predator-Prey Models with Constant Growth.* A simple expression for the growth rate of the prey has the form

$$r_1(t, x, y) = \alpha - \beta y, \quad \alpha, \beta > 0.$$

It allows the following interpretation: If no predator is present ($y = 0$), the prey evolves at a constant growth rate α. The presence of predators diminishes this growth rate by an amount which is proportional to the predator population. ("Many foxes devour many rabbits.")

We use a similar expression for the growth rate of the predator and set

$$r_2(t, x, y) = -\gamma + \delta x, \quad \gamma, \delta > 0,$$

that is, if no prey is available, the predator species becomes extinct at the constant rate γ. The presence of prey diminishes the death rate, in fact, it is proportional to the prey population.

Under these assumptions we obtain the special predator-prey equations (also known as *Volterra-Lotka equations*):

$$\begin{aligned} \dot{x} &= (\alpha - \beta y)x \\ \dot{y} &= (\delta x - \gamma)y \end{aligned} \qquad \alpha,\ \beta,\ \gamma,\ \delta > 0. \tag{12}$$

If we set

$$p(t) := (x(t), y(t)) \in \mathbb{R}^2$$

and

$$f : \mathbb{R}^2 \to \mathbb{R}^2, \quad (x, y) \mapsto ((\alpha - \beta y)x, (\delta x - \gamma)y),$$

this system can be written in the form

$$\dot{p} = f(p). \tag{13}$$

A solution of this differential equation is a path $p : I \subseteq \mathbb{R} \to \mathbb{R}^2$ such that its tangent vector $\dot{p}(t)$ at the point $p(t)$ is given by $f(p(t))$. By "attaching" the vector $f(p)$ at the point $p \in \mathbb{R}^2$, we obtain the *direction field* of differential equation (13). A solution of (13) then is a curve C in $\mathbb{R}^2$ (parametrized and oriented by $t \in I \subseteq \mathbb{R}$) such that at every point p on C, $f(p)$ is a tangent vector of C.

In the case of system (12), there are two *stationary points* (i.e., $p \in \mathbb{R}^2$ so that $f(p) = 0$), namely, $p_0 = (0, 0)$ and $p_1 := (\gamma/\delta, \alpha/\beta)$. The two coordinate axes and the two lines, $x = \gamma/\delta$ and $y = \alpha/\beta$, partition the plane into 9 subregions in which the tangent vectors $\dot{p} = (\dot{x}, \dot{y})$, of possible solution curves to (12), show the qualitative features as depicted in Figure 3.

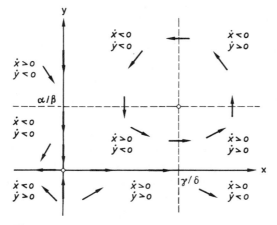

Figure 3: The direction field for system (12)

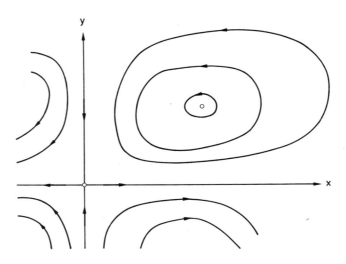

Figure 4: The phase portrait of system (12)

Later, we will see that a solution curve goes through every point in the plane and that the solution curves have the qualitative behavior as shown in Figure 4. From this *phase portrait* one reads off the "longtime behavior" of both populations, where, of course, in our case only the solution curves in the first quadrant are of interest, because negative populations do not exist. For every given initial state $(x_0, y_0) \in (0, \infty)^2$ satisfying $(x_0, y_0) \neq p_1$, the solution of the IVP

$$\dot{p} = f(p), \quad p(t_0) = (x_0, y_0),$$

lies on a closed curve around the stationary point p_1. Later, we will see that this means that $I = \mathbb{R}$, i.e., that the solution exists for all time and that it is periodic, that is, $p(t + \tau) = p(t)$ for some appropriate $\tau > 0$ and all $t \in \mathbb{R}$. This means that each population, $x(t)$ and $y(t)$, executes "oscillations" which, moreover, are "opposite" ("when the predator population increases, the prey population decreases, and conversely"). This characteristic, moreover, is *stable* with respect to small changes in the initial values (x_0, y_0).

If, on the other hand, $x_0 = 0$, that is, if initially there is no prey, then the predator species becomes extinct ($y(t) \rightarrow 0$ as $t \rightarrow \infty$). If no predators are present ($y_0 = 0$), the prey population grows indefinitely ($x(t) \rightarrow \infty$ as $t \rightarrow \infty$). This of course reflects the fact that in these cases the model turns into the one-species model dealt with in (1.1 a). In contrast to the case above, this characteristic behavior is *unstable*. A small change in initial values (e.g., a change from $x_0 = 0$ to $x_0 > 0$) results in a completely different longtime behavior.

(b) *Predator-Prey Models with Limited Growth.* As was done with the logistic equation of (1.1 b), we now modify the Volterra-Lotka system by "social friction terms," which, in particular, prevent the unlimited growth of the prey population in the absence of predators. We now consider the system

$$\dot{x} = (\alpha - \beta y)x - \lambda x^2$$
$$\dot{y} = (\delta x - \gamma)y - \mu y^2,$$

where α, β, γ, δ, λ, μ are positive constants, i.e., the system

$$\dot{x} = (\alpha - \beta y - \lambda x)x$$
$$\dot{y} = (\delta x - \gamma - \mu y)y. \tag{14}$$

In order to obtain the direction field, we observe that the vector field is parallel to the y-axis ($\dot{x} = 0$) along the straight line

$$L : \alpha - \beta y - \lambda x = 0$$

and that the vectors tangent to the solution curves are horizontal ($\dot{y} = 0$) along the straight line

$$M : \delta x - \gamma - \mu y = 0.$$

In Figure 5 we consider the case when the two lines, L and M, do not intersect in the first quadrant. System (14) has, in this case, the two stationary points,

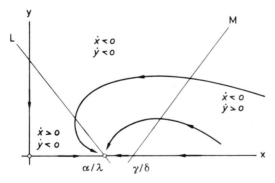

Figure 5: The direction field for (14) if L and M do not intersect in $\mathbb{R}_+^2$

$(0,0)$ and $(\alpha/\lambda, 0)$, in the region of interest $\mathbb{R}_+^2$. Later we will show that the solution curves exist for all time and that they have the qualitative features as indicated in Fig. 5. In particular, the predator species always becomes extinct and the system asymptotically approaches the equilibrium $(\alpha/\lambda, 0)$, which is stable and *"attracting"* if at the initial time t_0 a positive prey population x_0 is present. If $x_0 = 0$, the predator will die out, which again is an unstable case.

In the second case, both lines, L and M, intersect in the first quadrant. Then the point of intersection z is an additional stationary point and the direction field has the qualitative features indicated in Fig. 6. Later, we will see that there are basically two possibilities for the qualitative behavior of solution curves, where we will restrict ourselves to the interesting case $x_0 > 0, y_0 > 0$.

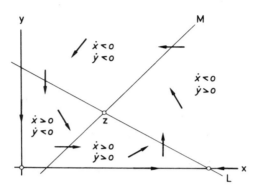

Figure 6: The direction field for (14) if L and M intersect in $\mathbb{R}^2_+$

In the first case z is a "*globally attracting*" stationary point. Every solution curve approaches the stationary point z as $t \to \infty$. In the case of Fig. 7, "the solutions spiral into the stationary point z as $t \to \infty$." In the long run, both

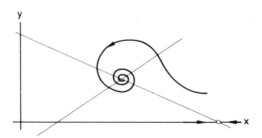

Figure 7: Case of a globally attracting stationary point

populations approach the "*equilibrium point*" $z = (z_1, z_2)$, in this case, as they execute opposing "damped oscillations" around z_1 and z_2, respectively. The approach can, however, also occur "*aperiodically*" (the case of a stable *node*). In the second case there exists a closed curve – a *limit cycle* – containing the stationary point z in its interior. Each solution curve starting in the interior winds infinitely often around this limit cycle and gets closer to it. The stationary point z is still a stable stationary point, but is no longer globally stable. Only those solution curves that start in the interior of the limit cycle can approach z in the long run. We therefore see that (once the assertions above have been proved) the predator-prey model, described by the differential equations (14), evolves in the long run either to an equilibrium state or a periodic population formation.

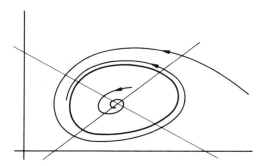

Figure 8: Case of a limit cycle

Remarks. (a) Here we neither go into the historical background of these models nor into the intricacies of interpretation. For the history (and anecdotes!) we refer to the book by M. Braun [1] and the literature cited therein.

Concerning the intricacies of these models, we only note that – despite their popularity in the "biomathematical literature" – these are only the simplest models which are capable of describing natural processes in a very confined space at best. Refined models must, in general, also take into account the spatial distribution of the species, as well as "migratory effects" which, for example, are of great importance for the spreading of epidemics. Problems such as these give rise to partial differential equations (in particular so-called *reaction-diffusion equations* (cf. section 4)). Their mathematical treatment is considerably more difficult and very little is known about them at the present time.

(b) All problems in this section are IVP's of the following type. A (smooth) function

$$f : \mathbb{R}^m \to \mathbb{R}^m$$

is given and we must find a (continuously differentiable) function $p : I \subseteq \mathbb{R} \to \mathbb{R}^m$ satisfying the differential equation

$$\dot{p}(t) = f(p(t)), \quad \forall t \in I,$$

and the initial condition

$$p(t_0) = p_0,$$

where I is some interval in $\mathbb{R}$ and $t_0 \in I$. In addition to the pure existence of solutions, the longtime behavior is of interest, particularly also in dependence on the initial values p_0 (the "stability problem").

Here we interpret f geometrically as a *vector field* on $\mathbb{R}^m$ ("at every point $p \in \mathbb{R}^m$ we attach the vector $f(p) \in \mathbb{R}^m$") and we seek curves C such that at every point $p \in C$, the vector $f(p)$ is a tangent vector to C. In general, f will be a vector field on a differentiable manifold M (i.e., f is a section of the tangent bundle $T(M)$) and we seek curves C in M such that at every point $p \in C$, the vector $f(p) \in T_p(M)$ is a tangent vector.

Problems

1. Verify the assertions in the text that the IVP's $\dot{p} = \alpha p$, $p(t_0) = p_0$, and $\dot{p} = \alpha p - \beta p^2$, $p(t_0) = p_0$, have unique solutions for every pair (t_0, p_0).

2. Let the one-parameter system

$$\dot{x} = 2x, \quad \dot{y} = \lambda y, \quad \lambda \in \mathbb{R},$$

be given. Determine all solutions and sketch the phase portraits for $\lambda = -1, 0, 1, 2$.

3. Determine all solutions of the system

$$\begin{bmatrix} \dot{x} \\ \dot{y} \end{bmatrix} = \begin{bmatrix} 5 & 3 \\ -6 & -4 \end{bmatrix} \begin{bmatrix} x \\ y \end{bmatrix}$$

and sketch the phase portrait. (*Hint*: Introduce new variables (ξ, η) by means of the transformation

$$\begin{bmatrix} \xi \\ \eta \end{bmatrix} = \begin{bmatrix} 2 & 1 \\ 1 & 1 \end{bmatrix} \begin{bmatrix} x \\ y \end{bmatrix}.\big)$$

Why have we introduced this transformation?

2. Variational Problems

(2.1) Geodesics. Let $M \subseteq \mathbb{R}^n$ be an m-dimensional C^2-manifold (e.g., the two-sphere $\mathbb{S}^2$ in $\mathbb{R}^3$) and let $A, B \in M$ denote two distinct points. We want to find a C^1-curve C in M which connects A and B and has minimal length $L(C)$.

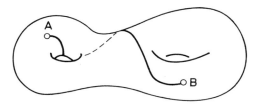

It is well-known that for every C^1-curve in $\mathbb{R}^n$ we have

$$L(C) = \int_\alpha^\beta |\dot{f}(t)|\,dt,$$

where $f : [\alpha, \beta] \subseteq \mathbb{R} \to \mathbb{R}^n$ is an arbitrary C^1–parametrization of C. For reasons of simplicity, we now assume that C is completely contained in some U, where (U, φ) is a *chart* on M. This means that $C \subseteq U$;

> U is open in M;
>
> $V := \varphi(U)$ is open in $\mathbb{R}^m$;
>
> $\varphi : U \to V$ is a homeomorphism;
>
> $g := \varphi^{-1} \in C^2(V, \mathbb{R}^n)$;
>
> $Dg(v) \in \mathcal{L}(\mathbb{R}^m, \mathbb{R}^n)$ is injective for all $v \in V$.

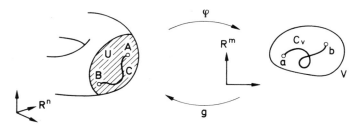

Then $v(t) := \varphi(f(t)), t \in [\alpha, \beta]$, defines a C^1-parametrized curve C_v in V connecting $a := \varphi(A)$ and $b := \varphi(B)$. Conversely, every C^1-curve connecting a and b in V defines a C^1-curve connecting A and B in U. Since $f(t) = g \circ \varphi(f(t)) = g \circ v(t)$, we have

$$\dot{f}(t) = Dg(v(t))\dot{v}(t),$$

and consequently

$$|\dot{f}(t)|^2 = (Dg(v(t))\dot{v}(t) \mid Dg(v(t))\dot{v}(t))_{\mathbb{R}^n}$$
$$= ([Dg(v(t))]^T Dg(v(t))\dot{v}(t) \mid \dot{v}(t))_{\mathbb{R}^m}.$$

Since Dg has the column representation $Dg = [D_1 g, \dots, D_m g]$, it follows that

$$[Dg]^T Dg = \begin{bmatrix} [D_1 g]^T \\ \vdots \\ [D_m g]^T \end{bmatrix} [D_1 g, \dots, D_m g] = [(D_i g \mid D_k g)]_{1 \le i, k \le m}.$$

Using the standard notation (of the Riemannian metric)

$$g_{ik}(v) := (D_i g(v) \mid D_k g(v)),$$

we thus obtain, in this special case, that

$$L(C) = \int_\alpha^\beta \sqrt{\sum_{i,k=1}^m g_{ik}(v(t)) \dot{v}^i(t) \dot{v}^k(t)} \ dt,$$

where $\dot{v} = dv/dt$ (is *not* the derivative with respect to arc length!) and $v = (v^1, \dots, v^m) \in V$.

Examples. (a) $M = \mathbb{S}^2$ in $\mathbb{R}^3$, *parametrized by spherical coordinates*: In this case we use the well-known chart which consists of

$$U := \mathbb{S}^2 \backslash (\mathbb{R}_+ \times \{0\} \times \mathbb{R})$$

and the parametrization

$$g : V := (0, 2\pi) \times \left(-\frac{\pi}{2}, \frac{\pi}{2}\right) \to \mathbb{R}^3, \quad (\varphi, \vartheta) \mapsto (\cos \varphi \cos \vartheta, \sin \varphi \cos \vartheta, \sin \vartheta)$$

(cf. Fig. 1).

In this case a simple calculation shows that

$$[g_{ik}(\varphi, \vartheta)] = \begin{bmatrix} \cos^2 \vartheta & 0 \\ 0 & 1 \end{bmatrix}.$$

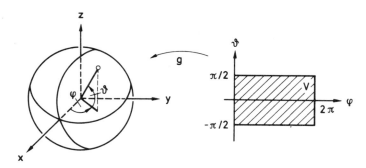

Figure 1: Parametrization of $\mathbb{S}^2$ by spherical coordinates

Hence in local spherical coordinates the length of the curve can be expressed as

$$L(C) = \int_\alpha^\beta \sqrt{\cos^2 \vartheta [\dot\varphi(t)]^2 + [\dot\vartheta(t)]^2}\ dt.$$

(b) In the trivial case when $M = \mathbb{R}^m$, we choose the natural chart $(\mathbb{R}^m, I_m)$, where $I_m := id_{\mathbb{R}^m}$ denotes the identity map $\mathbb{R}^m \to \mathbb{R}^m$. Then $g_{ik} = \delta_{ik}$, and thus we obtain the usual formula for the length of a curve in $\mathbb{R}^m$,

$$L(C) = \int_\alpha^\beta \sqrt{\sum_{i=1}^m [\dot v^i(t)]^2}\ dt = \int_\alpha^\beta |\dot v(t)|\, dt.$$

(c) If we consider a variational problem with *constraints*, by considering only those curves that lie entirely in U, we obtain the following problem

$$\int_\alpha^\beta \sqrt{\sum_{i,k=1}^m g_{ik}(v(t)) \dot v^i(t) \dot v^k(t)}\ dt \quad \Longrightarrow \quad \underset{\substack{v \in C^1([\alpha,\beta],V) \\ v(\alpha)=a,\ v(\beta)=b}}{\mathrm{Min}}.$$

If v is a solution of this problem, then $f := g \circ v$ is a solution of the extremal problem with constraints, a so-called *geodesic* on M. $\square$

(2.2) General Variational Problems. The variational problem above has the following form: We are given $\alpha, \beta \in \mathbb{R}$ with $\alpha < \beta$, an open set $W \subseteq \mathbb{R}^m \times \mathbb{R}^m$, points $a, b \in \mathbb{R}^m$, and a function

$$L \in C^1([\alpha, \beta] \times W, \mathbb{R}).$$

We then seek a function x from the class Z of all admissible functions such that the integral

$$\int_\alpha^\beta L(t, x(t), \dot x(t))\, dt$$

is a minimum, where

$$Z := \{\, x \in C^1([\alpha, \beta], \mathbb{R}^m) \mid |x(\alpha) = a,\ x(\beta) = b,\ (x(t), \dot x(t)) \in W,\ \forall t \in [\alpha, \beta] \,\}$$

(2.3) Remark. It follows immediately from the derivation in (2.1) that $g_{ik} = g_{ki}$ and

$$\sum_{i,k=1}^m g_{ik}(x) y^i y^k \ge 0, \quad \forall\, x \in V,\ \forall\, y \in \mathbb{R}^m.$$

Since $Dg(x)$ is injective, and since

$$\sum g_{ik}(x) y^i y^k = (Dg(x)y \mid Dg(x)y) = |Dg(x)y|^2,$$

it follows that equality holds if and only if $y = 0$. Therefore the symmetric matrix $[g_{ik}]$ is *positive definite*, and in this case we have $W = V \times (\mathbb{R}^m \setminus \{0\})$ and $L \in C^1([\alpha, \beta] \times W, \mathbb{R})$.

For a more abstract formulation of the variational problem above we need the following considerations.

(2.4) Lemma. $E := C^1([\alpha, \beta], \mathbb{R}^m)$ *is a normed vector space with norm*

$$\|x\|_{C^1} := \|x\|_C + \|\dot{x}\|_C = \max_{\alpha \le t \le \beta} |x(t)| + \max_{\alpha \le t \le \beta} |\dot{x}(t)|,$$

and

$$U_W := \{x \in E \mid (x(t), \dot{x}(t)) \in W, \quad \forall\, t \in [\alpha, \beta]\}$$

is open in E.

Proof. E is obviously a normed vector space. Let $x \in U_W$ be arbitrary. Since $M := \{(x(t), \dot{x}(t)) \mid t \in [\alpha, \beta]\}$ is compact, there exists some $r > 0$ such that

$$\operatorname{dist}(M, W^c) = \inf_{m \in M} \inf_{\xi \in W^c} |m - \xi| \ge 2r$$

(with the convention that $\operatorname{dist}(M, \emptyset) := \infty$).

For each $y \in E$ satisfying $\|x - y\|_{C^1} < r$ we have $y \in U_W$, i.e., U_W is open in E. $\qquad\square$

(2.5) Lemma. $C_0^1([\alpha, \beta], \mathbb{R}^m) := \{x \in E \mid x(\alpha) = x(\beta) = 0\}$ *is a closed vector subspace of E.*

If $\bar{x} \in E$ satisfies $\bar{x}(\alpha) = a$ and $\bar{x}(\beta) = b$, then

$$M := \{x \in E \mid x(\alpha) = a, \quad x(\beta) = b\} = \bar{x} + C_0^1([\alpha, \beta], \mathbb{R}^m),$$

i.e., M is a translate of a (closed) subspace of E.

Proof. The assertion is trivial. $\qquad\square$

Let now $U := M \cap U_W$ (i.e., $U = Z$) and

$$f(x) := \int_\alpha^\beta L(t, x(t), \dot{x}(t))dt, \quad \forall\, x \in U. \tag{1}$$

Then U is open in M and f is a real-valued function on U,

$$f : U \to \mathbb{R}.$$

The variational problem above now has the following simple abstract formulation: Find some $x \in U$ such that

$$f(x) \le f(y), \quad \forall\, y \in U,$$

that is, we seek a global minimum of the function f in U.

Thus we see that, by means of an appropriate interpretation (the paths of Z will be interpreted as points in some function space, namely E), the variational problem acquires a simple form which formally resembles the classical problem of minimizing a function of one variable. In the classical case of one real variable it is well-known that if the minimum is assumed at a point inside an open set, it must necessarily be a *critical point*, i.e., the derivative vanishes. If U is an open set in $\mathbb{R}^m$, then, in particular, all directional derivatives vanish at a critical point. We will see now that – again, in some appropriate interpretation – this criterion can be carried over verbatim to the abstract case. Since in our case U is only open in M, we may, of course, only consider those directions $v \in E$ for which $x_0 + tv$ lies in U for small $|t|$. This must be taken into account in the following definition of the directional derivative.

Let E be a normed vector space and let M denote a translate of a vector subspace of E (i.e., there exist a unique subspace V of E and some element $m \in E$ such that $M = m + V$. Here, of course, m is only determined modulo V, i.e., $m + V = m_1 + V$ if and only if $m - m_1 \in V$). Moreover, assume that U is open in M and let $f : U \to \mathbb{R}$ be given. For $u_0 \in M$ and every $v \in V$ we set

$$\delta f(u_0; v) := \lim_{t \to 0} \frac{f(u_0 + tv) - f(u_0)}{t}$$

whenever this limit exists. That is, $\delta f(u_0; v)$ is the *directional derivative* of f at u_0 in the direction v.

If this directional derivative exists *for every direction* $v \in V$, we set

$$\delta f(u_0) := \delta f(u_0; \cdot)$$

and call

$$\delta f(u_0) : V \to \mathbb{R}$$

the *first variation of f at the point u_0 with respect to the subspace V*.

(2.6) Remarks. (a) The directional derivative $\delta f(u_0; v)$ is *homogeneous* in the second variable, i.e.,

$$\delta f(u_0; \lambda v) = \lambda \delta f(u_0; v), \quad \forall \lambda \in \mathbb{R}.$$

Indeed, if $\lambda \neq 0$, we have

$$\delta f(u_0; \lambda v) = \lim_{t \to 0} \frac{f(u_0 + t\lambda v) - f(u_0)}{t} = \lambda \lim_{t \to 0} \frac{f(u_0 + \lambda t v) - f(u_0)}{\lambda t}$$

$$= \lambda \lim_{\tau \to 0} \frac{f(u_0 + \tau v) - f(u_0)}{\tau} = \lambda \delta f(u_0; v),$$

and if $\lambda = 0$, the assertion is trivial. Hence for every $v \in V \setminus \{0\}$ we have

$$\delta f(u_0; v) = \|v\| \, \delta f \left(u_0; \frac{v}{\|v\|} \right).$$

This shows that it would have sufficed to consider the directional derivative for the unit vectors $v \in V$ with $\|v\| = 1$. For many purposes, however, it is advantageous not to impose restrictions on the length of the direction vectors v.

(b) In the special case when U is open *in* E, i.e., if $M = E$, we may consider the case when $f : U \to \mathbb{R}$ is *differentiable* at u_0, that is to say, there exists some $Df(u_0) \in \mathcal{L}(E, \mathbb{R})$ such that

$$\lim_{h \to 0} \frac{f(u_0 + h) - f(u_0) - Df(u_0)h}{\|h\|} = 0.$$

It is easily seen in this case (as it is in $\mathbb{R}^m$!) that

$$\delta f(u_0)h = \delta f(u_0; h) = Df(u_0)h, \quad \forall h \in E,$$

that is, $\delta f(u_0) = Df(u_0)$. But the concept of the first variation is much weaker than the concept of the first derivative. For example, it is not required that the map $v \mapsto \delta f(u_0; v)$ is linear. $\qquad \square$

After these preliminaries we are ready to prove the following fundamental theorem. It generalizes the classical one-dimensional criterion which says that every local extremum must be a critical point.

(2.7) Theorem. *Let E be a normed vector space and let $M = m + V$ for some subspace V of E. Moreover, let U be open in M and assume that f is a real-valued function on U. If f has a local extremum at $u_0 \in U$ and if $\delta f(u_0)$ exists with respect to V, then*

$$\delta f(u_0) = 0,$$

that is, a necessary condition for the existence of a local extremum is the vanishing of the first variation.

Proof. For each $v \in V$, the function $t \mapsto \varphi(t) := f(u_0 + tv)$ is defined in a neighborhood of $0 \in \mathbb{R}$. At 0 it has a relative extremum and is differentiable at $0 \in \mathbb{R}$ with $\varphi'(0) = \delta f(u_0; v)$. Now the assertion follows from the classical one-dimensional criterion. $\qquad \square$

If $\delta f(u_0) = 0$, we say that u_0 *is a stationary point of the functional* f (i.e., of the real-valued function f). The point u_0 is also called a *critical point* of f.

In the following proposition we will explicitly determine the first variation for the concrete functional (1).

(2.8) Proposition. *Assume that $-\infty < \alpha < \beta < \infty$ and let W be open in $\mathbb{R}^m \times \mathbb{R}^m$. Moreover, let $L \in C^1([\alpha, \beta] \times W, \mathbb{R})$ and assume that M is a translate of a closed subspace of $C^1([\alpha, \beta], \mathbb{R}^m)$. Then*

$$U := \{x \in M \mid (x(t), \dot{x}(t)) \in W, \ \forall t \in [\alpha, \beta]\}$$

is open in M and the first variation of the functional

$$f(x) := \int_\alpha^\beta L(t, x(t), \dot{x}(t)) \, dt, \quad x \in U, \tag{2}$$

at $u_0 \in M$ with respect to $V := u_0 - M$ is given by

$$\delta f(u_0; v) = \int_\alpha^\beta \{D_2 L(t, u_0(t), \dot{u}_0(t))v(t) + D_3 L(t, u_0(t), \dot{u}_0(t))\dot{v}(t)\} \, dt.$$

Here $D_2 L$ denotes the derivative of the function (of m variables)

$$x \mapsto L(t, x, y) \quad \text{for fixed } (t, y),$$

and $D_3 L$ denotes the derivative of the function

$$y \mapsto L(t, x, y) \quad \text{for fixed } (t, x).$$

Proof. (i) With the notation of lemma (2.4) we have $U = U_W \cap M$. Hence U is open in M.

(ii) With $\varphi(s) := f(u_0 + sv)$ we have

$$\varphi(s) = \int_\alpha^\beta L(t, u_0(t) + sv(t), \dot{u}_0(t) + s\dot{v}(t))dt$$

for all $-\epsilon \le s \le \epsilon$ and some appropriate $\epsilon > 0$. That is, $\varphi(s)$ is defined by a parameter-dependent integral

$$\varphi(s) = \int_\alpha^\beta \psi(t, s)dt,$$

where

$$\psi(t, s) := L(t, u_0(t) + sv(t), \dot{u}_0(t) + s\dot{v}(t)).$$

Since ψ and $D_2\psi$ are continuous on $[\alpha, \beta] \times [-\epsilon, \epsilon]$, we may differentiate under the integral sign and hence we get

$$\delta f(u_0; v) = \varphi'(0) = \int_\alpha^\beta D_2\psi(t, 0) \, dt.$$

From this the assertion follows. □

(2.9) Corollary. *If, in addition, the map*

$$t \mapsto D_3 L(t, u_0(t), \dot{u}_0(t))$$

is continuously differentiable, then

$$\delta f(u_0; v) = \int_\alpha^\beta [D_2 L(t, u_0(t), \dot{u}_0(t)) - \frac{d}{dt} D_3 L(t, u_0(t), \dot{u}_0(t))]v(t) \, dt$$
$$+ D_3 L(t, u_0(t), \dot{u}_0(t))v(t)\big|_\alpha^\beta.$$

Proof. Use integration by parts. □

(2.10) Remark. For fixed (t, x, y), $D_2 L(t, x, y)$ and $D_3 L(t, x, y)$ are linear transformations from $\mathbb{R}^m$ into $\mathbb{R}$. With respect to the canonical basis they have the following Jacobian matrix representations:

$$D_2 L = \left[\frac{\partial L}{\partial x^1}, \ldots, \frac{\partial L}{\partial x^m} \right], \quad D_3 L = \left[\frac{\partial L}{\partial y^1}, \ldots, \frac{\partial L}{\partial y^m} \right].$$

For $v = (v^1, \ldots, v^m) \in \mathbb{R}^m$ we therefore have

$$\left[D_2 L - \frac{d}{dt} D_3 L \right] v = \sum_{i=1}^{m} \left(\frac{\partial L}{\partial x^i} - \frac{d}{dt} \frac{\partial L}{\partial y^i} \right) v^i. \tag{3}$$

If, as usual, we identify the dual space $(\mathbb{R}^m)' = \mathcal{L}(\mathbb{R}^m, \mathbb{R})$ with $\mathbb{R}^m$ by means of the Euclidean inner product

$$(x \mid y) := \sum_{i=1}^{m} x^i y^i,$$

then $D_2 L$ and $D_3 L$ can be identified with the gradient of L with respect to x and y, respectively. That is,

$$D_2 L = \text{grad}_x L := \left(\frac{\partial L}{\partial x^1}, \ldots, \frac{\partial L}{\partial x^m} \right)$$

and

$$D_3 L = \text{grad}_y L := \left(\frac{\partial L}{\partial y^1}, \ldots, \frac{\partial L}{\partial y^m} \right).$$

Then (3) becomes

$$\left[D_2 L - \frac{d}{dt} D_3 L \right] v = \left(\text{grad}_x L - \frac{d}{dt} \text{grad}_y L \mid v \right). \qquad \square$$

If u_0 is a critical point of the functional in (2), then $\delta f(u_0) = 0$. In the case of the variational problem with fixed boundaries and under the assumptions of corollary (2.9), the following lemma shows that the expression $D_2 L - \frac{d}{dt} D_3 L$ must vanish identically. The assertion of this lemma is not quite obvious because v must belong to the subspace $C_0^1([\alpha, \beta], \mathbb{R}^m)$ and cannot be chosen arbitrarily in $C^1([\alpha, \beta], \mathbb{R}^m)$.

(2.11) Lemma. *Let* $g \in C([\alpha, \beta], \mathbb{R}^m)$ *and assume that*

$$\int_{\alpha}^{\beta} (g(t) \mid v(t)) \, dt = 0, \quad \forall \, v \in C_0^1([\alpha, \beta], \mathbb{R}^m).$$

Then $g = 0$.

Proof. Let us assume that $g \neq 0$. Then there exists some $i = 1, \ldots, m$ such that $g^i \neq 0$. Since $g^i : [\alpha, \beta] \to \mathbb{R}$ is continuous, there exist some $x_0 \in (\alpha, \beta)$ and some $\epsilon > 0$ such that $U_\epsilon := (x_0 - \epsilon, x_0 + \epsilon)$ still lies entirely in $[\alpha, \beta]$ and so that

$g^i(t) \neq 0$ for all $t \in U_\epsilon$. Now we choose a function $v^i \in C^\infty(\mathbb{R}, \mathbb{R})$ which satisfies $\text{supp}(v^i) \subseteq U_\epsilon$ and $v^i > 0$ (cf. remark (2.12)). Then

$$v := (0, \ldots, 0, \underset{(i)}{v^i}, 0, \ldots, 0) \in C_0^1([\alpha, \beta], \mathbb{R}^m)$$

and

$$\int_\alpha^\beta (g(t) \mid v(t))\, dt = \int_\alpha^\beta g^i(t) v^i(t)\, dt = \int_{x_0-\epsilon}^{x_0+\epsilon} g^i(t) v^i(t)\, dt \neq 0,$$

since $g^i v^i$ is continuous, does not change its sign, and does not vanish identically. This contradiction implies the assertion. $\qquad\square$

(2.12) Remark. We know that the function

$$\varphi(t) := \begin{cases} e^{-1/t}, & \text{if } t > 0 \\ 0, & \text{if } t \leq 0 \end{cases}$$

is infinitely differentiable. Since also the function

$$\mathbb{R}^m \to \mathbb{R}, \qquad x \mapsto 1 - |x|^2,$$

is *smooth* (i.e., infinitely differentiable), it follows that the same is true for the composition of both functions. More precisely we have: If $c > 0$ and

$$\omega(x) := \begin{cases} c e^{1/(|x|^2-1)}, & \text{if } |x| < 1 \\ 0, & \text{if } |x| \geq 1, \end{cases}$$

then $\omega \in C^\infty(\mathbb{R}^m, \mathbb{R})$, $\omega \geq 0$, and

$$\text{supp}\,(\omega) := \overline{\{y \in \mathbb{R}^m \mid \omega(y) \neq 0\}} = \overline{\mathbb{B}}^m.$$

Now we choose $c := \int_{|x| \leq 1} e^{1/(|x|^2-1)}\, dx$ and for every $\epsilon > 0$ we set

$$\omega_\epsilon(x) := \epsilon^{-m} \omega(x/\epsilon), \qquad \forall\, x \in \mathbb{R}^m.$$

Then evidently we have

$$\begin{aligned} &\omega_\epsilon \in C^\infty(\mathbb{R}^m, \mathbb{R}), \quad \omega_\epsilon \geq 0, \quad \text{supp}\,(\omega_\epsilon) = \epsilon \overline{\mathbb{B}}^m, \\ &\omega_\epsilon(-x) = \omega_\epsilon(x), \quad \forall\, x \in \mathbb{R}^m \\ &\int_{\mathbb{R}^m} \omega_\epsilon(x)\, dx = 1. \end{aligned} \tag{4}$$

Every one-parameter family $\{\omega_\epsilon \mid \epsilon \in (0, \infty)\}$ of functions satisfying (4) is called a family of *mollifiers*.

If $f : \mathbb{R}^m \to \mathbb{K}$ is an arbitrary locally (Lebesque-) integrable function, then it is easily seen that the *convolution*

$$f_\epsilon(x) := \omega_\epsilon * f(x) := \int_{\mathbb{R}^m} \omega_\epsilon(x-y) f(y)\, dy = \int_{x+\epsilon\mathbb{B}^m} \omega_\epsilon(x-y) f(y)\, dy$$

exists for all $x \in \mathbb{R}^m$. From well-known results about parameter-dependent integrals and from the change-of-variables formula it easily follows that

$$\omega_\epsilon * f \in C^\infty(\mathbb{R}^m, \mathbb{K}), \quad \omega_\epsilon * f = f * \omega_\epsilon,$$

$$\operatorname{supp}(\omega_\epsilon * f) \subseteq \operatorname{supp}(f) + \epsilon \bar{\mathbb{B}}^m, \tag{5}$$

$$D^{(k)}(\omega_\epsilon * f) = (D^{(k)} \omega_\epsilon) * f,$$

where $D^{(k)}$ denotes a general kth order partial derivative.

If now $M \subseteq \mathbb{R}^m$ is measurable and $f : M \to \mathbb{K}$ is locally integrable, then the *natural extension* $\tilde{f} : \mathbb{R}^m \to \mathbb{K}$ is locally integrable, where we have set

$$\tilde{f}(x) := \begin{cases} f(x), & \text{if } x \in M \\ 0, & \text{if } x \notin M. \end{cases}$$

In this case we make the convention that by the convolution of f with the mollifier ω_ϵ, $\omega_\epsilon * f$, we mean the convolution $\omega_\epsilon * \tilde{f}$.

With this convention and the relation

$$f_\epsilon(x) - f(x) = f * \omega_\epsilon(x) - f(x) = \int_{\epsilon \bar{\mathbb{B}}^m} [f(x-y) - f(x)] \omega_\epsilon(y) \, dy$$

one easily derives the following *approximation*: *If $M \subseteq \mathbb{R}^m$ is measurable and $f \in C(M, \mathbb{K})$ is locally integrable, then $f_\epsilon := \omega_\epsilon * f$ converges to f as $\epsilon \to 0$, in fact, the convergence is uniform on compact subsets of M.*

The existence of a function v^i, as was needed in lemma (2.11), now follows immediately from these considerations. For this it suffices to take the convolution of the characteristic function of the interval $[x_0 - \epsilon/2, x_0 + \epsilon/2]$ with the mollifier $\omega_{\epsilon/2}$. $\qquad \square$

We now return to our original general variational problem and stipulate the following manner of writing: By the *variational problem with fixed end points*

$$\delta \int_\alpha^\beta L(t, x, \dot{x}) \, dt = 0, \quad x(\alpha) = a, \quad x(\beta) = b,$$

we symbolize the problem of finding a stationary point for the functional

$$f(x) := \int_\alpha^\beta L(t, x, \dot{x}) dt$$

which is defined on a subset of $C^1([\alpha, \beta], \mathbb{R}^m)$, determined by the constraints $x(\alpha) = a$, $x(\beta) = b$ (namely, $M = \bar{x} + C_0^1([\alpha, \beta], \mathbb{R}^m)$, where $\bar{x} \in C^1([\alpha, \beta], \mathbb{R}^m)$ satisfies $\bar{x}(\alpha) = a$, $\bar{x}(\beta) = b$), i.e., the problem

$$\delta f(x) = 0 \quad \text{with respect to} \quad C_0^1([\alpha, \beta], \mathbb{R}^m).$$

Every solution of this problem, that is, every stationary point of this functional is called an *extremal of the variational problem*.

(2.13) Theorem. *Let $-\infty < \alpha < \beta < \infty$ and assume that W is open in $\mathbb{R}^m \times \mathbb{R}^m$. Moreover, let $a, b \in \mathbb{R}^m$ and $L \in C^1([\alpha, \beta] \times W, \mathbb{R})$. If $u_0 \in C^1([\alpha, \beta], \mathbb{R}^m)$ is an extremal of the variational problem*

$$\delta \int_{\alpha}^{\beta} L(t, x, \dot{x}) \, dt = 0, \quad x(\alpha) = a, \quad x(\beta) = b,$$

and if the function $t \mapsto D_3 L(t, u_0(t), \dot{u}_0(t))$ is continuously differentiable, then u_0 satisfies the Euler equation

$$\frac{d}{dt} D_3 L(t, u_0(t), \dot{u}_0(t)) = D_2 L(t, u_0(t), \dot{u}_0(t))$$

and the constraints

$$u_0(\alpha) = a, \quad u_0(\beta) = b.$$

Proof. The assertion follows immediately from corollary (2.9), remark (2.10) and from lemma (2.11). □

It follows from remark (2.10) that the Euler equation is *a system of m implicit differential equations:*

$$\frac{d}{dt} \left[\frac{\partial L}{\partial y^i}(t, u_0^1(t), \dots, u_0^m(t), \dot{u}_0^1(t), \dots, \dot{u}_0^m(t)) \right]$$

$$= \frac{\partial L}{\partial x^i}(t, u_0^1(t), \dots, u_0^m(t), \dot{u}_0^1(t), \dots, \dot{u}_0^m(t)),$$

$i = 1, \dots, m$. Following classical usage, we set

$$L_{x^i} := \frac{\partial L}{\partial x^i}, \quad L_{\dot{x}^i} := \frac{\partial L}{\partial y^i}, \quad i = 1, \dots, m,$$

and

$$L_x := (L_{x^1}, \dots, L_{x^m}), \quad L_{\dot{x}} := (L_{\dot{x}^1}, \dots, L_{\dot{x}^m}).$$

Then the Euler equation becomes

$$\frac{d}{dt} L_{\dot{x}} = L_x,$$

that is,

$$\frac{d}{dt} L_{\dot{x}^i} = L_{x^i}, \quad i = 1, \dots, m,$$

where the argument $(t, u, \dot{u})$ has been suppressed.

(2.14) Remark. When we calculate the (total) derivative with respect to time (if this is permissible), we obtain

$$\frac{d}{dt} L_{\dot{x}^i} = L_{\dot{x}^i t} + \sum_{k=1}^{m} L_{\dot{x}^i x^k} \dot{x}^k + \sum_{k=1}^{m} L_{\dot{x}^i \dot{x}^k} \ddot{x}^k.$$

This makes it evident that the Euler equation is a *second order system*, i.e., the unknown function occurs with its second order derivative. If the matrix $[L_{\dot{x}^i \dot{x}^k}]_{1 \leq i,k \leq m}$ is invertible, then it is possible to solve for $\ddot{x}$ and we obtain an *explicit second order system* of the form

$$\ddot{x} = g(t, x, \dot{x}),$$

that is,

$$\ddot{x}^i = g^i(t, x^1, \ldots, x^m, \dot{x}^1, \ldots, \dot{x}^m), \quad i = 1, \ldots, m,$$

for some appropriate function

$$g : D \subseteq [\alpha, \beta] \times \mathbb{R}^m \times \mathbb{R}^m \to \mathbb{R}^m.$$

Thus, in this case the extremals are solutions of the *boundary value problem* (BVP)

$$\begin{cases} \ddot{x} = g(t, x, \dot{x}) & \text{in} \quad (\alpha, \beta) \\ x(\alpha) = a, \quad x(\beta) = b \end{cases}$$

$\square$

(2.15) Example. *Geodesics.* In this case it follows from (2.1) that the integrand of the variational problem with fixed end points is given by

$$L(t, x, \dot{x}) = \sqrt{\sum_{j,k=1}^{m} g_{jk}(x)\dot{x}^j \dot{x}^k} \ .$$

From this we obtain

$$L_{x^i} = \frac{1}{2L} \sum_{j,k=1}^{m} D_i g_{jk}(x)\dot{x}^j \dot{x}^k \tag{6}$$

and

$$L_{\dot{x}^i} = \frac{1}{L} \sum_{k=1}^{m} g_{ik}(x)\dot{x}^k \tag{7}$$

for $i = 1, \ldots, m$. Hence the Euler equation becomes:

$$\frac{d}{dt}\left(\frac{\sum_{k=1}^{m} g_{ik}(x)\dot{x}^k}{L} \right) = \frac{\sum_{j,k=1}^{m} D_i g_{jk}(x)\dot{x}^j \dot{x}^k}{2L}, \quad i = 1, \ldots, m. \tag{8}$$

Based on the derivation in (2.1) we have

$$L(t, x(t), \dot{x}(t)) = |(g \circ x)^{\cdot}(t)|, \quad \forall t \in [\alpha, \beta],$$

i.e., $L(t, x(t), \dot{x}(t))$ is the length of the tangent vector of the C^1-path $g \circ x : [\alpha, \beta] \to M$ at the point $g(x(t)) \in M$. The Euler equation was derived under the assumption that

$$(x(t), \dot{x}(t)) \in W = V \times (\mathbb{R}^m \setminus \{0\}), \quad \forall t \in [\alpha, \beta],$$

(cf. remark (2.3)). Since $Dg(x)$ is injective, it follows that $(g \circ x)^{\cdot}(t) = Dg(x(t))\dot{x}(t) \neq 0$ for all $t \in [\alpha, \beta]$, that is, the path $g \circ x$ is *regular*, i.e., it has a nonvanishing tangent everywhere. It is, however, well-known that the curve $C \subseteq M \subseteq \mathbb{R}^m$, described by this path, can be parametrized by arc length. This means that a parameter $s \in [\alpha', \beta'] \subseteq \mathbb{R}$ can be chosen such that $|(g \circ x)^{\cdot}(s)| = 1$ for all $s \in [\alpha', \beta']$. If we now assume that the parameter t has already been chosen so that it represents the arc length of the path $g \circ x$, then $L(t, x(t), \dot{x}(t)) = 1$ for all $t \in [\alpha, \beta]$ and the Euler equations (8) simplify to

$$\frac{d}{dt}\left(\sum_{k=1}^{m} g_{ik}(x)\dot{x}^k\right) = \frac{1}{2}\sum_{j,k=1}^{m} D_i g_{jk}(x)\dot{x}^j \dot{x}^k, \quad i = 1,\ldots,m. \tag{9}$$

In order to simplify the notation, we will now make use of the *summation convention*, that is, we omit the summation sign and agree to sum those indices from 1 to m that occur twice in a product. We also set

$$g_{jk,i} := D_i g_{jk}.$$

Then (9) takes on the form

$$\frac{d}{dt}(g_{ik}(x)\dot{x}^k) = \frac{1}{2} g_{jk,i}\dot{x}^j \dot{x}^k, \quad i = 1,\ldots,m. \tag{10}$$

Since the matrix $[g_{jk}]$ is positive definite, it is invertible. The inverse will be denoted by $[g^{jk}]$, i.e., we have

$$g^{jl}g_{lk} = \delta_k^j, \quad 1 \le j, k, \le m. \tag{11}$$

If we perform the differentiation in (10), we obtain

$$g_{ik}(x)\ddot{x}^k + g_{ik,j}\dot{x}^j \dot{x}^k - \frac{1}{2} g_{jk,i}\dot{x}^j \dot{x}^k = 0, \quad i = 1,\ldots,m.$$

Multiplying by g^{li}, summing over $i = 1,\ldots,m$, and using (11), it follows that

$$\ddot{x}^l + g^{li}(g_{ik,j} - \frac{1}{2} g_{jk,i})\dot{x}^j \dot{x}^k = 0, \quad l = 1,\ldots,m. \tag{12}$$

Because

$$\begin{aligned}
g_{ik,j}\dot{x}^j \dot{x}^k &= (D_j D_i g \mid D_k g)\dot{x}^j \dot{x}^k + (D_i g \mid D_j D_k g)\dot{x}^j \dot{x}^k \\
&= (D_k D_i g \mid D_j g)\dot{x}^j \dot{x}^k + (D_i g \mid D_k D_j g)\dot{x}^j \dot{x}^k = g_{ij,k}\dot{x}^j \dot{x}^k
\end{aligned} \tag{13}$$

(renaming of summation indices), we may write (12) in the symmetric form

$$\ddot{x}^l + \frac{1}{2} g^{li}(g_{ik,j} + g_{ij,k} - g_{jk,i})\dot{x}^j \dot{x}^k = 0, \quad l = 1,\ldots,m.$$

Under the assumptions above and using the *Christoffel symbols*

$$\Gamma_{ik}^l := \frac{1}{2} g^{lj}(g_{ij,k} + g_{jk,i} - g_{ki,j}), \quad i, k, l = 1,\ldots,m,$$

the Euler equation becomes:

$$\ddot{x}^l + \Gamma_{ik}^l \dot{x}^i \dot{x}^k = 0, \quad l = 1,\ldots,m, \tag{14}$$

(where we made use of the symmetry $g_{ij,k} = g_{ji,k}$).

The system of differential equations (14) is known as the *differential equation of geodesics (in local coordinates)*. If $x \in C^2([\alpha, \beta], V)$ is a solution of (14), then $g \circ x$ is called a *geodesic of M*, irrespective of whether or not x is a solution of the minimization problem with constraints (2.1).   □

(2.16) Remarks. (a) It should be noted that $\Gamma^l_{ik} \in C^{r-2}(V, \mathbb{R})$ if M is a C^r-manifold.

(b) We have derived the differential equation of geodesics under the assumption that $L = 1$, i.e., that the extremal of the variational problem already is parametrized by arc length. Conversely, it is a simple consequence of (13) that the following holds: *If x is a solution of the differential equation of geodesics, then $|(g \circ x)^{\cdot}(t)|$ is constant*, i.e., the geodesic of M is traversed at constant speed.

(c) For a general m-dimensional *Riemannian manifold* of class C^2 with metric $[g_{ik}]$, the differential equation of geodesics in local coordinates is also given by (14), where the Christoffel symbols are defined as above.   □

(2.17) Examples. (a) $M = \mathbb{R}^m$ *with the Euclidean metric*. In this case we may of course choose the natural chart $(\mathbb{R}^m, I_m)$. Then $g_{ik} = \delta_{ik}$ and consequently $\Gamma^l_{ik} = 0$. The differential equation of geodesics becomes

$$\ddot{x} = 0.$$

All solutions of this differential equation have the form

$$x(t) = at + b, \quad \text{where} \quad a, b \in \mathbb{R}^m.$$

Hence the geodesics in $\mathbb{R}^m$ are exactly the straight lines.

(b) $M = \mathbb{S}^2$, *parametrized by spherical coordinates.* From example (a) in (2.1) we obtain

$$[g^{ik}] = \begin{bmatrix} \frac{1}{\cos^2 \vartheta} & 0 \\ 0 & 1 \end{bmatrix}$$

and, in addition, we find that $g_{11,2} = -2 \cos \vartheta \sin \vartheta$ and that all other $g_{ij,k}$ vanish. From this we calculate the Christoffel symbols to be

$$\Gamma^1_{12} = \Gamma^1_{21} = -\tan \vartheta, \quad \Gamma^2_{11} = \sin \vartheta \cos \vartheta,$$

while all other Γ^l_{ik} vanish. Therefore the *geodesic differential equation in spherical coordinates* becomes:

$$\ddot{\varphi} - 2 \tan \vartheta \dot{\varphi} \dot{\vartheta} = 0$$
$$\dot{\vartheta} + \sin \vartheta \cos \vartheta \dot{\varphi}^2 = 0 \tag{15}$$

For the Euler equation of the variational problem

$$\delta \int_\alpha^\beta \sqrt{\cos^2 \vartheta \dot{\varphi}^2 + \dot{\vartheta}^2} \, dt = 0, \quad (\varphi(\alpha), \vartheta(\alpha)) = a, \quad (\varphi(\beta), \vartheta(\beta)) = b$$

we obtain

$$\frac{d}{dt} \left(\frac{\cos^2 \vartheta \dot{\varphi}}{L} \right) = 0, \quad \frac{d}{dt} \left(\frac{\dot{\vartheta}}{L} \right) = \frac{-\sin \vartheta \cos \vartheta \dot{\varphi}^2}{L},$$

where $L = \sqrt{\cos^2 \vartheta \dot{\varphi}^2 + \dot{\vartheta}^2}$. From this we again deduce (15) if $L = 1$.   □

(2.18) Remarks. (a) The Euler equation

$$\frac{d}{dt} L_{\dot{x}} = L_x \tag{16}$$

for the variational problem

$$\delta \int_\alpha^\beta L(t, x, \dot{x}) dt = 0, \quad x(\alpha) = a, \quad x(\beta) = b,$$

depends, of course, on the coordinates used. *However, the extremals are independent of the specific coordinates.* If one utilizes coordinates which are specially adapted to the problem, one obtains, in general, particularly simple differential equations. Since the Euler equation always has the form in (16), it is easily expressed in different coordinates. One only needs to perform a change of variables in the corresponding integral.

Hence for the geodesics in the plane, for example, we have

$$\delta \int_\alpha^\beta \sqrt{\dot{x}^2 + \dot{y}^2} \, dt = 0, \quad x(\alpha) = a, \; x(\beta) = b,$$

from which we deduce, as above, the differential equation of geodesics in Euclidean coordinates

$$\ddot{x} = 0, \quad \ddot{y} = 0.$$

Employing the polar coordinates $x = r \cos \varphi$, $y = r \sin \varphi$, the variational problem becomes

$$\delta \int_\alpha^\beta \sqrt{\dot{r}^2 + r^2 \dot{\varphi}^2} \, dt = 0, \quad (r(\alpha), \varphi(\alpha)) = a, \; (r(\beta), \varphi(\beta)) = b.$$

The corresponding Euler equations are

$$\frac{d}{dt} \left(\frac{\dot{r}}{L} \right) = \frac{r \dot{\varphi}^2}{L}, \quad \frac{d}{dt} \left(\frac{r^2 \dot{\varphi}}{L} \right) = 0,$$

from which we deduce the equations for geodesics in polar coordinates ($L = 1$)

$$\ddot{r} - r \dot{\varphi}^2 = 0, \quad (r^2 \dot{\varphi})^{\cdot} = 0.$$

This formulation clearly is considerably less well adapted to the problem than the formulation in Euclidean coordinates.

(b) One can, of course, also consider variational problems in several independent variables. If, for instance, $\Omega \subseteq \mathbb{R}^n$ is some domain, then the problem

$$\delta \int_\Omega L(x, u, \text{grad } u) dx = 0, \quad u \mid \partial \Omega = \varphi,$$

is the direct generalization of the one-dimensional variational problem with fixed endpoints above. For example, the question of finding a surface of smallest area, which passes through a given curve in space, leads to such a problem. In analogy with the above, one derives the Euler equation which, however, now is a *partial differential equation*.

(c) In the derivation of the Euler equation additional assumptions were made (e.g., that the extremal is twice differentiable) which require separate justifications. It is clear that every solution of the Euler equation is a critical point of the functional $\int_\alpha^\beta L(t, x, \dot{x})dt$, but the converse is in general not true.

(d) In the classical calculus of variations one analyzes the Euler equation, that is to say, one discusses when it can be solved and the nature of its solutions, and one tries to find further conditions which guarantee that, for instance, a solution of the Euler equation represents a local or global minimum of the variational problem.

In the case of higher dimensional variational problems, the Euler equations are partial differential equations which are considerably more difficult to analyze and about which considerably less is known. Such partial differential equations also occur in other connections. That is why in the modern theory of partial differential equations one takes the opposite course. The corresponding variational problem is studied directly (e.g., with the methods of nonlinear functional analysis) and one tries to prove the existence of critical points of the functional. In this way one obtains information about the existence of solutions of the partial differential equations.

(e) It is again emphasized that in deriving the Euler equation the existence of an extremal was assumed. In this entire section not a single existence statement was made. Proving the existence of solutions of the Euler equations, satisfying the boundary conditions above (as well as others), is the purpose of the theory of differential equations and will be discussed exhaustively.   □

Problems

1. Prove the assertions in remark (2.12).

2. Prove remark (2.16 b).

3. Let $-\infty < \alpha < \beta < \infty$ and assume that $W \subseteq \mathbb{R}^m \times \mathbb{R}^m$ is open. In addition, let $L \in C^2([\alpha, \beta], \times W, \mathbb{R})$ and $h_i \in C^1(\mathbb{R}^m, \mathbb{R})$, $i = \alpha, \beta$, be given. Show that every twice continuously differentiable solution of the *variational problem with free end points*

$$\delta \left[\int_\alpha^\beta L(t, x, \dot{x})\, dt + h_\alpha(x(\alpha)) + h_\beta(x(\beta)) \right] = 0$$

(i.e., we allow all C^1-paths $x : [\alpha, \beta] \to \mathbb{R}^m$ such that $(x(t), \dot{x}(t)) \in W$ for all $t \in [\alpha, \beta]$) must satisfy the Euler equation

$$\frac{d}{dt} L_{\dot{x}} = L_x$$

and appropriate boundary conditions (which?).

4. Prove that $C^1([\alpha, \beta], \mathbb{R}^m)$, $-\infty < \alpha < \beta < \infty$, is a Banach space with the usual norm $|| \cdot ||_{C^1}$. Is this also true with respect to the maximum norm $|| \cdot ||_C$?

3. Classical Mechanics

We consider a mechanical system with m degrees of freedom, which is described by the (generalized) coordinates

$$q = (q^1, \ldots, q^m).$$

The problem in mechanics is to determine the state $q(t)$ of the system at time t under the assumption that it is known at time t_0, that is, determine q as a function of t from a given $q(t_0)$.

We now make the fundamental *assumption*: The mechanical system can be completely described by the *kinetic energy*

$$T(t, q, \dot{q})$$

and the *potential energy*

$$U(t, q).$$

Then *Hamilton's principle of least action* says that: *The system "moves" from time t_0 to time t_1 in such a way that the* *action integral*

$$\int_{t_0}^{t_1} (T - U) dt$$

becomes stationary with respect to all (virtual) motions with the same initial and end points. In other words: The motion is an extremal of the variational problem

$$\delta \int_{t_0}^{t_1} (T - U) dt = 0, \quad q(t_0), q(t_1) \quad \text{fixed.}$$

In mechanics the integrand

$$L := T - U$$

is called the *Lagrangian*, the generalized coordinates q are elements of the *configuration space* M (in general an m-dimensional differentiable manifold), $\dot{q}$ are the (*generalized*) *velocities*, and the Euler equation of the variational problem above, i.e.,

$$\frac{d}{dt} \left(\frac{\partial L}{\partial \dot{q}} \right) = \frac{\partial L}{\partial q},$$

is called the *Euler-Lagrange equation*.

(3.1) Examples. (a)   In many important cases the kinetic energy is a (positive definite) quadratic form with respect to the velocities, i.e.,

$$T(t, q, \dot{q}) = \frac{1}{2}(A(t, q)\dot{q} \mid \dot{q}) = \frac{1}{2}\sum_{i,j=1}^{m} a_{ij}(t, q)\dot{q}^i \dot{q}^j,$$

where

$$A(t, q) = A(t, q)^T \in \mathcal{L}(\mathbb{R}^m), \quad \forall(t, q).$$

In this case we have

$$\frac{\partial L}{\partial \dot{q}} = A(t, q)\dot{q}.$$

In particular, if A is independent of q, i.e., if

$$T(t, q, \dot{q}) = Q(t, \dot{q}) := \frac{1}{2}(A(t)\dot{q} \mid \dot{q}),$$

the Euler-Lagrange equation becomes

$$\frac{d}{dt}(A(t)\dot{q}) = -\frac{\partial U}{\partial q} \quad (= -\operatorname{grad} U(q)). \tag{1}$$

If, moreover, A is constant, i.e., if $A(t) = A$ for all t, then (1) simplifies to

$$-A\ddot{q} = \operatorname{grad} U(q), \tag{2}$$

where $A = A^T \in \mathcal{L}(\mathbb{R}^m)$.

(b) *The equations of motion of n particles in a potential field.* In this case let M be an open subset of $\mathbb{R}^{3n}$. We decompose q in the form

$$q = (x_1, \dots, x_n),$$

where each $x_i \in \mathbb{R}^3$ denotes the position of the ith particle. Assume that the kinetic energy has the form

$$T(t, q, \dot{q}) = Q(\dot{q}) := \frac{1}{2}\sum_{i=1}^{m} m_i |\dot{x}_i|^2,$$

with the positive constant "masses" m_i. With $U(q) = U(x_1, \dots, x_n)$, the Euler-Lagrange equation becomes

$$-m_i \ddot{x}_i = \frac{\partial U}{\partial x_i}(x_1, \dots, x_n), \quad i = 1, \dots, n, \tag{3}$$

where $\frac{\partial U}{\partial x_i}$ denotes the gradient of U with respect to the variables $x_i \in \mathbb{R}^3$. This follows immediately from (2), since in this case

$$A = \operatorname{diag}[m_1, m_1, m_1, m_2, m_2, m_2, \dots, m_n, m_n, m_n].$$

In the special case when $n = 1$, (3) represents Newton's equations of motion of a particle in a potential field, which are well-known. According to Newton's "law of attraction," the potential for the classical three-body problem of celestial mechanics (sun-earth-moon) is given by

$$U(x_1, x_2, x_3) := \frac{m_1 m_2}{|x_1 - x_2|} + \frac{m_2 m_3}{|x_2 - x_3|} + \frac{m_3 m_1}{|x_3 - x_1|},$$

where U is defined wherever the denominators do not vanish, hence on an open subset of $\mathbb{R}^{3\cdot 3} = \mathbb{R}^9$. □

Next, we define the mechanical (total) *energy* E of the system by

$$E := T + U$$

and we easily obtain the following theorem.

(3.2) Conservation of Energy. *Assume that*

$$T(t, q, \dot{q}) = Q(\dot{q}) = \frac{1}{2}(A\dot{q} \mid \dot{q}), \quad \text{where} \quad A = A^T \in \mathcal{L}(\mathbb{R}^m),$$

and

$$U(t, q) = U_0(q).$$

Then the energy is constant along solutions of the Euler-Lagrange equation, that is, "the energy is a constant of motion."

Proof. We must show that the function $t \mapsto E(q(t), \dot{q}(t))$ is constant for every solution $t \mapsto q(t)$ of the Euler-Lagrange equation

$$-A\ddot{q} = \text{grad } U_0(q). \tag{4}$$

This follows from the relation

$$\frac{dE}{dt} = \frac{\partial E}{\partial q}\dot{q} + \frac{\partial E}{\partial \dot{q}}\ddot{q} = (\text{grad } U_0 \mid \dot{q}) + (A\dot{q} \mid \ddot{q})$$

$$\overset{(4)}{=} -(A\ddot{q} \mid \dot{q}) + (A\dot{q} \mid \ddot{q}) = 0,$$

which derives from the symmetry of A. □

(3.3) Remarks. (a) An energy preserving mechanical system is called a *conservative system*. For this reason, one also says that the differential equation

$$-A\ddot{x} = \text{grad } U(x), \quad x \in \mathbb{R}^m, \ A = A^T \in \mathcal{L}(\mathbb{R}^m),$$

is *conservative* (or forms a *conservative system*), even if, initially, it was not derived as the Euler-Lagrange equation of a mechanical system. By the preceding, one can always associate with it the "energy"

$$E(x, \dot{x}) = \frac{1}{2}(A\dot{x} \mid \dot{x}) + U(x), \quad x \in M,$$

which, in general, may have no physical meaning!

(b) If a function is constant along every solution of a differential equation, it is called a *first integral* of this equation. If F is a first integral of a second order differential equation, then *this means that the phase curve* $t \mapsto (x(t), \dot{x}(t))$ *in phase space* (that is the space $M \times \mathbb{R}^m$, if $M \subseteq \mathbb{R}^m$; in general, however, this is the tangent bundle $T(M)$ of the manifold M) *lies on a level set*

$$F^{-1}(c) = \{(x, \dot{x}) \mid F(x, \dot{x}) = c\}, \quad c \in \mathbb{R}.$$

This fact often provides important information about the qualitative characteristics of solutions of a differential equation.

(3.4) Examples. (a) Let $M \subseteq \mathbb{R}^m$ be open and assume that $A \in \mathcal{L}(\mathbb{R}^m)$ is *positive semi-definite*, i.e.,

$$A = A^T \quad \text{and} \quad (Ax \mid x) \geq 0, \quad \forall x \in \mathbb{R}^m.$$

Then for every solution $t \mapsto x(t)$ of the differential equation

$$-A\ddot{x} = \text{grad } U(x), \quad x \in M \subseteq \mathbb{R}^m,$$

and for all time t, we have the relation

$$U(x(t)) \leq E_0,$$

where

$$E_0 := \frac{1}{2}(A\dot{x}(t_0) \mid \dot{x}(t_0)) + U(x(t_0))$$

denotes the energy at some arbitrary time t_0 (i.e., *for all time the solution remains in the "potential well"* $\{x \in M \mid U(x) \leq E_0\}$).

Of course, this is an immediate result of the conservation of energy and the fact that

$$U(x(t)) \leq \frac{1}{2}(A\dot{x}(t) \mid \dot{x}(t)) + U(x(t)) = E(t) = E_0.$$

(b) *Phase portraits.* Let $V \subseteq \mathbb{R}$ be open and $f \in C^1(V, \mathbb{R})$. We then consider the differential equation (in one variable)

$$-\ddot{x} = f(x). \tag{5}$$

This equation has the form of (3) with the "potential"

$$U(x) = \int_{x_0}^{x} f(\xi)\, d\xi$$

for some arbitrary $x_0 \in V$. Consequently the corresponding "energy"

$$E(x, \dot{x}) = \frac{1}{2}\,\dot{x}^2 + U(x)$$

is a first integral.

The differential equation (5) is evidently equivalent to the system

$$\begin{aligned} \dot{x} &= y \\ \dot{y} &= -f(x), \end{aligned} \tag{6}$$

that is to say, a first order system in the *phase plane*,

$$\dot{u} = F(u), \tag{7}$$

where $u = (x, y)$ and $F(u) := (y, -f(x))$ (i.e., a system in the (x, y) plane, where y corresponds to the "velocity" of the motion).

Based on the conservation of energy, we know that every solution of (7) (if indeed there exists one) must lie in some level set $E^{-1}(c)$, $c \in \mathbb{R}$. For the analysis of the level sets of E the following observations are useful:

(α) *The critical points of the function*

$$(x, y) \longmapsto E(x, y) := \frac{y^2}{2} + U(x)$$

are exactly the points

$$(x, 0) \quad \text{such that} \quad U'(x) = 0,$$

that is, the equilibrium positions of equation (7) *all lie on the x-axis and are precisely the critical points of the potential U.*

(β) *The level sets are symmetric with respect to the x-axis. This of course follows from the fact that $E(x, -y) = E(x, y)$.*

(γ) *If W is open in $V \times \mathbb{R}$ and if W contains no critical point of E, then*

$$E^{-1}(c) \cap W = \{(x, y) \in W \mid E(x, y) = c\}, \quad c \in \mathbb{R},$$

is a one-dimensional C^2-manifold, hence it is a disjoint union of C^2-curves. In particular, if c is a regular value of E (i.e., if $DE(x, y) \neq 0, \forall (x, y) \in E^{-1}(c)$), then the level set $E^{-1}(c)$ is a disjoint union of C^2-curves in $\mathbb{R}^2$.

This is essentially a consequence of the implicit function theorem.

(δ) *If $(x, 0)$ is a regular point of E (i.e., if $DE(x, 0) \neq 0$), then the level curve through $(x, 0)$ intersects the x-axis orthogonally.*

Indeed, if $t \longmapsto (x(t), y(t))$ is a local parametrization of $E^{-1}(c)$ in some neighborhood of $(x, 0)$ (where $c := E(x, 0)$) satisfying $(x(0), y(0)) = (x, 0)$, then it follows from

$$\frac{dE}{dt}(x(t), y(t)) = y(t)\dot{y}(t) + U'(x(t))\dot{x}(t) = 0$$

that $\dot{x}(0) = 0$, because at $t = 0$ we have $y(0) = 0$ and $U'(x) = f(x) \neq 0$ (since $(x, 0)$ is a regular point). Therefore the tangent to $E^{-1}(c)$ at the point $(x, 0)$ is parallel to the y-axis.

(ε) *On a level set $E^{-1}(c)$, the smaller the value of y (the "velocity"), the larger the potential $U(x)$ must be, and conversely* (because $y^2/2 + U(x) = c$). For this it is useful to imagine a ball rolling in a frictionless "potential well" U (i.e., along the graph of U). When it rolls downward, its velocity increases, and when it rises, the velocity decreases. If at some time t_0 the ball is at rest ($y = 0$), then at no other time can it reach a position x for which $U(x) > U(x(t_0))$ holds.

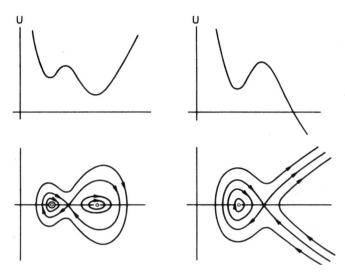

Figure 1: Phase portrait of $-\ddot{x} = f(x)$ with $f = U'$

After these general observations, it is easy to deduce the qualitative behavior of the level sets of E, and thereby (modulo the still missing existence and uniqueness theorems) the phase portrait of the differential equation (5), from the profile of the graph of U (cf. Fig. 1).

The level sets containing equilibrium points are called separatrices since they, in general, separate level sets of distinct topological structures. It will follow from the existence and uniqueness theorem that in the first example the separatrix consists of 3 parts and in the second example it consists of 4 parts, and that the equilibrium point cannot be reached from any point on one of the regular curves. In addition, we will show that the closed regular curves are traversed infinitely many times, which means that they are "trajectories" of a *periodic motion*. (One should illustrate this by means of the rolling ball model!)

(c) *The simple pendulum.* We consider a mass point of mass m in the gravitational field of the earth, suspended from a rigid and weightless rod of length l, which can rotate in a plane, free and without friction, about the point of suspension. It is clear that the mass point must move along a circle. Hence the system has one degree of freedom and can be completely described by the angle of rotation φ.

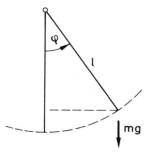

(One can also imagine a bead gliding along a frictionless circular wire of radius l). Here the configuration space M is the manifold S^1_l, the circle of radius l, which can of course be described by polar coordinates. Therefore $q = l\varphi$ and $\dot{q} = l\dot{\varphi}$ and the kinetic energy becomes

$$T = \frac{ml^2}{2}\dot{\varphi}^2.$$

If the scale for φ is chosen so that $\varphi = 0$ corresponds to the lowest position, then for the potential energy (as a function of φ, but not of q!) we obtain

$$U(\varphi) := mg(l - l\cos\varphi) = -mgl\cos\varphi + \text{constant}.$$

The resulting Euler-Lagrange equation, describing the motion of the mass point, becomes

$$\ddot{\varphi} = -\lambda\sin\varphi, \quad \text{where} \quad \lambda := g/l. \tag{8}$$

According to (b), the phase portrait of (8) has the qualitative features depicted in Fig. 2. Here the closed, regular level curves correspond to oscillations of the pendulum about its (stable) equilibrium position ($\varphi = 0$), while the curves outside the separatrices correspond to full rotations of the pendulum. The separatrices describe the limit case, whereby the pendulum approaches (as $t \to \infty$) the unstable equilibrium ($\varphi = \pi$).

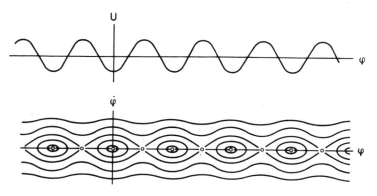

Figure 2: The phase portrait of $\ddot{\varphi} = -\lambda\sin\varphi$

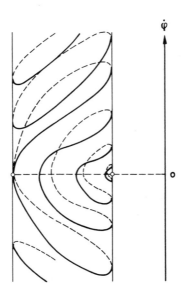

Figure 3: The phase portrait of the pendulum equation in the tangent bundle $T(\mathbb{S}^1)$

Equation (8) depends on the parameter λ. The only effect of a small change in λ (e.g., the length of the pendulum l) is either an expansion or a contraction of the phase portrait in the $\dot{\varphi}$-direction. No drastic changes ("catastrophes") occur with respect to the qualitative behavior, that is to say, *equation* (8) *is structurally stable with respect to small perturbations of* λ. In general, this is not the case, as the following example (d) shows.

The phase portrait in Fig. 2 is unsatisfactory since, for instance, the curves outside the separatrices correspond to full rotations of the pendulum, yet are not closed, even though they represent periodic motions. In addition to this, the closed curves around the points $2k\pi, k \in \mathbb{Z}\backslash\{0\}$, have no physical meaning. This, of course, stems from the fact that equation (8) describes a motion on a manifold, namely $\mathbb{S}^1$. If we are interested in global phenomena (as, for example, the rotation of the pendulum), then the motion should not be described in the phase plane, but rather on the tangent bundle of this manifold, $T(\mathbb{S}^1)$. We obtain the tangent bundle $T(\mathbb{S}^1)$ graphically by ("smoothly") attaching at every point x in $\mathbb{S}^1$ the tangent space $T_x(\mathbb{S}^1) \cong \mathbb{R}$. In this simple case, $T(\mathbb{S}^1)$ can be identified with $\mathbb{S}^1 \times \mathbb{R}$. Then the phase portrait of the pendulum equation (8) takes on the more satisfactory form depicted in Fig. 3 (which, of course, can be obtained from Fig. 2 by identifying $-\pi$ with π (mod 2π)).

(d) *A bifurcation problem.* We now consider a differential equation of the form $-\ddot{x} = f_\lambda(x)$, depending on a parameter λ, in particular, the equation

$$-\ddot{x} = x^3 - \lambda x, \quad x \in \mathbb{R}.$$

For $\lambda \leq 0$ and $\lambda > 0$, the potentials

$$U_\lambda(x) := \frac{x^4}{4} - \lambda\frac{x^2}{2}$$

have distinct qualitative features which are reflected in the qualitative differences of the phase portraits.

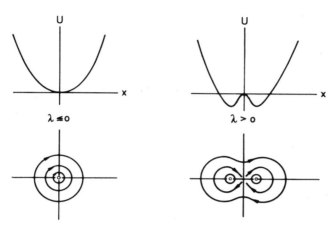

Figure 4: Typical phase portrait for $\lambda \leq 0$ and $\lambda > 0$

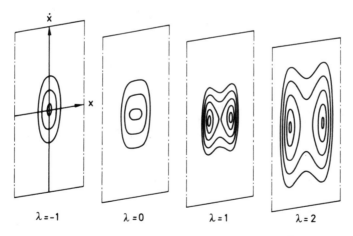

Figure 5: The phase portrait of equation $-\ddot{x} = x^3 - \lambda x$

At $\lambda = 0$ the equilibrium position $(x, y) = (0, 0)$ "bifurcates" into 3 equilibrium positions, namely $(0,0), (\pm\sqrt{\lambda}, 0)$, and from a single family of closed phase curves we get 3 such families, separated from each other by the separatrix. We call $\lambda = 0$ a *bifurcation point*. A better visual description is obtained if the (continuous) family of phase portraits is sketched in the (λ, x, y)-space, as indicated in Fig. 5.

The Legendre Transformation and Hamiltonian Systems

In the special case when

$$T(t, q, \dot{q}) = Q(\dot{q}) := \frac{1}{2} \sum_{i=1}^{n} m_i |\dot{x}_i|^2$$

$$U(t, q) = U(x_1, \ldots, x_n),$$

where $q := (x_1, \ldots, x_n) \in (\mathbb{R}^3)^n$, the Euler-Lagrange equations

$$m_i \ddot{x}_i = -\frac{\partial U}{\partial x_i}, \quad i = 1, \ldots, n$$

are obviously equivalent to the system

$$\begin{aligned} m_i \dot{x}_i &=: p_i \\ \dot{p}_i &= -\frac{\partial U}{\partial x_i} \end{aligned} \quad i = 1, \ldots, n. \tag{9}$$

If we set

$$p := (m_1 \dot{x}_1, \ldots, m_n \dot{x}_n) \in \mathbb{R}^{3n}$$

($p_i = m_i \dot{x}_i$ is the momentum of the ith mass point), and if we define the *Hamiltonian H* by

$$H(p, q) := \frac{1}{2} \sum_{i=1}^{n} \frac{|p_i|^2}{m_i} + U(q),$$

i.e., H is the total energy, expressed in space and momentum variables, then (9) can be written in the form

$$\dot{p} = -H_q, \quad \dot{q} = H_p. \tag{10}$$

Since the Lagrangian L is given by

$$L(q, \dot{q}) = \frac{1}{2} \sum_{i=1}^{n} m_i |\dot{q}_i|^2 - U(q),$$

the following relationship between the Hamiltonian H and the Lagrangian L evidently exists:

$$H(p, q) = (p \mid \dot{q}) - L(q, \dot{q}), \tag{11}$$

where

$$p = L_{\dot{q}}. \tag{12}$$

Here we solve equation (12) for $\dot{q}$ and use this result on the right-hand side of (11) to obtain the Hamiltonian as a function of p and q.

In what follows it will be shown *that for every variational problem, whose Lagrangian $L(t, q, \dot{q})$ is convex with respect to $\dot{q}$, the Euler equation is equivalent* *to a Hamiltonian system* of the form (10). Here one transforms the one system into the other by means of the so-called Legendre transformation. In this way it is possible to recast the, in general, implicit second order Euler equation into an equivalent explicit first order system of differential equations, which it is easier to work with in many investigations.

A version of Taylor's theorem is recalled in the next lemma. Here grad $f(x) = \nabla f(x) \in \mathbb{R}^n$ denotes the *gradient* of the function $f : U \subseteq \mathbb{R}^n \to \mathbb{R}$ at the point x. We call to mind that the connection between the gradient $\nabla f(x) \in \mathbb{R}^n$ and the (total) differential

$$df(x) = Df(x) \in \mathcal{L}(\mathbb{R}^n, \mathbb{R}) = (\mathbb{R}^n)'$$

is given by the formula

$$df(x)h = Df(x)h = (\nabla f(x) \mid h), \quad \forall\, h \in \mathbb{R}^n,$$

and that $\nabla f(x)$ is uniquely determined by $df(x)$ (and the inner product $(\cdot \mid \cdot)$). As is well-known, in Euclidean coordinates we have $\nabla f(x) = (D_1 f(x), \ldots, D_n f(x))$, where $D_i = \partial/\partial x^i$ denotes the partial derivative with respect to x^i.

If now $f : U \subseteq \mathbb{R}^n \to \mathbb{R}$ is twice differentiable, then

$$Ddf(x) \in \mathcal{L}(\mathbb{R}^n, (\mathbb{R}^n)'), \quad \forall\, x \in U,$$

because $df : U \to (\mathbb{R}^n)'$. We know that the map

$$\vartheta : \mathbb{R}^n \to (\mathbb{R}^n)', \quad y \mapsto (y \mid \cdot),$$

i.e.,

$$\vartheta(y)z = (y \mid z), \quad \forall\, y, z \in \mathbb{R}^n,$$

is a vector space isomorphism (and is often used to identify $(\mathbb{R}^n)'$ with $\mathbb{R}^n$). By means of this isomorphism we define the linear operator

$$D^2 f(x) := \vartheta^{-1} Ddf(x) \in \mathcal{L}(\mathbb{R}^n),$$

that is to say,

$$(D^2 f(x)y \mid z) = (Ddf(x)y)z, \quad \forall\, y, z \in \mathbb{R}^n.$$

With respect to the canonical basis of $\mathbb{R}^n$, $e_i = (\delta_{ij})_{1 \leq j \leq n}$, $i = 1, \ldots, n$, $D^2 f(x)$ can then be identified with the *Hessian*

$$[D_i D_j f(x)]_{1 \leq i, j \leq n}.$$

It follows from a theorem of H. A. Schwarz that $D^2 f(x)$ is *symmetric*, i.e.,

$$(D^2 f(x)y \mid z) = (D^2 f(x)z \mid y), \quad \forall\, y, z \in \mathbb{R}^n,$$

whenever $f \in C^2(U, \mathbb{R})$, that is, whenever f is twice continuously differentiable. (If $D^2 f(x)$ is identified with the Hessian, we know that this means that $[D^2 f(x)]^T = D^2 f(x)$.)

Finally, if $(V, (\cdot \mid \cdot))$ is an arbitrary (real) inner product space and if $A : V \to V$ is a linear transformation (a *linear operator*), then A is called *symmetric* if

$$(Ax \mid y) = (x \mid Ay), \quad \forall\, x, y \in V.$$

If A is symmetric and satisfies

$$(Ax \mid x) \geq 0, \quad \forall x \in V,$$

then A is called *positive semi-definite*. A is *positive definite* if A is symmetric and if there exists a constant $\alpha > 0$ such that

$$(Ax \mid x) \geq \alpha ||x||^2, \quad \forall x \in V, \tag{13}$$

where $||x|| := \sqrt{(x \mid x)}$ is the norm induced by the inner product. (In case $V = \mathbb{R}^n$, this is the Euclidean norm if, as usual, it is based on the Euclidean inner product.) Finally, we recall that *in the finite dimensional case relation* (13) *is equivalent to*

$$(Ax \mid x) > 0, \quad \forall x \in V \backslash \{0\}.$$

Moreover, in this case A is positive [semi-]definite if and only if all eigenvalues of A are positive [nonnegative].

(3.5) Lemma. *Let $U \subseteq \mathbb{R}^n$ be open and $f \in C^2(U, \mathbb{R})$. In addition, we assume that the line segment*

$$[\![x, y]\!] := \{x + t(y - x) \mid 0 \leq t \leq 1\}$$

is completely contained in U. Then

$$f(y) = f(x) + (\nabla f(x) \mid y - x) + \int_0^1 (1 - t)(D^2 f(x + t(y - x))(y - x) \mid y - x) \, dt.$$

Proof. Define $\varphi \in C^2([0, 1], \mathbb{R})$ by $\varphi(t) := f(x + t(y - x))$. Then we have $\varphi(0) = f(x)$, $\varphi(1) = f(y)$, as well as $\varphi'(t) = (\nabla f(x + t(y - x)) \mid y - x)$ and

$$\varphi''(t) = (D^2 f(x + t(y - x))(y - x) \mid y - x).$$

It follows from the fundamental theorem of calculus that

$$\varphi(1) = \varphi(0) + \int_0^1 \varphi'(t) dt.$$

Using integration by parts in the last integral ($u = \varphi', du = \varphi'' dt, dv = dt, v = -(1 - t)$), we get

$$\varphi(1) = \varphi(0) + \varphi'(0) + \int_0^1 (1 - t)\varphi''(t) dt,$$

from which the assertion follows by evaluating. $\square$

As is customary, a set $C \subseteq \mathbb{R}^n$ (more generally: a subset of a vector space V) is called *convex* if for every two points $x, y \in C$, the line segment $[\![x, y]\!]$ lies completely in C. If C is convex, then $f : C \to \mathbb{R}$ is called [*strictly*] *convex* if for every pair $x, y \in C$, the function

$$\varphi_{x,y} : [0, 1] \to \mathbb{R}, \quad t \mapsto f(x + t(y - x)),$$

is [*strictly*] convex. As a consequence, we immediately obtain from lemma (3.5) and from well-known theorems about convex functions of one variable the following corollary.

(3.6) Corollary. *Let $U \subseteq \mathbb{R}^n$ be open and convex and assume that $f \in C^2(U, \mathbb{R})$. Then f is convex if and only if $D^2 f(x)$ is positive semi-definite (for every $x \in U$). If $D^2 f(x)$ is positive definite for every $x \in U$, then f is strictly convex.*

The next lemma states a simple condition under which the map $\nabla f : \mathbb{R}^n \to \mathbb{R}^n$ is bijective.

(3.7) Lemma. *Let $f \in C^2(\mathbb{R}^n, \mathbb{R})$ and assume that $D^2 f$ is* <u>*uniformly positive definite*</u>*, i.e., there exists some $\alpha > 0$ such that*

$$(D^2 f(x)y \mid y) \geq \alpha |y|^2, \quad \forall x, y \in \mathbb{R}^n. \tag{14}$$

Then for every $y \in \mathbb{R}^n$, the equation

$$\nabla f(x) = y$$

has a unique solution

Proof. For $g(x) := f(x) - (y \mid x)$ we have

$$g \in C^2(\mathbb{R}^n, \mathbb{R}), \quad \nabla g(x) = \nabla f(x) - y, \quad D^2 g = D^2 f.$$

Consequently, it suffices to consider the case $y = 0$.

By corollary (3.6), f is strictly convex and hence it has at most one critical point (in fact, a minimum). Therefore the equation $\nabla f(x) = 0$ has at most one solution.

From lemma (3.5) and from (14) we derive the estimate

$$f(x) = f(0) + (\nabla f(0) \mid x) + \int_0^1 (1 - t)(D^2 f(tx)x \mid x) \, dt$$

$$\geq f(0) - |\nabla f(0)| \, |x| + \frac{\alpha}{2} |x|^2$$

for all $x \in \mathbb{R}^n$. This implies the existence of some $R > 0$ such that $f(x) \geq f(0)$ for all $|x| \geq R$. Therefore f must assume its minimum inside the ball $\mathbb{B}^n(0, R)$, if there is one. But $\bar{\mathbb{B}}^n(0, R)$ is compact and so $f|\bar{\mathbb{B}}^n(0, R)$ has a minimum, and by the preceding estimate this is the global minimum of f in $\mathbb{R}^n$. Hence there exists some $x \in \mathbb{B}^n(0, R)$ such that $\nabla f(x) = 0$. $\square$

Let now $f \in C(\mathbb{R}^n, \mathbb{R})$. Then the *Legendre transformation* of f,

$$f^* : \mathbb{R}^n \to \mathbb{R} \cup \{+\infty\},$$

is defined by

$$f^*(y) := \sup_{x \in \mathbb{R}^n} \{(y \mid x) - f(x)\}.$$

If f is convex and if, for example, the graph of f lies above the hyperplane $x \mapsto (y \mid x)$, then $-f^*(y)$ represents the minimal distance from the graph of f to this hyperplane (measured in the vertical direction).

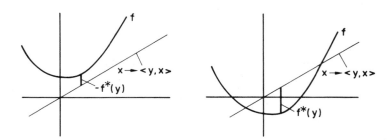

(3.8) Proposition. *Let $f \in C^2(\mathbb{R}^n, \mathbb{R})$ and assume that $D^2 f$ is uniformly positive definite. Then the following is true:*

(i) $f^*(y) = (y \mid x(y)) - f(x(y)), \quad \forall y \in \mathbb{R}^n,$

where $x(y)$ denotes the unique solution of the equation $\nabla f(x) = y$ (i.e., $x(y) = (\nabla f)^{-1}(y)$);

(ii) $f^* \in C^2(\mathbb{R}^n, \mathbb{R}), \quad f^*$ *is strictly convex and* $\nabla f^* = (\nabla f)^{-1}$;

(iii) $f(x) + f^*(y) \geq (y \mid x), \quad \forall\, x, y \in \mathbb{R}^n$
$\quad\quad f(x) + f^*(y) = (y \mid x) \Leftrightarrow x = x(y)$;

(iv) $f^{**} := (f^*)^* = f.$

Proof. (i) For $g(x) := f(x) - (y \mid x)$ we have:

$$g \in C^2(\mathbb{R}^n, \mathbb{R}), \quad \nabla g(x) = \nabla f(x) - y, \quad D^2 g = D^2 f.$$

Consequently g is strictly convex and at $x(y)$ it has a unique global minimum (cf. lemmas (3.6) and (3.7)). Now the assertion follows from

$$\min_{x \in \mathbb{R}^n} g(x) = g(x(y)) = -\max_{x \in \mathbb{R}^n} (-g(x)) = -f^*(y).$$

(ii)   With $h := \nabla f \in C^1(\mathbb{R}^n, \mathbb{R}^n), Dh$ can be identified with $D^2 f$. Since $D^2 f(x)$ is positive definite, we have $D^2 f(x) \in \mathcal{GL}(\mathbb{R}^n)$, i.e., $Dh(x)$ is invertible for every $x \in \mathbb{R}^n$. Hence the inverse function theorem implies that $x(\cdot) = h^{-1} \in C^1(\mathbb{R}^n, \mathbb{R}^n)$. From (i) we now deduce that $f^* \in C^1(\mathbb{R}^n, \mathbb{R})$. Because $\nabla f(x(y)) = y$, it follows that

$$Df^*(y) = (x(y) \mid \cdot) + (y \mid Dx(y)\cdot) - (\nabla f(x(y)) \mid Dx(y)\cdot) = (x(y) \mid \cdot).$$

This proves the relation $\nabla f^*(y) = x(y) = (\nabla f)^{-1}(y)$. Therefore $\nabla f^* \in C^1(\mathbb{R}^n, \mathbb{R}^n)$ and hence $f^* \in C^2(\mathbb{R}^n, \mathbb{R})$.

Now let $y, z \in \mathbb{R}^n$ be arbitrary and set

$$u := \nabla f^*(y) = (\nabla f)^{-1}(y), \quad v := \nabla f^*(z) = (\nabla f)^{-1}(z).$$

Then it follows from the mean value theorem that

$$(\nabla f^*(y) - \nabla f^*(z) \mid y - z) = (u - v \mid \nabla f(u) - \nabla f(v))$$

$$= (u - v \mid \int_0^1 D^2 f(v + t(u - v))(u - v)\,dt)$$

$$= \int_0^1 (D^2 f(v + t(u - v))(u - v) \mid u - v)\,dt \geq \alpha |u - v|^2,$$

hence

$$(\nabla f^*(y) - \nabla f^*(z) \mid y - z) > 0, \quad \forall\, y, z \in \mathbb{R}^n, \quad y \neq z. \tag{15}$$

Let now $\varphi(t) := f^*(z + t(y - z))$ for all $t \in [0, 1]$. Then $\varphi'(t) = (\nabla f^*(z + t(y - z)) \mid y - z)$, and for all $0 \leq t_1 < t_2 \leq 1$ we have

$$\varphi'(t_2) - \varphi'(t_1) = (t_2 - t_1)^{-1}(\nabla f^*(a) - \nabla f^*(b) \mid a - b),$$

where $a := z + t_2(y - z)$ and $b := z + t_1(y - z)$. Hence it follows from (15) that φ' is strictly increasing. Therefore φ is strictly convex and, since this holds for all $y, z \in \mathbb{R}^n$, f^* is strictly convex.

(iii)   is trivial.

(iv)   Since $f^* \in C^2(\mathbb{R}^n, \mathbb{R})$ is strictly convex, it follows that

$$f^{**}(x) = \max_{y \in \mathbb{R}^n} \{(x \mid y) - f^*(y)\}$$

holds if and only if there exists some $y = y(x) \in \mathbb{R}^n$ such that $\nabla f^*(y) = x$ (because this is the case if and only if the function $y \mapsto (x \mid y) - f^*(y)$ has a critical point at $y(x)$). By (ii), this is the case if and only if $\nabla f(x) = y$. Therefore $f^{**} : \mathbb{R}^n \to \mathbb{R}$ is well-defined and we have

$$f^{**}(x) = (x \mid y(x)) - f^*(y(x)) = (x \mid y(x)) - (y(x) \mid x) + f(x) = f(x)$$

for all $x \in \mathbb{R}^n$. $\square$

Evidently, the function $f : \mathbb{R}^n \to \mathbb{R}$ is convex if and only if the *epigraph*

$$\text{epi}(f) := \{(x, \xi) \in \mathbb{R}^n \times \mathbb{R} \mid \xi \geq f(x)\} \subseteq \mathbb{R}^n \times \mathbb{R}$$

is convex. Proposition (3.8 iii) shows that epi(f) lies above the hyperplane

$$x \mapsto (y \mid x) - f^*(y) \tag{16}$$

and that the hyperplane touches the graph of f exactly at the point $(x(y), f(x(y)))$, i.e., (16) is a *support hyperplane* to epi(f). From this it follows that the graph of f is the *envelope* of the family of hyperplanes $\{x \mapsto (y \mid x) - f^*(y) \mid y \in \mathbb{R}^n\}$.

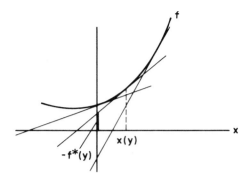

After these preliminaries we can prove the following general theorem.

(3.9) Theorem. *Suppose that $M \subseteq \mathbb{R}^n$ is open and that $-\infty < \alpha < \beta < \infty$. Moreover, let $L \in C^2((\alpha, \beta) \times M \times \mathbb{R}^n, \mathbb{R})$ and assume that for each fixed $(t_0, q_0, \dot{q}_0) \in (\alpha, \beta) \times M \times \mathbb{R}^n$,*

$$D_{\dot{q}}^2 L(t_0, q_0, \dot{q}_0) =: L_{\dot{q}\dot{q}}(t_0, q_0, \dot{q}_0) \in \mathcal{L}(\mathbb{R}^n)$$

is uniformly positive definite. Then the Euler equation

$$\frac{d}{dt}(L_{\dot{q}}) = L_q \tag{17}$$

is equivalent to the <u>*Hamiltonian system*</u>

$$\dot{p} = -H_q, \quad \dot{q} = H_p, \tag{18}$$

where the <u>*Hamiltonian*</u> *H is the Legendre transformation of the Lagrangian L with respect to the variable $\dot{q}$, that is to say,*

$$H(t, p, q) = (p \mid \dot{q}) - L(t, q, \dot{q}), \tag{19}$$

where

$$p = L_{\dot{q}}. \tag{20}$$

(Here $H_q := \nabla_q H$ denotes the gradient with respect to the variable q for fixed t, p, etc. In addition, equation (20) must be solved for $\dot{q}$: $\dot{q} = \dot{q}(t, q, p) = (L_{\dot{q}})^{-1}(p)$, where t and q are thought of as parameters. The result must then be inserted into the right-hand side of (19). By lemma (3.7), $p = L_{\dot{q}}$ can be solved uniquely for $\dot{q}$.)

Proof. "$\Longrightarrow$" It follows from (20) and from the Euler equation (17) that $\dot{p} = L_q$. By differentiating (19) with respect to q and using $p = L_{\dot{q}}$, we obtain

$$(H_q \mid h) = \left(p \mid \frac{\partial \dot{q}}{\partial q}h\right) - (L_q \mid h) - \left(L_{\dot{q}} \mid \frac{\partial \dot{q}}{\partial q}h\right) = -(L_q \mid h)$$

for all $h \in \mathbb{R}^n$. (Here of course $\frac{\partial \dot{q}}{\partial q}$ denotes the derivative of the function $q \mapsto$ $\dot{q}(t, q, p)$, i.e., $\frac{\partial \dot{q}}{\partial q} \in \mathcal{L}(\mathbb{R}^n)$.) Hence $L_q = -H_q$ and therefore $\dot{p} = -H_q$. Since proposition (3.8) implies $\nabla f^* = (\nabla f)^{-1}$, it follows that in our case we have $H_p = (L_{\dot{q}})^{-1}$, and therefore $\dot{q} = H_p$ now follows from (20).

"$\Longleftarrow$" Because $f^{**} = f$ (cf. proposition (3.8)), we have

$$L(t, q, \dot{q}) = (p \mid \dot{q}) - H(t, q, p),$$

where

$$\dot{q} = H_p, \quad \text{i.e.,} \quad p = p(t, q, \dot{q}) = (H_p)^{-1}(\dot{q}).$$

Hence proposition (3.8) implies that

$$p = (H_p)^{-1} = L_{\dot{q}}$$

(because $\nabla f^* = (\nabla f)^{-1}$), hence

$$\dot{p} = \frac{d}{dt}(L_{\dot{q}}) = -H_q, \tag{21}$$

where the last equality follows from (18). As above, we find that $L_q = -H_q$ and, using (21), the Euler equation (17) follows. $\qquad \square$

(3.10) Example. Let

$$L(t, q, \dot{q}) = T(t, q, \dot{q}) - U(t, q) = E_{\text{kin}} - E_{\text{pot}},$$

where

$$T(t, q, \dot{q}) = \frac{1}{2}(A(t, q)\dot{q} \mid \dot{q}), \quad \forall (t, q, \dot{q}) \in \mathbb{R} \times \mathbb{R}^n \times \mathbb{R}^n \tag{22}$$

and $A(t, q) \in \mathcal{L}(\mathbb{R}^n)$ is uniformly positive definite. Then the Hamiltonian H is given by

$$H(t, p, q) = (p \mid \dot{q}) - L(t, q, \dot{q}),$$

where

$$p = L_{\dot{q}} = A(t, q)\dot{q}.$$

We therefore have $\dot{q} = A(t, q)^{-1}p$ and hence

$$H(t, p, q) = (p \mid A^{-1}p) - \frac{1}{2}(AA^{-1}p \mid A^{-1}p) + U(t, q)$$

$$= \frac{1}{2}(p \mid A^{-1}p) + U(t, q) = E_{\text{kin}} + E_{\text{pot}}.$$

In other words, *if the kinetic energy (22) is given by a uniformly positive definite $A(t, q) \in \mathcal{L}(\mathbb{R}^n)$, then the Hamiltonian is the total energy, expressed in terms of the space and momentum variables.* $\qquad \square$

Let $U \subseteq \mathbb{R} \times \mathbb{R}^n \times \mathbb{R}^n$ be open and $H \in C^1(U, \mathbb{R})$. Then the (explicit) system of differential equations

$$\dot{x} = H_y, \quad \dot{y} = -H_x, \tag{23}$$

(where $H_x := \nabla_x H$, $H_y := \nabla_y H$ and $H = H(t, x, y)$) is called a *Hamiltonian system* with *Hamiltonian H*. If we set $z := (x, y) \in \mathbb{R}^n \times \mathbb{R}^n$ and

$$H_z := (H_x, H_y) = \nabla_z H,$$

then the Hamiltonian system (23) can be written in the form

$$\dot{z} = J H_z,$$

where

$$J := \begin{bmatrix} 0 & I_n \\ -I_n & 0 \end{bmatrix} \in \mathcal{L}(\mathbb{R}^n \times \mathbb{R}^n)$$

denotes the *symplectic normal form*.

(3.11) Proposition. *Let $U \subseteq \mathbb{R}^n \times \mathbb{R}^n$ be open and let $I \subseteq \mathbb{R}$ be an interval. Furthermore, let $H \in C^1(I \times U, \mathbb{R})$ and assume that $z(\cdot) \in C^1(I, U)$ is a solution of the Hamiltonian system*

$$\dot{z} = J H_z.$$

Then

$$\frac{d}{dt} H(t, z(t)) = \frac{\partial H}{\partial t}(t, z(t)), \quad \forall t \in I.$$

Proof. Applying the chain rule, we obtain

$$\frac{d}{dt} H = \frac{\partial H}{\partial t} + (H_z \mid \dot{z}) = \frac{\partial H}{\partial t} + (H_z \mid J H_z).$$

Evidently, J is antisymmetric ($J = -J^T$) and therefore the last term is zero.   □

(3.12) Corollary (*"Conservation of Energy"*).   *Let $U \subseteq \mathbb{R}^n \times \mathbb{R}^n$ be open and $H \in C^1(U, \mathbb{R})$. Then the Hamiltonian is a first integral of the autonomous Hamiltonian system*

$$\dot{z} = J \nabla H.$$

In this connection, a differential equation $\dot{x} = f(x)$ is called *autonomous* if f does not explicitly depend on the independent variable (i.e, on t).

(3.13) Remark. For $z := (x, y)$, $\zeta := (\xi, \eta) \in \mathbb{R}^n \times \mathbb{R}^n$ we set

$$[z, \zeta] := (z \mid J\zeta) = (x \mid \eta) - (y \mid \xi).$$

Then

$$[\cdot, \cdot] : \mathbb{R}^{2n} \times \mathbb{R}^{2n} \to \mathbb{R}$$

is an alternating bilinear form, that is to say,

$$[\cdot, \cdot] =: \omega \in \Omega^2(\mathbb{R}^{2n}).$$

This (constant) differential form is *nondegenerate*, that is,

$$(\omega(z, \zeta) = 0, \quad \forall\, z \in \mathbb{R}^{2n}) \Longleftrightarrow (\zeta = 0).$$

Every 2-form on $\mathbb{R}^{2n}$ has the canonical representation

$$\sum_{i,j=1}^{2n} a_{ij} dz^i \wedge dz^j.$$

Since

$$dz^i \wedge dz^j(z, \zeta) = z^i \zeta^j - z^j \zeta^i,$$

we easily derive the representation

$$\omega = dx \wedge dy := \sum_{i=1}^{n} dx^i \wedge dy^i.$$

Of course, we also have $d\omega = 0$, that is, ω is *closed*.

Now let M denote an arbitrary differentiable manifold and let $\omega \in \Omega^2(M)$ be a nondegenerate closed 2-form. Then ω is called a *symplectic form* on M and (M, ω) is a *symplectic manifold*.

If (M, ω) is a symplectic manifold, then every vector field $v \in \mathfrak{X}(M)$ on M defines a 1-form

$$\omega \lrcorner v \in \Omega^1(M),$$

i.e.,

$$(\omega \lrcorner v)(w) := \omega(v, w), \quad \forall\, w \in \mathfrak{X}(M),$$

and

$$\omega \lrcorner : \mathfrak{X}(M) \to \Omega^1(M)$$

is an isomorphism. Therefore, if $f \in C^1(M, \mathbb{R})$, there exists a unique vector field $X_f \in \mathfrak{X}M$ such that $(\omega \lrcorner v)(X_f) = df(v)$, that is,

$$\omega(v, X_f) = df(v), \quad \forall\, v \in \mathfrak{X}(M).$$

This vector field is called the *Hamiltonian vector field* on (M, ω) with Hamiltonian f. In particular, if $M = U \subseteq \mathbb{R}^{2n}$ is open and $\omega = dx \wedge dy$, then $X_f = J\,\text{grad}\, f$. (Observe the similarities between the definition of the gradient and that of the Hamiltonian vector field.)

For a more detailed exposition of Hamiltonian mechanics on manifolds we refer to the books by Arnold [1] and Abraham-Marsden [1]. □

Problems

1. Let $H \in C^1([\alpha, \beta] \times W, \mathbb{R})$, where $-\infty < \alpha < \beta < \infty$, and $W \subseteq \mathbb{R}^m \times \mathbb{R}^m$ is open. Prove that the function $(p_0, q_0) \in C^1([\alpha, \beta], W)$ is an extremal of the variational problem with partially fixed boundary

$$\delta \int_\alpha^\beta [(p \mid \dot{q}) - H(t, p, q)] \, dt = 0, \qquad q(\alpha) = a, \quad q(\beta) = b,$$

$(a, b \in \mathbb{R}^m$ fixed) if and only if (p_0, q_0) satisfies the Hamiltonian equations

$$\dot{p} = -H_q, \quad \dot{q} = H_p \quad \text{in} \quad (\alpha, \beta)$$

and the boundary conditions $q(\alpha) = a$, $q(\beta) = b$.

2. Let $|x|_p := (\sum_{i=1}^m |x^i|^p)^{1/p}, 1 < p < \infty$, denote the l_p-norm on $\mathbb{R}^m$. Calculate the Legendre transformation f^* of the function

$$f : \mathbb{R}^m \to \mathbb{R}, \qquad x \mapsto \frac{1}{p} |x|_p^p.$$

3. Let $f = U'$, where $U \in C^2((\alpha, \beta), \mathbb{R})$, and consider the differential equation

$$-\ddot{x} = f(x).$$

(a)   Sketch the phase portraits for the following potentials:

(i) $U(x) = -\frac{1}{x} + \frac{a}{x^2}, \quad x > 0, a > 0.$

(ii)

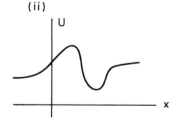

(iii)

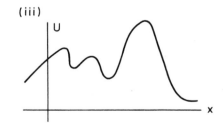

(b)   Prove that for the time it takes to get from point x_1 to point x_2 on the energy level curve $E = E_0$ (in one direction), we have:

$$t_2 - t_1 = \int_{x_1}^{x_2} \frac{dx}{\sqrt{2(E - U(x))}}.$$

Use this to calculate the period of the simple pendulum.

4. Consider the motion of a particle in a *central field*, that is, we suppose

$$-m\ddot{x} = \text{grad } U(x), \quad x \in \mathbb{R}^3 \backslash \{0\},$$

where $U(x) = U_0(|x|)$ and $U_0 \in C^2((0, \infty), \mathbb{R})$.

(a)   Prove that the *angular momentum* M relative to the point 0 is "conserved," where M is defined by the cross product

$$M := [x, m\dot{x}](:= x \times m\dot{x}).$$

(b)   Show that all orbits are planar (in a plane perpendicular to M).

(c)   Prove *Kepler's law*, which says that the radius vector "sweeps out equal area in equal time."

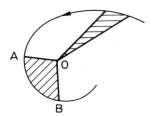

(*Hint for* (c): Use Stokes' theorem (which is also valid if there are "corners")

$$\int_{\triangle} dx \wedge dy = \frac{1}{2} \int_{\partial\triangle} x\, dy - y\, dx$$

to calculate the area of the planar "triangle" $\triangle := OAB$. Observe that the boundary integral equals $\frac{1}{2} \int_{\widehat{AB}} x\, dy - y\, dx$.)

4. Diffusion Problems

We consider a *perfect fluid moving* in a region $G \subseteq \mathbb{R}^3$. Let t denote the *time*, $\rho(x,t)$ the *density* at time t and at point x, and let $v(x,t)$ be the *velocity* at the point x and time t. Then

$$m_G(t) := \int_G \rho(x,t)\, dx \tag{1}$$

represents the *mass* in G at time t.

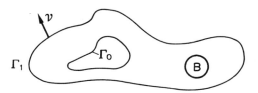

Now let B denote a subregion of G such that $\overline{B}$ is a compact C^2-manifold with boundary completely lying in G (e.g., a small ball). Then an approximation of the mass leaving B through the boundary ∂B during the time interval $[t, t + \Delta t]$ is given by the expression

$$\Delta t \int_{\partial B} (\rho v(x,t) \mid \nu)\, d\sigma(x) \tag{2}$$

(ν = outward unit normal along ∂B, $d\sigma$ = surface element of ∂B).

Indeed, the amount of fluid that flows through a small surface element $d\sigma$ during the time interval $[t, t + \Delta t]$ approximately fills a parallelepiped with height $h := (v\Delta t \mid \nu)$ and base $d\sigma$. The base is formed by a part of the tangent plane of ∂B of area $d\sigma$. Now (2) follows by taking the usual limits.

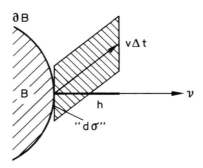

Therefore

$$-\int_{\partial B} (\rho v(\cdot, t) \mid \nu) d\sigma \tag{3}$$

represents the *increase in mass at time t due to the flow into the region B*. If we apply the divergence theorem to (3), we get

$$-\int_{\partial B} (\rho v(\cdot, t) \mid \nu)\, d\sigma = -\int_{B} \operatorname{div}(\rho v(\cdot, t))\, dx. \tag{4}$$

Furthermore, we assume that fluid is created (or destroyed) inside the region G, and we let

$$f(x, t)$$

denote the *source density* at the point x and time t. Then

$$\int_{B} f(x, t) dx$$

is the *amount of fluid that is being created* in B at time t (of course, it could also be negative if there are in fact *sinks*, i.e., if fluid is destroyed).

In summary, we find that the increase in mass over time inside B and at time t is given by

$$\frac{dm_B}{dt}(t) = -\int_{B} \operatorname{div}(\rho v(x, t))\, dx + \int_{B} f(x, t)\, dx \tag{5}$$

("conservation of mass"). Under mild regularity assumptions one can differentiate (1) inside the integral and thus obtain from (5)

$$\int_B \left[\frac{\partial \rho}{\partial t} + \operatorname{div}(\rho v) - f \right](x, t)\, dx = 0, \tag{6}$$

which holds for every "permissible" region $B \subseteq G$ (permissible here means that the divergence theorem can be applied and that $\overline{B} \subseteq G$).

We now set

$$\varphi(x, t) := \left[\frac{\partial \rho}{\partial t} + \operatorname{div}(\rho v) - f \right](x, t), \quad \forall (x, t) \in G \times \mathbb{R},$$

and assume that φ is continuous. Suppose t is fixed and let $x_0 \in G$ be an arbitrary point. It follows from the considerations above that

$$\frac{\int_B \varphi(x, t) dx}{\operatorname{vol}(B)} = 0$$

for every permissible region B. If we now consider only those permissible regions for which $x_0 \in B$, then a simple estimate shows that

$$\varphi(x_0, t) = \lim_{\substack{\operatorname{vol}(B) \to 0 \\ x_0 \in B}} \frac{\int_B \varphi(x, t)\, dx}{\operatorname{vol}(B)}.$$

We thus obtain from (6) (and the regularity assumptions above, which we assume hold in what follows) that the *transport equation*,

$$\frac{\partial \rho}{\partial t} + \operatorname{div}(\rho v) = f, \tag{7}$$

holds at every point $x \in G$ and for every time t.

If the flow is *source free*, then (7) can be simplified to

$$\frac{\partial \rho}{\partial t} = -\operatorname{div}(\rho v).$$

If, moreover, the density ρ is *independent of time and space*, then

$$\operatorname{div}(v) = 0,$$

that is, *an incompressible source free fluid is divergence free*.

In addition, we now make the assumption (which can be corroborated experimentally in many concrete cases) *that the fluid flows from regions of high density to areas of low density*, i.e., we assume that a *compensating process*, a *diffusion*, takes place.

In the simplest case we then have

$$\rho v = -a \operatorname{grad} \rho,$$

where $a > 0$ is a constant *diffusion coefficient*. However, in general we cannot expect the fluid to follow curves of steepest descent. The direction of the flow is rather dependent on the place, the time and the material (i.e., the fluid considered). That is, we will have

$$\rho v = -A \operatorname{grad} \rho, \tag{8}$$

where $A = [a^{ik}]_{1 \le i, k \le n}$ is a *diffusion matrix* depending on x, t and ρ, i.e., $A = A(x, t, \rho) \in \mathcal{L}(\mathbb{R}^3)$. The statement that the fluid flows from areas of high density to areas of lower density implies that

$$(\rho v \mid \operatorname{grad} \rho) \le 0$$

must hold, hence

$$(A \operatorname{grad} \rho \mid \operatorname{grad} \rho) \ge 0.$$

This is, of course, satisfied whenever $A(x, t, \rho) \in \mathcal{L}(\mathbb{R}^3)$ is *positive* [semi]*definite*, which is usually assumed.

The fundamental assumption that was used to derive (8) is known under different names (depending on the physical context), for example, *Fick's law* or *law of heat conduction*, etc.

It now follows from (7) and (8) that the *diffusion equation*

$$\frac{\partial \rho}{\partial t} - \operatorname{div}(A \operatorname{grad} \rho) = f \tag{9}$$

holds in G. In *Euclidean coordinates* (9) becomes (for general n)

$$\frac{\partial \rho}{\partial t} - \sum_{i,k=1}^{n} D_i(a^{ik} D_k \rho) = f,$$

where $D_i = \dfrac{\partial}{\partial x^i}$, $i = 1, \ldots, n$, i.e.,

$$\frac{\partial \rho}{\partial t} = \sum_{i,k=1}^{n} \frac{\partial}{\partial x^i}\left(a^{ik} \frac{\partial \rho}{\partial x^k} \right) + f.$$

Here we have a *second order partial differential equation*.

In the particular case when $A = aI_n$, $a > 0$, (9) simplifies to

$$\frac{\partial \rho}{\partial t} - a\Delta \rho = f, \tag{10}$$

where $\Delta \rho := \operatorname{div}(\operatorname{grad}\rho)$, i.e.,

$$\Delta \rho = \sum_{i=1}^{n} \frac{\partial^2 \rho}{\partial (x^i)^2}$$

in Euclidean coordinates. Equation (10) is called the (nonhomogeneous) *heat equation* and Δ is the *Laplace operator*.

We now must consider what will happen along the boundary of G, ∂G. There are several possibilities. To this end, we assume that ∂G consists of two disjoint parts,

$$\partial G = \Gamma_0 \cup \Gamma_1$$

(both are two-dimensional (in general: $(n - 1)$-dimensional) C^1-manifolds such that G lies locally on one side of ∂G. In other words: $\overline{G}$ is an n-dimensional C^1-manifold with boundary in $\mathbb{R}^n$).

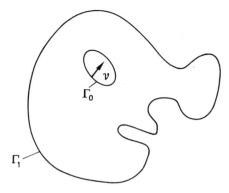

On Γ_0 the density ρ is kept at a given value φ (where, of course, φ may depend on (x, t), in general, e.g., $\varphi \in C(\Gamma_0 \times \mathbb{R}, \mathbb{R})$), i.e.,

$$\rho = \varphi \quad \text{on} \quad \Gamma_0 \tag{11}$$

We assume that the boundary part Γ_1 may possibly be permeable. Then (see above)

$$\Delta t (\rho v(x, t) \mid \nu)\, d\sigma$$

represents the amount of fluid streaming outward during the time interval $[t, t + \Delta t]$ and through the surface element $d\sigma$, that is to say,

$$(\rho v(x, t) \mid \nu)\, d\sigma$$

is the change of the amount of fluid over time evaluated at time t and at the point x, hence, in general, it is a function of position, time and density, i.e.,

$$(\rho v(x, t) \mid \nu(x)) = \psi(x, t, \rho).$$

In general, the function ψ is known, since it can be deduced from appropriate physical theories which can be confirmed experimentally. So, for instance, it is often meaningful to assume that the amount of outflowing fluid is proportional to the amount of fluid present at that time and place or, slightly more general, that

$$\psi(x, t, \rho) = \alpha(x, t)\rho - \beta(x, t). \tag{12}$$

Here α and β are *measures of the permeability of the boundary*.

From (8) we obtain (since $A = A^T$, which is always included in the definition of positive semi-definiteness!)

$$(\rho v \mid \nu) = -(A\, \mathrm{grad}\, \rho \mid \nu) = -(A\nu \mid \mathrm{grad}\, \rho),$$

and in Euclidean coordinates:

$$(\rho v \mid \nu) = -\sum_{i=1}^{n} \left(\sum_{k=1}^{n} a^{ik} \nu^k \right) \frac{\partial \rho}{\partial x^i}.$$

If $A = aI_n$ for $a > 0$, then $(A\nu \mid \mathrm{grad}\rho) = a(\nu \mid \mathrm{grad}\rho) = a\frac{\partial \rho}{\partial \nu}$, where $\partial / \partial \nu$ denotes the derivative in the direction of the outward normal. In general, $\nu_A := A\nu$ denotes the *conormal* with respect to A, and

$$\frac{\partial \rho}{\partial \nu_A} := (A\nu \mid \mathrm{grad}\rho)$$

is called the *conormal derivative* (on ∂G).

In summary, we have on Γ_1:

$$\frac{\partial \rho}{\partial \nu_A} = -\psi(x, t, \rho) \tag{13}$$

or, under assumption (12),

$$\frac{\partial \rho}{\partial \nu_A} + \alpha\rho = \beta. \tag{14}$$

In particular, if the boundary Γ_1 is *impermeable*, then

$$\frac{\partial \rho}{\partial \nu_A} = 0 \quad \text{on} \quad \Gamma_1.$$

Finally, we assume that we know the density distribution at time $t = t_0$:

$$\rho(x, t_0) = u_0(x), \quad \forall x \in G, \tag{15}$$

and that we are interested in knowing the density $\rho(x, t)$ at $x \in G$ and time $t > t_0$. That is to say, we seek a function $u : \overline{G} \times [t_0, \infty) \to \mathbb{R}$ which satisfies the following *initial-boundary value problem*:

$$\begin{cases} \dfrac{\partial u}{\partial t} - \mathrm{div}(A\,\mathrm{grad}\,u) = f & \text{in } G \times (t_0, \infty) \\[2mm] \qquad\qquad\qquad u = \varphi & \text{on } \Gamma_0 \\[2mm] \dfrac{\partial u}{\partial \nu_A} + \alpha u = \beta & \text{on } \Gamma_1 \\[2mm] \qquad\qquad u(\cdot, t_0) = u_0. \end{cases} \tag{16}$$

Here $A, f, \alpha, \beta, \varphi$ are known functions of (x, t) and possibly also of u and grad u, and for each fixed argument, A is a positive definite symmetric matrix.

(4.1) Remarks. (a) The "fluid" considered in the derivation of (16) need not be a real fluid at all. For example, u could be the temperature of some medium (then heat is the "fluid") or the population of some place-dependent population model (cf. section 1), i.e., a population model in which "migration" of the individuals is taken into account. In the latter case, the "source term" f, of course, represents the changes due to birth and death of the population at the position x and time t.

One can of course also consider systems with several populations, for example, *predator-prey systems subject to diffusion* (i.e., migration effects). We then obtain *systems* of *reaction-diffusion equations*:

$$
\begin{cases}
\dfrac{\partial u^i}{\partial t} - \operatorname{div}(A_i \operatorname{grad} u^i) = f(x, t, u^1, \dots, u^m) & \text{in } G \times (t_0, \infty), \\[2mm]
\qquad\qquad u^i = \varphi^i \quad \text{on } \Gamma_0, \\[2mm]
\qquad \dfrac{\partial u^i}{\partial \nu_{A_i}} + \alpha_i u^i = \beta_i \quad \text{on } \Gamma_1, \\[2mm]
\qquad\qquad u^i = u_0^i \quad \text{if } t = t_0,
\end{cases}
$$

$i = 1, \dots, m$ (no summation convention!).

Such reaction-diffusion equations arise, especially in chemistry, for the mathematical descriptions of chemical reactions in solutions, in which diffusion takes place. They are of much current interest and most ensuing mathematical questions (such as the global existence of solutions, stability, general longtime behavior, etc.) are still open.

(b) The initial-boundary value problem (16) is a problem from the theory of partial differential equations. Formally, problem (16) can also be written as an initial value problem for some ordinary differential equation. This cannot be done in $\mathbb{R}^n$, but *rather in some appropriate infinite dimensional space*. To this end we set

$$
v(t) := u(\cdot, t),
$$

that is, v is a function from $\mathbb{R}$ into some appropriate function space on G, $X(G)$ (e.g., $X(G) = C(G)$ or $X(G) = L^2(G)$). For reasons of simplicity we also assume that $A = A(x, u)$ and $f(x, u)$ do not depend on t explicitly. Moreover, assume that α, β and φ are independent of t. We set

$$
\operatorname{dom}(F) := \left\{ v \in X(G) \mid v \in C^2(G),\ v|\Gamma_0 = \varphi,\ \left(\frac{\partial}{\partial \nu_A} + \alpha \right) v = \beta \text{ on } \Gamma_1 \right\}
$$

and

$$
F(v) := \operatorname{div}(A \operatorname{grad} v) + f, \quad \forall v \in \operatorname{dom}(F).
$$

Then formally (16) has the structure

$$
\dot{v} = F(v), \quad v(t_0) = u_0.
$$

This approach to problem (16), that is to say, its treatment as an ordinary differential equation in an infinite dimensional Banach space, is extremely important and far-reaching. It is the basis of the functional analytic theory of evolution equations.

(c) Let $M \subseteq \mathbb{R}^n$ be an m-dimensional C^2-manifold and let $f \in C^2(M, \mathbb{R})$. It is known that in local coordinates we have

$$
(\operatorname{grad} f)^i = \sum_{j=1}^m g^{ij} \frac{\partial f}{\partial x^j}, \quad i = 1, 2, \dots, m, \tag{17}
$$

and

$$
\Delta_M f := \operatorname{div} \operatorname{grad} f = \frac{1}{\sqrt{g}} \sum_{i,j=1}^m \frac{\partial}{\partial x^i} \left(\sqrt{g}\, g^{ij} \frac{\partial f}{\partial x^j} \right). \tag{18}
$$

Here Δ_M denotes the Laplace-Beltrami operator on M,

$$[g^{ij}]_{1\leq i,j\leq m} = [g_{ij}]^{-1}_{1\leq i,j\leq m},$$

$$[g_{ij}]_{1\leq i,j\leq m} = \left[\left(\frac{\partial}{\partial x^i} \,\Big|\, \frac{\partial}{\partial x^j}\right)\right]_{1\leq i,j\leq m}$$

and $\sqrt{g} = \sqrt{\det[g_{ij}]}$. These relations also hold for every m-dimensional Riemannian manifold with metric $[g_{ij}]_{1\leq i,j\leq m}$ (see, e.g., Bröcker [1] or Abraham & Marsden [1]).

If we now assume that the matrix A in (16) is positive definite and only depends on $x \in G$, we may employ

$$[g_{ij}] := aA^{-1} = [a^{ij}]^{-1}a, \quad \text{where} \quad a := (\det A)^{1/(m-2)}, \tag{19}$$

as a Riemannian metric on G and thereby obtain the Riemannian manifold $M := (G, [g_{ik}])$. It follows that $\sqrt{g}[g^{ij}] = \frac{\sqrt{g}}{a}[a^{ij}]$ and, because

$$(\sqrt{g})^2 = \det[g_{ij}] = a^m \det A^{-1} = \frac{a^m}{\det A} = \frac{a^m}{a^{m-2}} = a^2,$$

we deduce from (18) that

$$\operatorname{div}(A\operatorname{grad}u) = \sum_{i,j=1}^{m} \frac{\partial}{\partial x^i}\left(a^{ij}\frac{\partial u}{\partial x^j}\right)$$

$$= a \cdot \frac{1}{\sqrt{g}} \sum_{i,j=1}^{m} \frac{\partial}{\partial x^i}\left(\sqrt{g}\, g^{ij}\frac{\partial u}{\partial x^j}\right)$$

$$= a(x)\Delta_M u,$$

where $a(x) = [\det A(x)]^{1/(m-2)}$.

Since $[g^{ij}] = \frac{1}{a}[a^{ij}]$, (17) implies that

$$\frac{\partial u}{\partial \nu_A} = \sum_{i,j=1}^{m} a^{ij}\frac{\partial u}{\partial x^j}\nu^i = a\sum_{i=1}^{m}\left(\sum_{j=1}^{m} g^{ij}\frac{\partial u}{\partial x^j}\right)\nu^i$$

$$= a(\operatorname{grad}_M u \mid \nu) = (\operatorname{grad}_M u \mid aA^{-1}A\nu)$$

$$= (\operatorname{grad}_M u \mid aA^{-1}\nu_A) = \sum_{i,j=1}^{m} g_{ij}\nu_A^i(\operatorname{grad}_M u)^j \tag{20}$$

$$= (\operatorname{grad}_M u \mid \nu_A)_M,$$

because, as we know,

$$(v \mid w)_M := \sum_{i,j=1}^{m} g_{ij}v^i w^j, \quad v, w \in T(M),$$

represents the inner product on the tangent bundle. Here the gradient $\operatorname{grad}_M f$ on the manifold M is defined, as usual, by

$$df(v) = (\text{grad}_M f \mid v)_M, \quad \forall v \in T(M).$$

Since

$$(v \mid \nu) = \left(v \mid aA^{-1}\left(\frac{1}{a}A\nu\right) \right) = \left(v \mid \frac{1}{a}\nu_A \right)_M$$

for all $v \in T(M) \cong M \times \mathbb{R}^m$, it follows that $\nu_A \in T(\partial M)^\perp$ is an outward pointing normal vector of length

$$\|\nu_A\|_M := \sqrt{(\nu_A \mid \nu_A)_M} = \sqrt{(A\nu \mid \nu)a} > 0.$$

Therefore (cf. (20))

$$\frac{\partial u}{\partial \nu_A} = (\text{grad}_M u \mid \nu_A)_M$$

is *the derivative of u on ∂M in the direction of the (not necessarily unit) outward normal vector ν_A.* Using the outward unit normal

$$\nu_M := \frac{\nu_A}{\|\nu_A\|_M}$$

on ∂M, we may now express the initial-boundary value problem (16) as an initial-boundary value problem on the Riemannian manifold M:

$$\begin{cases} \dfrac{\partial u}{\partial t} - a\Delta_M u & = & f & \text{in } M \times (t_0, \infty), \\ u & = & \varphi & \text{on } \Gamma_0, \\ \dfrac{\partial u}{\partial \nu_M} + \alpha_M u & = & \beta_M & \text{on } \Gamma_1, \\ u(\cdot, t_0) & = & u_0 & \text{on } M, \end{cases}$$

where $\alpha_M := \alpha/\|\nu_A\|_M$ and $\beta_M := \beta/\|\nu_A\|_M$. $\qquad\qquad\qquad\qquad\qquad\qquad\square$

Separation of Variables

We now consider the *homogeneous* initial-boundary value problem

$$\begin{array}{rcll} \dfrac{\partial u}{\partial t} - \text{div}(A\,\text{grad}u) & = & 0 & \text{in } G \times (0, \infty) \\ u & = & 0 & \text{on } \Gamma_0 \times (0, \infty) \\ \dfrac{\partial u}{\partial \nu_A} + \alpha u & = & 0 & \text{on } \Gamma_1 \times (0, \infty) \\ u(\cdot, 0) & = & u_0 & \text{on } G, \end{array} \qquad (21)$$

where we assume that A and α depend only on x and not on t. An important special case of (21) is illustrated by the ordinary *heat equation without external sources*, i.e., the problem

$$\begin{array}{rcll} \dfrac{\partial u}{\partial t} - \Delta u & = & 0 & \text{in } G \times (0, \infty) \\ u & = & 0 & \text{on } \partial G \times (0, \infty) \\ u(\cdot, 0) & = & u_0 & \text{on } G, \end{array}$$

where we have chosen the so-called *Dirichlet boundary condition* (i.e., the temperature is prescribed on ∂G).

To solve (21), we apply the *method of separation of variables* by finding solutions of the form

$$u(x, t) = v(x)w(t).$$

If we substitute this into the differential equation, we obtain (with $\cdot = d/dt$)

$$
\begin{aligned}
v\dot{w} - w\,\mathrm{div}(A\,\mathrm{grad}v) &= 0 \quad \text{in } G \times (0, \infty), \\
wv &= 0 \quad \text{on } \Gamma_0 \times (0, \infty), \\
w\left(\frac{\partial v}{\partial \nu_A} + \alpha v\right) &= 0 \quad \text{on } \Gamma_1 \times (0, \infty).
\end{aligned}
\tag{22}
$$

If for x we choose some fixed x_0 satisfying $v(x_0) \neq 0$, then it follows from (22) that

$$\dot{w} = -\lambda w, \quad t \geq 0, \tag{23}$$

must hold for some constant λ. From this we deduce the explicit form of w,

$$w(t) = ce^{-\lambda t}, \quad t \geq 0, \tag{24}$$

for some arbitrary constant $c \in \mathbb{R} \setminus \{0\}$. From (23), (24) and (22), we now obtain the following *boundary value problem* for the function v:

$$
\begin{aligned}
-\mathrm{div}(A\,\mathrm{grad}v) &= \lambda v \quad \text{in } G, \\
v &= 0 \quad \text{on } \Gamma_0, \\
\frac{\partial v}{\partial \nu_A} + \alpha v &= 0 \quad \text{on } \Gamma_1.
\end{aligned}
\tag{25}
$$

We now consider the simplest special case, namely, the case when $G = (a, b)$ is a bounded open interval in $\mathbb{R}$. Then $-\mathrm{div}(A\,\mathrm{grad}\,v)$ is a differential operator of the form

$$-(pv')', \quad ' := \frac{d}{dx},$$

for some $p \in C^1((a, b), \mathbb{R})$ satisfying $p > 0$ (which corresponds to the positive definiteness of A). With the *convention* that

$$\nu(a) := -1, \quad \nu(b) := 1,$$

the boundary conditions now become:

at $x = a$:
$$
\begin{cases}
v(a) = 0 \\
\quad\quad \text{or} \\
-p(a)v'(a) + \alpha v(a) = 0
\end{cases}
$$

and

at $x = b$:
$$
\begin{cases}
v(b) = 0 \\
\quad\quad \text{or} \\
p(b)v'(b) + \alpha v(b) = 0,
\end{cases}
$$

which is abbreviated to

$$Bv = 0$$

in the following. Thus, in this case, (25) takes on the form

$$\begin{cases} -(pv')' = \lambda v & \text{in } (a, b) \\ \quad Bv = 0 & \text{on } \partial(a, b) = \{a, b\}. \end{cases} \tag{26}$$

Boundary value problems of this type are called *Sturm-Liouville boundary value problems*. They represent special cases within the theory of elliptic boundary value problems.

At the moment we further specialize to $a = 0$, $b = \pi$ and $p = 1$, i.e., we analyze the particular problem

$$\begin{cases} -v'' = \lambda v & \text{in } (0, \pi), \\ v(0) = v(\pi) = 0. \end{cases} \tag{27}$$

One easily verifies that the function

$$v(x) := A \sin(\sqrt{\lambda}x) + B \cos(\sqrt{\lambda}x) \tag{28}$$

is a solution of the differential equation $-v'' = \lambda v$ for every choice of constants A and B. Later we will see that (28) already represents all solutions. The boundary conditions imply:

$$v(0) = 0 \quad \Rightarrow \quad B = 0$$

$$v(\pi) = 0 \quad \Rightarrow \quad \sin(\sqrt{\lambda}\pi) = 0 \quad \Rightarrow \quad \sqrt{\lambda} \in \mathbb{Z}.$$

We thus see that (27) has *nontrivial* solutions (that is, nonvanishing solutions) only if

$$\lambda \in \{n^2 \mid n \in \mathbb{N}^*\}.$$

If we now set

$$\lambda_n := n^2, \quad \forall n \in \mathbb{N}^*,$$

and

$$w_n(t) := c_n e^{-\lambda_n t}, \quad c_n \in \mathbb{R}, \ c_n \neq 0,$$

then it follows from (24) and the remarks above that, for each $n \in \mathbb{N}^*$, the function

$$u_n(x, t) := v_n(x)w_n(t) = c_n \sin(nx)e^{-n^2 t}$$

is a solution of the heat equation

$$\begin{cases} u_t - u_{xx} = 0 & \text{in } (0, \pi) \times (0, \infty) \\ u(0, t) = u(\pi, t) = 0, & \forall t \geq 0. \end{cases} \tag{29}$$

Since (29) clearly represents a linear problem, it follows that the sum of two solutions is also a solution, that is, the so-called *superposition principle* holds. By means of a purely formal superposition we therefore obtain a solution of the form

$$u(x, t) = \sum_{n=1}^{\infty} v_n(x)w_n(t) = \sum_{n=1}^{\infty} c_n \sin(nx)e^{-n^2 t}.$$

(This is indeed a solution if the convergence of the series is such that it can be differentiated "term-by-term," once with respect to t and twice with respect to x.)

To determine the unknown coefficients c_n, we must keep in mind that the initial condition $u(\cdot, 0) = u_0$ must also be satisfied, that is, we must have

$$u(x, 0) = \sum_{n=1}^{\infty} c_n \sin(nx) = u_0(x), \quad \forall x \in (0, \pi).$$

In other words: We must determine the coefficients c_n so that the function u_0 is represented by the *Fourier series*

$$\sum_{n=1}^{\infty} c_n \sin(nx)$$

(if at all possible).

We now suppose that, in fact, we have

$$u_0(x) = \sum_{n=1}^{\infty} c_n \sin(nx), \tag{30}$$

and that the convergence is such that the subsequent manipulations can be executed. We then multiply (30) by $\sin(mx)$, we integrate from 0 to π, and we employ the *orthogonality relations*

$$\int_0^{\pi} \sin(nx) \sin(mx)\, dx = \begin{cases} 0, & \text{if } n \neq m \\ \dfrac{\pi}{2}, & \text{if } n = m, \end{cases}$$

(which are easily verified). As a result, the *Fourier coefficients* c_n are given by the expressions

$$c_n = \frac{2}{\pi} \int_0^{\pi} \sin(nx) u_0(x)\, dx.$$

Using the shortened forms

$$(u \mid v) := \int_0^{\pi} u(x) v(x)\, dx$$

and

$$e_n(x) := \sqrt{\frac{2}{\pi}} \sin(nx), \quad \forall n \in \mathbb{N}^*,$$

we have

$$(e_n \mid e_m) = \delta_{nm}, \quad \forall n, m \in \mathbb{N}^*, \tag{31}$$

and

$$\sqrt{\frac{\pi}{2}} c_n = (u_0 \mid e_n), \quad \forall n \in \mathbb{N}^*.$$

Therefore the Fourier series has the form

$$\sum_{n=1}^{\infty} (u_0 \mid e_n) e_n,$$

and the (formal!) series

$$u(x, t) = \sum_{n=1}^{\infty} (u_0 \mid e_n) e_n(x) e^{-n^2 t}$$

(formally) represents a solution of the initial-boundary value problem

$$\begin{cases} u_t - u_{xx} &= 0 & \text{in } (0, \pi) \times (0, \infty), \\ u(0, t) = u(\pi, t) &= 0 & \forall t \geq 0 \\ u(x, 0) &= u_0(x) & \forall x \in (0, \pi). \end{cases}$$

Evidently, relations (31) mean exactly that $\{e_n \mid n \in \mathbb{N}\}$ is an *orthonormal set* in the Hilbert space $L^2(0, \pi)$. This observation suggests to give the general BVP (25) also a more abstract formulation. To this end, we consider an arbitrary (bounded) domain $G \subseteq \mathbb{R}^m$ and define a linear operator A on the Hilbert space $H := L^2(G)$, with the usual inner product

$$(u \mid v) := \int_G u(x)v(x)\, dx,$$

as follows:

$$\text{dom}(\mathsf{A}) := \{u \in C^2(\overline{G}) \mid \; u|_{\Gamma_0} = 0, \; \frac{\partial u}{\partial \nu_A} + \alpha u = 0 \text{ on } \Gamma_1\}$$

and

$$\mathsf{A}u := -\text{div}(A \,\text{grad} u), \quad \forall u \in \text{dom}(\mathsf{A}).$$

One verifies that

$$\mathsf{A} : \text{dom}(\mathsf{A}) \subseteq H \to H$$

is a linear *symmetric* operator, i.e., we have

$$(\mathsf{A}u \mid v) = (\mathsf{A}v \mid u), \quad \forall u, v \in \text{dom}(\mathsf{A}).$$

With these definitions, (25) assumes the simple form

$$\mathsf{A}v = \lambda v. \tag{32}$$

With reference to linear algebra, (32) is called an *eigenvalue problem*, and $\lambda \in \mathbb{R}$ is called an *eigenvalue* of A if there exists some $v \neq 0$ such that (32) is satisfied. Every such v is called an eigenfunction corresponding to the eigenvalue λ and the set

$$\sigma_p(\mathsf{A}) := \{\lambda \in \mathbb{R} \mid \lambda \text{ is an eigenvalue of } \mathsf{A} \}$$

is called the *point spectrum* of A.

Since (25) is a different formulation of (32), it is also called an *eigenvalue problem*.

The above considerations for the simplest special case (29) lead to the following questions:

(1) *What is* $\sigma_p(\mathsf{A})$?

Under very general conditions, it is shown in functional analysis that $\sigma_p(\mathsf{A})$ is an infinite set with no finite limit point, i.e.,

$$\sigma_p(\mathsf{A}) = \{\lambda_n \mid n \in \mathbb{N}\}.$$

(In the special case above we had $\lambda_n = n^2$.)

(2) *Can $u_0 \in H$ be represented as a Fourier series in terms of the eigenfunctions, i.e., do we have*

$$u_0 = \sum_{n=0}^{\infty} (u_0 \mid e_n) e_n, \quad \text{(33)}$$

where $Ae_n = \lambda_n e_n$? This is the so-called *completeness problem*.

(3) *Can the series*

$$\sum_{n=0}^{\infty} (u_0 \mid e_n) e_n(x) e^{-\lambda_n t}$$

be differentiated term-by-term, i.e., more precisely, does the operator $\frac{\partial}{\partial t} + A$ commute with $\sum_{n=0}^{\infty}$ (regularity problem)?

If all of this is true (and in the theory of partial differential equations it is shown in great generality that it is correct), the initial-boundary value problem (21), which can be formulated as an "abstract IVP" in the Hilbert space H,

$$\begin{cases} \dot{u} + Au &= 0, \quad \text{if } t > 0 \\ u(0) &= u_0, \end{cases} \quad \text{(34)}$$

has a solution given by

$$u(t, x) := \sum_{n=0}^{\infty} (u_0 \mid e_n) e_n(x) e^{-\lambda_n t}. \quad \text{(35)}$$

If, for the moment, we assume that (35) makes sense and that $\{e_n \mid n \in \mathbb{N}\}$ is an orthonormal set in the Hilbert space H, then it easily follows (at least formally) that

$$\|u(\cdot, t)\|_H^2 = (u(\cdot, t) \mid u(\cdot, t)) = \sum_{n=0}^{\infty} (u_0 \mid e_n)^2 e^{-2\lambda_n t}.$$

If we further assume that $\lambda_0 \leq \lambda_1 \leq \lambda_2 \leq \ldots$ holds (as is the case in the specific example above), we obtain the estimate

$$\|u(\cdot, t)\|_H \leq e^{-\lambda_0 t} \sqrt{\sum_{n=0}^{\infty} (u_0 \mid e_n)^2} = e^{-\lambda_0 t} \|u_0\|_H,$$

where the last equality follows (formally) from (33). So if we have $\lambda_0 > 0$ (cf. the preceding example), then

$$u(\cdot, t) \to 0 \quad \text{in } L^2(G) \quad \text{as } t \to \infty.$$

This is a statement about the longtime behavior of solutions of (34). It says that the "steady-state solution" $u = 0$ (i.e., the zero of $Au = 0$) is asymptotically stable. [In the case of a heat conduction model (with zero boundary conditions), this corresponds to the fact that over a long period of time the temperature of the object considered, decreases to zero if no heat sources are present and if the surface temperature is maintained at 0 (i.e., if all the heat that reaches the surface is "absorbed").]

The preceding formal considerations should give a taste of the functional analytic methods which can be employed treating problems in partial differential equations. It has been found most appropriate to formulate especially diffusion problems as abstract ordinary differential equations in suitable Hilbert or Banach spaces, as was indicated in (34). Here the fact that the resulting operator A can, in general, not be defined on all of H, that is, that we are dealing with an *unbounded* (i.e., not continuous) linear operator, represents fundamental difficulties. They can be surmounted by drawing upon more penetrating methods in functional analysis.

To demonstrate the method of separation of variables on a further example, we now consider the simplest model of *heat conduction in a homogeneous ball*. Hence we let $G := \mathbb{B}^3 := \{x \in \mathbb{R}^3 \mid |x| < 1\}$ and we seek one (or all!) solution(s) of the initial-boundary value problem

$$
\begin{cases}
\dfrac{\partial u}{\partial t} - \Delta u & = & 0 & \text{in } \mathbb{B}^3 \times (0, \infty), \\
u & = & 0 & \text{on } \mathbb{S}^2 \times (0, \infty), \\
u(\cdot, 0) & = & u_0 & \text{in } \mathbb{B}^3.
\end{cases}
$$

By separating the variables, $u = w(t)v(x)$, we obtain for v the *eigenvalue problem* (EVP)

$$
\begin{cases}
-\Delta v & = & \lambda v & \text{in } \mathbb{B}^3 \\
v & = & 0 & \text{on } \mathbb{S}^2 .
\end{cases}
\tag{36}
$$

Since

$$
v\Delta u = v\,\mathrm{div}(\mathrm{grad}\, u) = \mathrm{div}(v\,\mathrm{grad}\, u) - (\mathrm{grad}\, v \mid \mathrm{grad}\, u),
$$

under the assumption that $v \in C^2(\overline{\mathbb{B}}^3)$ is a solution of (36), and by multiplying the equation $-\Delta v = \lambda v$ by v and subsequent integration over $\mathbb{B}^3$, we obtain from (36) that

$$
\lambda \int_{\mathbb{B}^3} v^2 \, dx = \int_{\mathbb{B}^3} -v\Delta v \, dx = -\int_{\mathbb{B}^3} \mathrm{div}(v\,\mathrm{grad}\, v) \, dx + \int_{\mathbb{B}^3} |\mathrm{grad}\, v|^2 \, dx.
$$

From this, Stokes' theorem, and using the boundary condition $v = 0$, we deduce that

$$
\lambda \int_{\mathbb{B}^3} v^2 \, dx = -\int_{\mathbb{S}^2} v\frac{\partial v}{\partial \nu} \, d\sigma + \int_{\mathbb{B}^3} |\mathrm{grad}\, v|^2 \, dx = \int_{\mathbb{B}^3} |\mathrm{grad}\, v|^2 \, dx,
$$

and so $\lambda \geq 0$ follows. Hence we may set

$$
\lambda = \mu^2 \quad \text{for} \quad \mu \geq 0.
\tag{37}
$$

It is now reasonable to make use of the symmetry of the domain and express the Laplace operator in spherical coordinates (rather than Euclidean coordinates). For this we recall that (cf. remark (4.1 c)) *in local coordinates* the Laplace-Beltrami operator of an m-dimensional C^2-submanifold M of $\mathbb{R}^n$ has the form

$$
\Delta_M u = \mathrm{div}(\mathrm{grad}\, u) = \frac{1}{\sqrt{g}} \sum_{i,j=1}^{m} \frac{\partial}{\partial x^i} \left(\sqrt{g}\, g^{ij} \frac{\partial u}{\partial x^j} \right).
$$

(4.2) Examples. (a) *Parametrization of $\mathbb{B}^3$ by spherical coordinates.* Let

$$
V := (0, 1) \times (0, 2\pi) \times \left(-\frac{\pi}{2}, \frac{\pi}{2} \right) \to \mathbb{B}^3
$$
$$
g : (r, \varphi, \vartheta) \mapsto (r \cos \varphi \cos \vartheta, r \sin \varphi \cos \vartheta, r \sin \vartheta)
$$

be the local parametrization of the open subset

$$U := \mathbb{B}^3 \setminus (\mathbb{R}_+ \times \{0\} \times \mathbb{R})$$

of the ball $\mathbb{B}^3$ by spherical coordinates. Using the formulae from (4.1 c), a calculation shows that [since $g_{ij} = (D_i g \mid D_j g)$] we have

$$\Delta_{\mathbb{B}^3} = \frac{1}{r^2} \left[\frac{\partial}{\partial r} \left(r^2 \frac{\partial}{\partial r} \right) + \frac{1}{\cos^2 \vartheta} \frac{\partial^2}{\partial \varphi^2} + \frac{1}{\cos \vartheta} \frac{\partial}{\partial \vartheta} \left(\cos \vartheta \frac{\partial}{\partial \vartheta} \right) \right]. \tag{38}$$

(b) *Parametrization of $\mathbb{S}^2$ by spherical coordinates.* Let

$$V := (0, 2\pi) \times \left(-\frac{\pi}{2}, \frac{\pi}{2} \right) \to \mathbb{S}^2$$
$$g : (\varphi, \vartheta) \mapsto (\cos \varphi \cos \vartheta, \sin \varphi \cos \vartheta, \sin \vartheta)$$

be the local parametrization of $U := \mathbb{S}^2 \setminus (\mathbb{R}_+ \times \{0\} \times \mathbb{R})$ by spherical coordinates. Then we have

$$\Delta_{\mathbb{S}^2} = \frac{1}{\cos^2 \vartheta} \frac{\partial^2}{\partial \varphi^2} + \frac{1}{\cos \vartheta} \frac{\partial}{\partial \vartheta} \left(\cos \vartheta \frac{\partial}{\partial \vartheta} \right). \tag{39}$$

$\square$

In particular, by comparing (38) and (39) we see that

$$\Delta_{\mathbb{B}^3} = \frac{1}{r^2} \left[\frac{\partial}{\partial r} \left(r^2 \frac{\partial}{\partial r} \right) + \Delta_{\mathbb{S}^2} \right]. \tag{40}$$

(It is easily seen that (40) holds "globally," i.e., on all of $\mathbb{B}^3$, irrespective of the particular parametrization of $\mathbb{S}^2$.)

With (37) and (40), the EVP (36) now becomes:

$$-\frac{1}{r^2} \left[\frac{\partial}{\partial r} \left(r^2 \frac{\partial v}{\partial r} \right) + \Delta_{\mathbb{S}^2} v \right] = \mu^2 v \quad \text{in } \mathbb{B}^3 \setminus \{0\}$$
$$v = 0 \quad \text{on } \mathbb{S}^2,$$

which suggests the *separation of variables* $v(r, \omega) = u(r) w(\omega)$, where $(r, \omega) \in (0, 1) \times \mathbb{S}^2$. Substituting $v = uw$ into the differential equation, we get

$$w[(r^2 u')' + \mu^2 r^2 u] + u \Delta_{\mathbb{S}^2} w = 0 \quad \text{in } \mathbb{B}^3 \setminus \{0\}.$$

If we now consider some fixed $r_0 \in (0, 1)$ such that $u(r_0) \neq 0$, we obtain for w the EVP

$$-\Delta_{\mathbb{S}^2} w = \sigma w \quad \text{on } \mathbb{S}^2. \tag{41}$$

If w is a solution of (41), then u must clearly satisfy

$$(r^2 u')' + (\mu^2 r^2 - \sigma) u = 0 \quad \text{in } (0, 1). \tag{42}$$

Multiplying (41) by w and subsequently integrating over $\mathbb{S}^2$, we obtain from Stokes' theorem that

$$\sigma \int_{\mathbb{S}^2} w^2 = \int_{\mathbb{S}^2} |\text{grad } w|^2,$$

whenever $w \in C^2(\mathbb{S}^2)$ is a solution of (41).

Hence, again, we deduce that $\sigma \geq 0$ and so we may set

$$\sigma = \nu^2 \quad \text{for} \quad \nu \geq 0.$$

For u we therefore obtain the second order linear ordinary differential equation (where $x := r$)

$$(x^2 u')' + (\mu^2 x^2 - \nu^2)u = 0 \quad \text{in} \quad (0, 1).$$

The boundary condition

$$u(1) = 0$$

follows from the boundary condition in (36). Since the function $v(r, \omega) = u(r)w(\omega)$, which is only defined on $\mathbb{B}^3 \setminus \{0\}$, must be extended to a continuously differentiable function on $\mathbb{B}^3$ so that it indeed represents a solution of (36), we must require that $u'(0) = 0$. Hence for the radial part u we obtain the BVP

$$\begin{cases} (x^2 u')' + (\mu^2 x^2 - \nu^2)u &= 0 \quad \text{for } 0 < x < 1, \\ u'(0) &= 0, \\ u(1) &= 0. \end{cases}$$

The differential equation can be written in the equivalent form

$$u'' + \frac{2}{x}u' + \left(\mu^2 - \frac{\nu^2}{x^2}\right)u = 0. \tag{43}$$

This makes it evident that a singularity exists at $x = 0$. Differential equations of this type can be discussed in great detail with the methods of complex variables. Specifically, in the case of the equation above, we are dealing with a differential equation of the so-called *Fuchsian class* (see, e.g., Walter [1]).

In (43) we now introduce a new independent variable by means of the transformation

$$y(x) := \frac{u(x)}{\sqrt{x}}.$$

Then (43) transforms into the differential equation

$$y'' + \frac{1}{x}y' + \left[\mu^2 - \frac{\nu^2 + 1/4}{x^2}\right]y = 0, \tag{44}$$

the so-called *Bessel equation*. Here we must keep in mind that $\nu^2 = \sigma$ follows from the eigenvalue problem (41) and that it therefore can be viewed as known in (44). Thus (44) has the form

$$-y'' - \frac{1}{x}y' + \frac{s^2}{x^2}y = \mu^2 y, \tag{45}$$

where $s^2 := \nu^2 + 1/4 \in (0, \infty)$. Finally, we then obtain an *eigenvalue problem for a second order singular differential operator*, or more precisely, *for the Bessel differential operator*

$$B_s(y) := -\frac{d^2 y}{dx^2} - \frac{1}{x}\frac{dy}{dx} + \frac{s^2}{x^2}y,$$

that is to say, the EVP

$$\begin{cases} B_s(y) &= \mu^2 y & \text{in } (0,1), \\ y(1) &= 0, \\ \lim_{x \to 0}(\sqrt{x}\, y(x))' &= 0. \end{cases}$$

We refer to Triebel [1, § 27] for a functional analytic treatment of the Bessel differential operator (under slightly different boundary conditions). Triebel's book also contains a mathematically flawless presentation of separation of variables for many problems in mathematical physics (e.g., problems in quantum mechanics), as well as a detailed treatment of the EVP (41) for the Laplace-Beltrami operator on $\mathbb{S}^m$ (Triebel [1, § 31]).

Problems

1. Let $G \subseteq \mathbb{R}^n$ be open and bounded, and assume that $\overline{G}$ is a C^2-manifold such that $\partial G = \Gamma_0 \cup \Gamma_1$ and $\Gamma_0 \cap \Gamma_1 = \emptyset$. Moreover, let $A \in C^1(\overline{G}, \mathcal{L}(\mathbb{R}^n))$ be such that $A(x)$ is positive definite for all $x \in \overline{G}$,

$$\text{dom } (\mathbb{A}) := \{u \in C^2(\overline{G}, \mathbb{R}) \mid u = 0 \text{ on } \Gamma_0 \text{ and } \frac{\partial u}{\partial \nu_A} + \alpha u = 0 \text{ on } \Gamma_1\},$$

where ν_A denotes the conormal on ∂G with respect to A and $\alpha \in C(\Gamma_1, \mathbb{R})$, and

$$\mathbb{A}u := -\text{div } (A \text{ grad } u), \quad \forall u \in \text{dom } (\mathbb{A}).$$

Prove that $\mathbb{A}$ is symmetric with respect to the $L^2(G)$-inner product

$$(u \mid v) := \int_G u(x)v(x)\, dx,$$

that is, prove that

$$(\mathbb{A}u \mid v) = (u \mid \mathbb{A}v), \quad \forall u, v \in \text{dom } (\mathbb{A}).$$

2. Let $H := L^2(0, \pi)$ and

$$e_n(x) := \sqrt{\frac{2}{\pi}} \sin(nx), \quad \forall x \in [0, \pi], \quad \forall n \in \mathbb{N}^*.$$

Then $\{e_n \mid n \in \mathbb{N}^*\}$ is an orthonormal set in H, i.e., we have $(e_n \mid e_m) = \delta_{nm}, \quad \forall n, m \in \mathbb{N}^*$.

Show that:

(a) If

$$v \in C_0^2[0, \pi] := \{u \in C^2([0, \pi], \mathbb{R}) \mid u(0) = u(\pi) = 0\},$$

then the "Fourier coefficients" $(v \mid e_n)$ satisfy the estimate

$$|(v \mid e_n)| \leq \frac{\sqrt{2\pi}}{n^2} \max_{0 \leq x \leq \pi} |v''(x)|, \quad \forall n \in \mathbb{N}^*.$$

(b) If $v \in C_0^2[0, \pi]$, then the method of separation of variables is mathematically justified for the initial value problem of the equation of heat conduction

$$\begin{cases} u_t - u_{xx} & = & 0 & \text{in } (0, \pi) \times (0, \infty), \\ u(0, t) & = & u(\pi, t) = 0 & \text{for } t \geq 0, \\ u(x, 0) & = & v(x) & \text{for } x \in [0, \pi], \end{cases} \tag{46}$$

that is, the function

$$u(x, t) := \sum_{n=1}^{\infty} (v \mid e_n) e_n(x) e^{-n^2 t}, \quad (x, t) \in [0, \pi] \times \mathbb{R}_+,$$

is a (classical) solution of (46). (Use, without proof, the fact that the Fourier series

$$\sum_{n=1}^{\infty} (v \mid e_n) e_n$$

converges pointwise to v on $[0, 1]$.)

5. Elementary Integration Methods

In what follows we let $E = (E, |\cdot|)$ denote an arbitrary Banach space over $\mathbb{K} := \mathbb{R}$ or $\mathbb{C}$. In all cases of importance in applications, we will have $E = \mathbb{K}^m$. In addition, let $D \subseteq E$ be some domain and let $I \subseteq \mathbb{R}$ be an open interval. Finally, let $(t_0, x_0) \in I \times D$ and assume that $f \in C(I \times D, E)$. We consider the *initial value problem*

$$(IVP)_{(t_0, x_0)} \qquad\qquad \begin{cases} \dot{x} = f(t, x) \\ x(t_0) = x_0. \end{cases}$$

A function $u : J \to D$ is called a *solution* of $(IVP)_{(t_0, x_0)}$ if the following holds:

(i) $J \subseteq I$ is a perfect interval (i.e., $\mathrm{int}(J) \neq \emptyset$) such that $t_0 \in J$;

(ii) $u \in C^1(J, D)$;

(iii) $\dot{u}(t) = f(t, u(t)), \quad \forall t \in J$;

(iv) $u(t_0) = x_0$.

If the function $u : J \to D$ merely satisfies conditions (i)–(iii), then u is a *solution of the differential equation* $\dot{x} = f(t, x)$.

If $\tilde{u} : \tilde{J} \to D$ is a further solution of $(IVP)_{(t_0, x_0)}$ and if $\tilde{u} \supseteq u$, then $\tilde{u}$ is called an *extension* of u. We say that u is a *nonextendible solution* if u has no proper extension and if the interval J is a *maximal interval of existence* for $(IVP)_{(t_0, x_0)}$. If, finally, $J = I$, then the (clearly nonextendible) solution u is called a *global solution*.

A solution $u : J \to D$ of the differential equation $\dot{x} = f(t, x)$ is evidently a C^1-path in D, whose tangent vector at the point $u(t)$ is given by $f(t, u(t))$. The graph of the solution u, that is,

$$\Gamma := \operatorname{graph}(u) := \{(t, u(t)) \mid t \in J\} \subseteq \mathbb{R} \times E,$$

is called an *integral curve* of the differential equation $\dot{x} = f(t, x)$. If Γ is parametrized by $\psi : t \mapsto (t, u(t)), \ t \in J$, then the tangent vector to the path ψ at the point $(t, u(t))$ is given by the vector $(1, f(\psi(t))) = (1, f(t, u(t)))$.

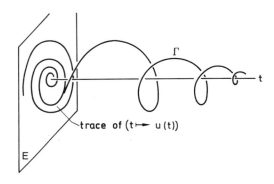

We briefly consider the one-dimensional special case $E = \mathbb{R}$. Then the triple $(t, x, f(t, x))$ is called a *line segment* and

$$\{(t, x, f(t, x)) \mid (t, x) \in I \times D\}$$

is the *direction field* of the differential equation $\dot{x} = f(t, x)$. Since, by the foregoing, $f(t, x)$ determines the slope of the tangent on Γ at (t, x), one can graphically illustrate the direction field by drawing a small line segment with slope $f(t, x)$ through the point $(t, x) \in I \times D$. This way one obtains an overview about the

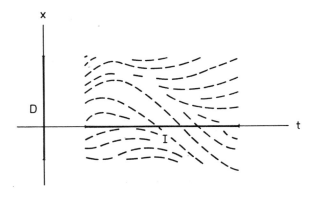

possible course of the integral curves.

Separable Equations

Let $E = \mathbb{R}$ and $f(t, x) = g(t)h(x)$, where $g \in C(I, \mathbb{R})$ and $h \in C(D, \mathbb{R})$. Then the IVP for a *first order separable differential equation* has the form

$$\dot{x} = g(t)h(x), \quad x(t_0) = x_0. \tag{1}$$

Let $x : J \to D$ now be a solution of (1) such that $h(x(t)) \neq 0, \ \forall t \in J$. Then dividing (1) by $h(x(t))$ and integrating from t_0 to $t \in J$, we obtain

$$\int_{t_0}^{t} g(\tau)\, d\tau = \int_{t_0}^{t} \frac{\dot{x}(\tau)\, d\tau}{h(x(\tau))} = \int_{x_0}^{x(t)} \frac{d\xi}{h(\xi)}. \tag{2}$$

Conversely, the following "*separable equation theorem*" states that solving the equation

$$\int_{x_0}^{x} \frac{d\xi}{h(\xi)} = \int_{t_0}^{t} g(\tau)\, d\tau \tag{3}$$

for x determines a solution of (1).

(5.1) Proposition. (a) *If $h(x_0) = 0$, then $x(t) = x_0, \ \forall t \in I$, is a global solution of (1).*

(b) *If $h(x_0) \neq 0$, there exists an open interval $J \subseteq I$ containing t_0 such that (1) has a unique solution on J ("local existence and uniqueness"). The solution can be obtained from (3) by solving for x.*

Proof. (a) is trivial.

(b) We define the unique function $G \in C^1(I, \mathbb{R})$ satisfying $G' = g$ and $G(t_0) = 0$ by $G(t) = \int_{t_0}^{t} g(\tau)\, d\tau$. In addition, let U denote the maximal interval in D such that $x_0 \in U$ and $h(\xi) \neq 0, \ \forall \xi \in U$. Then U is open in $\mathbb{R}$. For $x \in U$ we set

$$H(x) := \int_{x_0}^{x} \frac{d\xi}{h(\xi)}.$$

Then $H \in C^1(U, \mathbb{R})$, $H(x_0) = 0$ and $H' = 1/h$. Therefore H is a C^1-diffeomorphism from U onto some open interval $V \subseteq \mathbb{R}$ and $H(x_0) = 0 \in V$. Then $W := G^{-1}(V)$ is an open neighborhood of t_0. Let J denote the largest open interval in W containing t_0 (the connected component of $\{t_0\}$ in W).

Now equation (3) becomes $H(x) = G(t)$, and for every $t \in J$ this is equivalent to

$$x(t) = H^{-1}(G(t)).$$

Consequently $x \in C^1(J, \mathbb{R})$ and $x(t_0) = H^{-1}(G(t_0)) = x_0$. Differentiating $H(x(t)) = G(t)$, $t \in J$, implicitly and using $H' = 1/h$, we obtain

$$\dot{x}(t) = g(t)\, h(x(t)), \quad \forall t \in J.$$

Therefore $x := H^{-1} \circ G : J \to D$ solves the IVP (1).

If $\tilde{x} \in C^1(\tilde{J}, D)$ is some other solution of (1) satisfying $h(\tilde{x}(t)) \neq 0$, $\forall t \in \tilde{J}$, then $\tilde{x}(\tilde{J}) \subseteq U$, because $\tilde{x}(\tilde{J})$ is a connected subset of D containing x_0. Now (2) implies $H(\tilde{x}(t)) = G(t)$ for all $t \in \tilde{J}$, hence

$$\tilde{x}(t) = H^{-1}(G(t)) = x(t), \quad \forall t \in J \cap \tilde{J}.$$

From the definition of J we deduce that $\tilde{J} \subseteq J$, hence $J \cap \tilde{J} = \tilde{J}$ and therefore $x \supseteq \tilde{x}$. This proves the local uniqueness. $\qquad \square$

(5.2) Examples. (a) $\dot{x} = 1 + x^2$, $x(t_0) = x_0$. In this case we have $I = D = \mathbb{R}$, $g = 1$ and $h(x) = 1 + x^2 \neq 0$, $\forall x \in \mathbb{R}$. Therefore proposition (5.1) guarantees a unique solution for each (t_0, x_0), which can be obtained from

$$\int_{x_0}^{x} \frac{d\xi}{1 + \xi^2} = \int_{t_0}^{t} d\tau$$

by solving for x. It follows that

$$x(t) = \tan(1 - \alpha) \quad \text{for all } t \in \left(\alpha - \frac{\pi}{2}, \alpha + \frac{\pi}{2} \right) =: J(t_0, x_0),$$

where $\alpha := t_0 - \tan^{-1} x_0$.

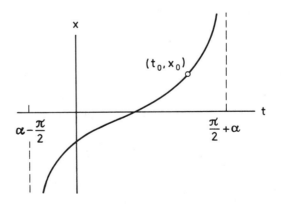

Evidently, this solution is *nonextendible, but not global*, although the right-hand side of the differential equation has the simplest form (a polynomial in x). Moreover, we see that exactly one integral curve of the differential equation $\dot{x} = 1 + x^2$ passes through each point (t_0, x_0) in the plane and all integral curves "run from $-\infty$ to ∞."

(b) $\dot{x} = f(x)$, *where* $f(x) = (\text{sign } x)\sqrt{|x|}$, $x \in \mathbb{R}$. In this case we have $I = D = \mathbb{R}$ again and the right-hand side is *odd*, i.e., we have $f(-x) = -f(x)$ for all $x \in \mathbb{R}$. This immediately implies that the direction field is symmetric with respect to the t-axis and that for every solution $t \mapsto x(t)$ also the "reflected" function $t \mapsto -x(t)$ is a solution. In addition, the differential equation has the trivial solution $x = 0$.

We now consider the IVP

$$\dot{x} = f(x), \quad x(t_0) = x_0. \tag{4}$$

If $x_0 > 0$, a solution of (4) can be calculated, using proposition (5.1), from the relation

$$\int_{x_0}^{x} \frac{d\xi}{\sqrt{\xi}} = \int_{t_0}^{t} d\tau,$$

and hence

$$x(t) = \left(\sqrt{x_0} + \frac{t - t_0}{2}\right)^2 \quad \text{if} \quad t > t_0 - 2\sqrt{x_0} =: \tau_0$$

(because if $t = \tau_0$, the term $\sqrt{x(t)}$ vanishes and therefore proposition (5.1 b) no longer applies). *Continuing this solution to the left by* 0 *gives the global solution*

$$\overline{x}(t) := \begin{cases} \left(\sqrt{x_0} + \dfrac{t - t_0}{2}\right)^2, & \text{if} \quad t \geq \tau_0, \\ 0, & \text{if} \quad t \leq \tau_0, \end{cases}$$

and clearly $\overline{x}$ is the unique, nonextendible solution of (4) for $x_0 > 0$.

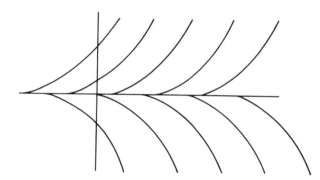

If $x_0 < 0$, we also obtain, based on the symmetry considerations from above, a unique, nonextendible solution through x_0, namely by reflecting $\tilde{x}$ with respect to the t-axis. *Hence for every* $(t_0, x_0) \in \mathbb{R}^2$ *such that* $x_0 \neq 0$, *the IVP* (4) *has a unique global solution*

$$x(t) = \begin{cases} \text{sign}\,(x_0) \left(\sqrt{|x_0|} + \dfrac{t - t_0}{2}\right)^2, & \text{if} \quad t \geq t_0 - 2\sqrt{|x_0|}, \\ 0, & \text{if} \quad t \leq t_0 - 2\sqrt{|x_0|}. \end{cases}$$

On the other hand, the IVP

$$\dot{x} = f(x), \quad x(t_0) = 0,$$

has for each $t_0 \in \mathbb{R}$, *in addition to the trivial solution* $x = 0$, *the infinite family of global solutions*

$$x_\alpha(t) := \begin{cases} \pm\dfrac{(t-\alpha)^2}{4}, & \text{if } t \geq \alpha, \\ 0, & \text{if } t \leq \alpha, \end{cases}$$

where α is an arbitrary real number satisfying $\alpha \geq t_0$.

The lack of uniqueness "occurs" at the point $x_0 = 0$. This is the only point at which f is not differentiable. In fact, f is not even *Lipschitz continuous* at 0, i.e., there exists no $\lambda \in \mathbb{R}_+$ such that

$$|f(x) - f(0)| \leq \lambda|x|$$

for all x in some neighborhood of 0. Later we will see that for Lipschitz continuous functions the IVP always has a unique solution.

(c) *The homogeneous first order linear differential equation.* Let $a \in C(I, \mathbb{R})$ and consider the IVP

$$\dot{x} = a(t)x, \quad x(t_0) = x_0, \tag{5}$$

for some arbitrary $(t_0, x_0) \in I \times \mathbb{R}$. If $x_0 \neq 0$, separation of variables gives us

$$\int_{x_0}^x \frac{d\xi}{\xi} = \int_{t_0}^t a(\tau)\,d\tau,$$

and thus

$$\ln\left|\frac{x}{x_0}\right| = \int_{t_0}^t a(\tau)\,d\tau,$$

or

$$\left|\frac{x(t)}{x_0}\right| = e^{\int_{t_0}^t a(\tau)\,d\tau}.$$

From this we read off that $x(t) \neq 0$ for all t in the interval of existence J. Consequently $J = I$ and, since $x(0) = x_0$, we have $|x(t)/x_0| = x(t)/x_0$. Hence for every $x_0 \neq 0$, the function

$$x(t) = x_0\, e^{\int_{t_0}^t a(\tau)\,d\tau}, \quad t \in I, \tag{6}$$

is the unique global solution of (5). If $x_0 = 0$, (5) has the trivial solution $x = 0$. It follows from (6) that no nontrivial solution meets the t-axis (i.e., vanishes). Consequently, the trivial solution is also unique. *Therefore* (6) *represents the unique global solution of the IVP* (5) *for each* $(t_0, x_0) \in I \times \mathbb{R}$.

It should be remarked that in the special case $a = $ constant, (6) becomes the solution $x(t) = x_0 e^{a(t-t_0)}$, already found in section 1.   □

1-Forms

Let I, D again denote open intervals in $\mathbb{R}$ and assume that $f \in C(I \times D, \mathbb{R})$. Then formally we can write the differential equation $dy/dx = f(x, y)$ as an "equation of differentials"

$$dy - f(x, y) \, dx = 0. \tag{7}$$

To make this more precise, we let

$$\Omega_k^1(M), \quad k \in \mathbb{N},$$

denote the set (the bundle) of all 1-*forms* of class C^k on (the manifold) M. If M is open in $\mathbb{R}^2$, we know that $\alpha \in \Omega_k^1(M)$ can be written in the form

$$\alpha = A dx + B dy,$$

where $A, B \in C^k(M, \mathbb{R})$. If $\varphi : V \to M$ is a C^1-path in M, then φ induces a 1-form on V, the *pull-back* $\varphi^* \alpha \in \Omega_0^1(V)$, which is defined by

$$\varphi^* \alpha(t) := [A(\varphi(t)) \, \dot{\varphi}^1(t) + B(\varphi(t)) \, \dot{\varphi}^2(t)] \, dt, \quad t \in V. \tag{8}$$

In particular, we have $\varphi^* \alpha = 0$ if and only if

$$A(\varphi(t)) \, \dot{\varphi}^1(t) + B(\varphi(t)) \, \dot{\varphi}^2(t) = 0, \quad t \in V. \tag{9}$$

Finally, a C^1-path $\varphi : V \to \mathbb{R}^2$ is *regular* if $\dot{\varphi}(t) \neq 0$ for all $t \in V$, and a *regular* C^1-*curve* Γ is an equivalence class of regular C^1-paths, where two regular C^1-paths $\varphi : V \to \mathbb{R}^2$ and $\psi : W \to \mathbb{R}^2$ are equivalent if and only if there exists some diffeomorphism $a : V \to W$ such that $\varphi = \psi \circ a$.

Now let

$$\alpha := dy - f dx \in \Omega_0^1(I \times D),$$

and assume that $u \in C^1(J, D)$ is a solution of the differential equation $\dot{y} = f(x, y)$. Then $\varphi(t) := (t, u(t)), t \in J$, defines a regular C^1-path on $I \times D$ for which we clearly have

$$\varphi^* \alpha(t) = [\dot{u}(t) - f(t, u(t))] \, dt = 0, \quad t \in J,$$

and therefore $\varphi^* \alpha = 0$.

Conversely, if $\varphi \in C^1(J, I \times D)$ is a regular C^1-path on $I \times D$ satisfying

$$\varphi^* \alpha = \left[\dot{\varphi}^2 - f(\varphi) \dot{\varphi}^1 \right] dt = 0, \tag{10}$$

we have $\dot{\varphi}^2 = f(\varphi^1, \varphi^2) \dot{\varphi}^1$. Suppose that for some $t \in J$ we have $\dot{\varphi}^1(t) = 0$. Then also $\dot{\varphi}^2(t) = 0$, which contradicts the regularity of φ. Therefore $\dot{\varphi}^1(t) \neq 0$ for all $t \in J$ and hence φ^1 is a C^1-diffeomorphism from J onto $V := \varphi^1(J)$. From (10) it follows, moreover, that

$$\frac{\dot{\varphi}^2(t)}{\dot{\varphi}^1(t)} = f(\varphi^1(t), \varphi^2(t)), \quad t \in J. \tag{11}$$

We now introduce a new variable x by $x := \varphi^1(t)$ and set $u(x) := \varphi^2 \left[\left(\varphi^1 \right)^{-1} (x) \right]$. From the chain rule it follows that

$$\frac{du}{dx} = \dot{\varphi}^2 \left[(\varphi^1)^{-1}(x) \right] \left[(\varphi^1)^{-1} \right]^{\cdot} (x) = \frac{\dot{\varphi}^2((\varphi^1)^{-1}(x))}{\dot{\varphi}^1((\varphi^1)^{-1}(x))}$$

and, using (11), we obtain

$$\frac{du}{dx} = f(x, u(x)), \quad x \in V.$$

We have shown:

(5.3) *Finding an integral curve for the differential equation*

$$y' = f(x, y)$$

is equivalent to the problem of finding a regular C^1-path $\varphi : J \to I \times D$ such that

$$\varphi^* \alpha = 0, \quad \text{where} \quad \alpha := dy - f dx \in \Omega_0^1(I \times D).$$

This motivates the following definition.

Definition. Let M be open in $\mathbb{R}^2$ and assume that $\alpha \in \Omega_0^1(M)$ is a continuous 1-form. A regular C^1-curve Γ in M is called a *solution (curve) of the equation* $\alpha = 0$ if there exists a regular C^1-parametrization $\varphi : J \to M$ of Γ such that $\varphi^* \alpha = 0$. The solution Γ *passes through the point* $(x_0, y_0) \in M$ if (x_0, y_0) lies in the image of Γ.

(5.4) Remarks. (a) The salient advantage of the definition above is the fact that for the solution curves of $\alpha = 0$ no specific parametrization stands out. In particular, the points on the curve in the (x, y) plane at which the tangents are vertical create no "difficulties" anymore. If, in fact, $\psi : \hat{J} \to M$ is some other regular parametrization of Γ, then there exists some C^1-diffeomorphism $h : J \to \hat{J}$ such that $\psi \circ \alpha = \varphi$. Then we have $\varphi^* \alpha = h^* \psi^* \alpha = 0$, which clearly implies $\psi^* \alpha = 0$.

(b) *If $\alpha \in \Omega_0^1(M)$ and if $h \in C(M, \mathbb{R})$ is such that $h(x, y) \neq 0, \forall (x, y) \in M$, then the equations $\alpha = 0$ and $h\alpha = 0$ have the same solution curves.*

Proof. Assume that $\varphi : J \to M$ is a regular C^1-parametrization of a solution curve Γ of $\alpha = 0$. It then follows immediately from $\varphi^* \alpha = 0$ that $\varphi^*(h\alpha) = (\varphi^* h)(\varphi^* \alpha) = (h \circ \varphi)\varphi^* \alpha = 0$. The converse also holds because $h(x, y) \neq 0, \quad \forall (x, y) \in M$. $\square$

(c) *Suppose $\alpha \in \Omega_0^1(M)$ is such that $\alpha(x_0, y_0) \neq 0$ for some $(x_0, y_0) \in M$. Then there exists an open neighborhood $U := I \times D$ of $(x_0, y_0) \in M$ and a differential equation defined on U and of the form*

$$y' = f(x, y) \quad or \quad x' = g(x, y),$$

which has the same solution curves in U as $\alpha = 0$.

Proof. Let $\alpha = A dx + B dy$. Then $(A(x_0, y_0), B(x_0, y_0)) \neq (0, 0)$. If $B(x_0, y_0) \neq 0$, then $B(x, y) \neq 0$ on some whole neighborhood U of (x_0, y_0). Consequently, $\alpha = 0$ and

$$\frac{1}{B}\alpha = \frac{A}{B} dx + dy = 0$$

have the same solution curves in U (by (b)). Now the assertion follows from (5.3). If $B(x_0, y_0) = 0$, we must have $A(x_0, y_0) \neq 0$ and so the assertion follows similarly.   □

(d) *The 1-form* $\alpha = A dx + B dy \in \Omega_0^1(M)$ *has the same solution curves as the system*

$$\dot{x} = B(x, y), \quad \dot{y} = -A(x, y).$$

Proof. Let $\varphi \in C^1(J, M)$ be regular. Then $\varphi^*\alpha = 0$ holds if and only if (9) is satisfied. The latter is the case if and only if there exists some $\lambda \in C(J, \mathbb{R} \setminus \{0\})$ such that

$$(\dot{\varphi}^1, \dot{\varphi}^2) = \lambda(B(\varphi), -A(\varphi)).$$

Let $t_0 \in J$ be arbitrary and set

$$a(t) := \int_{t_0}^{t} \lambda(\tau) \, d\tau, \quad t \in J.$$

Then a is a C^1-diffeomorphism from J onto $V := a(J)$ and for $(x(s), y(s)) := (\varphi^1 \circ a^{-1}(s), \varphi^2 \circ a^{-1}(s))$, $s = a(t) \in V$, we have

$$(\dot{x}(s), \dot{y}(s)) = \frac{1}{\dot{a}(t)}(\dot{\varphi}^1(t), \dot{\varphi}^2(t)) = (B(x(s), y(s)), -A(x(s), y(s)))$$

for all $s \in V$.   □

A 1-form $\alpha \in \Omega_0^1(M)$ is called *exact* if there exists some $f \in C^1(M, \mathbb{R})$ such that $df = \alpha$, and $\alpha \in \Omega_1^1(M)$ is *closed* if $d\alpha = 0$.

(5.5) Remarks. (a) *Since* $d^2 = 0$, *every exact form is closed.* It is well-known that the converse is false.

   (b) *The 1-form* $\alpha = A dx + B dy \in \Omega_1^1(M)$ *is closed if and only if the integrability condition*

$$\frac{\partial A}{\partial y} = \frac{\partial B}{\partial x}$$

is satisfied.

   In fact, from the rules for the exterior derivative we obtain

$$da = dA \wedge dx + dB \wedge dy$$
$$= \left(\frac{\partial A}{\partial x} dx + \frac{\partial A}{\partial y} dy\right) \wedge dx + \left(\frac{\partial B}{\partial x} dx + \frac{\partial B}{\partial y} dy\right) \wedge dy$$
$$= \left(-\frac{\partial A}{\partial y} + \frac{\partial B}{\partial x}\right) dx \wedge dy,$$

and so the assertion follows.   □

Recall that $c \in \mathbb{R}$ is a *regular value* of $f \in C^1(M, \mathbb{R})$ if $df(x, y) \neq 0$ for all $(x, y) \in f^{-1}(c)$. This is clearly equivalent to

$$\text{grad } f(x, y) \neq 0, \quad \forall (x, y) \in f^{-1}(c),$$

i.e., *the level set $f^{-1}(c)$ contains no critical point.*

(5.6) Proposition. *Let $\alpha \in \Omega_0^1(M)$ and $f \in C^1(M, \mathbb{R})$ be such that $df = \alpha$, and assume that $c \in \mathbb{R}$ is a regular value of f. Then the level set $f^{-1}(c)$ is a locally finite union of solution curves of the equation $\alpha = 0$. In addition, (the image of) every solution curve of $\alpha = 0$ lies in a level set of f.*
Proof. If c is a regular value, we know that $f^{-1}(c)$ is a one-dimensional C^1-manifold, and so, in particular, it has a regular local C^1-parametrization (implicit function theorem). If $\varphi : J \to M$ is a regular C^1-parametrization of a curve Γ such that image $(\Gamma) \subseteq f^{-1}(c)$, then $\varphi^* f = f \circ \varphi = c$ on J. Consequently

$$0 = d(\varphi^* f) = \varphi^* df = \varphi^* \alpha. \tag{12}$$

Therefore Γ is a solution curve of $\alpha = 0$.

Conversely, if φ is a regular C^1-parametrization of a solution curve Γ of $\alpha = 0$, then (12) holds (read from right to left!). Therefore $\varphi^* f = f \circ \varphi$ is constant on J. $\qquad\square$

A function $f \in C^1(M, \mathbb{R})$ is called an *integral of equation $\alpha = 0$*, or an *antiderivative* of $\alpha \in \Omega_0^1(M)$ if $\alpha = df$.

(5.7) Corollary. *If $f \in C^1(M, \mathbb{R})$ is an antiderivative of $\alpha \in \Omega_0^1(M)$, then the regular segments of the level sets represent all regular solutions of $\alpha = 0$.*

Here we understand by a *regular segment of a level set* $f^{-1}(c)$ the set $f^{-1}(c) \backslash K$, where

$$K := \{(x, y) \in M \mid df(x, y) = 0\},$$

that is, the critical points are removed from the level set.

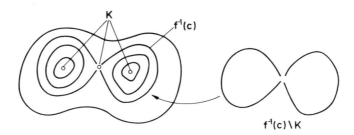

We know from calculus that the converse of (5.5 a) holds if M is a sufficiently "nice" domain. More precisely, we have the following theorem.

(5.8) Theorem. *Assume that $M \subseteq \mathbb{R}^2$ is a simply connected domain and let $\alpha \in \Omega_1^1(M)$. Then α is exact if and only if α is closed.*

For a proof of this theorem we refer to the literature (e.g., Fleming [1], CH 8.10). In what follows we will only consider the special case when M is an open rectangle in $\mathbb{R}^2$, but we will assume less regularity.

(5.9) Theorem. *Let $M \subseteq \mathbb{R}^2$ be an open rectangle and $\alpha = A\,dx + B\,dy \in \Omega_0^1(M)$. In addition, we assume that the partial derivatives $A_y := \partial A/\partial y$ and B_x exist and are continuous on M. If the integrability condition $A_y = B_x$ is satisfied, then α is exact.*

Proof. First we assume that $f \in C^1(M, \mathbb{R})$ satisfies $df = \alpha$. Then we have $f_x = A$ and $f_y = B$. Integrating the first of these equations from x_0 to x, for fixed y, we obtain

$$f(x, y) = \int_{x_0}^{x} A(t, y)\,dt + h(y), \tag{13}$$

where $h(y)$ is some integration constant. Since $f \in C^1(M, \mathbb{R})$ and $A_y \in C(M, \mathbb{R})$, it follows that $h \in C^1((\alpha, \beta), \mathbb{R})$, where $M = (a, b) \times (\alpha, \beta)$. By differentiating (13) with respect to y, we now obtain

$$B(x, y) = f_y(x, y) = \int_{x_0}^{x} A_y(t, y)\,dt + h'(y),$$

and thus

$$h'(y) = B(x, y) - \int_{x_0}^{x} A_y(t, y)\,dt. \tag{14}$$

Because

$$\frac{\partial}{\partial x}\left[B(x, y) - \int_{x_0}^{x} A_y(t, y)\,dt\right] = B_x(x, y) - A_y(x, y) = 0,$$

the right-hand side of (14) is just a function of $y \in (\alpha, \beta)$. Consequently h can be determined by integration:

$$h(y) = \int_{y_0}^{y} [B(x, s) - \int_{x_0}^{x} A_y(t, s)\,dt]\,ds$$

for some arbitrary $y_0 \in (\alpha, \beta)$. Finally, we obtain

$$f(x, y) = \int_{x_0}^{x} A(t, y)\,dt + \int_{y_0}^{y} [B(x, s) - \int_{x_0}^{x} A_y(t, s)\,dt]\,ds \tag{15}$$

for some arbitrary $(x_0, y_0) \in M$. Therefore, if $df = \alpha$, then f has necessarily the form (15).

We now *define* $f \in C^1(M, \mathbb{R})$ by formula (15) for some arbitrary $(x_0, y_0) \in M$. One immediately sees that $df = \alpha$. □

(5.10) Remarks. (a) Under the assumptions of theorem (5.9), two antiderivatives of α can clearly only differ by a constant.

(b) We explicitly note *that the proof of theorem (5.9) is constructive*, that is, it contains an explicit technique to find an antiderivative of α (and therefore all).

(c) Based on proposition (5.6), the integration problem of equation $\alpha = 0$ can be considered solved if an antiderivative can be found. Then, if a solution of $\alpha = 0$ is sought passing through (x_0, y_0), the *general solution* is obtained in the implicit form

$$f(x, y) = c,$$

where the "integration constant" c is determined by the condition $f(x_0, y_0) = c$.   □

(5.11) Examples. (a) We want to determine the general solution of the equation

$$\alpha := (y \cos x + 2xe^y) \, dx + (\sin x + x^2 e^y + 2) \, dy = 0.$$

Evidently $\alpha = A dx + B dy \in \Omega^1_\infty(\mathbb{R}^2)$ and, since

$$A_y = \cos x + 2xe^y = B_x,$$

α is closed and hence exact (by theorem (5.9)).

If $f \in C^1(\mathbb{R}^2, \mathbb{R})$ is an antiderivative, then

$$f_x = y \cos x + 2xe^y, \qquad f_y = \sin x + x^2 e^y + 2. \tag{16}$$

Hence by integrating the first equation with respect to x we obtain

$$f(x, y) = y \sin x + x^2 e^y + h(y).$$

Differentiating with respect to y and using the second equation in (16), we obtain

$$\sin x + x^2 e^y + h'(y) = f_y = \sin x + x^2 e^y + 2,$$

hence $h'(y) = 2$ and therefore $h(y) = 2y$. Consequently

$$f(x, y) := y \sin x + x^2 e^y + 2y$$

is an antiderivative of α and so the general solution of the equation $\alpha = 0$ is given by

$$y \sin x + x^2 e^y + 2y = c, \quad c \in \mathbb{R}.$$

(b) Let I, D be open intervals in $\mathbb{R}$ and let $g \in C(I, \mathbb{R})$, $h \in C(D, \mathbb{R})$ be such that $h(x) \neq 0$ for all $x \in D$. It follows from (5.3) that the differential equation

$$\dot{x} = g(t)h(x) \tag{17}$$

has the same solution curves as the 1-form

$$\alpha := dx - g(t)h(x) \, dt,$$

and so it has, by (5.4 b), the same solution curves as the 1-form

$$\beta := \frac{1}{h}\alpha = \frac{1}{h}dx - g\,dt \in \Omega_0^1(D \times I).$$

This form clearly satisfies the integrability condition $A_t = (1/h)_t = 0 = B_x = (-g)_x$. Hence theorem (5.9) and its method of proof is applicable. So there exists some $f \in C^1(D \times I, \mathbb{R})$ such that $f_x = 1/h$ and $f_t = -g$. Consequently

$$f(x,t) = \int_{x_0}^x \frac{d\xi}{h(\xi)} + a(t), \quad x_0 \in D \text{ is arbitrary,}$$

from which we deduce

$$a' = f_t = -g.$$

Therefore

$$a(t) = -\int_{t_0}^t g(\tau)\,d\tau, \quad t_0 \in I \text{ is arbitrary,}$$

and hence

$$f(x,t) = \int_{x_0}^x \frac{d\xi}{h(\xi)} - \int_{t_0}^t g(\tau)\,d\tau, \quad (x_0, t_0) \in D \times I \text{ is arbitrary.}$$

The general solution of the differential equation (17) is given implicitly by

$$f(x,t) = \int_{x_0}^x \frac{d\xi}{h(\xi)} - \int_{t_0}^t g(\tau)\,d\tau = c, \quad c \in \mathbb{R}.$$

Since $f(x_0, t_0) = 0$, we obtain the solution of the initial value problem

$$\dot{x} = g(t)h(x), \quad x(t_0) = x_0,$$

in the implicit form

$$\int_{x_0}^x \frac{d\xi}{h(\xi)} = \int_{t_0}^t g(\tau)\,d\tau.$$

This is the formula we have obtained in proposition (5.1) by separation of variables. It follows that the method of separation of variables is contained, as a special case, in the integration problem of 1-forms. $\square$

In example (5.11 b) we have changed the 1-form $\alpha := dx - g(t)h(x)\,dt$, which is not closed, to the closed 1-form $\beta := \frac{1}{h}\alpha$, which, according to (5.4 b), has the same solution curves as α. This trick, of general importance, is known as the method of integrating factors. More precisely: Let $\alpha \in \Omega_0^1(M)$ and $h \in C(M, \mathbb{R})$ be such that $h(x,y) \neq 0$, $\forall(x,y) \in M$. Then if $h\alpha$ is exact, h is an *integrating factor for* α.

(5.12) Remarks. (a) *If h is an integrating factor for α and if f is an antiderivative of $h\alpha$, then the general solution of the equation $\alpha = 0$ is given implicitly by $f(x,y) = c$, $c \in \mathbb{R}$.*

Proof. This follows from (5.4 b) and (5.6). $\square$

(b) If $\alpha \in \Omega_1^1(M)$ and if $h \in C^1(M, \mathbb{R})$ is an integrating factor for α, then $h\alpha$ is exact, hence closed. From (5.5 b) we deduce the integrability condition

$$(hA)_y = (hB)_x,$$

that is,

$$Ah_y - Bh_x + (A_y - B_x)h = 0, \tag{18}$$

where $\alpha = A\,dx + B\,dy$. Therefore theorems (5.8) and (5.9) imply the following: *If M is simply connected, and if $h \in C^1(M, \mathbb{R})$ satisfies the first order partial differential equation* (18) *and*

$$h(x, y) \neq 0, \quad \forall (x, y) \in M,$$

then h is an integrating factor for $\alpha = A\,dx + B\,dy \in \Omega_1^1(M)$. This is also true if M is a rectangle and if for $\alpha \in \Omega_0^1(M)$ the derivatives A_y and B_x exist and are continuous on M.

Hence we have "reduced" the integration problem of the equation $\alpha = 0$ to the more difficult problem of finding a solution of the partial differential equation (18). Fortunately, it is often possible to find a solution of (18) in the special form $h = h(x)$, or $h = h(y)$, or $h = h(xy)$, etc.   □

(5.13) Examples. (a) *The Volterra-Lotka Equations* $\dot{x} = (\alpha - \beta y)x$, $\dot{y} = (\delta x - \gamma)y$. We are interested in the biologically relevant case $x > 0$, $y > 0$. By remark (5.4 d), this system is equivalent to the problem of integrating the 1-form

$$\omega := (\gamma - \delta x)y\,dx + (\alpha - \beta y)x\,dy.$$

Since ω is not closed, we will try to find an integration factor of the form $h = h(xy)$. From (18) and by means of an easy calculation we obtain the relation

$$xy[(\gamma - \delta x) - (\alpha - \beta y)]\dot{h} + [(\gamma - \delta x) - (\alpha - \beta y)]h = 0.$$

This implies the equation $t\dot{h} + h = 0$ (where $t := xy$), which has the solution $h(t) = t^{-1}$ (e.g., by separation of variables). The 1-form

$$\frac{1}{xy}\omega = \left(\frac{\gamma}{x} - \delta\right)dx + \left(\frac{\alpha}{y} - \beta\right)dy$$

is exact in $(0, \infty)^2$, and by the method of example (5.11 a) we find that the general solution of the equation $\omega = 0$ – and thereby of the Volterra-Lotka system – is given by

$$\alpha \ln y + \gamma \ln x - \beta y - \delta x = \text{constant},$$

which has the equivalent form

$$y^\alpha e^{-\beta y} x^\gamma e^{-\delta x} = \text{constant}.$$

The function

$$F(x, y) := \alpha \ln y + \gamma \ln x - \beta y - \delta x, \quad x, y > 0,$$

clearly has a maximum at the point $(\gamma/\delta, \alpha/\beta)$, but no other critical points. We also have $F(x, y) \to -\infty$ as $(x, y) \to \partial M$. From this it follows that the level curves $\{(x, y) \in M \mid F(x, y) = c\}$ are closed curves whose interior contains the critical point $(\gamma/\delta, \alpha/\beta)$. (A mathematically correct proof of the intuitively clear fact, that the level set corresponding to some fixed value c consists of a single closed curve, follows from corollary (24.22), which will be proved later.) This then shows that the Volterra-Lotka system indeed has the qualitative features in M as depicted in Figure 4 of section 1, i.e., that all noncritical solutions in M are periodic.

(b) *The Nonhomogeneous First Order Linear Differential Equation.* Let I be an open interval in $\mathbb{R}$ and let $a, b \in C(I, \mathbb{R})$. We consider the differential equation

$$\dot{x} = a(t)x + b(t). \tag{19}$$

According to (5.3), the integration problem for (19) is equivalent to the problem of integrating the 1-form $\alpha := dx - [ax + b]\, dt \in \Omega_0^1(M)$, where $M := \mathbb{R} \times I$. This form is not closed. Trying to find an integration factor of the form $h = h(t)$, (18) reduces to the ordinary differential equation

$$\dot{h} = -ah, \tag{20}$$

which, by (5.2 c), has the solution

$$h(t) = e^{-\int_{t_0}^{t} a(\tau)\, d\tau}, \qquad t_0 \in I. \tag{21}$$

By (5.12 b), h is an integrating factor for α, hence $h\alpha$ is exact. Following the methods of example (5.11 a), we find the general solution of (19) in the form $f(x, y) = c$, where

$$f(x, t) = h(t)x - \int_{t_0}^{t} h(\tau)\, b(\tau)\, d\tau, \qquad \forall t \in I,$$

and $t_0 \in I$ is arbitrary. If $x_0 \in \mathbb{R}$ is fixed, then, by solving the implicit equation

$$f(x, t) = f(x_0, t_0) = h(t_0)x_0$$

for x, we obtain the function

$$x(t) = \frac{h(t_0)}{h(t)} x_0 + \int_{t_0}^{t} \frac{h(\tau)}{h(t)} b(\tau)\, d\tau.$$

It follows from (21) that

$$x(t) = e^{\int_{t_0}^{t} a(\tau)\, d\tau} x_0 + \int_{t_0}^{t} e^{\int_{s}^{t} a(\tau)\, d\tau} b(s)\, ds, \qquad t \in I,$$

is a solution of the IVP

$$\dot{x} = a(t)x + b(t), \qquad x(t_0) = x_0. \tag{22}$$

If $y \in C^1(I, \mathbb{R})$ is some other solution of (22), that is to say, if

$$\dot{y} = a(t)y + b(t), \qquad y(t_0) = x_0,$$

then, by taking the difference, it follows that the function $u := x - y \in C^1(I, \mathbb{R})$ is a solution of the homogeneous IVP

$$\dot{u} = a(t)u, \qquad u(t_0) = 0. \tag{23}$$

According to (5.2 c), (23) has the unique solution $u = 0$. Therefore $x = y$, i.e., (22) can be solved uniquely. With this we have proved the following theorem:

(5.14) Theorem. *Let $I \subseteq \mathbb{R}$ be an open interval and let $a, b \in C(I, \mathbb{R})$. Then for every $(t_0, x_0) \in I \times \mathbb{R}$, the nonhomogeneous first order linear differential equation*

$$\dot{x} = a(t)x + b(t), \quad x(t_0) = x_0,$$

has a unique global solution x. It is given by the formula

$$x(t) = U(t, t_0)x_0 + \int_{t_0}^{t} U(t, s)b(s) \, ds, \quad t \in I, \tag{24}$$

where

$$U(t, s) := e^{\int_s^t a(\tau) \, d\tau}, \quad \forall s, t \in I. \tag{25}$$

(5.15) Remarks. (a) For fixed $t, t_0 \in I$, the mapping $x_0 \mapsto U(t, t_0)x_0$ is clearly a linear transformation from $\mathbb{R}$ into itself. Hence we may think of $U(t, t_0)$ as a linear operator on $\mathbb{R}$. It then follows from (25) (using $\mathcal{L}(\mathbb{R}) \cong \mathbb{R}$) that

(i) $U \in C^1(I \times I, \mathcal{L}(\mathbb{R}))$;

(ii) $U(s, s) = \mathrm{id}_{\mathcal{L}(\mathbb{R})}, \quad \forall s \in I$;

(iii) $U(t, \tau)U(\tau, s) = U(t, s), \quad \forall s, t, \tau \in I$;

(iv)   for each $s \in I$, $u(\cdot) := U(\cdot, s) \in C^1(I, \mathcal{L}(\mathbb{R}))$ is the unique solution of the homogeneous IVP $\dot{u} = a(t)u$, $u(s) = \mathrm{id}_{\mathcal{L}(\mathbb{R})}(= 1)$;

(v)   if a is constant, then $U(t, s) = V(t - s)$, where $V(t) = e^{at}$.

From (iii) we deduce, in particular, that

$$U(t, s) = [U(s, t)]^{-1}, \quad \forall s, t \in I.$$

If E is an arbitrary Banach space and if

$$U : I \times I \to \mathcal{L}(E)$$

is a mapping with properties (i) – (iii) (where of course $\mathcal{L}(\mathbb{R})$ must be replaced by $\mathcal{L}(E)$), then U is called an *evolution operator*. In particular, if $I = \mathbb{R}$ and $U(t, s) = V(t-s)$, where $V : I \to \mathcal{L}(E)$, then conditions (ii) and (iii) reduce to

(ii) $V(0) = \mathrm{id}_{\mathcal{L}(E)}$;

(iii) $V(t)V(s) = V(t - s), \quad \forall s, t \in \mathbb{R}$.

The mapping $V : \mathbb{R} \to \mathcal{L}(E)$, in this case, is a *representation of the additive group* $(\mathbb{R}, +)$ *on the Banach algebra* $\mathcal{L}(E)$. Such representations occur in the theory of partial differential equations such as, for example, in the functional analytic treatment of initial value problems for the wave equation or the (time dependent) Schrödinger equation. In these cases one can show that, under an appropriate interpretation, formula (24) represents the solution of such initial value problems. Moreover, we will come across solution formula (24) again when we treat the initial value problems for linear systems of differential equations (cf. CH. III).

(b) The standard *trick* used to solve the IVP

$$\dot{x} = a(t)x + b(t), \quad x(t_0) = x_0, \tag{26}$$

is the so-called method of *variation of constants*. Here one first determines an arbitrary solution of the homogeneous equation

$$\dot{x} = a(t)x, \tag{27}$$

such as, for instance,

$$u(t) := e^{\int_{t_0}^{t} a(\tau)\,d\tau}. \tag{28}$$

Then one tries to find a solution of (26) of the form $x(t) = c(t)u(t)$ for some unknown function c (the "variable constant"). Substituting this into differential equation (26), we obtain

$$\dot{c}u + \dot{u}c = acu + b,$$

and so, using (27), we have $\dot{c} = b/u$. By integrating we obtain from this, and using (28) and the initial condition, formula (24). For this reason, the representation (24) is also called the *variation of constants formula*. □

Change of Variables

An important technique for solving differential equations, which is also of great theoretical significance, is the *introduction of new dependent variables*. To illustrate this, we consider the differential equation

$$\dot{x} = f(t, x), \quad x(t_0) = x_0, \tag{29}$$

where $f \in C(I \times D, E)$. In addition, we consider a mapping

$$\varphi \in C^1(I \times D, E),$$

with the property that for every $t \in I$,

$$\varphi_t := \varphi(t, \cdot) : D \to E$$

is a C^1-*diffeomorphism on* E (i.e., $M_t := \varphi_t(D)$ is an open subset of E, $\varphi_t : D \to M_t$ is bijective, and φ_t and φ_t^{-1} are continuously differentiable). If $u : J \to D$ is a solution of (29), then for the function transformed by φ,

$$v(t) := \varphi_t(u(t)) = \varphi(t, u(t)), \quad t \in J,$$

we clearly have

$$\dot{v}(t) = D\varphi_t(u(t))\dot{u}(t) + \frac{\partial\varphi}{\partial t}(t, u(t))$$

$$= D\varphi_t(u(t))f(t, u(t)) + \frac{\partial\varphi}{\partial t}(t, u(t)).$$

Since $u(t) = \varphi_t^{-1}(v(t))$, we therefore have

$$\dot{v}(t) = D\varphi_t(\varphi_t^{-1}(v(t)))f(t, \varphi_t^{-1}(v(t))) + \frac{\partial\varphi}{\partial t}(t, \varphi_t^{-1}(v(t))).$$

If we also observe that

$$D\varphi_t(\varphi_t^{-1}(y)) = [D(\varphi_t^{-1})(y)]^{-1},$$

we have shown that:

u : J → E is a solution of the differential equation $\dot{x} = f(t, x)$ if and only if $v := \varphi_t(u) : J → E$ is a solution of the transformed differential equation

$$\dot{y} = [D\varphi_t^{-1}(y)]^{-1}f(t, \varphi_t^{-1}(y)) + \frac{\partial\varphi}{\partial t}(t, \varphi_t^{-1}(y)). \tag{30}$$

(5.16) Remark. In the special case

$$\dot{x} = f(x), \quad \varphi = \varphi(x),$$

the transformed equation becomes

$$\dot{y} = [D\varphi^{-1}(y)]^{-1}f(\varphi^{-1}(y)). \tag{31}$$

Here $\varphi : D → M$ is a C^1-diffeomorphism from the open set $D \subseteq E$ onto the open set $M \subseteq E$. In the special case $E = \mathbb{R}^m$, we can interpret D and M as m-dimensional manifolds, and their tangent bundles $T(D)$ and $T(M)$ can be identified with $D \times \mathbb{R}^m$ and $M \times \mathbb{R}^m$, respectively.

We now define the vector field $X \in \mathfrak{X}(D)$ on D by

$$X(x) := (x, f(x)) \in T_x(D), \quad \forall x \in D.$$

The diffeomorphism $\varphi^{-1} : M → D$ induces the isomorphism $T\varphi^{-1} : T(M) → T(D)$, where

$$T\varphi^{-1}(y, \eta) = (\varphi^{-1}(y), D\varphi^{-1}(y)\eta) \in T_{\varphi^{-1}(y)}(D)$$

for all $(y, \eta) \in T_y(M)$. With the aid of $T\varphi^{-1}$ we can "pull back" the vector field X on D to obtain the vector field $(\varphi^{-1})^*X \in \mathfrak{X}(M)$, where, as we know, $(\varphi^{-1})^*X$ is defined by

$$(\varphi^{-1})^*X(y) = (T_y\varphi^{-1})^{-1}X(\varphi^{-1}(y)), \quad \forall y \in M.$$

Hence we have

$$(\varphi^{-1})^*X(y) = (y, [D\varphi^{-1}(y)]^{-1}f(\varphi^{-1}(y))), \quad \forall y \in M.$$

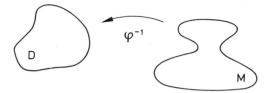

Consequently, the right-hand side of the transformed differential equation (31) represents exactly the principal part of the vector field $(\varphi^{-1})^* X$, pulled back from D to M by means of φ^{-1}.   □

(5.17) Examples. (a) *Homogeneous Differential Equations.* Assume that $D \subseteq \mathbb{R}$ is an open interval and let $f \in C(D, \mathbb{R})$. Then a differential equation of the form

$$\dot{x} = f\left(\frac{x}{t}\right) \tag{32}$$

is called *homogeneous.* It is reasonable to introduce the new variable $y := x/t$. Then in this case we have $I = (0, \infty)$ and $\varphi(t, x) = x/t$. One easily calculates that the transformed differential equation is given by

$$\dot{y} = \frac{1}{t}[f(y) - y].$$

Consequently, we have reduced the integration problem of (32) to that of a separable equation, which was treated earlier.

(b) *Bernoulli Differential Equations.* By this we mean the differential equations

$$\dot{x} = a(t)x + b(t)x^{\alpha},$$

where $a, b \in C(I, \mathbb{R})$ and $\alpha \in \mathbb{R}$, $\alpha \neq 1$. In this case $D = (0, \infty)$, and we introduce the new variable $y := x^{1-\alpha}$. Then

$$\varphi : (0, \infty) \to (0, \infty), \quad x \mapsto x^{1-\alpha},$$

is a diffeomorphism and $\varphi^{-1}(y) = y^{1/(1-\alpha)}$. One verifies that in this case the transformed equation takes on the form

$$\dot{y} = (1 - \alpha)(a(t)y + b(t)).$$

Therefore Bernoulli differential equations can be transformed into linear differential equations, and so we can regard the integration problem as completely solved.   □

The preceding examples show that sometimes it is possible to recast an ordinary differential equation into a simple form by means of a clever transformation, and then solve it. It is, in general, not possible to develop a systematic theory, but rather, one is dependent on ingenious techniques. For a multitude of tricks and for a collection of (more or less explicitly) solvable ordinary differential equations we refer to Kamke [1]. In what follows, our goal is not to solve particular equations explicitly, but to build a mathematical theory which allows us to understand those phenomena that occur in differential equations. Such insight is, in general, significantly more valuable than to obtain closed forms for individual solutions, which often do not contain very much information.

(5.18) Remark. Let $D \subseteq \mathbb{R}^m$ be open and assume that $I \subseteq \mathbb{R}$ is an open interval. Moreover, let $g \in C(I \times D, \mathbb{R})$. Then

$$x^{(m)} = g(t, x, \dot{x}, \ldots, x^{(m-1)}) \tag{33}$$

represents an mth *order ordinary differential equation in explicit form*. More precisely, by (33) we understand the problem of finding functions $u : J \to \mathbb{R}$ – *solutions of* (33) – such that the following holds:

(i) $u \in C^m(J, \mathbb{R})$;

(ii) $(u(t), \dot{u}(t), \ldots, D^{m-1}u(t)) \in D, \quad \forall t \in J$;

(iii) $D^m u(t) = g(t, u(t), \ldots, D^{m-1}u(t)), \quad \forall t \in J$.

Theoretically it is now extremely important that *differential equation* (33) *is equivalent to a first order system in explicit form, that is to say, to a differential equation of the form*

$$\dot{y} = f(t, y), \quad f \in C(I \times D, \mathbb{R}^m),$$

namely the system

$$\dot{y}_1 = y_2$$
$$\dot{y}_2 = y_3$$
$$\vdots$$
$$\dot{y}_{m-1} = y_m$$
$$\dot{y}_m = g(t, y_1, \ldots, y_m).$$

Hence we have $y := (y_1, \ldots, y_m) \in \mathbb{R}^m$ and

$$f(t, y_1, \ldots, y_m) := (y_2, y_3, \ldots, y_m, g(t, y_1, y_2, \ldots, y_m)).$$

The assertion now follows immediately from the "transformation" $x = y_1$.   □

Problems

1. Determine the solutions of the initial value problems

$$y' = -\frac{2y}{x} + 4x, \quad y(1) = y_0,$$

for $y_0 = 1$ and $y_0 = 2$.

2. Let $\alpha = A\,dx + B\,dy$ be a continuously differentiable 1-form on $\mathbb{R}^2$. Show that:

If $(A_y - B_x)/B$ is a function of x only and defined on some simply connected domain $U \subseteq \mathbb{R}^2$, then there exists an integrating factor $h = h(x)$ on U. Determine h explicitly.

3. Determine the solution curves of

(a) $(3xy + y^2) \, dx + (x^2 + xy) \, dy = 0$;

(b) $y \, dx + (2xy - e^{-2y}) \, dy = 0$.

4. Determine the solution curves of

$$(ax + y) \, dx - x \, dy = 0, \quad a \in \mathbb{R},$$

and sketch the solutions near the critical point $(0,0)$. (*Hint*: Show, by substituting $x = e^t$ in the solutions found, that the solution curves can be continuously extended into the point $(0,0)$ and determine the limit positions of the tangents at the point $(0,0)$.)

5. Determine the solution curves for the predator-prey system

$$\dot{x} = ax - bxy, \quad \dot{y} = -cy + dxy, \quad a, b, c, d > 0,$$

in implicit form. (*Hint*: Reduce it to an integration problem of an appropriate 1-form $\alpha = 0$.)

6. Determine the solutions of the differential equation

$$y' = \frac{x + y}{x - y}$$

using:

(a) the substitution $z = y/x$;

(b) the introduction of polar coordinates;

and sketch the solution curves.

7. (a) Show that the *Riccati differential equation*

$$y' + p(x)y + r(x)y^2 = q(x),$$

$p, q, r \in C(I, \mathbb{R})$, can be reduced to a Bernoulli differential equation by setting $y = u + v$ if a particular solution u is known.

(b) Guess a solution of the differential equation

$$y' - (1 - 2x)y + y^2 = 2x \tag{34}$$

and then determine the general solution of (34).

8. Determine the general solution of the differential equation

$$y'(x + x^2 y) = y - xy^2$$

using the transformation $z := \varphi(x, y) = xy$.

9. Solve the initial value problem

$$y' = y^2 - x^2, \quad y(0) = 1, \tag{35}$$

by seeking a solution in the form of a *power series* $y(x) = \sum_{k=0}^{\infty} a_k x^k$. Formally substitute this series into (35) and compare coefficients (of powers of the same order). Then show (by induction) that the coefficients satisfy the estimate $|a_k| \leq 1$. Use this to deduce that the formal solution found is indeed a solution of (35) in the interval $-1 < x < 1$.

Chapter II
Existence and Continuity Theorems

In this chapter we will prove the fundamental existence theorem for ordinary differential equations, the Cauchy-Peano theorem. This local result will be extended to a global existence and uniqueness theorem under somewhat stronger conditions. Theorems about the continuous and differentiable dependence of the solutions on all the data, including parameters, will be proved. These theorems are fundamental for the qualitative study of ordinary differential equations.

Autonomous differential equations generate (local) flows. Because such flows, and in particular also semiflows, appear in other connections (e.g. partial differential equations), we will study the fundamental properties of flows in metric spaces.

The proofs are written – whenever possible – in such a way that they can be extended to the infinite dimensional case. Minor necessary modifications will be pointed out at the appropriate places. Ordinary differential equations in Banach spaces play a role in nonlinear functional analysis – in particular in connection with variational methods.

6. Preliminaries

We begin with a fundamental inequality.

(6.1) Gronwall's Lemma. *Let J be an interval in $\mathbb{R}$, $t_0 \in J$, and a, β, $u \in C(J, \mathbb{R}_+)$. If we assume that*

$$u(t) \leq a(t) + \left| \int_{t_0}^{t} \beta(s)u(s)\,ds \right|, \quad \forall t \in J, \tag{1}$$

then it follows that

$$u(t) \leq a(t) + \left| \int_{t_0}^{t} a(s)\beta(s)e^{\left| \int_{s}^{t} \beta(\sigma)\,d\sigma \right|}\,ds \right|, \quad \forall t \in J. \tag{2}$$

Proof. With $v(t) := \int_{t_0}^{t} \beta(s)u(s)\,ds$ it follows from (1) that

$$\dot{v}(t) = \beta(t)u(t) \leq a(t)\beta(t) + \mathrm{sgn}(t - t_0)\beta(t)v(t), \quad \forall t \in J.$$

Multiplying this inequality by

$$\gamma(t) := \exp\left\{-\left|\int_{t_0}^t \beta(s)\, ds\right|\right\} = \exp\left\{-\int_{t_0}^t \text{sgn}(s - t_0)\beta(s)\, ds\right\},$$

we obtain $\gamma\dot{v} \le a\beta\gamma - \dot{\gamma}v$, and so $(\gamma v)^{\cdot} - a\beta\gamma \le 0$. Now integrating and using $v(t_0) = 0$, we get:

$$\text{sgn}(t - t_0)v(t) \le \text{sgn}(t - t_0)\int_{t_0}^t a\beta\gamma\, ds/\gamma(t)$$

$$= \left|\int_{t_0}^t [a(s)\beta(s)\gamma(s)/\gamma(t)]\, ds\right|, \quad \forall t \in J.$$

From (1) and the definition of γ it follows that

$$u(t) \le a(t) + \text{sgn}(t - t_0)v(t)$$

$$\le a(t) + \left|\int_{t_0}^t a(s)\beta(s)\exp\left\{\left|\int_s^t \beta(\sigma)\, d\sigma\right|\right\}\, ds\right|, \quad \forall t \in J,$$

which is the estimate for u, as claimed. $\qquad\qquad\square$

(6.2) Corollary. *Let* $a(t) = a_0(|t - t_0|)$, *where* $a_0 \in C(\mathbb{R}_+, \mathbb{R}_+)$ *is a monotone increasing function, and assume that*

$$u(t) \le a(t) + \left|\int_{t_0}^t \beta(s)u(s)\, ds\right|, \quad \forall t \in J.$$

Then we obtain the estimate

$$u(t) \le a(t)e^{\left|\int_{t_0}^t \beta(s)\, ds\right|}, \quad \forall t \in J.$$

Proof. Since $a(s) \le a(t)$ for $|s - t_0| \le |t - t_0|$, it follows from (2) that

$$u(t) \le a(t)\left[1 + \left|\int_{t_0}^t \beta(s)\exp\{|\int_s^t \beta(\sigma)\, d\sigma|\}\, ds\right|\right]$$

$$= a(t)\left[1 + \text{sgn}(t - t_0)\int_{t_0}^t \beta(s)\exp\{\text{sgn}(t - t_0)\int_s^t \beta(\sigma)\, d\sigma\}\, ds\right]$$

$$= a(t)\exp\left\{\text{sgn}(t - t_0)\int_{t_0}^t \beta(\sigma)\, d\sigma\right\}, \quad \forall t \in J,$$

which is the assertion. $\qquad\qquad\square$

Lipschitz continuous functions play an important role in the theory of (ordinary) differential equations. For this reason we want to study this class of functions more carefully and explain their relation to the continuously differentiable functions.

Assume that X and Y are metric spaces and let T be a topological space. A function $f : T \times X \to Y$ is called *uniformly Lipschitz continuous with respect to* $x \in X$, if there exists a constant $\lambda \in \mathbb{R}_+$ such that

$$d(f(t,x), f(t,\overline{x})) \leq \lambda d(x, \overline{x}), \quad \forall x, \overline{x} \in X, \ \forall t \in T.$$

Each $\lambda \in \mathbb{R}_+$ with this property is called a *Lipschitz constant* for f. (Of course d denotes the respective metrics in X and Y.)

The function $f : T \times X \to Y$ is called (locally) *Lipschitz continuous with respect to* $x \in X$, if every point $(t_0, x_0) \in T \times X$ has a neighborhood $U \times V$ in $T \times X$ such that $f|(U \times V)$ is uniformly Lipschitz continuous with respect to $x \in V$. Finally, we set

$$C^{0,1^-}(T \times X, Y) := \{ f : T \times X \to Y \mid f \in C(T \times X, Y)$$
$$\text{and } f \text{ is Lipschitz continuous with respect to } x \in X \}.$$

If T is a single point, and therefore $f : X \to Y$, then we suppress the phrase "with respect to $x \in X$." We then set

$$C^{1^-}(X, Y) := \{ f : X \to Y \mid f \text{ is Lipschitz continuous} \}.$$

Of course we have

$$C^{1^-}(X, Y) \subseteq C(X, Y),$$

and by definition

$$C^{0,1^-}(T \times X, Y) \subseteq C(T \times X, Y).$$

Finally, if X and Y are open subsets of the Banach spaces E and F, respectively, then $C^{0,1}(T \times X, Y)$ denotes the set of all continuous functions $f : T \times X \to Y$ which have continuous partial derivatives with respect to $x \in X$. That is,

$$C^{0,1}(T \times X, Y) := \{ f \in C(T \times X, Y) \mid D_2 f \in C(T \times X, \mathcal{L}(E, F)) \}.$$

With this notation we obtain the following elementary but important theorem.

(6.3) Proposition. *Let E and F be Banach spaces with $D \subseteq E$ open and let T be an arbitrary topological space. Then*

$$C^{0,1}(T \times D, F) \subseteq C^{0,1^-}(T \times D, F).$$

In particular, we have

$$C^1(D, F) \subseteq C^{1^-}(D, F),$$

that is, every continuously differentiable function is Lipschitz continuous.

Proof. Let $(t_0, x_0) \in T \times D$ and $f \in C^{0,1}(T \times D, F)$ be arbitrary. Then there exists a neighborhood $U \times V$ of (t_0, x_0) in $T \times D$ such that

$$\|D_2 f(t,x) - D_2 f(t_0, x_0)\| \le 1, \quad \forall (t,x) \in U \times V.$$

With $m := 1 + \|D_2 f(t_0, x_0)\|$ we then have

$$\|D_2 f(t,x)\| \le m < \infty, \quad \forall (t,x) \in U \times V.$$

Without loss of generality we may assume that V is convex. From the mean value theorem we then obtain the estimate

$$\|f(t,x) - f(t,\overline{x})\| \le \sup_{0 \le s \le 1} \|D_2 f(t, \overline{x} + s(x - \overline{x}))\| \|x - \overline{x}\| \le m\|x - \overline{x}\|$$

for all $(t,x), (t,\overline{x}) \in U \times V$, which is our assertion.   □

The following proposition has significant technical importance. In particular, it says that *every Lipschitz continuous function defined on compact subsets is uniformly Lipschitz continuous.*

(6.4) Proposition. *Let X and Y be metric spaces and let T be a compact topological space. Suppose that $K \subseteq X$ is compact and $f \in C^{0,1^-}(T \times X, Y)$. Then there exists an open neighborhood W of K in X such that $f|(T \times W)$ is uniformly Lipschitz continuous with respect to $x \in W$.*

Proof. By assumption, for every $(t,x) \in T \times X$ there exists an open neighborhood $U_t \times V_x$ of (t,x) in $T \times X$ and some constant $\lambda(t,x) \in \mathbb{R}_+$ such that

$$d(f(\overline{t}, \overline{x}), f(\overline{t}, \overline{\overline{x}})) \le \lambda(t,x) d(\overline{x}, \overline{\overline{x}})$$

for all $(\overline{t}, \overline{x}), (\overline{t}, \overline{\overline{x}}) \in U_t \times V_x$. Without loss of generality we may assume that

$$V_x = \mathbb{B}(x, \epsilon(x)) := \{y \in X \mid d(y,x) < \epsilon(x)\}$$

holds for some suitable $\epsilon(x) > 0$. Since $T \times K$ is compact, there exist $(t_i, x_i) \in T \times K$, $i = 1, \ldots, m$, with

$$T \times K \subseteq \bigcup_{i=1}^{m} U_{t_i} \times \mathbb{B}(x_i, \epsilon(x_i)/2).$$

Consequently

$$W := \bigcup_{i=1}^{m} \mathbb{B}(x_i, \epsilon(x_i)/2)$$

is an open neighborhood of K in X.

First we will show that the set $f(T \times W)$ has a finite diameter. To this end, let $(t,x), (s,y) \in T \times W$ be arbitrary. Then there exist indices $i, j \in \{1, \ldots, m\}$

with $(t, x) \in U_{t_i} \times V_{x_i}$ and $(s, y) \in U_{t_j} \times V_{x_j}$. From this, from the compactness of $T \times T$, and from the continuity of $(t, s) \mapsto d(f(t, x_i), f(s, x_j))$ it follows that

$$d(f(t, x), f(s, y))$$
$$\leq d(f(t, x), f(t, x_i)) + d(f(t, x_i), f(s, x_j)) + d(f(s, x_j), f(s, y))$$
$$\leq \lambda(t_i, x_i)\epsilon(x_i) + \max_{(s,t)\in T\times T} d(f(t, x_i), f(s, x_j)) + \lambda(t_j, x_j)\epsilon(x_j)$$
$$=: M_{ij} < \infty.$$

With $M := \max\{M_{ij} \mid 1 \leq i, j \leq m\}$ we have

$$\mathrm{diam}(f(T \times W)) \leq M < \infty.$$

With

$$\delta := \min\{\epsilon(x_1), \dots, \epsilon(x_m)\}/2 > 0$$

it follows that

$$\lambda := \max\{\lambda(t_1, x_1), \dots, \lambda(t_m, x_m), \delta^{-1}\mathrm{diam}(f(T \times W))\} \in \mathbb{R}_+$$

is well defined.

Now let $(t, x), (t, y) \in T \times W$ be arbitrary. Then there is some $i \in \{1, \dots, m\}$ so that $(t, x) \in U_{t_i} \times \mathbb{B}(x_i, \epsilon(x_i)/2)$. If $d(x, y) < \delta$, then

$$d(y, x_i) \leq d(y, x) + d(x, x_i) < \delta + \epsilon(x_i)/2 \leq \epsilon(x_i)$$

and thus $y \in V_{x_i} = \mathbb{B}(x_i, \epsilon(x_i))$. Therefore $(t, y) \in U_{t_i} \times V_{x_i}$ and

$$d(f(t, x), f(t, y)) \leq \lambda(t_i, x_i)d(x, y) \leq \lambda d(x, y).$$

If, on the other hand, $d(x, y) \geq \delta$, then

$$d(f(t, x), f(t, y)) \leq \mathrm{diam} f(T \times W) = [\delta^{-1}\mathrm{diam} f(T \times W)]\delta \leq \lambda d(x, y).$$

Therefore $d(f(t, x), f(t, y)) \leq \lambda d(x, y)$ for all $(t, x), (t, y) \in T \times W$. $\qquad\square$

After these preparations we return again to the differential equations. For the rest of this section we make the following stipulations:

$J \subseteq \mathbb{R}$ *is an open interval, E is an arbitrary Banach space (over $\mathbb{K}$), $D \subseteq E$ is open and $f \in C(J \times D, E)$.*

A function $u : J_u \to D$ is called a *solution of the differential equation*

$$\dot{x} = f(t, x) \tag{3}$$

if the following holds:

(i) $J_u \subseteq J$ is a perfect interval (i.e., $\mathrm{int}(J_u) \neq \emptyset$);
(ii) $u \in C^1(J_u, D)$;
(iii) $\dot{u}(t) = f(t, u(t)), \quad \forall t \in J_u$.

If $\epsilon > 0$, then $u : J_u \to D$ is called an ϵ-*approximate solution* of (3) if the following holds:

(i) $J_u \subseteq J$ is a perfect interval;

(ii) $u \in C(J_u, D)$, and u is piecewise continuously differentiable (i.e., J_u can be written as a finite union of perfect subintervals $I_1, \ldots, I_m$, so that u is continuously differentiable on each $\overline{I_k}$);

(iii) For every subinterval $I \subseteq J_u$ on which u is continuously differentiable we have

$$\|\dot{u}(t) - f(t, u(t))\| \le \epsilon, \quad \forall t \in I.$$

(6.5) Remarks. (a) *Let J_u be a perfect subinterval of J and let $u : J_u \to D$. Then u is a solution of the differential equation $\dot{x} = f(t, x)$ if and only if $u \in C(J_u, D)$ and*

$$u(t) = u(t_0) + \int_{t_0}^{t} f(s, u(s)) \, ds, \quad \forall t \in J_u, \tag{4}$$

where $t_0 \in J_u$ is arbitrary.

This is an immediate consequence of the fundamental theorem of calculus. The trivial fact that the *integral equation* (4) is equivalent to the differential equation (3) is of great theoretical importance. It permits us, in fact, "to work in the space of continuous functions" without regard to differentiability questions.

(b)   *Let $u : J_u \to D$ be an ϵ-approximate solution of the differential equation $\dot{x} = f(t, x)$. Then we have*

$$\left\| u(t) - u(t_0) - \int_{t_0}^{t} f(s, u(s)) \, ds \right\| \le \epsilon |t - t_0|, \quad \forall t \in J_u,$$

where $t_0 \in J_u$ is arbitrary.

Proof. We consider the case $t > t_0$. (The case $t < t_0$ is treated analogously.) There exists a decomposition $t_0 =: s_0 < s_1 < \cdots < s_m := t$ with $u|[s_i, s_{i+1}] \in C^1([s_i, s_{i+1}], D)$ for $i = 0, \ldots, m - 1$. By the fundamental theorem of calculus we have

$$u(s_{i+1}) - u(s_i) = \int_{s_i}^{s_{i+1}} \dot{u}(s) \, ds.$$

From this it follows that

$$\left\| u(s_{i+1}) - u(s_i) - \int_{s_i}^{s_{i+1}} f(s, u(s)) \, ds \right\| \le \int_{s_i}^{s_{i+1}} \|\dot{u}(s) - f(s, u(s))\| \, ds \le \epsilon(s_{i+1} - s_i),$$

for $i = 0, 1, \ldots, m - 1$. Now the assertion follows since

$$u(t) - u(t_0) - \int_{t_0}^{t} f(s, u(s)) \, ds = \sum_{i=0}^{m-1} \left[u(s_{i+1}) - u(s_i) - \int_{s_i}^{s_{i+1}} f(s, u(s)) \, ds \right]. \quad \square$$

The following simple estimate will play a fundamental role later on.

(6.6) Lemma. *Let $f : J \times D \to E$ be uniformly Lipschitz continuous with respect to $x \in D$ and with Lipschitz constant λ. If $u : J_u \to D$ and $v : J_v \to D$ are, respectively, ϵ_1- and ϵ_2-approximate solutions of $\dot{x} = f(t, x)$, then for every $t_0 \in J_u \cap J_v$ we have:*

$$\|u(t) - v(t)\| \le \{\|u(t_0) - v(t_0)\| + (\epsilon_1 + \epsilon_2)|t - t_0|\}e^{\lambda|t - t_0|},$$

for all $t \in J_u \cap J_v$.

Proof. Using (6.5 b) and the identity

$$u(t) - v(t) = \left[u(t) - u(t_0) - \int_{t_0}^t f(s, u(s))\, ds\right]$$
$$- \left[v(t) - v(t_0) - \int_{t_0}^t f(s, v(s))\, ds\right] + [u(t_0) - v(t_0)]$$
$$+ \int_{t_0}^t [f(s, u(s)) - f(s, v(s))]ds,$$

it follows that

$$\|u(t) - v(t)\| \le (\epsilon_1 + \epsilon_2)|t - t_0| + \|u(t_0) - v(t_0)\|$$
$$+ \lambda \left|\int_{t_0}^t \|u(s) - v(s)\|\, ds\right|$$

for all $t \in J_u \cap J_v$. Now the assertion follows from corollary (6.2). $\square$

As a first application we prove the following *uniqueness theorem* "for Lipschitz continuous right-hand sides."

(6.7) Theorem. *Let $f \in C^{0,1-}(J \times D, E)$ and assume that $u : J_u \to D$ and $v : J_v \to D$ are solutions of $\dot{x} = f(t, x)$ with $u(t_0) = v(t_0)$, for some $t_0 \in J_u \cap J_v$. Then $u = v$ holds identically on $J_u \cap J_v$.*

Proof. It suffices to prove the assertion for every compact perfect subinterval $I \subseteq J_u \cap J_v$ with $t_0 \in I$. Since $K := u(I) \cup v(I) \subseteq D$ is compact, there exists an open neighborhood W of K in D such that $f \mid (I \times W)$ is uniformly Lipschitz continuous with respect to $x \in W$ (cf. proposition (6.4)). Now the assertion follows from lemma (6.6). $\square$

Problems

1. Show that, under the assumptions of lemma (6.6), the estimate

$$\|u(t) - v(t)\| \leq \|u(t_0) - v(t_0)\| e^{\lambda |t - t_0|} + \frac{\epsilon_1 + \epsilon_2}{\lambda} \left[e^{\lambda |t - t_0|} - 1 \right]$$

holds for all $t, t_0 \in J_u \cap J_v$. Also show that this estimate is sharper than the one in lemma (6.6).

2. Show that the estimate in problem 1 is sharp, that is, it cannot, in general, be improved.

3. Give an example to show that the inclusions

$$C^1(D, F) \subseteq C^{1^-}(D, F) \subseteq C(D, F)$$

are proper ($D \subseteq E$ is open; E, F are Banach spaces).

7. Existence Theorems

In this section we let $J \subseteq \mathbb{R}$ be an open interval, $E = (E, |\cdot|)$ be a finite dimensional Banach space over $\mathbb{K}$, $D \subseteq E$ be open and $f \in C(J \times D, E)$.

Moreover let $(t_0, x_0) \in J \times D$ and let the constants $a, b > 0$ be fixed so that

$$[t_0 - a, t_0 + a] \subseteq J \quad \text{and} \quad \overline{\mathbb{B}}(x_0, b) \subseteq D,$$

and set $R := [t_0 - a, t_0 + a] \times \overline{\mathbb{B}}(x_0, b)$.

(7.1) Lemma. *Let $M := \max |f(R)|$ and $\alpha := \min(a, b/M)$. Then for every $\epsilon > 0$ there exists an ϵ-approximate solution*

$$u \in C([t_0 - \alpha, t_0 + \alpha], \overline{\mathbb{B}}(x_0, b))$$

of $\dot{x} = f(t, x)$ with $u(t_0) = x_0$ and

$$|u(t) - u(s)| \leq M|t - s|, \quad \forall t, s \in [t_0 - \alpha, t_0 + \alpha].$$

Proof. Since $f|R$ is uniformly continuous, there exists some $\delta > 0$ such that

$$|f(t, x) - f(\bar{t}, \bar{x})| \leq \epsilon, \quad \forall (t, x), (\bar{t}, \bar{x}) \in R$$

with $|t - \bar{t}| \leq \delta$ and $|x - \bar{x}| \leq \delta$. We now partition the interval $[t_0 - \alpha, t_0 + \alpha]$ into subintervals

$$t_0 - \alpha =: t_{-n} < t_{-n+1} < \cdots < t_{-1} < t_0 < t_1 < \cdots < t_n := t_0 + \alpha,$$

such that

$$\max_{i=-n+1,\ldots,n} |t_{i-1} - t_i| \leq \min(\delta, \delta/M)$$

holds.

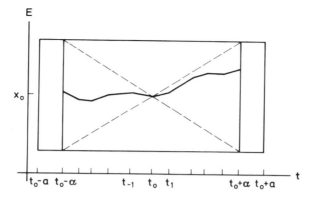

We then define inductively a polygonal curve, a so-called *Euler polygon*, by

$$u(t) := \begin{cases} u(t_i) + (t - t_i)f(t_i, u(t_i)), & \text{if } i \geq 0 \\ u(t_{i+1}) + (t - t_{i+1})f(t_{i+1}, u(t_{i+1})), & \text{if } i \leq -1, \end{cases}$$

where $t_i \leq t \leq t_{i+1}$. One easily verifies that u is defined on all of $[t_0 - \alpha, t_0 + \alpha]$ and that

$$u \in C([t_0 - \alpha, t_0 + \alpha], \overline{\mathbb{B}}(x_0, b)),$$

as well as $|u(t) - u(s)| \leq M|t - s|$, holds for all $s, t \in [t_0 - \alpha, t_0 + \alpha]$. Moreover, we evidently have

$$\dot{u}(t) = f(t_i, u(t_i))$$

for all $t \in [t_i, t_{i+1}] \cap [t_0, \infty)$ and all $t \in [t_{i-1}, t_i] \cap (-\infty, t_0]$, and also

$$|u(t) - u(t_i)| \leq \delta$$

for all $t \in [t_i, t_{i+1}] \cap [t_0, \infty)$ and all $t \in [t_{i-1}, t_i] \cap (-\infty, t_0]$. From these facts it follows easily that u is an ϵ-approximate solution of $\dot{x} = f(t, x)$. $\qquad\square$

For every $\epsilon > 0$, this lemma furnishes an ϵ-approximate solution on the fixed interval $[t_0 - \alpha, t_0 + \alpha]$. Now, if we knew that for some sequence $\epsilon_n \to 0$ the sequence of ϵ_n-approximate solutions, (u_{ϵ_n}), would converge uniformly to a function $u \in C([t_0 - \alpha, t_0 + \alpha], E)$, then a simple limit argument would show that u is a solution of the IVP $\dot{x} = f(t, x)$, $x(t_0) = x_0$. Such a convergent subsequence exists, if the set of ϵ-approximate solutions is relatively compact in $C([t_0 - \alpha, t_0 + \alpha], E)$. Hence we need a criterion for the compactness of subsets of $C([t_0 - \alpha, t_0 + \alpha], E)$. Such a criterion is furnished by the following Arzéla-Ascoli theorem. To this end, recall that for every compact topological space K and every Banach space F the space $C(K, F)$ is a Banach space with respect to the *sup-norm*

$$\|f\|_C := \max_{x \in K} \|f(x)\|.$$

(7.2) Lemma. *(Arzéla-Ascoli): Let K be a compact metric space and let F be an arbitrary Banach space. Moreover, let $\mathcal{M} \subseteq C(K, F)$. Then $\mathcal{M}$ is relatively compact (i.e., $\overline{\mathcal{M}}$ is compact) if and only if the following holds:*

(i) $\mathcal{M}$ *is equicontinuous, that is, for every $y \in K$ and every $\epsilon > 0$ there exists a neighborhood V of y in K such that*

$$\|f(x) - f(y)\| < \epsilon, \quad \forall x \in V, \ \forall f \in \mathcal{M};$$

(ii) $\mathcal{M}(y) := \{f(y) \mid f \in \mathcal{M}\}$ *is relatively compact in F for every $y \in K$.*

If F is finite dimensional, then $\mathcal{M}$ is precompact if and only if $\mathcal{M}$ is equicontinuous and bounded.

Proof. If F is an arbitrary Banach space, we refer to Lang [1] for a simple proof. Proofs which apply to more general spaces can be found, for example, in Dugundji [1] or Schubert [1].

When $\mathcal{M}$ is relatively compact, then consequently $\mathcal{M}$ is bounded. If $\mathcal{M}$ is bounded (in $C(K, F)$ of course), then evidently $\mathcal{M}(y)$ is bounded in F for every $y \in K$. Therefore $\mathcal{M}(y)$ is relatively compact when F is finite dimensional (and hence isomorphic to $\mathbb{K}^{\dim(F)}$). With this the last assertion follows from the general case. $\square$

After these preliminaries we can now easily prove the following fundamental existence theorem.

(7.3) Theorem. *(Cauchy-Peano): Assume that $f \in C(J \times D, E)$. Then the IVP*

$$\dot{x} = f(t, x), \quad x(t_0) = x_0,$$

has at least one solution u on $[t_0 - \alpha, t_0 + \alpha]$ with $u([t_0 - \alpha, t_0 + \alpha]) \subseteq \overline{\mathbb{B}}(x_0, b)$.

Proof. For each $n \in \mathbb{N}^*$, lemma (7.1) implies the existence of a $\frac{1}{n}$-approximate solution u_n on $\overline{J}_\alpha := [t_0 - \alpha, t_0 + \alpha]$ such that $u_n(\overline{J}_\alpha) \subseteq \overline{\mathbb{B}}(x_0, b)$ and

$$|u_n(t) - u_n(s)| \leq M|s - t|, \quad \forall s, t \in \overline{J}_\alpha. \tag{1}$$

From (1) follows, in particular, that the set

$$\mathcal{M} := \{u_n \mid n \in \mathbb{N}^*\} \subseteq C(\overline{J}_\alpha, E)$$

is equicontinuous. Moreover, it follows from (1) that

$$|u_n(t)| \leq |u_n(t_0)| + M|t - t_0| \leq |x_0| + b,$$

for all $n \in \mathbb{N}^*$ and all $t \in \overline{J}_\alpha$. Consequently $\mathcal{M}$ is bounded in $C(\overline{J}_\alpha, E)$ and by lemma (7.2) $\mathcal{M}$ is precompact in $C(\overline{J}_\alpha, E)$. Therefore there exists some

$u \in C(\overline{J}_\alpha, E)$ and a subsequence (u_{n_k}) of (u_n) such that $u_{n_k} \to u$ in $C(\overline{J}_\alpha, E)$, as $k \to \infty$. So (u_{n_k}) converges uniformly to u on $\overline{J}_\alpha$. By (6.5 b) we have that

$$\left| u_{n_k}(t) - x_0 - \int_{t_0}^t f(s, u_{n_k}(s))\, ds \right| \leq \frac{1}{n_k} |t - t_0|,$$

for all $t \in \overline{J}_\alpha$ and all $k \in \mathbb{N}$. Since the convergence is uniform, it follows that we can take the limit under the integral and we get

$$u(t) - x_0 - \int_{t_0}^t f(s, u(s))\, ds = 0, \quad \forall t \in \overline{J}_\alpha.$$

The assertion now follows from this and (6.5 a). □

From theorem (7.3) and theorem (6.7) we immediately obtain the following proposition.

(7.4) Local Existence and Uniqueness Theorem. *Assume that* $f \in C^{0,1-}(J \times D, E)$. *Then the IVP*

$$\dot{x} = f(t, x), \quad x(t_0) = x_0$$

has a unique solution u *on* $[t_0 - \alpha, t_0 + \alpha]$.

(7.5) Remarks. (a) The solution of the IVP in theorem (7.3) is, in general, not unique, as is shown by example (5.2b).

(b) The method used in the above proof is also numerically useful. The Euler polygons can very simply be obtained by an algorithm which can easily be programmed on a computer. In general, however, only the convergence of a subsequence can be guaranteed. A better result is obtained if

$f|R$ is uniformly Lipschitz continuous with respect to $x \in \overline{B}(x_0, b)$. *Then the entire sequence of* ϵ_n*-approximate solutions,* (u_{ϵ_n}), *converges uniformly on* $\overline{J}_\alpha := [t_0 - \alpha, t_0 + \alpha]$, *as* $n \to \infty$, *to the unique solution* u *of the IVP*

$$\dot{x} = f(t, x), \quad x(t_0) = x_0,$$

whenever $\epsilon_n \to 0$. *We have the error estimate*

$$|u_{\epsilon_n}(t) - u(t)| \leq \epsilon_n |t - t_0| e^{\lambda |t - t_0|},$$

where λ *denotes a Lipschitz constant.*

In fact, from lemma (6.6) it follows that

$$|u_{\epsilon_n}(t) - u_{\epsilon_m}(t)| \leq (\epsilon_n + \epsilon_m) \alpha e^{\lambda \alpha}, \quad \forall t \in \overline{J}_\alpha,$$

and for all $n, m \in \mathbb{N}$. Hence (u_{ϵ_n}) is a Cauchy sequence in the Banach space $C(\overline{J}_\alpha, E)$.

(c) The above error estimate shows that the method of Euler polygons is not very well suited to numerically approximate the solution over a large time interval. One develops methods (e.g., multistep methods) in the theory of "numerical integration of ordinary differential equations" which are better suited for these purposes.   □

The central result of this section is the following theorem.

(7.6) Global Existence and Uniqueness Theorem. *Assume that*

$$f \in C^{0,1-}(J \times D, E).$$

Then for every $(t_0, x_0) \in J \times D$ *there exists a unique nonextendible solution*

$$u(\cdot, t_0, x_0) : J(t_0, x_0) \to D$$

of the IVP

$$\dot{x} = f(t, x), \quad x(t_0) = x_0. \tag{2}$$

The maximal interval of existence $J(t_0, x_0)$ *is open:*

$$J(t_0, x_0) = (t^-(t_0, x_0), t^+(t_0, x_0)),$$

and we either have

$$t^- := t^-(t_0, x_0) = \inf J, \quad resp. \quad t^+ := t^+(t_0, x_0) = \sup J,$$

or

$$\lim_{t \to t^\pm} \min \left\{ \operatorname{dist}(u(t, t_0, x_0), \partial D), |u(t, t_0, x_0)|^{-1} \right\} = 0.$$

(Here of course we mean the limit as $t \to t^-$ when $t^- > \inf J$, and $t \to t^+$ when $t^+ < \sup J$, respectively. Moreover, we use the *convention*: dist $(x, \emptyset) = \infty$.)

Proof. Let $(t_0, x_0) \in J \times D$ be fixed. By theorem (7.4) there exists some $\alpha > 0$ such that the IVP (2) has a unique solution u on $\overline{J}_\alpha := [t_0 - \alpha, t_0 + \alpha]$. Again, by theorem (7.4) there exists some $\beta > 0$ such that the IVP

$$\dot{x} = f(t, x), \quad x(t_0 + \alpha) = u(t_0 + \alpha),$$

has a unique solution v on $\overline{J}_{\alpha,\beta} := [t_0 + \alpha - \beta, t_0 + \alpha + \beta]$. Now, using theorem (6.7), we have $u = v$ on $\overline{J}_\alpha \cap \overline{J}_{\alpha,\beta}$. It then follows that the function

$$u_1 := \begin{cases} u, & \text{on } \overline{J}_\alpha \\ v, & \text{on } \overline{J}_{\alpha,\beta}, \end{cases}$$

defined on $\overline{J}_\alpha \cup \overline{J}_{\alpha,\beta}$, is a solution of the IVP (2) and is a proper extension of u. Since a similar argument can be made at $t_0 - \alpha$, we see that u can be properly extended to the right and to the left.

We now set

$$t^+ := t^+(t_0, x_0) := \sup\{\, \beta \in \mathbb{R} \mid (2) \text{ has a solution on } [t_0, \beta]\,\}$$

and

$$t^- := t^-(t_0, x_0) := \inf\{\, \gamma \in \mathbb{R} \mid (2) \text{ has a solution on } [\gamma, t_0]\,\}.$$

Then, based upon the uniqueness theorem (6.7), there exists a unique solution

$$u := u(\cdot, t_0, x_0) : J(t_0, x_0) := (t^-, t^+) \to D$$

of (2) so that u cannot be extended. In particular, it follows that $J(t_0, x_0)$ is open, because otherwise we can apply the above argument to extend u past either t^+ or t^-.

Consider the case $t^+ < \sup J$, and assume there exist some $\epsilon > 0$ and a sequence $t_i \to t^+$ such that $t_i < t^+$ and

$$|u(t_i)| \le 1/2\epsilon \quad \text{and} \quad \operatorname{dist}(u(t_i), \partial D) \ge 2\epsilon, \quad \forall i \in \mathbb{N}. \tag{3}$$

Without loss of generality we may assume that $\epsilon^2 \le 1/2$. Moreover, let

$$M := \max\{|f(t, x)| \mid t_0 \le t \le t^+, \ |x| \le 1/\epsilon, \ \operatorname{dist}(x, \partial D) \ge \epsilon\}$$

and $0 < \delta < \epsilon/M$. We then have

$$\begin{aligned} &|u(t_i + s)| < 1/\epsilon \text{ and } \operatorname{dist}(u(t_i + s), \partial D) > \epsilon \\ &\text{for all } i \in \mathbb{N} \text{ and } 0 \le s \le \min\{\delta, t^+ - t_i\}. \end{aligned} \tag{4}$$

Indeed, if (4) were false, there would exist some $k \in \mathbb{N}$ and some $\beta \in (0, \min\{\delta, t^+ - t_k\}]$ with $|u(t_k + s)| \le 1/\epsilon$ and $\operatorname{dist}(u(t_k + s), \partial D) \ge \epsilon$ for $0 \le s \le \beta$ and either

$$|u(t_k + \beta)| = 1/\epsilon \quad \text{or} \quad \operatorname{dist}(u(t_k + \beta), \partial D) = \epsilon.$$

We then would have

$$|f(t_k + s, u(t_k + s))| \le M, \quad \text{for all } 0 \le s \le \beta,$$

and therefore (cf. 6.5 a)

$$|u(t_k + \beta) - u(t_k)| \le \int_{t_k}^{t_k + \beta} |f(s, u(s))|\, ds \le \beta M \le \delta M < \epsilon.$$

Consequently, we would get

$$|u(t_k + \beta)| < |u(t_k)| + \epsilon \le (1/2\epsilon) + \epsilon \le 1/\epsilon,$$

since $\epsilon \le 1/2\epsilon$ (because $\epsilon^2 \le 1/2$), and

$$\operatorname{dist}(u(t_k + \beta), \partial D) \ge \operatorname{dist}(u(t_k), \partial D) - |u(t_k + \beta) - u(t_k)| > 2\epsilon - \epsilon = \epsilon.$$

But this contradicts the choice of β.

Because of (4) we obtain, for all $i \in \mathbb{N}$ with $t^+ - t_i \le \delta$, the estimate

$$|u(t) - u(s)| \leq \left| \int_s^t |f(\tau, u(\tau))| \, d\tau \right| \leq M|t - s|, \quad \forall s, t \in [t_i, t^+). \tag{5}$$

If now (t'_k) is an arbitrary sequence such that $t'_k < t^+$ and $t'_k \to t^+$, then (5) shows that $(u(t'_k))$ is a Cauchy sequence in E. Therefore the limit

$$y := \lim_{k \to \infty} u(t'_k)$$

exists and $y \in D$, since $\mathrm{dist}(u(t), \partial D) \geq \epsilon$ for t close to t^+. It also follows from (5) that the limit

$$\lim_{k \to \infty} \int_{t_0}^{t'_k} f(s, u(s)) \, ds$$

exists. If now (s_k) is some other sequence with $s_k \to t^+$ and $s_k < t^+$, then it follows similarly that

$$\lim_{k \to \infty} u(s_k) = z \in D.$$

Hence (5) implies that

$$|y - z| = \lim_{k \to \infty} |u(t'_k) - u(s_k)| \leq M \lim_{k \to \infty} |t'_k - s_k| = 0.$$

From this we obtain

$$y = \lim_{t \to t^+} u(t),$$

and a similar argument shows that

$$\lim_{t \to t^+} \int_{t_0}^t f(s, u(s)) ds = \int_{t_0}^{t^+} f(s, u(s)) ds,$$

i.e., the improper integral on the right converges. If we now set

$$v(t) := \begin{cases} u(t), & \text{for } t^- < t < t^+ \\ y, & \text{for } t = t^+, \end{cases}$$

we see that

$$v \in C((t^-, t^+], D)$$

and

$$v(t) = x_0 + \int_{t_0}^t f(s, v(s)) \, ds, \quad \forall t \in (t^-, t^+].$$

Therefore v is a solution of the IVP (2) on the interval $(t^-, t^+]$, which contradicts the choice of t^+. This shows that (3) cannot hold. We therefore have

$$\lim_{t \to t^+} \min\{\mathrm{dist}(u(t), \partial D), |u(t)|^{-1}\} = 0.$$

The argument at the point t^- is similar. $\qquad \square$

(7.7) Corollary. *Let* $f \in C^{0,1-}(J \times D, E)$ *and*

$$\gamma^+(t_0, x_0) := \{u(t, t_0, x_0) \mid t \in [t_0, t^+(t_0, x_0))\}.$$

(a) *If* $\gamma^+(t_0, x_0)$ *is bounded, we either have* $t^+ = \sup J$ *or* $\mathrm{dist}(u(t, t_0, x_0), \partial D) \to 0$
as $t \to t^+$.

(b) *If* $\gamma^+(t_0, x_0)$ *is contained in a compact subset of* D, *then* $t^+ = \sup J$.

Similar assertions apply to t^- *and*

$$\gamma^-(t_0, x_0) := u\left((t^-, t_0], t_0, x_0\right).$$

The above result can be expressed somewhat imprecisely as: *either the solution exists for all time, or it approaches the boundary of D* (where the boundary of D includes the "point at infinity" ($|x| = \infty$)).

A useful criterion, implying the boundedness of all solutions of the differential equations (for finite time), is given by the following proposition. Example (5.2 a) shows that it cannot be improved significantly.

(7.8) Proposition. *Assume there exist* $\alpha, \beta \in C(J, \mathbb{R}_+) \cap L^1(J, \mathbb{R})$ *such that*

$$|f(t, x)| \leq \alpha(t)|x| + \beta(t), \quad \forall(t, x) \in J \times D, \tag{6}$$

(i.e., f is <u>*linearly bounded*</u> *w.r.t.* $x \in D$). *Then every solution of* $\dot{x} = f(t, x)$ *is* bounded.

Proof. Let $u : J_u \to D$ be a solution of $\dot{x} = f(t, x)$. Then it follows from (6) and remark (6.5 a) that

$$|u(t)| \leq |u(t_0)| + \left| \int_{t_0}^t \beta(s)\, ds \right| + \left| \int_{t_0}^t \alpha(s)|u(s)|\, ds \right|, \quad \forall t \in J_u.$$

The assertion is now a simple consequence of Gronwall's lemma (6.1). $\square$

We obtain now easily the following fundamental *global existence and uniqueness theorem for linear differential equations* by applying the above results.

(7.9) Theorem. *Let* $A \in C(J, \mathcal{L}(E))$ *and* $b \in C(J, E)$. *Then the linear (nonhomogeneous) IVP*

$$\dot{x} = A(t)x + b(t), \quad x(t_0) = x_0,$$

has a unique global solution for every $(t_0, x_0) \in J \times E$.

Proof. We set $f(t, x) := A(t)x + b(t)$ and choose a fixed $(s, y) \in J \times E$. Moreover, we choose some $\delta > 0$ so that $[s - \delta, s + \delta] \subseteq J$. Then for all $(t, x) \in J \times E$ with

$|t - s| \leq \delta$ we have:

$$|f(s, y) - f(t, x)| \leq |A(s) - A(t)||y|$$

$$+ \left(\max_{|\tau - s| \leq \delta} |A(\tau)| \right) |x - y| + |b(s) - b(t)|.$$

This shows that $f \in C(J \times E, E)$. Also $D_2 f(t, x) = A(t)$ and therefore $D_2 f \in C(J \times E, \mathcal{L}(E))$. It follows that

$$f \in C^{0,1}(J \times E, E) \subseteq C^{0,1^-}(J \times E, E)$$

(cf. proposition (6.3)). Finally, f is linearly bounded because

$$|f(t, x)| \leq |A(t)||x| + |b(t)|, \quad \forall (t, x) \in J \times E.$$

Now the assertion follows from proposition (7.8) and theorem (7.6).   □

(7.10) Remarks. (a) The Cauchy-Peano theorem is wrong if dim $E = \infty$. For a counter-example we refer to Deimling [1]. *If a and b are chosen so small that $f|R$ is bounded, then the local existence and uniqueness theorem (7.4) remains also true in case the Banach space E is infinite dimensional.* (The boundedness of $f|R$ in this case can no longer be deduced from the compactness, but rather, it follows from the continuity.) From the uniform Lipschitz continuity of $f|R$ with respect to $x \in \overline{\mathbb{B}}(x_0, b)$ and from the compactness of $[t_0 - a, t_0 + a]$, one can easily deduce that $f|R$ is uniformly continuous. Then lemma (7.1) remains correct and the limit can be taken (without compactness), as in remark (7.5 b). For a different proof, one which is based on the historically important *Picard-Lindelöf iteration* and also applies to the infinite dimensional case, we refer to the problems at the end of this section.

(b) *In the infinite dimensional case the global existence theorem (7.6) remains true with the same proof if one makes the additional assumption: f is bounded on bounded subsets of D which have a positive distance from ∂D.* One should note that the latter assumption is *only* used to say something about the behavior of $u(\cdot, t_0, x_0)$ as $t \to t^{\pm}$.

(c) Based on (b), one easily verifies that theorem (7.9) remains also true in case dim $E = \infty$.

(d) Ordinary differential equations in infinite dimensional Banach spaces play a role in some areas of *nonlinear functional analysis*. For further details in the case dim $E = \infty$, we refer to the books by Deimling [1] and Martin [1].

Problems

1. *Banach Fixed Point Theorem.* Let X be a complete metric space and $f : X \to X$ a contraction, i.e., there exists some $\alpha \in (0, 1)$ such that

$$d(f(x), f(y)) \leq \alpha d(x, y), \quad \forall x, y \in X.$$

Prove that:

(i) f has a unique fixed point $\overline{x} = f(\overline{x})$.
(ii) $\overline{x}$ can be calculated by iteration, i.e., for every $x_0 \in X$, the sequence (x_n) such that

$$x_{n+1} := f(x_n), \quad n \in \mathbb{N},$$

   converges to $\overline{x}$.
(iii) We have the error estimate

$$d(x_n, \overline{x}) \le \frac{\alpha^n}{1 - \alpha} d(x_1, x_0).$$

(*Hint:* Show that (x_n) is a Cauchy sequence.)

2. *Continuous Parameter Dependence.* Let X be a complete metric space and Λ a topological space. Assume that $f : \Lambda \times X \to X$ satisfies:

(i)   There exists some $\alpha \in (0, 1)$ such that $d(f(\lambda, x), f(\lambda, y)) \le \alpha d(x, y)$ for all $x, y \in X$
       and $\lambda \in \Lambda$.
(ii) $f(\cdot, x) : \Lambda \to X$ is continuous for every $x \in X$.

Show that: For every $\lambda \in \Lambda$, the function $f(\lambda, \cdot) : X \to X$ has a unique fixed point $x(\lambda)$
and $x(\cdot) \in C(\Lambda, X)$.

3. *The Picard-Lindelöf Theorem.* Let both $J \subseteq \mathbb{R}$ and $D \subseteq E$ be open, where E
is an arbitrary (not necessarily finite dimensional) Banach space. Moreover, let $f \in$
$C(J \times D, E), (t_0, x_0) \in J \times D$, and $a, b, \lambda, M \in \mathbb{R}$ be such that

$$R := [t_0 - a, t_0 + a] \times \overline{\mathbb{B}}(x_0, b) \subseteq J \times D,$$
$$\|f(t, x) - f(t, y)\| \le \lambda \|x - y\|, \quad \forall (t, x), (t, y) \in R,$$
$$\|f(t, x)\| \le M, \quad \forall (t, x) \in R.$$

Finally, let

$$\alpha := \min\left(a, \frac{b}{M}, \frac{1}{2\lambda}\right)$$

and $I := [t_0 - \alpha, t_0 + \alpha]$. Then the IVP

$$\dot{x} = f(t, x), \quad x(t_0) = x_0,$$

has a unique solution $u : I \to \overline{\mathbb{B}}(x_0, b)$ and it can be "calculated" by the following scheme:

$$u_{m+1}(t) = x_0 + \int_{t_0}^{t} f(s, u_m(s))ds, \quad m \in \mathbb{N}, \ t \in I,$$

where $u_0 \in C(I, \overline{\mathbb{B}}(x_0, b))$, with $u_0(t_0) = x_0$, is arbitrary.

Here the sequence (u_m) converges uniformly on I as $m \to \infty$.

(*Hint:* Show that the Banach Fixed Point theorem can be applied to

$$T : X \to C(I, E), \quad \text{with} \quad Tv(t) := x_0 + \int_{t_0}^{t} f(s, v(s))\, ds, \quad t \in I,$$

where $X := \{ v \in C(I, E) \mid v(t_0) = x_0,\ \|v - v_0\|_C \leq b,\ v_0(t) := x_0,\ \forall t \in I \}$.)

4. Solve the initial value problem in $\mathbb{R}$,

$$\dot{x} = ax, \quad x(0) = 1, \quad a \in \mathbb{R}, \tag{7}$$

with (a) the Euler polygon method and (b) the Picard-Lindelöf iteration.

(a) Let $t > 0$ and partition the interval $[0, t]$ into m equal subintervals. What is the value of the Euler polygon u_m corresponding to this subdivision at the point t, i.e., what is $u_m(t)$? What happens to $u_m(t)$ as $m \to \infty$, and what can one say when $t < 0$?

(b) Using the "initial function" $u_0(t) = 1$, calculate explicitly the mth iteration u_m of the solution to (7) by the Picard-Lindelöf iteration method. On which interval does the iteration converge?

8. Continuity Theorems

In what follows, we always let J denote an open interval in $\mathbb{R}$ and D an open subset of a finite dimensional Banach space E. Moreover, let Λ denote a locally compact metric space and assume that

$$f \in C(J \times D \times \Lambda, E)$$

is Lipschitz continuous with respect to $x \in D$.

Then, for every $\lambda \in \Lambda$ and $(\tau, \xi) \in J \times D$, theorem (7.6) implies that there exists a unique noncontinuable solution (for short: *the solution*)

$$u(\cdot, \tau, \xi, \lambda) : J(\tau, \xi, \lambda) \to D$$

of the *parameter-dependent* initial value problem

$$\dot{x} = f(t, x, \lambda), \quad x(\tau) = \xi.$$

Here

$$J(\tau, \xi, \lambda) := (t^-(\tau, \xi, \lambda), t^+(\tau, \xi, \lambda)) \subseteq J$$

is the maximal interval of existence (which is open by theorem (7.6)). We denote by

$$\mathcal{D}(f, \Lambda) := \{ (t, \tau, \xi, \lambda) \in J \times J \times D \times \Lambda \mid t \in J(\tau, \xi, \lambda) \}$$

the domain of the function

$$u : (t, \tau, \xi, \lambda) \mapsto u(t, \tau, \xi, \lambda) \in D.$$

We also set $\mathcal{D}(f) := \mathcal{D}(f, \Lambda)$, whenever Λ contains only one element.

The main goal of this section is to show that $\mathcal{D}(f, \Lambda)$ is open in $J \times J \times D \times \Lambda$, hence open in $\mathbb{R} \times \mathbb{R} \times E \times \Lambda$, and that $u : \mathcal{D}(f, \Lambda) \to D$ is continuous. To show this we need a few preliminaries.

(8.1) Lemma. *Let X and Y be topological spaces, and assume that A, B and C are metric spaces.*

(i) *If $f : X \times Y \to A$ satisfies*
    (a) $f(x, \cdot) : Y \to A$ *is continuous for every $x \in X$,*
    (b) $f(\cdot, y) : X \to A$ *is continuous, uniformly with respect to $y \in Y$, then f is continuous.*

(ii) $f : A \times B \to C$ *is uniformly Lipschitz continuous if and only if f is uniformly Lipschitz continuous with respect to x and with respect to y.*

(iii) $C^{1-}(B, C) \circ C^{1-}(A, B) \subseteq C^{1-}(A, C)$; *that is, the composition of Lipschitz continuous functions is Lipschitz continuous.*

Proof. (i) Let $(x_0, y_0) \in X \times Y$ and $\epsilon > 0$ be arbitrary. Because of (b), there exists a neighborhood U of x_0 in X such that

$$d(f(x, y), f(x_0, y)) < \epsilon/2, \quad \forall (x, y) \in U \times Y,$$

and because of (a), there exists a neighborhood V of y_0 in Y such that

$$d(f(x_0, y), f(x_0, y_0)) < \epsilon/2, \quad \forall y \in V.$$

Consequently we have

$$d(f(x, y), f(x_0, y_0)) \leq d(f(x, y), f(x_0, y)) + d(f(x_0, y), f(x_0, y_0)) < \epsilon$$

for all $(x, y) \in U \times V$.

(ii) The necessity of the above condition is clear. That this condition is also sufficient follows from the inequality

$$d(f(a, b), f(\overline{a}, \overline{b})) \leq d(f(a, b), f(\overline{a}, b)) + d(f(\overline{a}, b), f(\overline{a}, \overline{b})),$$

which holds for all $(a, b), (\overline{a}, \overline{b}) \in A \times B$.

(iii) Let $f \in C^{1-}(A, B)$ and $g \in C^{1-}(B, C)$, and let $x_0 \in A$ be arbitrary. Then there exists a neighborhood V of $y_0 := f(x_0)$ in B and some $\lambda \in \mathbb{R}_+$ with

$$d(g(y), g(\overline{y})) \leq \lambda d(y, \overline{y}), \quad \forall y, \overline{y} \in V.$$

Moreover, there exists a neighborhood U of x_0 in A and a constant $\mu \in \mathbb{R}_+$ such that $f(U) \subseteq V$ and

$$d(f(x), f(\overline{x})) \leq \mu d(x, \overline{x}), \quad \forall x, \overline{x} \in U.$$

Hence, for $x, \overline{x} \in U$, we have

$$d(g(f(x)), g(f(\overline{x}))) \leq \lambda d(f(x), f(\overline{x})) \leq \lambda \mu d(x, \overline{x}),$$

which proves the assertion. $\square$

First we will prove a local continuity result.

(8.2) Lemma. *For every* $(\overline{\tau}, \overline{\xi}, \overline{\lambda}) \in J \times D \times \Lambda$ *there exists a neighborhood* $I \times V \times W$ *of* $(\overline{\tau}, \overline{\xi}, \overline{\lambda})$ *in* $J \times D \times \Lambda$ *such that*

$$u \in C^{1-,0}([I \times I \times V] \times W, D).$$

Proof. There exist numbers $a, b, \mu > 0$ and a compact neighborhood W of $\overline{\lambda}$ in Λ such that

$$A := [\overline{\tau} - 2a, \overline{\tau} + 2a] \times \overline{\mathbb{B}}(\overline{\xi}, 2b) \times W \subseteq J \times D \times \Lambda$$

and

$$|f(t, x, \lambda) - f(t, y, \lambda)| \leq \mu |x - y|, \quad \forall (t, x, \lambda), (t, y, \lambda) \in A. \tag{1}$$

Hence, for every

$$(\tau, \xi, \lambda) \in B := [\overline{\tau} - a, \overline{\tau} + a] \times \overline{\mathbb{B}}(\overline{\xi}, b) \times W,$$

we have the inclusion

$$[\tau - a, \tau + a] \times \overline{\mathbb{B}}(\xi, b) \times W \subseteq A.$$

We now set

$$M := \sup_{(\tau, \xi, \lambda) \in A} |f(\tau, \xi, \lambda)| \quad \text{and} \quad \alpha := \min(a, b/M). \tag{2}$$

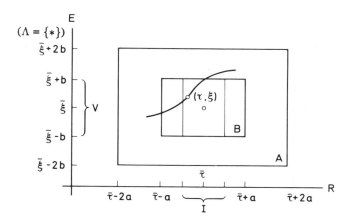

Then, by theorem (7.3), there exists the solution $u(\cdot, \tau, \xi, \lambda)$ on $[\tau - \alpha, \tau + \alpha]$ for every $(\tau, \xi, \lambda) \in B$.

With

$$I := [\overline{\tau} - \alpha/2, \overline{\tau} + \alpha/2] \quad \text{and} \quad V := \overline{\mathbb{B}}(\overline{\xi}, b)$$

we then have

$$I \times I \times V \times W \subseteq \mathcal{D}(f, \Lambda).$$

From (6.5 a) and (2) it follows that

$$|u(s, \tau, \xi, \lambda) - u(t, \tau, \xi, \lambda)| \le M|s - t| \tag{3}$$

for all $s, t \in I$ and $(\tau, \xi, \lambda) \in I \times V \times W$. From Lemma (6.6) and (1) we obtain the estimate

$$|u(t, \tau, \xi, \lambda) - u(t, \tau, \eta, \lambda)| \le e^{\alpha \mu}|\xi - \eta| \tag{4}$$

for all $\xi, \eta \in V$ and $(t, \tau, \lambda) \in I \times I \times W$.

The uniqueness of the solution implies that

$$u(t, \sigma, \xi, \lambda) = u(t, \tau, u(\tau, \sigma, \xi, \lambda), \lambda)$$

for all $\sigma, \tau \in I$ and $(t, \xi, \lambda) \in I \times V \times W$. Hence (4) implies that

$$|u(t, \tau, \xi, \lambda) - u(t, \sigma, \xi, \lambda)| \le |\xi - u(\tau, \sigma, \xi, \lambda)|e^{\alpha \mu},$$

and so from $u(\sigma, \sigma, \xi, \lambda) = \xi$ and (3) we finally get

$$|u(t, \tau, \xi, \lambda) - u(t, \sigma, \xi, \lambda)| \le Me^{\alpha \mu}|\tau - \sigma| \tag{5}$$

for all $\sigma, \tau, \in I$ and $(\tau, \xi, \lambda) \in I \times V \times W$.

From (3), (4), (5) and from lemma (8.1 ii) it now follows that for every $\lambda \in W$

$$u(\cdot, \cdot, \cdot, \lambda) \text{ is uniformly Lipschitz continuous on } I \times I \times V, \text{ and} \tag{6}$$
the Lipschitz constant is independent of $\lambda \in W$.

To simplify, we set $u := u(\cdot, \tau, \xi, \lambda)$ and $v := v(\cdot, \tau, \xi, \nu)$, where $(\tau, \xi) \in I \times V$ and $\lambda, \nu \in W$. We then obtain from

$$u(t) - v(t) = \int_{\tau}^{t} [f(s, u(s), \lambda) - f(s, v(s), \nu)]ds$$

$$= \int_{\tau}^{t} [f(s, u(s), \lambda) - f(s, u(s), \nu)]ds + \int_{\tau}^{t} [f(s, u(s), \nu) - f(s, v(s), \nu)]ds,$$

with $T := [\min\{\tau, t\}, \max\{\tau, t\}]$, the estimate

$$|u(t) - v(t)| \le \alpha \max_{s \in T} |f(s, u(s), \lambda) - f(s, u(s), \nu)| + \left| \int_{\tau}^{t} \mu |u(s) - v(s)|ds \right|$$

for all $\lambda, \nu \in W$ and $(t, \tau, \xi) \in I \times I \times V$. Here we note that, based on theorem (7.3), we have

$$u(s), v(s) \in \overline{\mathbb{B}}(\xi, b) \subseteq \overline{\mathbb{B}}(\overline{\xi}, 2b)$$

for all $s \in T$, $\lambda, \nu \in W$ and $(\tau, \xi) \in I \times V$, and hence the Lipschitz estimate (1) can be applied. Therefore corollary (6.2), with $\lambda, \nu \in W$ and $(t, \tau, \xi) \in I \times I \times V$, implies the estimate

$$|u(t, \tau, \xi, \lambda) - u(t, \tau, \xi, \nu)| \leq \alpha e^{\alpha\mu} \delta(\lambda, \nu),$$

where

$$\delta(\lambda, \nu) := \max_{s \in T} |f(s, u(s, \tau, \xi, \lambda), \lambda) - f(s, u(s, \tau, \xi, \lambda), \nu)|.$$

Since $C := T \times u(T, \tau, \xi, \lambda) \times W$ is compact for every fixed $(\tau, \xi, \lambda) \in I \times V \times W$, it follows that $f|C$ is uniformly continuous. Therefore $\delta(\lambda, \nu) \to 0$ as $\nu \to \lambda$; i.e., for fixed $(t, \tau, \xi) \in I \times I \times V$, the function $u(t, \tau, \xi, \cdot) : W \to D$ is continuous. From (6) and lemma (8.1 i) it follows that $u \in C(I \times I \times V \times W, D)$. $\qquad \square$

After these preliminaries we can prove the following basic *global continuity theorem for parameter-dependent initial value problems.*

(8.3) Theorem. $\mathcal{D}(f, \Lambda)$ *is open in* $J \times J \times D \times \Lambda$ *and* $u \in C(\mathcal{D}(f, \Lambda), D)$. *Moreover,*

$$u(\cdot, \cdot, \cdot, \lambda) \in C^{1-}(\mathcal{D}(f(\cdot, \cdot, \lambda)), D)$$

for all $\lambda \in \Lambda$.

Proof. Let $(t^*, \tau_0, \xi_0, \lambda_0) \in \mathcal{D}(f, \Lambda)$ be arbitrary. It follows from lemma (8.2) that for every $(\overline{\tau}, \overline{\xi}) \in J \times D$ there exists a neighborhood $\tilde{I}_{\overline{\tau}} \times \tilde{V}_{\overline{\xi}} \times \tilde{W}_{(\overline{\tau}, \overline{\xi})}$ of $(\overline{\tau}, \overline{\xi}, \lambda_0)$ in $J \times D \times \Lambda$ such that

$$u \in C^{1-,0}([\tilde{I}_{\overline{\tau}} \times \tilde{I}_{\overline{\tau}} \times \tilde{V}_{\overline{\xi}}] \times \tilde{W}_{(\overline{\tau}, \overline{\xi})}, D). \tag{7}$$

By making appropriate changes we obtain a neighborhood $I_{\overline{\tau}} \times V_{\overline{\xi}} \times W_{(\overline{\tau}, \overline{\xi})}$ of $(\overline{\tau}, \overline{\xi}, \lambda_0)$, such that $u(t, \tau, \xi, \lambda) \in \tilde{V}_{\overline{\xi}}$ for all $(t, \tau, \xi, \lambda) \in I_{\overline{\tau}} \times I_{\overline{\tau}} \times V_{\overline{\xi}} \times W_{(\overline{\tau}, \overline{\xi})}$.

Since

$$K := \{(t, u(t, \tau_0, \xi_0, \lambda_0)) \mid t \in [\tau_0, t^*]\} \subseteq J \times D$$

is compact (Here we consider the case $t^* \geq \tau_0$. The case $t^* \leq \tau_0$ is treated similarly.), there exist $(\overline{\tau}_i, \overline{\xi}_i) \in K$, $i = 1, \ldots, m$, with $K \subseteq \cup_{i=1}^m (I_i \times V_i)$, where we have set $I_i := I_{\overline{\tau}_i}$ and $V_i := V_{\overline{\xi}_i}$. By making the intervals I_i sufficiently small, we may assume that $I_i = [t_i, t_{i+1}]$, $i = 1, \ldots, m$, with $t_1 < \tau_0 < t_2 < \cdots < t_m < t^* < t_{m+1}$. In addition, set $W := \cap_{i=1}^m W_{(\overline{\tau}_i, \overline{\xi}_i)}$.

For $i = 1, \ldots, m$ let

$$Z_i := \{(t, u(t, t_i, \xi, \lambda), \lambda) \mid t \in I_i, \ \xi \in V_i, \ \lambda \in W\} \subseteq I_i \times D \times W$$

be a so-called "solution cylinder in the interval I_i," and let $Z_i(t)$ denote the projection of Z_i along $t \in I_i$, i.e.,

$$Z_i(t) := \{(\xi, \lambda) \in D \times \Lambda \mid (t, \xi, \lambda) \in Z_i\}.$$

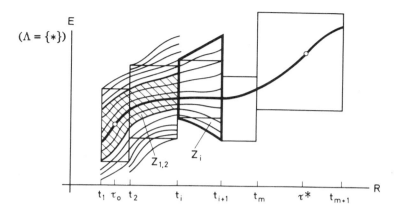

From $u(t, t_i, u(t_i, t, \xi, \lambda), \lambda) = \xi$ and (7) it follows that

$$g := u(t, t_i, \cdot, \cdot) \times id_W : V_i \times W \to Z_i(t)$$

is a homeomorphism with inverse

$$g^{-1} = u(t_i, t, \cdot, \cdot) \times id_W.$$

In particular, $V_1 \times W$ and $Z_1(t_2)$ are homeomorphic, and therefore

$$B_2 := Z_1(t_2) \cap (V_2 \times W)$$

is a neighborhood of $(u(t_2, \tau_0, \xi_0, \lambda_0), \lambda_0)$. We now can replace $Z_1 \cup Z_2$ by the well-defined solution cylinder in the interval $I_1 \cup I_2$,

$$Z_{1,2} := \{(t, u(t, t_2, \xi, \lambda), \lambda) \mid t \in I_1 \cup I_2, \ (\xi, \lambda) \in B_2\}.$$

Repeating this argument with $Z_{1,2}$ instead of Z_1, etc., we obtain inductively a solution cylinder Z in the interval $[t_1, t_{m+1}]$,

$$Z = \{(t, u(t, t_m, \xi, \lambda), \lambda) \mid t \in [t_1, t_{m+1}], \ (\xi, \lambda) \in B_m\},$$

where B_m is an appropriate neighborhood of $(u(t_m, \tau_0, \xi_0, \lambda_0), \lambda)$ in $D \times \Lambda$.
Moreover, all projections

$$Z(t_i), \quad i = 1, \ldots, m+1,$$

of Z are homeomorphic and we have

$$Z(t_i) \subset V_i \times W, \quad i = 1, \ldots, m.$$

We now set

$$Z[t_i, t_{i+1}] := \{(t, u(t, t_m, \xi, \lambda), \lambda) \mid t \in [t_i, t_{i+1}], (\xi, \lambda) \in B_m\}.$$

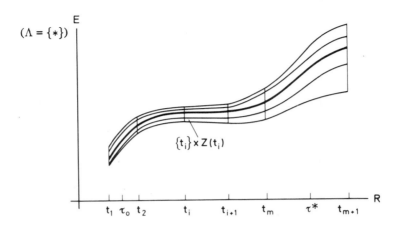

Then we have

$$Z[t_i, t_{i+1}] = \{(t, u(t, t_i, \xi, \lambda), \lambda) \mid t \in [t_i, t_{i+1}], (\xi, \lambda) \in Z(t_i)\},$$

so in particular, $Z[t_i, t_{i+1}] \subset [t_i, t_{i+1}] \times V_i \times W$. It follows from lemma (8.2) that the function

$$h_i : [t_i, t_{i+1}] \times V_i \times W \to \{t_i\} \times V_i \times W, \qquad (t, \xi, \lambda) \mapsto (t_i, u(t_i, t, \xi, \lambda), \lambda)$$

is continuous. Since, based on uniqueness,

$$Z[t_i, t_{i+1}] = h_i^{-1}(\{t_i\} \times Z(t_i))$$

and since we may assume, without loss of generality, that $Z(t_i)$ is open, we have that $Z[t_i, t_{i+1}]$ is open in $[t_i, t_{i+1}] \times D \times \Lambda$. This implies that Z is open in $[t_1, t_{m+1}] \times D \times \Lambda$; hence it is a neighborhood of $\{(t, u(t, \tau_0, \xi_0, \lambda_0), \lambda_0) \mid t_1 \le t \le t_{m+1}\}$. Consequently, there exists a neighborhood $\hat{I}_1 \times \hat{A}_1 \times \hat{W}$ of $(\tau_0, \xi_0, \lambda_0)$ in $J \times D \times W$ with $\hat{I}_1 \times \hat{A}_1 \times \hat{W} \subseteq Z$.

From the construction of Z it follows that the functions

$$
\begin{aligned}
\varphi_1 \quad &: \hat{I}_1 \times \hat{A}_1 \times \hat{W} \to Z(t_2), (\tau, \xi, \lambda) \mapsto (u(t_2, \tau, \xi, \lambda), \lambda) \\
\varphi_2 \quad &: Z(t_2) \to Z(t_3), \qquad (\xi, \lambda) \quad \mapsto (u(t_3, t_2, \xi, \lambda), \lambda) \\
&\;\;\vdots \\
\varphi_{m-1} &: Z(t_{m-1}) \to Z(t_m), \quad (\xi, \lambda) \quad \mapsto (u(t_m, t_{m-1}, \xi, \lambda), \lambda) \\
\varphi_m \quad &: I_m \times Z(t_m) \to V_m, \quad (t, \xi, \lambda) \mapsto u(t, t_m, \xi, \lambda)
\end{aligned}
$$

are well-defined. We obtain from lemma (8.2) that all these functions are Lipschitz continuous for each fixed λ.

Since

$$u(t, \tau, \xi, \lambda) = u(t, t_m, u(t_m, \tau, \xi, \lambda), \lambda)$$

for all $t \in I_m$ and $(\tau, \xi, \lambda) \in \hat{I}_1 \times \hat{A}_1 \times \hat{W}$, it follows by induction that

$$u(t, \tau, \xi, \lambda) = \varphi_m(t, \varphi_{m-1} \circ \cdots \circ \varphi_1(\tau, \xi, \lambda)) \tag{8}$$

for all $(t, \tau, \xi, \lambda) \in I_m \times \hat{I}_1 \times \hat{A}_1 \times \hat{W}$. Consequently, we have that

$$U := I_m \times \hat{I}_1 \times \hat{A}_1 \times \hat{W} \subseteq \mathcal{D}(f, \Lambda),$$

and, since U is a neighborhood of $(t^*, \tau_0, \xi_0, \lambda_0)$, that $\mathcal{D}(f, \Lambda)$ is open. The assertions about continuity now follow from (8) and lemma (8.1 iii).   □

If one assumes that f is Lipschitz continuous with respect to $\lambda \in \Lambda$, one obtains, by appropriate modifications of the above proof, that u is Lipschitz continuous in all variables. But instead of going through these modifications, we will deduce this result from theorem (8.3) by a simple trick, for the case that Λ is an open subset of a finite dimensional Banach space.

(8.4) Theorem. *Assume that Λ is open in a finite dimensional Banach space F, that $M := D \times \Lambda$, and that $f \in C^{0,1^-}(J \times M, E)$. Then*

$$u \in C^{1^-}(\mathcal{D}(f, \Lambda), D).$$

Proof. Clearly M is open in $G := E \times F$ and

$$g := (f, 0) \in C^{0,1^-}(J \times M, G).$$

Moreover, the IVP

$$\dot{x} = f(t, x, \lambda), \quad x(\tau) = \xi, \tag{9}$$

is evidently equivalent to the parameter-independent IVP

$$\dot{z} = g(t, z), \quad z(\tau) = (\xi, \lambda). \tag{10}$$

By theorem (8.3), the solution $v : \mathcal{D}(g) \to M$ of (10) is Lipschitz continuous. From

$$v(t, \tau, (\xi, \lambda)) = (u(t, \tau, \xi, \lambda), \lambda)$$

we get that

$$\mathcal{D}(g) = \mathcal{D}(f, \Lambda),$$

and hence the assertion follows.   □

(8.5) Remarks. (a) Because of remark (7.10 b), *all results of this section remain true if* $\dim E = \infty$.

(b) One easily verifies *that Λ may be an arbitrary metric space if $\lambda \mapsto f(t, x, \lambda)$ is continuous, uniformly for (t, x) in compact subsets of $J \times D$.*   □

Problems

1. Verify that remarks (8.5) are correct.

2. Prove lemma (8.2) with the aid of problems 2 and 3 of section 7.

9. Differentiability Theorems

In this section we assume that E is a real Banach space, and that Λ is an open subset of a finite dimensional Banach space F. Assuming that the spaces are real is no loss of generality, since we can identify E with $\mathbb{K}^n$ and $\mathbb{C}^n$ always with $\mathbb{R}^{2n}$ (by separating the real and imaginary parts).

We now assume that $f \in C^{0,1}(J \times (D \times \Lambda), E)$ and that the solution $u : \mathcal{D}(f, \Lambda) \to D$ is differentiable. If we differentiate the equations

$$\frac{\partial u}{\partial t}(t, \tau, \xi, \lambda) = f(t, u(t, \tau, \xi, \lambda), \lambda) \tag{1}$$

$$u(\tau, \tau, \xi, \lambda) = \xi,$$

with respect to τ, ξ and λ, and if we interchange the derivatives with $\partial/\partial t$, we (formally) obtain for all $(\tau, \xi, \lambda) \in J \times D \times \Lambda$:

(a) the function

$$J(\tau, \xi, \lambda) \to E, \quad t \mapsto \frac{\partial u}{\partial \tau}(t, \tau, \xi, \lambda) \tag{2}$$

satisfies the IVP

$$\begin{cases} \dot{y} = D_2 f(t, u(t, \tau, \xi, \lambda), \lambda)y \\ y(\tau) = -f(\tau, \xi, \lambda); \end{cases} \tag{3}$$

(b) the function

$$J(\tau, \xi, \lambda) \to \mathcal{L}(E), \quad t \mapsto \frac{\partial u}{\partial \xi}(t, \tau, \xi, \lambda) \tag{4}$$

satisfies the IVP

$$\begin{cases} \dot{z} = D_2 f(t, u(t, \tau, \xi, \lambda), \lambda)z \\ z(\tau) = id_E; \end{cases} \tag{5}$$

(c) the function

$$J(\tau, \xi, \lambda) \to \mathcal{L}(F, E), \quad t \mapsto \frac{\partial u}{\partial \lambda}(t, \tau, \xi, \lambda) \tag{6}$$

satisfies the IVP

$$\begin{cases} \dot{v} = D_2 f(t, u(t, \tau, \xi, \lambda), \lambda)v + D_3 f(t, u(t, \tau, \xi, \lambda), \lambda) \\ v(\tau) = 0. \end{cases} \tag{7}$$

Here we have set

$$\frac{\partial u}{\partial \xi} := D_3 u : \mathcal{D}(f, \Lambda) \to \mathcal{L}(E)$$

and

$$\frac{\partial u}{\partial \lambda} := D_4 u : \mathcal{D}(f, \Lambda) \to \mathcal{L}(F, E).$$

The goal of this section is to show that these formal operations are justified.

(9.1) Remarks. (a) If $g \in C^{0,1}(J \times D, E)$, and if $v : J_v \to D$ is any solution of the differential equation $\dot{y} = g(t, y)$, then the linear differential equation

$$\dot{z} = D_2 g(t, v(t))z$$

is called the *variational equation along the solution* v (or the *linearization at* v).

(b) Let $f \in C^{0,1}(J \times (D \times \Lambda), E)$ and set

$$\mu := (\tau, \xi, \lambda) \in J \times D \times \Lambda =: M,$$

and

$$A(t, \mu) := D_2 f(t, u(t, \tau, \xi, \lambda), \lambda),$$

as well as

$$b(t, \mu) := D_3 f(t, u(t, \tau, \xi, \lambda), \lambda).$$

Since by theorem (8.3) we have

$$u \in C(\mathcal{D}(f, \Lambda), D),$$

it follows that

$$A(t, \mu) \in \mathcal{L}(E)$$

and

$$b(t, \mu) \in \mathcal{L}(F, E)$$

are continuous in (t, μ). Therefore (3), (5) and (7) represent parameter-dependent linear IVP, namely the problems

$$\dot{y} = A(t, \mu)y, \quad y(\tau) = -f(\mu), \tag{3}$$

$$\dot{z} = A(t, \mu)z, \quad z(\tau) = id_E, \tag{5}$$

and

$$\dot{v} = A(t, \mu)v + b(t, \mu), \quad v(\tau) = 0. \tag{7}$$

Here (3) is an IVP in E, (5) is an IVP in the finite dimensional Banach space $\mathcal{L}(E)$, and (7) is an IVP in the finite dimensional Banach space $\mathcal{L}(F, E)$. The parameter space M is open in $\mathbb{R} \times E \times F$ and so it is locally compact. By theorem (7.9), every one of these IVP's has a unique global solution, i.e., for each fixed $\mu \in M$ the solution exists on all of $J(\mu) = J(\tau, \xi, \lambda)$. The solutions of these linear IVP's are, by theorem (8.3), also continuous in all variables. $\square$

(9.2) Theorem. *Let $f \in C^{0,1}(J \times (D \times \Lambda), E)$. Then*

$$u \in C^1(\mathcal{D}(f, \Lambda), D),$$

and $\frac{\partial u}{\partial \tau}$, $\frac{\partial u}{\partial \xi}$ and $\frac{\partial u}{\partial \lambda}$ are the solutions of the linearized IVP's (3), (5) and (7), respectively. Moreover, the "mixed second order partial derivatives"

$$\frac{\partial^2 u}{\partial \tau \partial t} = \frac{\partial^2 u}{\partial t \partial \tau}, \quad \frac{\partial^2 u}{\partial \xi \partial t} = \frac{\partial^2 u}{\partial t \partial \xi}, \quad \frac{\partial^2 u}{\partial \lambda \partial t} = \frac{\partial^2 u}{\partial t \partial \lambda} \tag{8}$$

exist and are continuous.

Proof. (a) If we prove the first part of the assertion, it follows from remark (9.1 b) that

$$\frac{\partial^2 u}{\partial t \partial \tau} = \frac{\partial}{\partial t}\left(\frac{\partial u}{\partial \tau}\right), \quad \frac{\partial^2 u}{\partial t \partial \xi}, \quad \frac{\partial^2 u}{\partial t \partial \lambda}$$

exist and are continuous. Since the function

$$(\tau, \xi, \lambda) \mapsto f(t, u(t, \tau, \xi, \lambda), \lambda)$$

is continuously differentiable, it follows by differentiating (1) that also the other mixed derivatives exist and that the equalities hold, as claimed. It therefore suffices to prove the first part of the assertion.

(b) By considering the "extended IVP"

$$\dot{z} = g(t, z), \quad z(\tau) = (\xi, \lambda),$$

where

$$g := (f, 0) \in C^{0,1}(J \times (D \times \Lambda), E \times F)$$

(cf. the proof of theorem (8.4)), we may assume, without loss of generality, that $\Lambda = \{0\}$.

(c) Let $(t, \tau, \xi) \in \mathcal{D}(f)$ and $\epsilon > 0$ be such that we have $\{(t, \tau)\} \times \mathbb{B}(\xi, \epsilon) \subseteq \mathcal{D}(f)$. Moreover, let $v(t, \tau, \xi, h) := u(t, \tau, \xi + h) - u(t, \tau, \xi)$ and

$$B(t, \tau, \xi, h) := \int_0^1 D_2 f(t, u(t, \tau, \xi) + sv(t, \tau, \xi, h)) \, ds,$$

where $h \in \mathbb{B}(\xi, \epsilon)$. (Since v is continuous, we may assume that ϵ is chosen so small that $u + sv \in D$ for all $0 \leq s \leq 1$.) It now follows from the mean value theorem that

$$\frac{\partial v}{\partial t}(t, \tau, \xi, h) = \frac{\partial u}{\partial t}(t, \tau, \xi + h) - \frac{\partial u}{\partial t}(t, \tau, \xi) =$$

$$f(t, u(t, \tau, \xi + h)) - f(t, u(t, \tau, \xi)) = B(t, \tau, \xi, h)v(t, \tau, \xi, h).$$

We also have

$$v(\tau, \tau, \xi, h) = \xi + h - \xi = h.$$

Therefore $v(\cdot, \tau, \xi, h)$ is the solution of the parameter-independent linear IVP

$$\dot{z} = B(t, \tau, \xi, h)z, \quad z(\tau) = h \tag{9}$$

in E, where $B(t, \tau, \xi, h) \in \mathcal{L}(E)$ depends continuously on its arguments. It follows from theorem (7.9) and theorem (8.3) that the linear IVP in $\mathcal{L}(E)$,

$$\dot{C} = B(t, \tau, \xi, h)C, \quad C(\tau) = id_E, \tag{10}$$

has a unique solution $C(t, \tau, \xi, h) \in \mathcal{L}(E)$, which also depends continuously on its arguments. Clearly $t \mapsto C(t, \tau, \xi, h)h$ is a solution of (9), and so by uniqueness we have

$$v(t, \tau, \xi, h) = C(t, \tau, \xi, h)h.$$

From this it follows that

$$|u(t, \tau, \xi + h) - u(t, \tau, \xi) - C(t, \tau, \xi, 0)h| \leq |C(t, \tau, \xi, h) - C(t, \tau, \xi, 0)|_{\mathcal{L}(E)}|h|$$
$$= o(|h|)$$

as $h \to 0$, and this implies the continuous differentiability of $\xi \mapsto u(t, \tau, \xi)$. We also have $D_3 u(t, \tau, \xi) = C(t, \tau, \xi, 0)$, and this function is a solution (as a function of t) of the IVP (10) with $h = 0$. Now because

$$B(t, \tau, \xi, 0) = D_2 f(t, u(t, \tau, \xi)) = A(t, \mu),$$

where $\mu := (\tau, \xi)$, it is a solution of the IVP (5).

(d) For $\sigma \in \mathbb{R}$ sufficiently close to zero we set

$$w(t, \tau, \xi, \sigma) := u(t, \tau + \sigma, \xi) - u(t, \tau, \xi).$$

As in (c), we have that

$$\frac{\partial w}{\partial t}(t, \tau, \xi, \sigma) = D(t, \tau, \xi, \sigma)w(t, \tau, \xi, \sigma),$$

where

$$D(t, \tau, \xi, \sigma) = \int_0^1 D_2 f(t, u(t, \tau, \xi) + sw(t, \tau, \xi, \sigma))\, ds.$$

Again, the mean value theorem implies that

$$w(\tau, \tau, \xi, \sigma) = u(\tau, \tau + \sigma, \xi) - \xi$$
$$= -[u(\tau + \sigma, \tau + \sigma, \xi) - u(\tau, \tau + \sigma, \xi)]$$
$$= -\sigma \int_0^1 \frac{\partial u}{\partial t}(\tau + s\sigma, \tau + \sigma, \xi) \, ds$$
$$= -\sigma \int_0^1 f(\tau + s\sigma, u(\tau + s\sigma, \tau + \sigma, \xi)) \, ds$$
$$= -\sigma f(\tau, \xi) + r(\tau, \xi, \sigma)\sigma,$$

where

$$r(\tau, \xi, \sigma) := \int_0^1 [f(\tau, \xi) - f(\tau + s\sigma, u(\tau + s\sigma, \tau + \sigma, \xi))] \, ds.$$

Hence $w(\cdot, \tau, \xi, \sigma)$ is a solution of the parameter-dependent linear IVP

$$\dot{z} = D(t, \tau, \xi, \sigma)z, \quad z(\tau) = -\sigma f(\tau, \xi) + r(\tau, \xi, \sigma)\sigma, \tag{11}$$

in E.

As in (c), we have that the parameter-dependent linear IVP in $\mathcal{L}(E)$

$$\dot{X} = D(t, \tau, \xi, \sigma)X, \quad X(\tau) = id_E,$$

has a unique solution $X(t, \tau, \xi, \sigma)$ which depends continuously on all its arguments. Uniqueness again implies that

$$w(t, \tau, \xi, \sigma) = X(t, \tau, \xi, \sigma)[-f(\tau, \xi) + r(\tau, \xi, \sigma)]\sigma.$$

And therefore

$$u(t, \tau + \sigma, \xi) - u(t, \tau, \xi) + X(t, \tau, \xi, 0)f(\tau, \xi)\sigma$$
$$= \{[X(t, \tau, \xi, 0) - X(t, \tau, \xi, \sigma)]f(\tau, \xi) + X(t, \tau, \xi, \sigma)r(\tau, \xi, \sigma)\} \sigma$$
$$= o(\sigma)$$

as $\sigma \to \infty$. Hence the function $\tau \mapsto u(t, \tau, \xi)$ is continuously differentiable and we have

$$D_2 u(t, \tau, \xi) = -X(t, \tau, \xi, 0)f(\tau, \xi).$$

Thus $D_2 u$ is the solution of the IVP (3) because $D(t, \tau, \xi, 0) = D_2 f(t, u(t, \tau, \xi)) = A(t, \mu)$. $\qquad \square$

The proof above shows that theorem (9.2) remains true if f depends continuously, but not necessarily differentiably, on an *additional* parameter $\mu \in M$. Then the derivatives $\partial u/\partial \tau$, $\partial u/\partial \xi$, $\partial u/\partial \lambda$ exist and are continuous in $(t, \tau, \xi, \lambda, \mu)$. In particular, we have the following result.

(9.3) Corollary. *Let M be a locally compact metric space and assume that $f \in C(J \times D \times \Lambda \times M, E)$ is continuously differentiable in $x \in D$ and $\lambda \in \Lambda$. Then the solution $u \in C(\mathcal{D}(f, \Lambda \times M), D)$ is continuously differentiable in τ, ξ and λ, and*

$\partial u/\partial \tau$, $\partial u/\partial \xi$ and $\partial u/\partial \lambda$ are the solutions of the linearized IVP's (3), (5) and (7), respectively, which now of course also depend on $\mu \in M$. Moreover, the mixed partial derivatives (8) exist and are continuous.

Based on corollary (9.3), it is easy now to deduce the existence of higher order derivatives of u. Here one always starts with the linearized IVP's (3), (5) or (7) and then applies induction. To simplify matters, we consider here only the case of differentiability with respect to the initial condition $\xi \in D$. We leave it up to the reader to follow this example, and state and prove the corresponding theorems for higher order derivatives of u with respect to τ and λ. Statements about higher order derivatives of u with respect to t follow directly from equation (1).

(9.4) Proposition. *Let M be a locally compact metric space and assume that $f \in C(J \times D \times M, E)$ is a C^m-function in the variable $x \in D$. Then the solution $u \in C(\mathcal{D}(f, M), D)$ is of class C^m as a function of the initial value $\xi \in D$.*

Proof. Corollary (9.3) implies that $\partial u/\partial \xi$ exists, is continuous, and is the solution of the IVP

$$\dot{z} = A(t, \tau, \xi, \mu)z, \quad z(\tau) = id_E, \tag{12}$$

in $\mathcal{L}(E)$, where

$$A(t, \tau, \xi, \mu) = D_2 f(t, u(t, \tau, \xi, \mu), \mu). \tag{13}$$

We now set

$$\hat{E} := \mathcal{L}(E), \quad \hat{D} := \mathcal{L}(E), \quad \hat{\Lambda} := D, \quad \hat{M} := J \times M,$$

and define $\hat{f} \in C(J \times \hat{D} \times \hat{\Lambda} \times \hat{M}, \hat{E})$ by

$$\hat{f}(t, \hat{x}, \hat{\lambda}, \hat{\mu}) := A(t, \tau, \hat{\lambda}, \mu)\hat{x},$$

where $\hat{\mu} = (\tau, \mu)$. Then (12) becomes

$$\dot{\hat{x}} = \hat{f}(t, \hat{x}, \hat{\lambda}, \hat{\mu}), \quad \hat{x}(\tau) = id_E. \tag{14}$$

If f is twice differentiable in x, it follows from corollary (9.3) and from (13) that $\hat{f}$ is continuously differentiable as a function of $\hat{x}$ and $\hat{\lambda}$. Hence we may apply corollary (9.3) to the IVP (14) and deduce that the solution $\partial u/\partial \xi$ of (14) is continuously differentiable with respect to $\hat{\lambda} = \xi$. Therefore u is a C^2-function of ξ. We now apply this reasoning again to the corresponding linearization of (14) and obtain the assertion by induction. $\square$

We now prove the following important theorem as a simple consequence of proposition (9.4).

(9.5) Theorem. *Assume that $f \in C^m(J \times D \times \Lambda, E)$ for some $m \in \mathbb{N}^* \cup \{\infty\}$. Then $u \in C^m(\mathcal{D}(f, \Lambda), D)$.*

Proof. We set

$$\hat{D} := J \times D \times \Lambda \subseteq \mathbb{R} \times E \times F = \hat{E}$$

and

$$\hat{f} := (1, f, 0) \in C^m(\hat{D}, \hat{E}).$$

Then the parameter-dependent nonautonomous IVP

$$\dot{x} = f(t, x, \lambda), \quad x(\tau) = \xi, \tag{15}$$

is equivalent to the parameter-independent autonomous IVP

$$\dot{y} = \hat{f}(y), \quad y(\tau) = (\tau, \xi, \lambda) \in \hat{E}. \tag{16}$$

Proposition (9.4) implies that the solution of (16), and therefore the solution of (15), is of class C^m as a function of (τ, ξ, λ). From

$$\frac{\partial u}{\partial t}(t, \tau, \xi, \lambda) = f(t, u(t, \tau, \xi, \lambda), \lambda) \tag{17}$$

it follows that $u \in C^1(\mathcal{D}(f, \Lambda), D)$. Hence (17) can be differentiated with respect to t, which implies that $\partial^2 u / \partial t^2$ exists and is continuous. It follows from the proof of proposition (9.4) that every kth order derivative of u with respect to (τ, ξ, λ), $k \leq m$, satisfies a differential equation which is obtained by differentiating (17) k-times. We can again apply the reasoning carried out above to this "differentiated" differential equation and thereby obtain the assertion by induction.   □

(9.6) Remarks.   (a) If one does not reduce the proof of theorem (9.5), as above, to proposition (9.4) but instead follows along the lines of the proof of proposition (9.4), one easily sees that *the assertion of theorem (9.5) remains correct if f only satisfies that all derivatives $D_t^i D_x^j D_\lambda^k f$, where $0 \leq i+j+k \leq m$ and $0 \leq i \leq m-1$, exist and are continuous, that is, it suffices to assume that "f is a C^{m-1}-function with respect to $t \in J$."*

(b) *The results of this section remain true if either* $\dim(E) = \infty$ *or* $\dim(F) = \infty$.

Indeed, for theorem (9.2) we only need to verify, by remark (8.5), that the mappings $(t, \mu, y) \mapsto A(t, \mu)y$ and $(t, \mu, z) \mapsto A(t, \mu)z + b(t, \mu)$ are continuous in μ, uniformly for (t, y) and (t, z) in compact sets, respectively. This is easily seen to be true. Based on (8.5 b) one obtains corollary (9.3) and proposition (9.4) also in case that M is an arbitrary metric space and that f is continuous in $\mu \in M$, uniformly in (t, x, λ) in compact subsets of $J \times D \times \Lambda$. The proof of theorem (9.5) remains unchanged.

For a different proof of theorem (9.5) in the infinite dimensional case we refer to Lang [1].   □

Problems

1. Verify the assertions in remark (9.6 a).

2. Verify the assertions in remark (9.6 b).

3. Study the proofs in Lang [1].

10. Flows

In this section let E *denote a finite dimensional Banach space over* $\mathbb{K}$, *let* D *be an open subset of* E *and let* $f \in C^{1-}(D, E)$. We now consider the *autonomous* differential equation

$$\dot{x} = f(x) \tag{1}$$

in D.

(10.1) Remarks. (a) It is appropriate to *interpret* f *as a vector field on* D, i.e., "we attach the vector $f(x) \in E$ to every point $x \in D$." More precisely: f will be identified with the vector field $X : D \to T(D) = D \times E$, given by:

$$x \mapsto (x, f(x)).$$

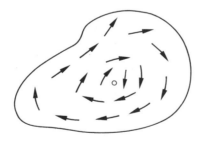

A solution of (1) can then be interpreted as a path in D, which has at every point $x \in D$ the tangent vector $f(x)$ (or more accurately $(x, f(x))$).

(b) If $u : J_u \to D$ is a solution of the IVP

$$\dot{x} = f(x), \quad x(\sigma) = \xi,$$

then, for each $s \in \mathbb{R}$, the function

$$v : J_u - s \to D, \quad t \mapsto u(t + s)$$

is a solution of the IVP

$$\dot{x} = f(x), \qquad x(\sigma - s) = \xi.$$

This follows from

$$\dot{v}(t) = \dot{u}(t + s) = f(u(t + s)) = f(v(t))$$

and $v(\sigma - s) = u(\sigma) = \xi$. Consequently we have for *the solution* of

$$\dot{x} = f(x), \qquad x(\tau) = \xi, \tag{2}$$

that is, for the uniquely determined noncontinuable solution

$$u(\cdot, \tau, \xi) : J(\tau, \xi) = (t^-(\tau, \xi), t^+(\tau, \xi)) \to D,$$

and for every $\sigma \in \mathbb{R}$, that uniqueness implies

$$u(t, \tau, \xi) = u(t + \sigma - \tau, \sigma, \xi)$$

for all

$$t \in J(\tau, \xi) = J(\sigma, \xi) + \tau - \sigma.$$

In particular we have

$$u(t, \tau, \xi) = u(t - \tau, 0, \xi), \qquad \forall\, t \in J(\tau, \xi),$$

and

$$J(\tau, \xi) = J(0, \xi) + \tau = (t^-(0, \xi) + \tau, t^+(0, \xi) + \tau)$$

for every $\tau \in \mathbb{R}$ and $\xi \in D$.

For $\xi \in D$ we now set

$$t^\pm(\xi) := t^\pm(0, \xi), \qquad J(\xi) := J(0, \xi),$$

that is,

$$J(\xi) = (t^-(\xi), t^+(\xi)),$$

and

$$\varphi(t, \xi) := u(t, 0, \xi), \qquad \forall\, t \in J(\xi).$$

We summarize:

For every $\xi \in D$,

$$\varphi(\cdot, \xi) : J(\xi) = (t^-(\xi), t^+(\xi)) \to D$$

is the (unique nonextendible) solution of the IVP

$$\dot{x} = f(x), \qquad x(0) = \xi.$$

Then, for each $\tau \in \mathbb{R}$, the (unique nonextendible) solution of the IVP

$$\dot{x} = f(x), \qquad x(\tau) = \xi,$$

is given by

$$t \mapsto \varphi(t - \tau, \xi), \qquad \forall\, t \in J(\xi) + \tau = (t^-(\xi) + \tau, t^+(\xi) + \tau).$$

(c) Let $J \subseteq \mathbb{R}$ be an open interval and $f \in C(J \times M, E)$. Then the nonautonomous (or time dependent) IVP

$$\dot{x} = f(t, x), \quad x(\tau) = \xi,$$

is equivalent to the autonomous IVP

$$\dot{x} = f(t, x), \quad x(\tau) = \xi,$$
$$\dot{t} = 1, \quad t(\tau) = \tau,$$

in $\mathbb{R} \times E$. Here of course $g := (1, f) \in C(J \times D, \mathbb{R} \times E)$ (see, for example, the proof of theorem (9.5)). The "time" t plays, in general, a special role in nonautonomous problems (e.g., in sections 7 and 8 it suffices to assume that $f \in C^{0,1-}(J \times D, E)$ instead of $f \in C^{1-}(J \times D, E)$), so that it cannot always be recommended to transform nonautonomous problems into those of autonomous type. $\qquad \square$

We now let $\Omega := \Omega(f)$ denote the domain of the function φ, i.e.,

$$\Omega := \Omega(f) := \{(t, x) \in \mathbb{R} \times D \mid t^-(x) < t < t^+(x), x \in D\}.$$

Then it follows from theorem (8.3) that:

(i) $\Omega(f)$ is open in $\mathbb{R} \times D$,
(ii) $\varphi : \Omega(f) \to D$ is Lipschitz continuous.

Moreover, it follows from theorem (9.5) (in case $\mathbb{K} = \mathbb{R}$) that:
(ii') If $f \in C^m(D, E)$, $m \geq 1$, then

$$\varphi \in C^m(\Omega(f), D).$$

Finally, we have that

(iii) $\varphi(0, \cdot) = id_D$

and, by the uniqueness of solutions,

(iv) $\varphi(t, \varphi(s, x)) = \varphi(t + s, x), \quad \forall \, x \in D,$

holds for all $s \in J(x)$ and $t \in J(\varphi(s, x))$.

Maps with properties (i)–(iv) above also occur in other contexts (e.g., in certain partial differential equations or integral equations). For this reason the following definition is meaningful:

Let M be a metric space and for every $x \in M$, let $J(x) := (t^-(x), t^+(x))$ be an open interval in $\mathbb{R}$ with $0 \in J(x)$. Moreover, set

$$\Omega := \bigcup_{x \in M} J(x) \times \{x\},$$

and let

$$\varphi : \Omega \to M$$

be a mapping with the following properties:

(i) Ω is open in $\mathbb{R} \times M$;
(ii) $\varphi : \Omega \to M$ is continuous;
(iii) $\varphi(0, \cdot) = id_M$;
(iv)   for each $x \in M$, $s \in J(x)$ and $t \in J(\varphi(s, x))$ we have $s + t \in J(x)$ and
       $\varphi(t, \varphi(s, x)) = \varphi(s + t, x)$.

Then φ is called a *flow* on M (or a (local) *dynamical system* on M). For each
$x \in M$ we call $t^-(x)$ and $t^+(x)$ the *negative*, respectively, *positive escape time of*
x. If $\Omega = \mathbb{R} \times M$, that is, if $t^-(x) = -\infty$ and $t^+(x) = \infty$ for all $x \in M$, then φ is
called a *global flow* (or *global dynamical system*).

   If φ is a given flow on M and if no confusion can be expected, we often set

$$t \cdot x := \varphi(t, x), \quad \forall\ (t, x) \in \Omega, \tag{3}$$

and for $R \times A \subseteq \Omega$ we set:

$$R \cdot A := \{t \cdot x \mid t \in R, \quad x \in A\}$$

with $R \cdot x := R \cdot \{x\}$ and $t \cdot A := \{t\} \cdot A$.

(10.2) Remarks. (a)   Condition (iv) of the above definition takes the following simple
form in the notation of (3): if $x \in M, s \in J(x)$ and $t \in J(s \cdot x)$, then $s + t \in J(x)$ and
the following is true

$$t \cdot s \cdot x := t \cdot (s \cdot x) = (t + s) \cdot x.$$

Also $0 \cdot x = x$ for all $x \in M$. If φ is a global flow, then for all $x \in M$, $s, t \in \mathbb{R}$ we have

$$t \cdot s \cdot x = (t + s) \cdot x,$$

i.e., the map $\mathbb{R} \times M \to M$, $(t, x) \mapsto t \cdot x$, is a continuous (left) *group action on* M of the
additive group $(\mathbb{R}, +)$.

(b)   Some authors apply the term flow (or dynamical system) exclusively to global flows
(and consider only those, which is, however, an unsatisfactory restriction).

(c)   If we replace in the definition of flows the set of "time parameters" $\mathbb{R}$ by $\mathbb{R}_+$ (that is,
if we replace the group $(\mathbb{R}, +)$ by the semigroup $(\mathbb{R}_+, +)$), we obtain a *semiflow*. Such semiflows are important for parabolic differential equations (e.g., reaction-diffusion equations),
or functional differential equations, which in general can be solved only in the "positive
time direction."

(d)   If M is a C^m-manifold ($m \geq 1$), for example, an open subset of E, and if we replace
(ii) in the above definition by

(ii$'$) $\varphi \in C^m(\Omega, M)$,

we obtain a C^m-flow (analogously: semiflow). $\qquad\qquad\qquad\qquad\qquad\qquad\qquad\Box$

(10.3) Theorem. *If $f \in C^{1-}(D, E)$, then the solution*

$$\varphi : \Omega(f) \to D$$

of the IVP

$$\dot{x} = f(x), \quad x(0) = \xi,$$

is a flow on D, the flow induced by f. If $\mathbb{K} = \mathbb{R}$ and $f \in C^m(D, E), m \geq 1$, then the flow is of class C^m.

If φ is a C^{2-}-flow on D (i.e., if $\varphi \in C^1$ and $D\varphi \in C^{1-}$), then it is induced by the vector field

$$D_1\varphi(0, \cdot) \in C^{1-}(D, E).$$

Proof. The first part of the theorem is a restatement of the considerations above.
  Now, let φ be a C^{2-}-flow on D. Then, for each $(t, x) \in \Omega$ and all sufficiently small $|h|$ (because Ω is open), we have for $u(t) := \varphi(t, x)$ that:

$$u(t + h) - u(t) = \varphi(h, u(t)) - u(t) = \varphi(h, u(t)) - \varphi(0, u(t)),$$

and hence $\dot{u}(t) = D_1\varphi(0, u(t))$. $\qquad\qquad\qquad\qquad\qquad\qquad\qquad$ □

(10.4) Remarks. (a) Evidently, it suffices that $D_1\varphi(0, \cdot)$ exists and is Lipschitz continuous.

(b) If the flow φ is induced by a vector field f, then f is called the *(infinitesimal) generator of the flow*. Since

$$f(x) = \lim_{t \to 0} \frac{t \cdot x - x}{t}, \quad \forall\, x \in D,$$

it follows that the generator is uniquely determined by the flow. A corresponding relation also holds for semiflows. In that case one only considers the right-hand limit. □

Let X be a topological space. Then a function $g : X \to [-\infty, \infty] =: \bar{\mathbb{R}}$ is called *lower semicontinuous* at $x_0 \in X$ if for every $\xi < g(x_0)$ there exists a neighborhood U of x_0 such that $g(x) > \xi$ for all $x \in U$. A function $g : X \to \bar{\mathbb{R}}$ is called *lower semicontinuous* if it is lower semicontinuous at every point, and g is called *upper semicontinuous* if $-g$ is lower semicontinuous.

(10.5) Lemma. *Let φ be a flow on M. Then*

(i) $t^+, -t^- : M \to (0, \infty]$ *are lower semicontinuous;*
(ii) *for all $(t, x) \in \Omega$ we have:*

$$J(t \cdot x) = J(x) - t.$$

Proof. (i) Let $(t, x) \in \Omega$. Since Ω is open, there exists a neighborhood U of x in M and some $\epsilon > 0$ such that $(t - \epsilon, t + \epsilon) \times U \subseteq \Omega$. Thus we have $t^-(y) \leq t - \epsilon < t < t + \epsilon \leq t^+(y)$ for all $y \in U$.

(ii) For each $s \in J(t \cdot x)$ we have $s + t \in J(x)$ by the definition of a flow. Hence $J(t \cdot x) \subseteq J(x) - t$.

Assume that $t^+(t \cdot x) < t^+(x) - t$. Then, in particular, $t^+(t \cdot x) < \infty$. Moreover, let $t_j \in J(t \cdot x)$ be such that $t_j \to t^+(t \cdot x)$. Then we have $t_j + t \in J(x)$ and by continuity we get

$$x_j := t_j \cdot (t \cdot x) = (t_j + t) \cdot x \to (t^+(t \cdot x) + t) \cdot x =: y.$$

Hence it follows from (i) and the inclusion already proved that

$$0 < t^+(y) \leq \liminf_{j \to \infty} t^+(x_j) \leq \liminf_{j \to \infty}(t^+(t \cdot x) - t_j) = 0.$$

This is a contradiction and therefore $t^+(t \cdot x) = t^+(x) - t$. Since a similar proof applies to t^-, we have proved the lemma.   □

Now let φ be a flow on M and $t \cdot x := \varphi(t, x)$. Then, for each $x \in M$, the function

$$\varphi_x := \varphi(\cdot, x) : J(x) \to M, \quad t \mapsto t \cdot x,$$

is called the *flow line through* x and M is the *phase space* of the flow. For each $x \in M$ we call

$$\gamma^+(x) := [0, t^+(x)) \cdot x = \{t \cdot x \mid 0 \leq t < t^+(x)\},$$

and

$$\gamma^-(x) := (t^-(x), 0] \cdot x,$$

and

$$\gamma(x) := (t^-(x), t^+(x)) \cdot x = \gamma^+(x) \cup \gamma^-(x)$$

the *positive semiorbit,* the *negative semiorbit,* and the *orbit* (or the *trajectory*) *through* x, respectively.

(10.6) Remarks. (a) The orbit $\gamma(x)$ is the image of the flow line $\varphi_x : J(x) \to M$. Hence the flow line represents a parametrization of the trajectory. This furnishes an orientation for $\gamma(x)$, namely, the "direction in which the flow line runs through the orbit." Graphically one can interpret the function $t \mapsto t \cdot x$ as the movement of the point x in time. For this reason we will usually furnish orbits with an arrow when represented graphically (cf. the figures in section 1).

(b) *The trajectories partition the phase space, i.e., each point of M is contained in exactly one trajectory.*

This partition of M is called the *phase portrait* of the flow (or of the vector field, if the flow is induced by a vector field).

Indeed, it is clear that $M = \bigcup_{x \in M} \gamma(x)$, i.e., every point is contained in a trajectory. If $\gamma(x) \cap \gamma(y) \neq \emptyset$, there exist some $s \in J(x)$ and $t \in J(y)$ with $s \cdot x = t \cdot y$. So, by (10.5 ii) we have that $-s \in J(x) - s = J(s \cdot x) = J(t \cdot y)$, and hence $x = (-s) \cdot s \cdot x = (t - s) \cdot y$. It follows that $\tau \cdot x = \tau \cdot (t-s) \cdot y = (t + \tau - s) \cdot y$ holds for every $\tau \in J(x) = J((t-s) \cdot y)$, i.e., $\gamma(x) \subseteq \gamma(y)$. For reasons of symmetry we also have $\gamma(y) \subseteq \gamma(x)$ and thus $\gamma(x) = \gamma(y)$, i.e., two trajectories are either identical or disjoint.   □

A point $x \in M$ is called a *critical point* or *stationary point* or *equilibrium point* of the flow φ if $t \cdot x = x$ for all $t \in J(x)$.

(10.7) Proposition. *The following statements* (i)–(vi) *are equivalent:*

(i) x *is a critical point;*
(ii) $\gamma(x) = \{x\}$;
(iii) $\gamma^+(x) = \{x\}$;
(iv) $\gamma^-(x) = \{x\}$;
(v) $[a, b] \cdot x = \{x\}$, *for some* $a, b \in J(x)$ *with* $a < b$;
(vi)   *there exists a sequence* $t_j \in J(x)$ *with* $t_j > 0$ *and* $\lim_{j \to \infty} t_j = 0$, *such that* $t_j \cdot x = x$ *for all* $j \in \mathbb{N}$.

If x is a critical point, then $J(x) = \mathbb{R}$.

Proof. First we will show:

$$\begin{aligned} &\text{If } t \cdot x = x \text{ for some } t \neq 0, \text{ then } J(x) = \mathbb{R} \text{ and } (nt) \cdot x = x \\ &\text{for all } n \in \mathbb{Z}. \end{aligned} \tag{4}$$

From (10.5 ii) follows that $J(x) = J(t \cdot x) = J(x) - t$, which is only possible if $J(x) = \mathbb{R}$, since $t \neq 0$.

From $t \cdot x = x$ it follows by induction that $(nt) \cdot x = x$ for all $n \in \mathbb{N}$. Since $(-t) \cdot x = (-t) \cdot (t \cdot x) = x$, the assertion follows now for all $n \in \mathbb{Z}$.

From (4) it follows, in particular, that $J(x) = \mathbb{R}$ in all cases (i)–(vi).

(vi) $\Rightarrow$ (i): If $t = nt_j$ for some $n \in \mathbb{Z}$ and $j \in \mathbb{N}$, it follows from (4) that $t \cdot x = x$. Assume now that $t \neq nt_j$ for all $(n, j) \in \mathbb{Z} \times \mathbb{N}$. Then for every $j \in \mathbb{N}$ there exists some $n_j \in \mathbb{Z}$ such that $n_j t_j \leq t < (n_j + 1)t_j$, i.e., $0 \leq t - n_j t_j < t_j$. Therefore $n_j t_j \to t$ as $j \to \infty$ and hence $x = (n_j t_j) \cdot x \to t \cdot x$, thus $x = t \cdot x$.

The remaining implications are now trivial.   □

(10.8) Proposition. *Assume that φ is a flow on D induced by the vector field $f \in C^{1-}(D, E)$. Then the critical points are exactly the zeros of f.*

Proof. If x is a critical point, then

$$f(x) = \lim_{t \to 0} \frac{t \cdot x - x}{t}$$

implies that $f(x) = 0$ (cf. (10.4 b)). Conversely, if $f(\bar{x}) = 0$, then the function $u : \mathbb{R} \to D$, $t \mapsto \bar{x}$, is a solution of the IVP $\dot{x} = f(x)$, $x(0) = \bar{x}$. Uniqueness now implies that $\bar{x} = u(t) = t \cdot \bar{x}$ for all $t \in J(x)$. $\qquad \square$

A point $x \in M$ is called a *periodic point* if there exists some $T \neq 0$ such that

$$(t + T) \cdot x = t \cdot x, \quad \forall\ t \in J(x).$$

Every $T \neq 0$ with this property is called a *period* of x. If x is periodic, we also call the orbit $\gamma(x)$ and the flow line $\varphi(\cdot, x)$ *periodic*.

(10.9) Proposition. *Let φ be a flow on M.*

(a) $x \in M$ *is periodic if and only if there exists some $T \neq 0$ with $T \cdot x = x$.*

(b) *If x is periodic, then $J(x) = \mathbb{R}$.*

(c) *If x is periodic, but not a critical point, then there exists a smallest positive period T of x, the* minimal (or fundamental) period*, and $T\mathbb{Z}^* = \{Tn \mid n \in \mathbb{Z}^* := \mathbb{Z} \setminus \{0\}\}$ is exactly the set of all periods of x.*

Proof. (a) If x is periodic with period T, then $T \cdot x = x$ holds trivially. Conversely, if $T \cdot x = x$ for some $T \neq 0$, then we have

$$t \cdot x = t \cdot (T \cdot x) = (t + T) \cdot x$$

for all $t \in J(x) = J(T \cdot x)$.

(b) This follows from statement (4) in the proof of proposition (10.7).

(c) Set

$$P := \{\tau \in \mathbb{R} \mid \tau \cdot x = x\}.$$

Then P is closed in $\mathbb{R}$ and $\{0\} \neq P \neq \mathbb{R}$. For $\tau_1, \tau_2 \in P$ we have $\tau_1 \cdot x = x = \tau_2 \cdot x$, hence $x = (-\tau_1) \cdot \tau_1 \cdot x = (-\tau_1) \cdot \tau_2 \cdot x = (\tau_2 - \tau_1) \cdot x$ and consequently $\tau_2 - \tau_1 \in P$. If we choose $\tau_2 = 0$, then it follows that $-\tau_1 \in P$. If we replace τ_1 by $-\tau_1$, then it follows that $\tau_1 + \tau_2 \in P$. Therefore P is a closed subgroup of $\mathbb{R}$.

Since $P \neq \mathbb{R}$, it contains a smallest element $T > 0$. Because otherwise we could find for each $\epsilon > 0$ some $\tau \in P$ with $0 < \tau < \epsilon$, and thus for every $t \in \mathbb{R}$ we then could find some $m \in \mathbb{Z}$ so that $|t - m\tau| < \epsilon$. Hence we would have $P = \bar{P} = \mathbb{R}$.

Now we show that $P = T\mathbb{Z}$. If this were not so, there would exist some $t \in P \backslash T\mathbb{Z}$ which we may assume, without loss of generality, to be positive. Then there exists some $n \in \mathbb{N}$ such that $nT < t < (n+1)T$ and thus $0 < t - nT < T$. Since P is a subgroup of $\mathbb{R}$ we get $t - nT \in P$, which contradicts the definition of T.   □

(10.10) Remark. In part (c) of the proof above we have shown: *If G is a closed subgroup of $(\mathbb{R}, +)$, then either $G = \{0\}, G = \mathbb{R}$ or G is infinite cyclic.*   □

In summary, we have proved the following statements about the phase portrait of a flow which we will formulate for the corresponding flow lines.

(10.11) Corollary. *Let φ be a flow on M and let $x \in M$ be arbitrary. Then for the flow line $\varphi_x : J(x) \to M$ one of the following statements holds:*

(i)   *φ_x is constant. This is the case if and only if x is a critical point.*
(ii)   *φ_x is periodic with positive minimal period T.*
(iii) *φ_x is injective.*

If $M = D$, and if φ is induced by a vector field $f \in C^{1-}(D, E)$, then (i) *holds if and only if $f(x) = 0$. In cases* (ii) *and* (iii) *the flow line is regular, i.e., we have $\dot{\varphi}_x(t) \neq 0, \quad \forall\, t \in J(x)$.*

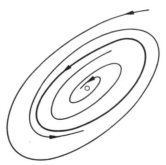

There exist exactly three types of trajectories: (i) stationary points, (ii) periodic orbits which are not stationary points, hence closed (regular) curves, and (iii) "open" (regular) curves, i.e., (regular) simple curves. (Here "regular" refers to the case when $M = D$.) We also know that, in cases (i) and (ii), the flow lines "exist for all time," i.e., $t^+(x) = \infty$ and $t^-(x) = -\infty$. Whenever the flow is induced by a vector field $f \in C^{1-}(D, E)$, we know from corollary (7.7) that $t^+(x) = \infty$ or $t^-(x) = -\infty$ holds if, respectively, either $\gamma^+(x)$ or $\gamma^-(x)$ is contained in a compact set. The following proposition shows that this is true for every flow.

(10.12) Proposition. *Let φ be a flow on M. If $t^+(x) < \infty$ [resp. $t^-(x) > -\infty$], then for every compact set $K \subseteq M$ there exists some $t_K \in J(x)$ such* *that $t \cdot x \notin K$ for all $t > t_K$ [resp. $t < t_K$], that is, <u>every point $x \in M$ with finite positive [resp. negative] escape time leaves every compact set completely, as $t \to t^+(x)$ [resp. $t \to t^-(x)$].</u>*

Proof. Assume that $t^+(x) < \infty$, and let K be a compact set so that there exists a sequence $t_j \to t^+(x)$ with $t_j \cdot x \in K$ for all $j \in \mathbb{N}$. Then the sequence $(t_j \cdot x)_{j \in \mathbb{N}}$ has an accumulation point $y \in K$. By lemma (10.5 i) there exists a neighborhood V of y in M and some $\delta > 0$ so that $t^+(z) > \delta$ for all $z \in V$. There exists some $k \in \mathbb{N}$ such that $t_k + \delta > t^+(x)$ and $t_k \cdot x \in V$, since $t_j + \delta > t^+(x)$ for $j \in \mathbb{N}$ sufficiently large and since y is an accumulation point of $(t_j \cdot x)$. Hence from lemma (10.5 ii) follows that

$$t^+(x) = t^+(t_k \cdot x) + t_k > \delta + t_k > t^+(x),$$

which is impossible. The case $t^-(x) > -\infty$ is proved analogously. $\qquad\square$

(10.13) Corollary. *If $\gamma^+(x)$ [resp. $\gamma^-(x)$] is relatively compact, then $t^+(x) = \infty$ [resp. $t^-(x) = -\infty$]. If M is compact, then φ is a global flow, i.e., $J(x) = \mathbb{R}$ for all $x \in M$.*

In what follows we set

$$\varphi^t(x) := \varphi(t, x) = t \cdot x, \quad \forall\, (t, x) \in \Omega,$$

and

$$\Omega_t := \{x \in M \mid (t, x) \in \Omega\},$$

for all $t \in \mathbb{R}$. *Then Ω_t is open in M (possibly empty) and*

$$\varphi^t \in C(\Omega_t, M), \quad \forall\, t \in \mathbb{R}. \tag{5}$$

We also have

$$\Omega_0 = M \quad \text{and} \quad \varphi^0 = id_M,$$

and, if φ is a global flow, then

$$\varphi^t \circ \varphi^s = \varphi^s \circ \varphi^t = \varphi^{s+t}, \quad \forall\, s, t \in \mathbb{R}.$$

In particular, it follows from this that the map $\varphi^t : M \to M$ is a homeomorphism with $(\varphi^t)^{-1} = \varphi^{-t}$ for each $t \in \mathbb{R}$. In the following theorem we will generalize this fact to general flows.

(10.14) Theorem. *Let φ be a flow on M. Then φ^t is a homeomorphism from Ω_t onto Ω_{-t} and $(\varphi^t)^{-1} = \varphi^{-t}$ for each $t \in \mathbb{R}$.*

Proof. It follows from lemma (10.5 ii) that $J(t \cdot x) = J(x) - t$. Hence, for every $t \in \mathbb{R}$ and $x \in \Omega_t$ we have $-t \in J(t \cdot x)$ and so $t \cdot x \in \Omega_{-t}$. Thus it follows that $(-t) \cdot (t \cdot x) = (t - t) \cdot x = x$ and therefore $\varphi^{-t}(\varphi^t(x)) = x$ for all $x \in \Omega_t$. If we replace t by $-t$, we obtain $\varphi^t(\varphi^{-t}(y)) = y$ for all $y \in \Omega_{-t}$. Hence φ^t is a bijection from Ω_t onto Ω_{-t} with $(\varphi^t)^{-1} = \varphi^{-t}$. Now the assertion follows from (5).   □

(10.15) Example. Suppose $M = \mathbb{R}$, and let φ denote the flow induced by the vector field $x \mapsto x^2$ on $\mathbb{R}$. Therefore $\varphi_y := \varphi(\cdot, y)$ is the solution of the IVP $\dot{x} = x^2$, $x(0) = y$. By separating the variables one immediately calculates that

$$t \cdot x = \begin{cases} \dfrac{1}{1/x - t}, & \text{if } x \neq 0 \\ 0, & \text{if } x = 0 \end{cases}$$

and

$$t^-(x) = \begin{cases} -\infty, & \text{if } x \geq 0 \\ \dfrac{1}{x}, & \text{if } x < 0 \end{cases} \qquad t^+(x) = \begin{cases} \dfrac{1}{x}, & \text{if } x > 0 \\ \infty, & \text{if } x \leq 0 \end{cases}$$

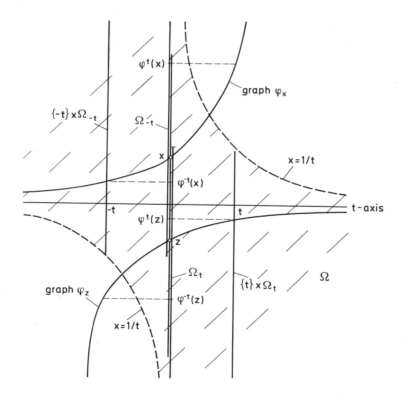

Therefore $\Omega \subseteq \mathbb{R} \times \mathbb{R}$ is the domain between the two branches of the hyperbola $x = 1/t$. For $t > 0$ we have $\Omega_t = (-\infty, 1/t)$ and $\Omega_{-t} = (-1/t, \infty)$. For each $x \in \Omega_t$ we obtain, from the projection of the point $(t, t \cdot x)$ on the graph φ_x onto the vertical axis, that $\varphi^t(x) \in \Omega_{-t}$.

In particular, we see that the only point x in M for which $J(x) = \mathbb{R}$ holds is the stationary point $x = 0$.

The homeomorphism $\varphi^t : \Omega_t \to \Omega_{-t}$ fixes the stationary point 0 and is orientation preserving from the interval $(-\infty, 0)$ onto $(-1/t, 0)$ and from the interval $(0, 1/t)$ onto $(0, \infty)$. The phase portrait of the flow consists of exactly three orbits, namely, $(-\infty, 0)$, $\{0\}$, $(0, \infty)$.

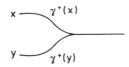

When $x < 0$, the positive semitrajectory $\gamma^+(x)$ lies in the compact set $[x, 0] \subseteq M$ and when $x > 0$, the negative semitrajectory $\gamma^-(x)$ lies in the compact set $[0, x]$. Even without the explicit representation of $t^{\pm}(x)$ we can deduce from this that $t^+(x) = \infty$ if $x < 0$, and $t^-(x) = -\infty$ if $x > 0$ (cf. corollary (10.13)).   □

(10.16) Remarks. (a) Lemma (10.5), proposition (10.7) and proposition (10.12) also hold for semiflows if the obvious modifications are made, because only the semigroup $(\mathbb{R}_+, +)$ is used in those (for semiflows meaningful parts of the) proofs. But for semiflows we have in general no "negative uniqueness," that is, it can happen that two semitrajectories through distinct points $x, y \in M$ intersect without being periodic.

$$x \underset{y}{\overset{\gamma^+(x)}{\underset{\gamma^+(y)}{\rightrightarrows}}}$$

It follows that for semiflows we cannot, in general, partition the phase space into disjoint semitrajectories. (For concrete examples of these situations we refer to Hale [2].)

(b) By remark (7.10 b), theorem (10.3) and proposition (10.8) remain true in case $\dim(E) = \infty$. In this case, if f is also bounded on bounded sets which have a positive distance from ∂D, the global existence and uniqueness theorem (7.6) contains a more precise statement than proposition (10.12) about the asymptotic behavior of the flow lines φ_x.   □

Problems

1. Show that the following rules define global flows on M and sketch the corresponding phase portraits.

(a) $M = \mathbb{R}^2$, $t \cdot (x, y) = (e^t x, e^t y)$

(b) $M = \mathbb{R}^2$, $t \cdot (x, y) = (e^t x, e^{-t} y)$
(c) $M = \mathbb{C}$, $t \cdot z = e^{it} z$
(d) $M = \mathbb{C}$, $t \cdot z = e^{(i-1)t} z$
(e) $M = \mathbb{R}^2$, $t \cdot (x, y) = (x, tx + y)$
(f) $M = \mathbb{R}^2$, $t \cdot (x, y) = (e^{-t} x, (tx + y) e^{-t})$

2. Let φ and ψ be global flows on the metric spaces M and N, respectively. Then a global flow $\varphi \times \psi$ on $M \times N$, the *product flow*, is defined by

$$t \cdot (x, y) := (t \cdot x, t \cdot y), \quad \forall \, (x, y) \in M \times N.$$

(a) Prove this assertion.

(b) Describe the orbits of the product flow on the cylinder $\mathbb{R} \times \mathbb{S}^1$ defined by the two flows

$$t \cdot x = e^t x, \quad x \in \mathbb{R},$$

and

$$t \cdot z = e^{i\alpha t} z, \quad z \in \mathbb{S}^1 \cong \{z \in \mathbb{C} \mid |z| = 1\},$$

where $\alpha \in \mathbb{R}$ is arbitrary.

(c) Describe the orbits of the product flow on the torus $T^2 := \mathbb{S}^1 \times \mathbb{S}^1$ defined by the two flows

$$t \cdot z = e^{2\pi i t} z, \quad t \cdot z = e^{2\pi i \alpha t} z, \quad z \in \mathbb{S}^1,$$

where $\alpha \in \mathbb{R}$. That is, show that, for $\alpha \in \mathbb{Q}$, every point is periodic and, for $\alpha \in \mathbb{R} \backslash \mathbb{Q}$, every trajectory is dense in T^2.

(*Hint for* $\alpha \in \mathbb{R} \backslash \mathbb{Q}$: observe first that $G := \{e^{2\pi i \alpha m} \mid m \in \mathbb{Z}\}$ is an infinite multiplicative subgroup of the group $\mathbb{S}^1$ (with respect to multiplication in $\mathbb{C}$), which is dense in $\mathbb{S}^1$.)

3. Determine the flow on $M = \mathbb{R}$, induced by the vector field $x \mapsto |x| x$. What are Ω and $J(x)$ for $x \in \mathbb{R}$? Determine the phase portrait of this flow. What is $\Omega_t := \{x \in M \mid (t, x) \in \Omega\}$ for $t \in \mathbb{R}$? Describe the action of the maps $\varphi^t : \Omega_t \to M$ and $\varphi^{-t} : \Omega_{-t} \to M$, where $\varphi^t(x) := t \cdot x$ for all $(t, x) \in \Omega$.

4. Let φ be a flow on the metric space M. Prove that:

(a) The set of all critical points is closed.

(b) $x \in M$ is critical if and only if every neighborhood of x contains a semitrajectory.

(c) Assume that for $x, y \in M$ we have $t \cdot y \to x$ as $t \to t^+(y)$, or as $t \to t^-(y)$. Then x is a critical point.

(*Hints:* (b) proof by contradiction. (c) follows from (b).)

5. Let X be a topological space and $f : X \to \bar{\mathbb{R}}$. Prove that:

(i)   The following are equivalent:
   (a) f is lower semicontinuous ($l.s.c.$).
   (b) $\forall \, \xi \in \mathbb{R}$ we have that $f^{-1}((\xi, \infty])$ is open in X.
   (c) $\forall \xi \in \mathbb{R}$ we have that $f^{-1}([-\infty, \xi])$ is closed in X.

(ii)   If $f_\alpha : X \to \bar{\mathbb{R}}$, $\alpha \in A$, is an arbitrary nonempty family of lower semicontinuous functions, then

$$\sup_\alpha f_\alpha : X \to \bar{\mathbb{R}} \quad [\text{where } (\sup_\alpha f_\alpha)(x) := \sup_\alpha f_\alpha(x)]$$

   is $l.s.c.$

(iii) f is $l.s.c.$ if and only if the *epigraph*

$$\mathrm{epi}(f) := \{(x, \xi) \in X \times \mathbb{R} \mid \xi \geq f(x)\}$$

   is closed.

(iv) $A \subseteq X$ is open if and only if the characteristic function χ_A is $l.s.c.$

(v)   If X is a metric space (more generally: if X is first countable), then f is $l.s.c.$ if and only if

$$f(x_0) \leq \liminf_{k \to \infty} f(x_k),$$

   for every sequence (x_k) such that $x_k \to x_0$.

6. Let X be a compact metric space and let $f : X \to X$ be such that

$$d(f(x), f(y)) < d(x, y), \quad \forall \, x \neq y.$$

Show that:

(i) f has a unique fixed point.
(ii)   Give an example of a mapping $f : [0, 1] \to [0, 1]$ with the properties above and such that f is not an α-contraction (i.e., globally Lipschitz continuous with Lipschitz constant α) for any $\alpha \in [0, 1)$.

(*Hint for* (i): Consider the function $x \mapsto d(x, f(x))$.)

7. Assume that A is a compact subset of a metric space M and let φ be a semiflow on M. Also assume that:

$$d(t \cdot x, t \cdot y) < d(x, y), \quad \forall x, y \in A, \ x \neq y, \ t > 0,$$

and

$$t \cdot x \in A, \quad \forall x \in A, \ \forall t \geq 0.$$

Show that:

(i)   There exists a unique critical point a in A.
(ii)   For every $x \in A$ we have $t \cdot x \to a$ as $t \to \infty$.

(*Hint:* (i) Problem 6. (ii) Consider the function $t \mapsto d(t \cdot x, a)$.)

8. Let $f \in C^{1-}(\mathbb{R}^m, \mathbb{R}^m)$ be such that

$$\langle f(x) - f(y), x - y \rangle > 0, \quad \forall\, x \neq y,$$

and assume that there exists some $R > 0$ with

$$\langle f(x), x \rangle > 0, \quad \forall |x| = R.$$

Then f has a unique zero x_0 and $|x_0| < R$.

(*Hint:* Apply problem 7 to the flow on $\mathbb{R}^m$, induced by $-f$.)

Chapter III
Linear Differential Equations

In this chapter we will treat the initial value problem of first order linear differential equations in a finite dimensional Banach space. In particular, we will completely solve the case of constant coefficient equations. The theory developed here is the foundation for the local analysis near critical points of general nonlinear systems.

Next to the classical results, we will also carry through the topological classification of hyperbolic linear flows, whereby the meaning of "saddle points" will be particularly emphasized. In the last section of this chapter we will briefly derive the most essential results for higher order differential equations by reducing them to first order systems.

11. Nonautonomous Linear Differential Equations

In this section let J denote an open interval in $\mathbb{R}$ and let $E = (E, |\cdot|)$ be a finite dimensional Banach space over $\mathbb{K}$. In addition, let

$$A \in C(J, \mathcal{L}(E)) \quad \text{and} \quad b \in C(J, E).$$

Then the differential equation

$$\dot{x} = A(t)x + b(t) \tag{1}$$

is called a *nonhomogeneous* (if $b \neq 0$), respectively, *homogeneous* (if $b = 0$) *first order linear differential equation in* E.

(11.1) Remark. If $E = \mathbb{K}^m$, and if we identify $A(t)$ with its matrix representation with respect to the canonical basis of $\mathbb{K}^m$,

$$A(t) = [a_{ij}(t)]_{1 \leq i,j \leq m},$$

then (1) can be written in the form:

$$\dot{x}^1 = a_{11}(t)x^1 + \cdots + a_{1m}(t)x^m + b^1(t)$$

$$\vdots$$

$$\dot{x}^m = a_{m1}(t)x^1 + \cdots + a_{mm}(t)x^m + b^m(t),$$

or

$$\begin{bmatrix} \dot{x}^1 \\ \vdots \\ \dot{x}^m \end{bmatrix} = \begin{bmatrix} a_{11}(t) & \cdots & a_{1m}(t) \\ \vdots & & \vdots \\ a_{m1}(t) & \cdots & a_{mm}(t) \end{bmatrix} \begin{bmatrix} x^1 \\ \vdots \\ x^m \end{bmatrix} + \begin{bmatrix} b^1(t) \\ \vdots \\ b^m(t) \end{bmatrix},$$

where we have identified $x = (x^1, \cdots, x^m) \in \mathbb{K}^m$ with the column vector $[x^1, \cdots, x^m]^T$. $\qquad\square$

Homogeneous Equations

Based on theorem (7.9), we already know that for each $(t_0, x_0) \in J \times E$ the IVP

$$\dot{x} = A(t)x + b(t),$$

$$x(t_0) = x_0,$$

has a unique global solution

$$u(\cdot, t_0, x_0) : J \to E.$$

Moreover, by theorem (8.3) we have

$$u \in C(J \times J \times E, E).$$

From this the following fundamental theorem follows immediately.

(11.2) Theorem. *The set of all solutions of the equation*

$$\dot{x} = A(t)x \tag{2}$$

forms a vector subspace V of $C^1(J, E)$ of dimension $m := \dim(E)$. For each fixed $t_0 \in J$, the mapping

$$\xi \mapsto u(\cdot, t_0, \xi) \tag{3}$$

defines an isomorphism from E onto V.

Proof. It follows from the unique solvability that for $\lambda, \mu \in \mathbb{K}$ and $\xi, \eta \in E$ we have

$$u(\cdot, t_0, \lambda\xi + \mu\eta) = \lambda u(\cdot, t_0, \xi) + \mu u(\cdot, t_0, \eta),$$

because each side is a solution of (2) with initial value $\lambda\xi + \mu\eta$ at $t = t_0$. Thus mapping (3) is linear. Again based on uniqueness, it follows from $u(\cdot, t_0, \xi) = 0$ that $\xi = 0$. Therefore the mapping in (3) is injective, hence a vector space isomorphism from E onto its image V. $\qquad\square$

(11.3) Remarks. (a) *Every linear combination of solutions of (2) is again a solution of (2).*

(b) *If $u \in C^1(J, E)$ is a solution of (2), and if $u(t_0) = 0$ for some $t_0 \in J$, then $u = 0$.*

(c) *There exist exactly* $m := \dim(E)$ *linearly independent solutions* $x_1, \ldots, x_m \in C^1(J, E)$ *of* (2). *Every set* $\{x_1, \ldots, x_m\}$ of m linearly independent solutions of (2) is called a *fundamental set of solutions* of (2).

(d) Assume that $E = \mathbb{K}^m$ and let $\{x_1, \ldots, x_m\}$ be a fundamental set of solutions of (2). Then we call the matrix X with columns $x_1, \ldots, x_m$, that is,

$$X(t) := [x_1(t), \ldots, x_m(t)],$$

a *fundamental matrix*. If, in addition, we have $X(t_0) = id_{\mathbb{K}^m}$, that is, $x_j^i(t_0) = \delta_{ij}$, $1 \le i, j \le m$, then $X_{t_0} := X$ is called the *special fundamental matrix of* (2) *at time* t_0. The special fundamental matrix X_{t_0} is the unique global solution of the homogeneous linear IVP in $\mathcal{L}(\mathbb{K}^m)$

$$\dot{X} = A(t)X, \quad X(t_0) = id_{\mathbb{K}^m} =: I_m,$$

where $\mathcal{L}(\mathbb{K}^m)$ has been identified via the canonical basis for $\mathbb{K}^m$ with the space of all $(m \times m)$-matrices over $\mathbb{K}$, $\mathbb{M}^m(\mathbb{K})$.

If X_τ *is the special fundamental matrix of* (2) *at time* τ, *then for each* $\xi \in \mathbb{K}^m$ *the unique solution* $u(\cdot, \tau, \xi) \in C^1(J, \mathbb{K}^m)$ *of the* IVP

$$\dot{x} = A(t)x, \quad x(\tau) = \xi,$$

is given by

$$u(t, \tau, \xi) = X_\tau(t)\xi.$$

In particular, the space of all solutions of (2) *is*

$$V = \{X_\tau(\cdot)\xi \mid \xi \in \mathbb{K}^m\} \subseteq C^1(J, \mathbb{K}^m). \qquad \square$$

In general, we call every solution of

$$\dot{X} = A(t)X \quad \text{in} \quad \mathbb{M}^m(\mathbb{K}) \tag{4}$$

a *solution matrix* of the homogeneous differential equation

$$\dot{x} = A(t)x \quad \text{in} \quad \mathbb{K}^m. \tag{5}$$

If X is a solution matrix of (5) (this is evidently the case if and only if every column vector of X is a solution of (5)), then the function

$$J \to \mathbb{K}, \quad t \mapsto W(t) := \det(X(t))$$

is called the *Wronskian* of the solution matrix $X = [x_1, \cdots, x_m]$ or of the set of solutions of (5), $\{x_1, \cdots, x_m\}$.

Liouville's Theorem

(11.4) Proposition *(Liouville). Let X be a solution matrix of the homogeneous linear differential equation*

$$\dot{x} = A(t)x \quad in \quad \mathbb{K}^m.$$

Then the Wronskian W of X is a solution of the homogeneous linear differential equation

$$\dot{y} = \operatorname{trace}(A(t))y \quad in \quad \mathbb{K}. \tag{6}$$

Hence we have

$$W(t) = W(\tau)e^{\int_\tau^t \operatorname{trace}(A(s))\,ds}, \quad \forall\ t,\tau \in J. \tag{7}$$

Proof. Since the determinant is an (alternating) m-linear function

$$\underbrace{\mathbb{K}^m \times \ldots \times \mathbb{K}^m}_{m} \to \mathbb{K},$$

it follows that

$$\dot{W}(t) = (\det(X))\dot{}(t)$$
$$= \sum_{j=1}^{m} \det[x_1(t), \cdots, x_{j-1}(t), \dot{x}_j(t), x_{j+1}(t), \cdots, x_m(t)],$$

for all $t \in J$.   Therefore

$$\dot{W} = \sum_{j=1}^{m} \det[x_1, \cdots, x_{j-1}, Ax_j, x_{j+1}, \cdots, x_m]. \tag{8}$$

If now $X = X_\tau$ is a special fundamental matrix at time $\tau \in J$, in other words, if $x_j(\tau) = e_j$, where $e_j^i = \delta_{ij}$, $1 \le i, j \le m$, it follows from (8) that

$$\dot{W}_\tau(\tau) = \operatorname{trace}(A(\tau)) = \operatorname{trace}(A(\tau))W_\tau(\tau), \tag{9}$$

where $W_\tau := \det X_\tau$ [because $Ae_j = j$th column of A and $W_\tau(\tau) = 1$]. If X is an arbitrary solution matrix, it follows from (11.3 d) that

$$X(t) = X_\tau(t)C, \quad \forall\ t \in J,$$

for some appropriate $C \in \mathbb{M}^m(\mathbb{K})$. Hence we get from (9) that

$$\dot{W}(\tau) = \dot{W}_\tau(\tau)\det C = \operatorname{trace}(A(\tau))W_\tau(\tau)\det C$$
$$= \operatorname{trace}(A(\tau))W(\tau).$$

Since this holds for every $\tau \in J$, we have shown that W is a solution of (6). Now one verifies that (7) is a solution of (6) and the assertion follows from the uniqueness (also see example (5.2c), where however only the real case was treated).   □

(11.5) Corollary. *The Wronskian of a solution matrix of $\dot{x} = A(t)x$ either is identically zero or else never vanishes. A set of solutions $\{x_1, \cdots, x_m\}$ is a fundamental set of solutions if and only if the corresponding Wronskian is different from zero.*

In order to give an application of Liouville's theorem, we need the following elementary proposition.

(11.6) Proposition. *Assume that M is a compact metric space and let $f : M \to \bar{\mathbb{R}}$ be lower semicontinuous. Then f assumes its minimum somewhere, that is, there exists some $m \in M$ such that $f(m) \le f(x)$ for all $x \in M$.*

Proof. Let (x_j) be a *minimizing sequence*, i.e., $f(x_j) \to a := \inf(f) \in \bar{\mathbb{R}}$. Since M is compact, hence sequentially compact, there exists a subsequence (y_k) of (x_j) and some $m \in M$ so that $y_k \to m$. Thus it follows from the lower semicontinuity (cf. problem (10.5)) that

$$f(m) \le \liminf_{k \to \infty} f(y_k) = \lim_{k \to \infty} f(y_k) = a,$$

which implies the assertion. □

(11.7) Remark. The proposition above is one of the fundamental existence theorems in the calculus of variations, because the functions occurring there are in many cases lower semicontinuous, but not continuous. However, the assumption that M is a compact metric space is too restrictive for many applications (in particular in the area of differential equations). But the proof above shows that it suffices to assume that M is *sequentially compact*. This is in general, that is, if M is not first countable, a much weaker assumption than compactness. The proof above also shows that it suffices to assume that f is "sequentially lower semicontinuous." Here we call a function $f : X \to \bar{\mathbb{R}}$, defined on an arbitrary topological space X, *sequentially lower semicontinuous at the point x* if for every sequence (x_k) in X with $x_k \to x$ we have:

$$f(x) \le \liminf_{k \to \infty} f(x_k).$$

The function f is called *sequentially lower semicontinuous* if f is sequentially lower semicontinuous at every point. Evidently, every lower semicontinuous function is also sequentially lower semicontinuous (cf. problem (10.5)), but if X is not first countable, the converse is in general not true.

We actually have now proved the following *fundamental existence theorem of the calculus of variations:*

Let X be a sequentially compact space and assume that $f : X \to \bar{\mathbb{R}}$ is sequentially lower semicontinuous. Then f assumes its minimum somewhere. □

We now come to the announced application of proposition (11.4).

(11.8) Theorem *(Liouville). Let $M \subseteq \mathbb{R}^m$ be open and $f \in C^1(M, \mathbb{R}^m)$. More-over, let φ denote the flow on M induced by f and let $K \subseteq M$ be compact. Finally, set $t_K^+ := \min_{x \in K} t^+(x)$ and $t_K^- := \max_{x \in K} t^-(x)$, and for every $t \in (t_K^-, t_K^+)$ let*

$$V(t) := \text{vol}_m(t \cdot K) := \int_{t \cdot K} dx^1 \wedge \cdots \wedge dx^m$$

denote the oriented m-dimensional volume of $t \cdot K$. Then

$$\dot{V}(t) = \int_{t \cdot K} \text{div } f \, dx.$$

Proof. Since $t^+, -t^- : M \to (0, \infty]$ are lower semicontinuous (lemma (10.5 i)), it follows from proposition (11.6) that $t_K^{\pm}$ are well-defined and we have $-\infty \le t_K^- < 0 < t_K^+ \le \infty$. By the definition of $t_K^{\pm}$ we have

$$K \subseteq \Omega_t = \{x \in M \mid (t, x) \in \Omega\}$$

for $t_K^- < t < t_K^+$. Theorem (10.14) implies that φ^t is a homeomorphism from Ω_t onto Ω_{-t}, with $(\varphi^t)^{-1} = \varphi^{-t}$. Since by theorem (10.3) the flow φ is continuously differentiable, it follows that

$$\varphi^t : \Omega_t \to \Omega_{-t}$$

is a C^1-diffeomorphism.

Let now $t \in (t_k^-, t_k^+)$ be fixed and let $\omega := dx^1 \wedge \cdots \wedge dx^m$ denote the canonical volume element on $\mathbb{R}^m$. Since

$$t \cdot K = \varphi^t(K)$$

is compact, $t \cdot K$ is measurable and has finite volume $V(t)$. And because $t \cdot K \subseteq \Omega_{-t}$, we have

$$V(t) = \int_{t \cdot K} \omega = \int_K (\varphi^t)^* \omega.$$

For every $h \in \mathbb{R}$ with $t_K^- < t + h < t_K^+$ we get

$$\varphi^{t+h}(x) = \varphi^h \circ \varphi^t(x), \quad \forall \ x \in K,$$

and for every two open sets U, V in $\mathbb{R}^m$ and every C^1-mapping

$$g : U \to V, \quad x \mapsto y = g(x),$$

we get

$$g^* dy^1 \wedge \cdots \wedge dy^m = \det(Dg) dx^1 \wedge \cdots \wedge dx^m.$$

From this it follows that

$$V(t + h) = \int_K (\varphi^h \circ \varphi^t)^* \omega = \int_K (\varphi^t)^* (\varphi^h)^* \omega$$

$$= \int_{t \cdot K} (\varphi^h)^* \omega = \int_{t \cdot K} \det(D\varphi^h) \, dx.$$

Now choose some $\epsilon > 0$ so that $t_K^- < t - \epsilon < t + \epsilon < t_K^+$ and set

$$g(s, x) := \det[D_2\varphi(s, x)], \quad \forall \ (s, x) \in [-\epsilon, \epsilon] \times (t \cdot K).$$

Then $g : [-\epsilon, \epsilon] \times (t \cdot K) \to \mathbb{R}$ is continuous and

$$V(t + s) = \int_{t \cdot K} g(s, x) dx, \quad \forall \ s \in [-\epsilon, \epsilon].$$

If we can show that

$$D_1 g : [-\epsilon, \epsilon] \times (t \cdot K) \to \mathbb{R}$$

exists and is continuous, then the compactness of $[-\epsilon, \epsilon] \times (t \cdot K)$ and the theorem about the differentiability of parameter-dependent integrals imply that $\dot{V}(t)$ exists and that

$$\dot{V}(t) = \int_{t \cdot K} D_1 g(0, x) dx. \tag{10}$$

Because

$$\dot{\varphi}_x(t) = f(\varphi_x(t)), \quad \forall \ t \in J(x),$$
$$\varphi_x(0) = x,$$

it follows from theorem (9.2) that the function

$$J(x) \to \mathcal{L}(\mathbb{R}^m), \quad t \mapsto D_2\varphi(t, x),$$

is the solution of the variational equation

$$\dot{X} = Df(\varphi(t, x))X \quad \text{in} \ \ \mathcal{L}(\mathbb{R}^m)$$

for each fixed $x \in M$.   Hence we obtain from proposition (11.4) (if, as usual, we identify $\mathcal{L}(\mathbb{R}^m)$ with $\mathsf{M}^m(\mathbb{R})$)

$$\det(X)^{\cdot}(t) = \text{trace}[Df(\varphi(t, x))] \det X(t), \quad \forall \ t \in J(x).$$

Thus for all $(s, x) \in [-\epsilon, \epsilon] \times (t \cdot K)$ we have:

$$D_1 g(s, x) = \text{trace}[Df(\varphi(s, x))]g(s, x).$$

Now $D_2\varphi(0, x) = id_{\mathbb{R}^m}$ implies that $\varphi(0, x) = x$ and so $g(0, x) = 1$.   Therefore we have that $D_1 g$ is continuous and

$$D_1 g(0, x) = \text{trace}[Df(x)] = \sum_{i=1}^m D_i f^i(x) = \text{div} \ f(x)$$

for all $x \in t \cdot K$.   The assertion now follows from (10). $\square$

(11.9) Corollary. *If $f \in C^1(M, \mathbb{R}^m)$ is* <u>*divergence free*</u>*, i.e., if* div $f = 0$*, then the flow induced by f is* <u>*volume preserving,*</u> *that is, for every compact set $K \subseteq M$ we have*

$$\mathrm{vol}_m(K) = \mathrm{vol}_m(t \cdot K)$$

for all t with $t_K^- < t < t_K^+$.

(11.10) Examples. (a) *Assume that $M \subseteq \mathbb{R}^{2m}$ is open and let $H \in C^2(M, \mathbb{R})$. Then the flow induced by the Hamiltonian vector field*

$$\dot{x} = H_y, \quad \dot{y} = -H_x, \quad (x, y) \in \mathbb{R}^m \times \mathbb{R}^m,$$

is volume preserving.

Proof. This follows from

$$\mathrm{div}(H_y, -H_x) = \sum_{i=1}^m \left[\frac{\partial^2 H}{\partial x^i \partial y^i} - \frac{\partial^2 H}{\partial y^i \partial x^i} \right] = 0. \qquad \square$$

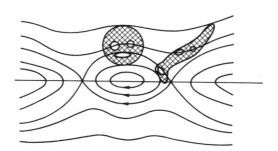

(b) *Assume that $A \in \mathcal{L}(\mathbb{R}^m)$ and let φ denote the global flow on $\mathbb{R}^m$ induced by the linear vector field*

$$(x \mapsto Ax) \in C^\infty(\mathbb{R}^m, \mathbb{R}^m).$$

Then for every compact set $K \subseteq \mathbb{R}^m$ we have

$$\mathrm{vol}_m(t \cdot K) = e^{t \, \mathrm{trace}(A)} \mathrm{vol}_m(K), \quad \forall t \in \mathbb{R}.$$

Proof. For $f(x) := Ax$ we have div $f = \mathrm{trace}(A)$. Therefore theorem (11.8) implies that

$$\dot{V}(t) = \int_{t \cdot K} \mathrm{trace}(A) \, dx = \mathrm{trace}(A) V(t), \quad \forall t \in \mathbb{R},$$

where $V(t) = \mathrm{vol}_m(t \cdot K)$ and $V(0) = \mathrm{vol}_m(K)$. $\qquad \square$

(11.11) Remarks. (a) Since the m-dimensional Lebesgue measure λ_m is *regular*, it follows, in particular, that

$$\lambda_m(A) = \sup\{\lambda_m(K) \mid K \subseteq A, \quad K \text{ compact}\}$$

for every Lebesgue measurable set $A \subseteq \mathbb{R}^m$ with finite measure. This implies that Liouville's theorem (11.8) can be extended to arbitrary Lebesgue measurable sets A which have finite measure and for which $t \cdot A$ is well-defined (cf. the proof of theorem (23.11)).

(b) Liouville's theorem plays an important role in a branch of statistical mechanics (from which mathematical *ergodic theory* has developed), in particular, in connection with example (11.10).   □

Nonhomogeneous Equations

We now turn to the nonhomogeneous differential equation

$$\dot{x} = A(t)x + b(t), \quad t \in J, \tag{11}$$

and immediately obtain the following theorem.

(11.12) Theorem. *The set of all solutions of the nonhomogeneous equation* (11) *forms the affine subspace*

$$v + V$$

of $C^1(J, E)$, *where* $v \in C^1(J, E)$ *is an arbitrary solution of the nonhomogeneous equation and* $V \subseteq C^1(J, E)$ *is the solution space of the corresponding homogeneous equation* $\dot{x} = A(t)x$.

Proof. If u and v are arbitrary solutions of (11), then clearly $u - v \in V$.   □

In order to effectively integrate the nonhomogeneous equation, we will try to transform it into a fairly simple form by means of a substitution. Since the linearity of the equation suggests a linear transformation (cf. section 5 for general transformations), we set

$$x = B y,$$

where

$$B \in C^1(J, \mathcal{GL}(E))$$

and

$$\mathcal{GL}(E) := \{B \in \mathcal{L}(E) \mid B^{-1} \in \mathcal{L}(E)\}.$$

Hence the transformed equation becomes:

$$\dot{y} = (B^{-1}x)^{\cdot} = (B^{-1})^{\cdot}x + B^{-1}\dot{x}$$
$$= (B^{-1})^{\cdot}B y + B^{-1}A B y + B^{-1}b.$$

From $BB^{-1} = id_E$ we get

$$\dot{B}B^{-1} + B(B^{-1})^{\cdot} = 0,$$

thus

$$(B^{-1})^{\cdot} = -B^{-1}\dot{B}B^{-1}$$

and therefore

$$\dot{y} = B^{-1}[AB - \dot{B}]y + B^{-1}b. \tag{12}$$

We now identify (by introducing a basis) E with $\mathbb{K}^m$ and $\mathcal{L}(E)$ with $\mathbb{M}^m(\mathbb{K})$, and we let B be the special fundamental matrix X_τ of the homogeneous equation $\dot{x} = A(t)x$ at time $\tau \in J$. Then the transformed equation (12) takes on the simple form

$$\dot{y} = X_\tau^{-1}(t)b(t).$$

A solution of this equation is the function

$$y_\tau(t) = \int_\tau^t X_\tau^{-1}(s)b(s)\,ds, \quad \forall\, t \in J.$$

Hence

$$v_\tau(t) := X_\tau(t) \int_\tau^t X_\tau^{-1}(s)b(s)\,ds, \quad t \in J,$$

is a solution of the nonhomogeneous equation (11). By theorem (11.12) and remark (11.3 d) we therefore get that all solutions of (11) have the form

$$u(t) = X_\tau(t)\xi + \int_\tau^t X_\tau(t)X_\tau^{-1}(s)b(s)\,ds, \quad t \in J, \tag{13}$$

where ξ runs through $\mathbb{K}^m$.

Let now X be an arbitrary fundamental matrix of the equation $\dot{x} = A(t)x$. Then the function

$$U(t, s) := X(t)X^{-1}(s), \quad t, s \in J, \tag{14}$$

is well-defined by corollary (11.5) and

$$U \in C^1(J \times J, \mathcal{L}(E)).$$

Moreover, we have

$$D_1U(t, s) = \dot{X}(t)X^{-1}(s) = A(t)X(t)X^{-1}(s) = A(t)U(t, s)$$

and

$$U(s, s) = id_E. \tag{15}$$

Hence for every $\tau \in J$, $U(\cdot, \tau)$ is the unique global solution of the IVP

$$\dot{Y} = A(t)Y, \quad Y(\tau) = id_E,$$

in $\mathcal{L}(E)$, that is, $U(\cdot, \tau)$ is the special fundamental matrix of $\dot{x} = A(t)x$ at time τ:

$$U(t, \tau) = X_\tau(t), \quad \forall\, t, \tau \in J. \tag{16}$$

We deduce from (14) (or from the unique solvability of the IVP above) that for $s, t, \sigma \in J$ we have:

$$U(t, \sigma)U(\sigma, s) = X(t)X^{-1}(\sigma)X(\sigma)X^{-1}(s) \tag{17}$$
$$= X(t)X^{-1}(s) = U(t, s).$$

Hence from (15) we get

$$U(t, s)U(s, t) = id_E$$

and therefore

$$U(s, t) = [U(t, s)]^{-1}, \quad \forall\, s, t \in J.$$

In particular, it follows from (16) that

$$X_\tau(t)^{-1} = U(\tau, t)$$

and hence

$$X_\tau(t)X_\tau^{-1}(s) = U(t, \tau)U(\tau, s) = U(t, s).$$

We now can write (13) in the form

$$u(t) = U(t, \tau)\xi + \int_\tau^t U(t, s)b(s)\, ds$$

and we have proved the following theorem.

(11.13) Theorem (*Variation of Constants*). *The unique global solution, $u(\cdot, \tau, \xi)$, of the IVP*

$$\dot{x} = A(t)x + b(t), \quad x(\tau) = \xi,$$

where $\tau \in J$ and $\xi \in E$, is given by the formula

$$u(t, \tau, \xi) = U(t, \tau)\xi + \int_\tau^t U(t, s)b(s)ds, \quad t \in J. \tag{18}$$

Here $U(\cdot, \tau)$ is the global solution of the IVP in $\mathcal{L}(E)$:

$$\dot{X} = A(t)X, \quad X(\tau) = id_E,$$

for all $\tau \in J$. In particular, the <u>evolution operator</u> satisfies

$$U \in C^1(J \times J, \mathcal{L}(E))$$

and

(i) $U(s, s) = id_E, \quad \forall\, s \in J,$
(ii) $U(t, \tau)U(\tau, s) = U(t, s), \quad \forall\, \tau, t, s \in J.$

If we identify E with $\mathbb{K}^m$ and $\mathcal{L}(E)$ with $\mathbb{M}^m(\mathbb{K})$, then $U(\cdot, \tau)$ is the special fundamental matrix of the homogeneous equation $\dot{x} = A(t)x$ at time $\tau \in J$.

(11.14) Remarks.   (a) If X_τ is the special fundamental matrix of the equation $\dot{x} = A(t)x$, it follows from remark (11.3) that $X_\tau(t)\xi$, with $\xi \in \mathbb{K}^m$, is the "general solution" of the homogeneous equation. If we set $y_\tau(t) = X_\tau(t)\xi(t)$, where $\xi \in C^1(J, \mathbb{K}^m)$, then

$$\dot{y}_\tau = \dot{X}_\tau\xi + X_\tau\dot{\xi} = Ay_\tau + X_\tau\dot{\xi}.$$

So if y_τ is a solution of the nonhomogeneous equation $\dot{y} = Ay + b$, then $b = X_\tau\dot{\xi}$ or $\dot{\xi} = X_\tau^{-1}b$, and therefore

$$y_\tau(t) = X_\tau(t) \int_\tau^t X_\tau^{-1}(s)b(s)\, ds, \quad \forall\ t \in J.$$

This is the expression also found above by means of a transformation and it explains the term "variation of constants."

(b) From theorem (11.12) *one obtains the general solution of the nonhomogeneous equation*

$$\dot{x} = A(t)x + b(t) \tag{19}$$

by adding an arbitrary solution of (19) – *a so-called particular solution* – *to the general solution of the homogeneous equation.* Here we mean by the general solution the solution $u(\cdot, \tau, \xi)$ in which the initial conditions are not specified, that is, they are "free parameters." Since

$$t \longmapsto \int_\tau^t U(t, s)b(s)\, ds, \quad t \in J,$$

is a particular solution of (19), and since $U(\cdot, \tau)\xi$ represents the general solution of the homogeneous equation $\dot{x} = A(t)x$, the statements above are confirmed by formula (18).

(c) If $m = 1$, then the evolution operator is clearly given by

$$U(t, s) = e^{\int_s^t a(\sigma)d\sigma}, \quad t, s \in J,$$

(cf. theorem (5.14), where of course only the case $\mathbb{K} = \mathbb{R}$ was considered). $\square$

Let E' denote the dual space of E. As is well-known, for every $B \in \mathcal{L}(E)$ the dual operator $B' \in \mathcal{L}(E')$ is defined by

$$\langle B'e', e \rangle = \langle e', Be \rangle, \quad \forall\ e' \in E', e \in E$$

where $\langle \cdot, \cdot \rangle : E' \times E \to \mathbb{K}$ denotes the duality pairing. We now define the *dual linear differential equation* corresponding to the homogeneous linear equation $\dot{x} = A(t)x$ by

$$\dot{y} = -A'(t)y \quad \text{in } E'.$$

The following theorem establishes a connection between the corresponding evolution operators.

(11.15) Proposition.   *If $U(t, \tau)$ is the evolution operator of the equation $\dot{x} = A(t)x$, then*

$$V(t, \tau) := U'(\tau, t)$$

is the evolution operator of the dual linear equation

$$\dot{y} = -A'(t)y,$$

and we have:

$$U'(\tau, t) = [U'(t, \tau)]^{-1}.$$

Proof. From the differentiability theorem (9.2) it follows that: $D_2 U(\cdot, \tau) \in C^1(J, \mathcal{L}(E))$ and, for every $\tau \in J$, $D_2 U(\cdot, \tau)$ is the solution of the IVP

$$\dot{z} = A(t)z, \quad z(\tau) = -A(\tau) \quad \text{in} \quad \mathcal{L}(E).$$

The unique solution of this equation is given by $-U(\cdot, \tau)A(\tau)$ and this implies that

$$D_2 U(t, \tau) = -U(t, \tau)A(\tau), \quad \forall \ t, \tau \in J.$$

By taking the dual we get from this that

$$[D_2 U(t, \tau)]' = D_2 U'(t, \tau) = -A'(\tau)U'(t, \tau), \quad \forall \ t, \tau \in J,$$

and $U'(t, t) = id_{E'}$, hence

$$V(t, \tau) = U'(\tau, t).$$

The assertion now follows since one easily verifies that for every $B \in \mathcal{GL}(E)$ the relation $(B^{-1})' = (B')^{-1}$ holds. We deduce from theorem (11.13) that $U(t, \tau)U(\tau, t) = id_E$, hence $U(\tau, t) = [U(t, \tau)]^{-1}$.   □

(11.16) Remarks. (a) Let $(\cdot \mid \cdot)$ be an inner product on E. Then we define a mapping $\vartheta : E \to E'$ by

$$\langle \vartheta(x), y \rangle := (y \mid x), \quad \forall \, x, y \in E.$$

Clearly we have

$$\vartheta(\alpha x + y) = \bar{\alpha}\vartheta(x) + \vartheta(y), \quad \forall \, \alpha \in \mathbb{K}, \quad x, y \in E,$$

that is, the mapping ϑ is *conjugate linear*. If $\mathbb{K} = \mathbb{R}$, then ϑ is linear. If $\vartheta(x) = 0$, that is, if $(y \mid x) = 0$ for all $y \in E$, then $x = 0$. Therefore ϑ is injective and hence – since E is finite dimensional and thus $\dim E = \dim E' - \vartheta$ is bijective. Finally, one verifies that $|\vartheta(x)| = |x|$, i.e., $\vartheta : E \to E'$ is a conjugate linear isometry, the so-called <u>duality mapping</u>. *In particular, if $\mathbb{K} = \mathbb{R}$, then ϑ is a norm preserving isomorphism from E onto E'* (i.e., an isometric isomorphism).

As is well-known, for each $A \in \mathcal{L}(E)$ we define the *adjoint* of A, $A^* \in \mathcal{L}(E)$, by

$$(Ay \mid x) = (y \mid A^*x), \quad \forall \, x, y \in E.$$

Therefore

$$\langle \vartheta(x), Ay \rangle = (Ay \mid x) = (y \mid A^*x) = \langle \vartheta(A^*x), y \rangle, \quad \forall \, x, y \in E,$$

which implies that $A'\vartheta(x) = \vartheta(A^*x)$ for all $x \in E$ and thus

$$A'\vartheta = \vartheta A^* \quad \text{or} \quad A^* = \vartheta^{-1} A'\vartheta. \tag{20}$$

In other words: the diagram

commutes. This implies, for example, *that the equation*

$$A'x' = y' \quad \text{in} \quad E'$$

is equivalent to the equation

$$A^*x = y \quad \text{in} \quad E.$$

To see this, it suffices to set $x' = \vartheta(x)$ and $y' = \vartheta(y)$. One can in fact avoid employing dual spaces and dual linear operators by considering the adjoints. This has the advantage that one can work entirely within the space E. When $\mathbb{K} = \mathbb{R}$, this is often expressed by identifying E' with E via the duality mapping ϑ. In this case one does not distinguish between the duality pairing $\langle \cdot, \cdot \rangle$ and the inner product $(\cdot \mid \cdot)$, or between A' and A^*. For the sake of greater clarity it is preferable not to use this identification.

We want to point out explicitly that *the adjoint $A^* \in \mathcal{L}(E)$ of $A \in \mathcal{L}(E)$ depends on the choice of the inner product $(\cdot \mid \cdot)$ in E.* This is also a reason why we prefer to work with the dual operator in order to obtain invariant statements.

(b) Let $\{e_1, \cdots, e_m\}$ be a basis for E and let $\{e'_1, \cdots, e'_m\}$ denote the corresponding *dual basis,* that is,

$$\langle e'_j, e_k \rangle = \delta_{jk}, \quad j, k = 1, \ldots, m.$$

Moreover, let $[a_{jk}]$ and $[b_{jk}]$ denote the matrix representations of $A \in \mathcal{L}(E)$ and $A' \in \mathcal{L}(E')$, respectively. We then have $Ae_k = \sum_j a_{jk} e_j$ and $A'e'_k = \sum_j b_{jk} e'_j$, and hence $a_{jk} = \langle e'_j, Ae_k \rangle$ and

$$b_{jk} = \langle A'e'_k, e_j \rangle = \langle e'_k, Ae_j \rangle = a_{kj}, \quad \forall j, k = 1, \ldots, m.$$

Consequently, we have

$$[b_{jk}] = [a_{jk}]^T,$$

i.e., *the matrix representing the dual operator A' with respect to the dual basis is the transpose of the matrix representing A.*

Let now $(\cdot \mid \cdot)$ be an inner product on E and assume that $\{e_1, \ldots, e_m\}$ is an orthonormal basis, that is,

$$(e_j \mid e_k) = \delta_{jk}, \quad j, k = 1, \ldots, m.$$

Moreover, let $[a^*_{jk}]$ denote the matrix representation of the adjoint $A^* \in \mathcal{L}(E)$ of A with respect to this basis. It then follows from $A^* e_k = \sum_j a^*_{jk} e_j$ and $A^{**} = A$, as above, that

$$a^*_{jk} = (A^* e_k \mid e_j) = (e_k \mid A e_j) = \overline{(A e_j \mid e_k)} = \overline{a_{kj}}.$$

Hence

$$[a^*_{jk}] = [a_{jk}]^* := \overline{[a_{jk}]^T},$$

that is, *the matrix representing the adjoint A^* with respect to an orthonormal basis is the conjugate transpose of A.*

Let $E = \mathbb{K}^m$ and assume that $(\cdot \mid \cdot)$ is the Euclidean inner product on $\mathbb{K}^m$,

$$(x \mid y) = \sum_{j=1}^m x^j \overline{y^j}.$$

Then the standard basis $e_1 = (1, 0, \cdots, 0), \ldots, e_m = (0, \ldots, 0, 1)$ is an orthonormal basis. If we identify E' with $\mathbb{K}^m$, that is, if we set

$$\langle x, y \rangle := \sum_{j=1}^m x^j y^j, \quad \forall\, x, y \in \mathbb{K}^m,$$

then the standard basis of $\mathbb{K}^m = E'$ is also the dual basis of the standard basis of $E = \mathbb{K}^m$. We therefore obtain the following matrix representations with respect to the standard basis of $\mathbb{K}^m$ and the resulting identifications: The matrix of the dual operator A' of $A \in \mathcal{L}(\mathbb{K}^m)$ is the transpose of the matrix representing A, and the matrix of the adjoint A^* is the conjugate transpose of the matrix representing A.

(c) Let $(\cdot \mid \cdot)$ be a fixed inner product on E. Then one defines the *adjoint linear differential equation* corresponding to the equation $\dot{x} = A(t)x$ by

$$\dot{y} = -A^*(t)y.$$

In contrast to the dual differential equation, this is an equation in E (which depends on the choice of the inner product, however). □

Problems

1. Let $J \subseteq \mathbb{R}$ be an open interval and $A \in C(J, \mathcal{L}(E))$. Moreover, let $(\cdot \mid \cdot)$ be an inner product on E. Prove that:

(i) If $U(t, \tau)$ is the evolution operator of the equation $\dot{x} = A(t)x$, then

$$W(t, \tau) := U^*(\tau, t) = [U^{-1}(t, \tau)]^*$$

is the evolution operator of the adjoint equation

$$\dot{y} = -A^*(t)y.$$

(ii)   If u is a solution of $\dot{x} = A(t)x$ and if v is a solution of the corresponding adjoint equation, the inner product $(u \mid v)$ is constant.

(iii)   If $A(t) = -A^*(t)$ holds $\forall\, t \in J$, that is, if $A(t)$ is *skew-hermitian*, then the function

$$V(x) := |x|^2, \quad x \in E,$$

is a first integral of $\dot{x} = A(t)x$. Therefore all solution curves of $\dot{x} = A(t)x$ lie on cylinders in $\mathbb{R} \times E$ with $\mathbb{R} \times \{0\}$ (i.e., the "time-axis") as center-axis.

2. Let $A \in C(\mathbb{R}, E)$. Prove that: If $A(t)$ and $\int_0^t A(\tau)\,d\tau$ commute for each $t \in \mathbb{R}$, then the solution operator of $\dot{x} = A(t)x$ satisfies

$$U(t,0) = \exp\left\{ \int_0^t A(\tau)\,d\tau \right\}.$$

3. Let the system $\dot{x} = A(t)x + b(t)$ on $(0, \infty)$ be given, where

$$A(t) = \begin{bmatrix} 0 & 1 \\ -\frac{2}{t^2} & \frac{2}{t} \end{bmatrix} \quad \text{and} \quad b(t) = \begin{bmatrix} t^4 \\ t^3 \end{bmatrix}.$$

Determine the solution for the initial value

$$x(2) = \begin{bmatrix} 1 \\ 4 \end{bmatrix}.$$

4.   Let $\mathsf{J} \in \mathcal{L}(\mathbb{R}^{2m})$ denote the symplectic normal form (cf. Section 3) and let $A \in C(J, \mathcal{L}(\mathbb{R}^{2m}))$ be such that $A(t) = [A(t)]^*$ for all $t \in J$. Prove that: If X is a special fundamental matrix of the equation $\dot{x} = \mathsf{J}A(t)x$, then

$$[X(t)]^* \mathsf{J} X(t) = \mathsf{J}, \quad \forall\, t \in J,$$

that is, $X(t)$ is *symplectic* for each $t \in J$.

12. Autonomous Linear Differential Equations

In this section E again denotes a finite dimensional Banach space over $\mathbb{K}$ and

$$A \in \mathcal{L}(E).$$

We now consider the autonomous linear homogeneous differential equation

$$\dot{x} = A\,x \tag{1}$$

in E.

(12.1) Remarks.   (a) If we identify E with $\mathbb{K}^m$ and $\mathcal{L}(E)$ with $\mathbb{M}^m(\mathbb{K})$, then (1) represents a "shorthand form" for a homogeneous system of m first order ordinary differential equations *with constant coefficients*.

(b)   Since the vector field

$$f : E \to E, \quad x \mapsto Ax,$$

is infinitely often differentiable, it induces a global C^∞-flow φ on E, by theorem (10.3) and theorem (7.9). $\square$

The Exponential Function

It follows from theorem (11.13) and remark (10.1 b) that the flow φ is given by

$$\varphi(t, x) = U(t, 0)x, \quad (t, x) \in \mathbb{R} \times E, \tag{2}$$

where $U(t) := U(t, 0)$ is the global solution of the linear IVP

$$\dot{X} = AX, \quad X(0) = id_E, \tag{3}$$

in $\mathcal{L}(E)$. With $I := id_E$, (3) is equivalent to the integral equation

$$X(t) = I + \int_0^t AX(\tau)\,d\tau = I + A \int_0^t X(\tau)\,d\tau, \quad t \in \mathbb{R},$$

which suggests the following iteration (cf. problem (7.3)):

$$X_0 := I$$

$$X_1(t) = I + A \int_0^t X_0(\tau)\,d\tau = I + tA$$

$$X_2(t) = I + A \int_0^t X_1(\tau)\,d\tau = I + tA + \frac{t^2}{2}A^2 \tag{4}$$

$$\vdots$$

$$X_{n+1}(t) = I + A \int_0^t X_n(\tau)\,d\tau = \sum_{k=0}^{n+1} \frac{t^k}{k!}A^k.$$

The series

$$\sum_{k=0}^{\infty} \frac{t^k}{k!} A^k \tag{5}$$

in $\mathcal{L}(E)$ is bounded in norm by the series

$$\sum_{k=0}^{\infty} \frac{|t|^k}{k!} \|A\|^k = e^{|t|\,\|A\|}, \quad t \in \mathbb{R},$$

in $\mathbb{R}$. By the Weierstrass M-test it now follows that (5) *converges in* $\mathcal{L}(E)$ *absolutely and uniformly on every compact interval of* $\mathbb{R}$. We now *define* $e^{tA} \in \mathcal{L}(E)$ by

$$e^{tA} := \sum_{k=0}^{\infty} \frac{t^k}{k!} A^k, \quad \forall\, t \in \mathbb{R}.$$

It follows from the uniform convergence on compact intervals that the function

$$U : \mathbb{R} \to \mathcal{L}(E), \quad t \mapsto e^{tA},$$

is continuous. It also follows from the local uniform convergence that we may take the limit in (4) (for fixed $t \in \mathbb{R}$) and obtain

$$U(t) = I + \int_0^t AU(s)ds, \quad \forall\, t \in \mathbb{R}.$$

Therefore U is the unique global solution of the IVP

$$\dot{X} = AX, \quad X(0) = I. \tag{6}$$

in $\mathcal{L}(E)$. From this the following theorem follows.

(12.2) Theorem. *For every $A \in \mathcal{L}(E)$, the function*

$$U : \mathbb{R} \to \mathcal{L}(E), \quad t \mapsto e^{tA}$$

is a C^∞–group homomorphism from the additive group $(\mathbb{R}, +)$ into the multiplicative group $\mathcal{GL}(E)$, i.e., we have

$$U(t + s) = U(t)U(s), \quad \forall\, s, t \in \mathbb{R}.$$

Proof. Since A is constant, it follows from remark (10.1 b) that the global solution of the autonomous IVP

$$\dot{x} = Ax, \quad x(\tau) = \xi,$$

$u(\cdot, \tau, \xi)$, satisfies the relation

$$u(t, \tau, \xi) = u(t - \tau, 0, \xi), \quad \forall\, t, \tau \in \mathbb{R}, \xi \in E.$$

Hence from theorem (11.13) we get that

$$U(t, \tau)\xi = U(t - \tau, 0)\xi, \quad \forall\, t, \tau \in \mathbb{R}, \xi \in E,$$

and consequently

$$U(t, \tau) = U(t - \tau, 0) = U(t - \tau), \quad \forall\, t, \tau \in \mathbb{R},$$

where the last equality follows from the unique solvability of the IVP (6). Now theorem (11.13 ii) implies that

$$U(t)U(s) = U(t,0)U(s,0) = U(t+s,s)U(s,0)$$
$$= U(t+s,0) = U(t+s)$$

for all $s,t \in \mathbb{R}$.

It follows from

$$U(t)U(-t) = U(-t)U(t) = U(0) = I$$

that $U(t) \in \mathcal{GL}(E)$ and $U(t)^{-1} = U(-t)$ for all $t \in \mathbb{R}$. Therefore U is a group homomorphism from $(\mathbb{R},+)$ into $\mathcal{GL}(E)$. Since $U \in C(\mathbb{R}, \mathcal{L}(E))$ and $\dot{U} = AU$, we obtain by induction that $U \in C^\infty(\mathbb{R}, \mathcal{L}(E))$.   □

(12.3) Remarks. (a) For $A \in \mathcal{L}(E)$ and $t,s \in \mathbb{R}$ we have:

(i) $e^{0A} = I, \quad e^{(t+s)A} = e^{tA}e^{sA}$.

(ii) $e^{tA} \in \mathcal{GL}(E)$ and $(e^{tA})^{-1} = e^{-tA}$.

(iii) $(e^{tA})^{\cdot} = Ae^{tA} = e^{tA}A$.

(iv) $\|e^{tA}\| \le e^{|t|\,\|A\|}$.

(b) *For the global flow φ on E, induced by the constant vector field $A \in \mathcal{L}(E)$, we have*

$$\varphi(t,x) = e^{tA}x, \quad \forall\,(t,x) \in \mathbb{R} \times E.$$

(c) Let $A \in \mathcal{L}(E)$ and $b \in C(J,E)$, where $J \subseteq \mathbb{R}$ is an open interval. Then the global solution of the nonhomogeneous IVP

$$\dot{x} = Ax + b(t), \quad x(\tau) = \xi, \quad \tau \in J, \xi \in E,$$

is given by the "variation of constants formula"

$$u(t,\tau,\xi) = e^{(t-\tau)A}\xi + \int_{\tau}^{t} e^{(t-s)A}b(s)\,ds, \quad t \in J.$$

(d) If $A \in \mathcal{L}(E)$, then the group homomorphism $t \mapsto e^{tA}$ is often called *"the one-parameter group of A in $\mathcal{L}(E)$."*   □

In the following proposition we collect the *rules for the exponential function in $\mathcal{L}(E)$*.

(12.4) Proposition. *Let $A, B \in \mathcal{L}(E)$.*

(i) *If $AB = BA$, then*

$$Ae^{B} = e^{B}A \quad and \quad e^{A+B} = e^{A}e^{B}.$$

(ii) *For $B \in \mathcal{GL}(E)$ we have*

$$e^{BAB^{-1}} = Be^{A}B^{-1}.$$

Proof. (i) For the functions $X, Y \in C^\infty(\mathbb{R}, \mathcal{L}(E))$, defined by $X(t) := Ae^{tB}$ and $Y(t) := e^{tB}A$, we have $X(0) = Y(0) = A$ and

$$\dot{X}(t) = ABe^{tB} = BAe^{tB} = BX(t),$$

as well as

$$\dot{Y}(t) = Be^{tB}A = BY(t)$$

for all $t \in \mathbb{R}$. Hence it follows from the unique solvability of the IVP $\dot{X} = BX$, $X(0) = A$, in $\mathcal{L}(E)$ that $X = Y$ and in particular that $X(1) = Y(1)$.

Similarly, set

$$U(t) := e^{t(A+B)} \quad \text{and} \quad V(t) := e^{tA}e^{tB}, \quad \forall t \in \mathbb{R}.$$

Then we have $U(0) = V(0) = I$ and

$$\dot{U}(t) = (A + B)U(t), \quad \forall t \in \mathbb{R},$$

as well as

$$\dot{V}(t) = A\,e^{tA}e^{tB} + e^{tA}B\,e^{tB}.$$

Using what we have already proved, we get

$$\dot{V}(t) = (A + B)V(t), \quad \forall t \in \mathbb{R},$$

and hence, again using the uniqueness theorem, the assertion follows.

(ii) With $X(t) := e^{tBAB^{-1}}$ and $Y(t) := Be^{tA}B^{-1}$ we have $X(0) = Y(0) = I$ and

$$\dot{X}(t) = BAB^{-1}X(t), \quad \forall t \in \mathbb{R},$$

as well as

$$\dot{Y}(t) = BAe^{tA}B^{-1} = BAB^{-1}Y(t), \quad \forall t \in \mathbb{R}.$$

From this it follows again that $X = Y$. $\qquad\square$

Solution Formulae

By setting $y = Bx$, where $B \in \mathcal{GL}(E)$, we transform the differential equation $\dot{x} = Ax$ into the equation

$$\dot{y} = BAB^{-1}y,$$

which has the fundamental matrix

$$e^{tBAB^{-1}} = Be^{tA}B^{-1}, \quad \forall t \in \mathbb{R}.$$

For the fundamental matrix of $\dot{x} = Ax$ we therefore have the relation

$$e^{tA} = B^{-1}e^{tBAB^{-1}}B, \quad \forall t \in \mathbb{R}.$$

In order to calculate the fundamental matrix of $\dot{x} = Ax$, we will bring the operator A into a fairly simple form by means of an appropriate substitution (which is equivalent to a change of basis).

Assume that E is the direct sum of the subspaces $E_1, \ldots, E_k$, that is,

$$E = E_1 \oplus \ldots \oplus E_k. \tag{7}$$

Then every $x \in E$ has a unique representation

$$x = x_1 + \ldots + x_k, \quad x_j \in E_j, \quad j = 1, \ldots, k,$$

and the canonical projections (corresponding to the direct sum) are defined by

$$P_j : E \to E_j, \quad x \mapsto x_j, \quad j = 1, \ldots, k.$$

Here we clearly have

$$P_j P_i = \delta_{ji} P_j, \quad i, j = 1, \ldots, k, \tag{8}$$

and

$$\sum_{j=1}^{k} P_j = I. \tag{9}$$

Conversely, if $P_1, \ldots, P_k \in \mathcal{L}(E)$ are operators satisfying (8) and (9), and if we set $E_j := P_j(E)$, we have

$$E = E_1 \oplus \ldots \oplus E_k,$$

and the operators P_j are the corresponding canonical projections.

Assume now that $A \in \mathcal{L}(E)$ is such that $A(E_j) \subseteq E_j$ for all $j = 1, \ldots, k$. Then we say that the direct sum (7) *decomposes* (or *reduces*) the endomorphism A, and clearly A is completely determined by the *parts* $A_j \in \mathcal{L}(E_j)$, where $A_j x := Ax$ for all $x \in E_j$. We write

$$A = A_1 \oplus \ldots \oplus A_k \tag{10}$$

and call A the *direct sum* of the operators $A_1, \ldots, A_k$.

(12.5) Lemma. *Assume* $A \in \mathcal{L}(E)$ *and let* $E = E_1 \oplus \ldots \oplus E_k$, *with the corresponding projections* $P_j, j = 1, \ldots, k$. *Then the direct sum* $E = E_1 \oplus \ldots \oplus E_k$ *decomposes* A *if and only if*

$$AP_j = P_j A, \quad j = 1, \ldots, k.$$

Proof. Assume that A is decomposed and let $x \in E$ be arbitrary. Then $P_j x \in E_j$ and consequently we have $AP_j x \in E_j$, hence $P_i AP_j x = \delta_{ij} AP_j x$ for all $j = 1, \ldots, k$. Summing up and using (9), we obtain

$$AP_i x = \sum_{j=1}^{k} \delta_{ij} AP_j x = \sum_{j=1}^{k} P_i AP_j x = P_i A \left(\sum_{j=1}^{k} P_j x \right) = P_i Ax.$$

Hence we have $AP_i = P_i A$ for all $i = 1, \ldots, k$.

Conversely, if A commutes with every P_j, it follows that

$$Ax = AP_j x = P_j Ax \in E_j,$$

where $x \in E_j$ and therefore $x = P_j x$. Hence $A(E_j) \subseteq E_j$ for all $j = 1, \dots, k$. $\qquad \square$

(12.6) Corollary. *Assume that the operator $A \in \mathcal{L}(E)$ is decomposed by the direct sum $E = E_1 \oplus \dots \oplus E_k$. Then this direct sum also decomposes e^{tA} for every $t \in \mathbb{R}$, that is, <u>the flow induced by</u> A <u>is decomposed by</u> $E = E_1 \oplus \dots \oplus E_k$.*

Proof. Lemma (12.5) implies that for $P_j, j = 1, \dots, k$, we have $AP_j = P_j A$. Hence from proposition (12.4) it follows that

$$P_j e^{tA} = e^{tA} P_j, \quad j = 1, \dots, k,$$

and now the assertion follows from lemma (12.5). $\qquad \square$

Now, let $\mathbb{K} = \mathbb{C}$ and $A \in \mathcal{L}(E)$, and assume that $\lambda_1, \dots, \lambda_k$ are the distinct *eigenvalues* of A, i.e., $\lambda_1, \dots, \lambda_k$ are the distinct (complex) roots of the *characteristic polynomial*

$$\det (A - \lambda) = 0, \quad \text{where } A - \lambda := A - \lambda I.$$

The set of all eigenvalues of A is called the *spectrum* of $A, \sigma(A)$, that is,

$$\sigma(A) = \{\lambda_j \mid j = 1, \dots, k\}.$$

Clearly we have:

$$\lambda \in \sigma(A) \Leftrightarrow \ker (A - \lambda) \neq \{0\} \Leftrightarrow \operatorname{im} (A - \lambda) \neq E,$$

because E is finite dimensional. Let $m_1, \dots, m_k$ denote the multiplicities of the roots $\lambda_1, \dots, \lambda_k$ of the characteristic polynomial. Then we have

$$\det (A - \lambda) = (-1)^m (\lambda - \lambda_1)^{m_1} \dots (\lambda - \lambda_k)^{m_k},$$

where $m = m_1 + \dots + m_k$, and m_j is called the *algebraic multiplicity* of the eigenvalue λ_j of A. In linear algebra it is shown that the *generalized eigenspace* of A corresponding to λ_j,

$$E_j := ker[(A - \lambda_j)^{m_j}],$$

has dimension m_j and that E has the direct sum decomposition

$$E = E_1 \oplus \dots \oplus E_k. \tag{11}$$

In this connection we call the canonical projections $P_j : E \to E_j$ the *eigenprojections* corresponding to the eigenvalues λ_j (see, for example, the first chapter of Kato [1] for an elegant and short treatment of the material required here, or see R. Walter [2]).

It is immediately evident that $A(E_j) \subseteq E_j$ for all $j = 1, \ldots, k$. Hence the direct sum (11) induces a decomposition of the operator A, i.e., $A = A_1 \oplus \cdots \oplus A_k$, and by corollary (12.6) we have

$$e^{tA} = e^{tA_1} \oplus \cdots \oplus e^{tA_k}.$$

Therefore it suffices to determine e^{tA_j} for every $j = 1, \ldots, k$.

Assume now that some $j \in \{1, \ldots, k\}$ is fixed and set

$$X := E_j, \quad \lambda := \lambda_j, \quad N := (A - \lambda)|X,$$

that is,

$$A_j := \lambda + N.$$

From the definition of X and N it clearly follows that

$$N^{m_j} = 0,$$

that is, N is *nilpotent*, the *eigennilpotent of A corresponding to the eigenvalue* λ_j.

Since N is nilpotent, we know from linear algebra that X can be written as a direct sum

$$X = X_1 \oplus \ldots \oplus X_s$$

with the following properties:

(i) $N(X_i) \subseteq X_i$, $i = 1, \ldots, s$, i.e., the decomposition reduces N,

(ii) Every X_i has a basis $\{u_{i,1}, \ldots, u_{i,q_i}\}$ such that $N u_{i,1} = 0$ and

$$N u_{i,r} = u_{i,r-1}, \quad 2 \leq r \leq q_i.$$

With respect to this basis the operator $A|X_i$ is represented by the $q_i \times q_i$ *Jordan matrix*

$$\begin{bmatrix} \lambda & 1 & 0 & \ldots & 0 & 0 \\ 0 & \lambda & 1 & \ldots & 0 & 0 \\ 0 & 0 & \lambda & \ldots & 0 & 0 \\ \vdots & \vdots & \vdots & \ddots & \vdots & \vdots \\ 0 & 0 & 0 & \ldots & \lambda & 1 \\ 0 & 0 & 0 & \ldots & 0 & \lambda \end{bmatrix}.$$

From (i) and (12.6) it follows that

$$e^{tA_i} = e^{tA|X_1} \oplus \cdots \oplus e^{tA|X_s},$$

and so it suffices to determine $e^{tA|X_i}$.

It follows from (ii) that

$$N^n u_{i,r} = u_{i,r-n}, \quad 1 \leq r \leq q_i, \quad 0 \leq n \leq r - 1 \tag{12}$$

and

$$N^r u_{i,r} = 0, \quad 1 \le r \le q_i. \tag{13}$$

Using this, the definition of $e^{tA}|X_i$ and proposition (12.4), we obtain for

$$x = \sum_{r=1}^{q_i} \alpha^r u_{i,r} \in X_i, \quad \alpha^r \in \mathbb{C},$$

that

$$e^{tA}x = e^{tA|X_i}x = e^{t\lambda}e^{tN}x$$

$$= e^{\lambda t} \sum_{r=1}^{q_i} \alpha^r \sum_{n=0}^{\infty} \frac{t^n}{n!} N^n u_{i,r}$$

$$= e^{\lambda t} \sum_{r=1}^{q_i} \alpha^r \sum_{n=0}^{r-1} \frac{t^n}{n!} u_{i,r-n}$$

$$= e^{\lambda t} \alpha^1 u_{i,1}$$

$$+ e^{\lambda t} \alpha^2 [u_{i,2} + t u_{i,1}]$$

$$+ e^{\lambda t} \alpha^3 [u_{i,3} + t u_{i,2} + \frac{t^2}{2} u_{i,1}]$$

$$\vdots$$

$$+ e^{\lambda t} \alpha^{q_i} [u_{i,q_i} + t u_{i,q_i-1} + \cdots + \frac{t^{q_i-1}}{(q_i-1)!} u_{i,1}].$$

In summary, we have proved the following theorem:

(12.7) Theorem. *Assume that $\mathbb{K} = \mathbb{C}$ and $A \in \mathcal{L}(E)$, and let $\lambda_1, \ldots, \lambda_k$ denote the distinct eigenvalues of A with algebraic multiplicities $m_1, \ldots, m_k$, respectively. For every $j = 1, \ldots, k$, there exist exactly m_j linearly independent solutions of the homogeneous differential equation*

$$\dot{x} = Ax \quad \text{in} \quad E, \tag{14}$$

of the form

$$x_{j,s}(t) = e^{\lambda_j t} p_{j,s-1}(t), \quad t \in \mathbb{R}, \quad 1 \le s \le m_j,$$

where $p_{j,\nu}(t)$ denotes a polynomial in t with coefficients in E and degree $\le \nu$. The set of all these solutions forms a fundamental set of solutions for (14).

If A is semisimple *(i.e., diagonalizable), then (14) has a fundamental set of solutions of the form*

$$\{e^{\lambda_j t} y_{j,s} \mid 1 \le s \le m_j, \quad 1 \le j \le k\},$$

where the $y_{j,s} \in E$ are linearly independent vectors.

(12.8) Remarks. (a) If $\lambda \in \sigma(A)$, we call

$$\dim [\ker (A - \lambda)]$$

the *geometric multiplicity* of the eigenvalue λ. Evidently, *the geometric multiplicity is at most equal to the algebraic multiplicity*, and A is called *semisimple* if these two multiplicities agree for all $\lambda \in \sigma(A)$. The considerations above show that *A is semisimple if and only if all eigennilpotents vanish, that is, if and only if the matrix representing A with respect to some appropriate basis is diagonal* (i.e., A is diagonalizable).

It is well-known that every hermitian operator $(A = A^*)$ (with respect to some inner product on E) or, more generally, every *normal* operator $A \in \mathcal{L}(E)$ is semisimple. Here we call $A \in \mathcal{L}(E)$ normal if $AA^* = A^*A$.

(b) If A is semisimple, then the matrix representing A with respect to some appropriate basis has the form

$$A = \mathrm{diag}\,[\lambda_1, \ldots, \lambda_1, \lambda_2, \ldots, \lambda_2, \ldots, \lambda_k, \ldots, \lambda_k].$$

Here λ_j occurs exactly m_j times. With respect to this basis the equation

$$\dot{x} = A\,x$$

simply becomes

$$\dot{x}^1 = \mu_1 x^1,$$
$$\vdots$$
$$\dot{x}^m = \mu_m x^m,$$

where $\mu_l = \lambda_j$ for $m_1 + \cdots + m_{j-1} < l \leq m_1 + \cdots + m_j$. This fact is reflected in the second part of theorem (12.7).   □

(12.9) Example. Let $E = \mathbb{C}^2$ and consider the system

$$\dot{x} = x - y$$
$$\dot{y} = 4x - 3y. \tag{15}$$

Then

$$A = \begin{bmatrix} 1 & -1 \\ 4 & -3 \end{bmatrix}$$

and

$$\det (A - \lambda) = (\lambda + 1)^2.$$

Hence A has a single eigenvalue, namely $\lambda = -1$, with algebraic multiplicity 2. Since

$$A - \lambda = A + 1 = \begin{bmatrix} 2 & -1 \\ 4 & -2 \end{bmatrix},$$

one immediately sees that

$$\begin{bmatrix} 1 \\ 2 \end{bmatrix} \in \ker(A + 1).$$

Hence

$$t \mapsto \begin{bmatrix} 1 \\ 2 \end{bmatrix} e^{-t}, \quad t \in \mathbb{R},$$

represents a solution of (15). By theorem (12.7), a second linearly independent solution has the form

$$\begin{bmatrix} x(t) \\ y(t) \end{bmatrix} = \begin{bmatrix} a + b\,t \\ c + d\,t \end{bmatrix} e^{-t}.$$

For this to be a solution, the constants must be determined such that

$$\begin{bmatrix} \dot{x}(t) \\ \dot{y}(t) \end{bmatrix} = \begin{bmatrix} b - a - b\,t \\ d - c - d\,t \end{bmatrix} e^{-t} = A \begin{bmatrix} a + b\,t \\ c + d\,t \end{bmatrix} e^{-t}, \quad \forall\, t \in \mathbb{R}.$$

Hence we must have

$$A \begin{bmatrix} a \\ c \end{bmatrix} = \begin{bmatrix} b - a \\ d - c \end{bmatrix} \quad \text{and} \quad A \begin{bmatrix} b \\ d \end{bmatrix} = - \begin{bmatrix} b \\ d \end{bmatrix}.$$

The second equation is exactly the eigenvalue equation and thus it has the solution $b = 1, d = 2$. Consequently the first equation now becomes

$$a - c = 1 - a$$
$$4a - 3c = 2 - c,$$

which clearly has the solution $a = 0, c = -1$. Hence a second linearly independent solution is given by

$$t \mapsto \begin{bmatrix} t \\ -1 + 2t \end{bmatrix} e^{-t}. \qquad \qquad \square$$

Assume now that $\mathbb{K} = \mathbb{R}$. Then the *complexification* of $E = (E, |\cdot|)$, $E_{\mathbb{C}}$, is a normed complex vector space defined as follows: On $E \times E$ define multiplication by complex scalars $\gamma := \alpha + i\beta \in \mathbb{C} = \mathbb{R} + i\mathbb{R}$ by

$$\gamma z := (\alpha x - \beta\, y, \beta x + \alpha\, y), \quad \forall\, z = (x, y) \in E \times E,$$

and a norm by

$$|z|_{E_{\mathbb{C}}} := \max_{0 \le \varphi \le 2\pi} |x \cos \varphi + y \sin \varphi|.$$

If E has dimension m (over $\mathbb{R}$), one easily verifies that with this multiplication $(E \times E, |\cdot|_{E_{\mathbb{C}}}) =: E_{\mathbb{C}}$ is a normed complex vector space of dimension m (over $\mathbb{C}$!). Moreover, we have

$$1(x, 0) = (x, 0)$$

and

$$i(x, 0) = (0, x),$$

as well as

$$|(x, 0)|_{E_{\mathbb{C}}} = |x|, \quad \forall\, x \in E.$$

Consequently, we may identify E with $E \times \{0\}$ in $E_{\mathbb{C}}$ and thus every element $z = (x, y)$ can be written uniquely in the form

$$z = x + iy.$$

For this reason we simply write

$$E_{\mathbb{C}} = E + iE.$$

Now, let $A \in \mathcal{L}(E)$. Then the *complexification* of A, $A_{\mathbb{C}}$, is defined by

$$A_{\mathbb{C}}(x + iy) := A\,x + iA\,y, \quad \forall\, x + iy \in E_{\mathbb{C}}.$$

Clearly $A_{\mathbb{C}}$ is an endomorphism of $E_{\mathbb{C}}$, i.e., $A_{\mathbb{C}} \in \mathcal{L}(E_{\mathbb{C}})$, and we have

$$|A_{\mathbb{C}}|_{\mathcal{L}(E_{\mathbb{C}})} = |A|_{\mathcal{L}(E)}. \tag{16}$$

Indeed, from

$$|A_{\mathbb{C}}(x + i\,y)|_{E_{\mathbb{C}}} = \max_{0 \le \varphi \le 2\pi} |A\,x\,\cos\,\varphi + A\,y\,\sin\,\varphi| \le |A|_{\mathcal{L}(E)}|x + i\,y|_{E_{\mathbb{C}}}$$

it follows that $|A_{\mathbb{C}}|_{\mathcal{L}(E_{\mathbb{C}})} \le |A|_{\mathcal{L}(E)}$. Conversely, if $x \in E \subseteq E_{\mathbb{C}}$, then

$$|A\,x| = |A_{\mathbb{C}}x|_{E_{\mathbb{C}}} \le |A_{\mathbb{C}}|_{\mathcal{L}(E_{\mathbb{C}})}|x|$$

and thus $|A|_{\mathcal{L}(E)} \le |A_{\mathbb{C}}|_{\mathcal{L}(E_{\mathbb{C}})}$. This shows that the norms are equal, as claimed. In what follows we will – if no confusion can be expected – denote the norms in $E_{\mathbb{C}}$ and $\mathcal{L}(E_{\mathbb{C}})$ by $|\cdot|$.

Now clearly

$$A_{\mathbb{C}}^n = (A^n)_{\mathbb{C}}, \quad \forall\, n \in \mathbb{N},$$

and consequently

$$(e^{tA})_{\mathbb{C}} = e^{tA_{\mathbb{C}}}, \quad \forall\, t \in \mathbb{R}.$$

By separating into real and imaginary parts, we thus obtain a fundamental set of solutions of the differential equation

$$\dot{x} = A\,x \quad \text{in} \quad E$$

from a fundamental set of solutions of the (complex) differential equation

$$\dot{z} = A_{\mathbb{C}}z \quad \text{in} \quad E_{\mathbb{C}}. \tag{17}$$

It follows from theorem (12.7) that every component of a solution of (17) is a linear combination of functions of the form

$$t^n e^{\lambda t}$$

with coefficients in $E_{\mathbb{C}}$, where $\lambda \in \sigma(A_{\mathbb{C}})$, $n < m(\lambda)$, and $m(\lambda)$ denotes the algebraic multiplicity of λ. From the equation

$$A_{\mathbb{C}}z = \lambda z$$

it follows that

$$A_{\mathbb{C}}\bar{z} = \overline{A_{\mathbb{C}}z} = \overline{\lambda z} = \bar{\lambda}\bar{z}.$$

Hence, if λ is an eigenvalue of $A_{\mathbb{C}}$, then so is $\bar{\lambda}$.

Let

$$\lambda = \alpha + i\omega \in \sigma(A_{\mathbb{C}}),$$

then

$$e^{\lambda t} = e^{\alpha t}[\cos\,(\omega t) + i\,\sin\,(\omega t)].$$

If we separate every term of the form

$$ct^n e^{\lambda t} = (a + ib)t^n e^{\alpha t}(\cos (\omega t) + i \, \sin (\omega t)),$$

where $c = a + ib \in E_{\mathbb{C}}$, into real and imaginary parts, then the following theorem follows immediately from theorem (12.7).

(12.10) Theorem. *Assume that $\mathbb{K} = \mathbb{R}$ and $A \in \mathcal{L}(E)$, and let u be a solution of the homogeneous differential equation*

$$\dot{x} = A x \quad in \; E.$$

Then u is a linear combination of functions of the form

$$t^k e^{\alpha t} \cos(\omega t)a, \quad t^l e^{\alpha t} \sin(\omega t)b, \quad a, b \in E,$$

where $\lambda := \alpha + i\omega$ runs through all the eigenvalues of $A_{\mathbb{C}}$ with $\omega \geq 0$ and

$$k, l \leq m(\lambda) - 1.$$

Here $m(\lambda)$ denotes the algebraic multiplicity of the eigenvalue λ of $A_{\mathbb{C}}$.

In order to simplify the exposition, *we will henceforth use the convention that by an eigenvalue of $A \in \mathcal{L}(E)$ we mean an eigenvalue of $A_{\mathbb{C}}$*, i.e.,

$$\sigma(A) := \sigma(A_{\mathbb{C}}).$$

This means that *we allow complex eigenvalues also in the real case*. This convention is meaningful because

$$\det(A_{\mathbb{C}} - \lambda) = \det(A - \lambda)$$

for all $\lambda \in \mathbb{R}$. Indeed, if $\{a_1 \ldots, a_m\}$ is a basis of E, then clearly $\{a_1 + i \, 0, \ldots, a_m + i \, 0\}$ is a basis of $E_{\mathbb{C}}$, and the matrix representations of A and $A_{\mathbb{C}}$ with respect to these two bases are the same.

Stability

As a simple application of theorem (12.7) we obtain the following important result.

(12.11) Stability criterion. *For $A \in \mathcal{L}(E)$ we have*

$$\lim_{t \to \infty} e^{tA} = 0 \; in \; \mathcal{L}(E) \Leftrightarrow Re \, \lambda < 0, \quad \forall \lambda \in \sigma(A),$$

that is, all eigenvalues of A lie in the negative open half plane of the complex plane if and only if every solution u of $\dot{x} = Ax$ in E satisfies:

$$\lim_{t \to \infty} u(t) = 0.$$

Proof. If $\mathbb{K} = \mathbb{R}$, then

$$(e^{tA})_{\mathbb{C}} = e^{tA_{\mathbb{C}}}, \quad \forall\, t \in \mathbb{R},$$

and, because $\max\{|x|, |y|\} \le |x + iy| \le 2\max\{|x|, |y|\}$ for $x + iy \in E + iE$, we have

$$(e^{tA})_{\mathbb{C}} \to 0 \ \text{ in } \ \mathcal{L}(E_{\mathbb{C}}) \Leftrightarrow e^{tA} \to 0 \ \text{ in } \ \mathcal{L}(E).$$

Therefore it suffices to consider the case $\mathbb{K} = \mathbb{C}$.

Assume that $Re\,\lambda < 0$ for all $\lambda \in \sigma(A)$, and let u be an arbitrary solution of $\dot{x} = Ax$. By theorem (12.7), u is a linear combination of functions of the form

$$t^n e^{\lambda t} y, \quad \lambda \in \sigma(A), \quad y \in E.$$

From

$$|t^n e^{\lambda t} y| = t^n e^{t Re\,\lambda} |y|, \quad t \ge 0,$$

it follows that $u(t) \to 0$ as $t \to \infty$.

Conversely, if $Re\,\lambda \ge 0$ for some $\lambda \in \sigma(A)$, then

$$u(t) := e^{\lambda t} y, \quad y \in \ker(A - \lambda), \quad y \ne 0,$$

is a solution of $\dot{x} = Ax$ and

$$\lim_{t \to \infty} |u(t)| = |y| \lim_{t \to \infty} e^{t Re\,\lambda} \ne 0.$$

It is clear that e^{tA} converges to zero in $\mathcal{L}(E)$ as $t \to \infty$ if and only if every solution u of $\dot{x} = Ax$ converges to zero in E as $t \to \infty$. Hence everything is proved. $\qquad\square$

(12.12) Corollary 1. *Let $A \in \mathcal{L}(E)$. Then $Re\,\lambda < 0$ for all $\lambda \in \sigma(A)$ if and only if every solution $u \ne 0$ of $\dot{x} = Ax$ satisfies*

$$\lim_{t \to -\infty} |u(t)| = \infty.$$

Proof. Assume that $u(t) = e^{tA} x$, where $x \ne 0$. For $t \le 0$ we have $u(t) = e^{-|t|A} x$ and hence

$$|x| = |e^{|t|A} u(t)| \le |e^{|t|A}| |u(t)|.$$

By (12.11) we have that

$$\lim_{t \to \infty} |e^{|t|A}| = 0$$

holds if and only if $Re\,\lambda < 0$ for all $\lambda \in \sigma(A)$. Therefore the assertion follows. $\square$

(12.13) Corollary 2. *Every solution $u \ne 0$ of $\dot{x} = Ax$ in E satisfies*

$$\lim_{t \to \infty} |u(t)| = \infty$$

if and only if $Re\,\lambda > 0$ for all $\lambda \in \sigma(A)$.

Proof. For $u(t) = e^{tA}x$ and $t > 0$ we have

$$u(t) = e^{-t(-A)}x.$$

Because

$$\lambda \in \sigma(A) \Leftrightarrow -\lambda \in \sigma(-A),$$

the assertion now follows from (12.12).   □

Analogously we obtain the following criterion.

(12.14) Boundedness criterion. *Let $A \in \mathcal{L}(E)$. Then every solution u of $\dot{x} = A x$ remains bounded as $t \to \infty$ if and only if the following holds:*

(i) $Re\,\lambda \leq 0, \quad \forall \lambda \in \sigma(A).$
(ii)   *Every $\lambda \in \sigma(A)$ with $Re\,\lambda = 0$ is a semisimple eigenvalue, i.e., the geometric and algebraic multiplicities agree.*

Proof. Again, it suffices to consider the case $\mathbb{K} = \mathbb{C}$. If conditions (i) and (ii) are satisfied, then every solution u is a linear combination of functions of the form

$$t^n e^{\lambda t}y, \quad \lambda \in \sigma(A), y \in E, \tag{18}$$

where $n = 0$ when $Re\,\lambda = 0$. From this the boundedness of u follows immediately.

Conversely, the proof of theorem (12.7) shows that for every $\lambda \in \sigma(A)$ there exists a solution of the form

$$e^{\lambda t}(y_1 + t\,y_2 + \ldots + t^{n-1}y_n),$$

where the vectors $y_1, \ldots, y_n \in \mathbb{C}^m$ are linearly independent. Here we may choose $n > 1$ if and only if λ is not semisimple. This shows that conditions (i) and (ii) are also necessary.   □

(12.15) Remark. The last proof shows that the solutions of $\dot{x} = A x$, corresponding to $Re\,\lambda \leq 0$, have at most *polynomial growth*. In fact, there are unbounded solutions if and only if A has an eigenvalue λ which is not semisimple and $Re\,\lambda = 0$.   □

Problems

1. Let

$$J = \begin{bmatrix} 0 & I_m \\ -I_m & 0 \end{bmatrix} \in \mathcal{L}(\mathbb{R}^m \times \mathbb{R}^m)$$

denote the symplectic normal form. Calculate e^J.

2. Assume that $A \in \mathcal{L}(\mathbb{C}^m)$ is semisimple and let $\lambda_1, \ldots, \lambda_k \in \mathbb{C}$ be the distinct eigenvalues of A. Calculate the eigenvalues of e^A.

3. Prove that

$$\det e^A = e^{\text{trace}(A)},$$

where $A \in \mathcal{L}(\mathbb{K}^m)$.

4. Calculate the general solution of the system:

$$\dot{x} = 4x + 9y + 3e^t$$
$$\dot{y} = -x - 2y - e^t.$$

5. Calculate the general solution of the system:

$$\dot{x} = y + z$$
$$\dot{y} = z + x$$
$$\dot{z} = x + y.$$

What is the behavior of the solutions as $t \to \pm\infty$?

13. The Classification of Linear Flows

In this section we will investigate in some important special cases the behavior of the *linear flow* $e^{tA}x$, $A \in \mathcal{L}(E)$, near the critical point $x = 0$. These investigations play a fundamental role for the stability theory of nonlinear equations $\dot{x} = f(x)$.

Planar Linear Flows

First we consider the *two-dimensional system*

$$\dot{x} = Ax \quad \text{in} \quad \mathbb{R}^2, \tag{1}$$

or in component form:

$$\dot{x}^1 = ax^1 + bx^2 \qquad a, b, c, d \in \mathbb{R}. \tag{2}$$
$$\dot{x}^2 = cx^1 + dx^2$$

From the preceding section we know that the solutions of (1) are characterized by $\sigma(A)$ and the algebraic multiplicities of the eigenvalues. It is therefore reasonable to consider, instead of (1), the transformed equation $(x = Py)$

$$\dot{y} = By,$$

where $B = P^{-1}AP$ for some appropriate $P \in \mathcal{GL}(\mathbb{R}^2)$. We must distinguish between various cases:

Case 1: A has real nonvanishing eigenvalues of opposite sign. In this case A is semisimple and so there exists some $P \in \mathcal{GL}(\mathbb{R}^2)$ such that

$$B = P^{-1}AP = \begin{bmatrix} \lambda & 0 \\ 0 & \mu \end{bmatrix}, \quad \lambda < 0 < \mu.$$

Thus the flow $e^{tB}y$, induced by B, is given (in the y-coordinates!) by

$$t \mapsto (e^{\lambda t}y^1, e^{\mu t}y^2).$$

In the (y^1, y^2) plane the phase portrait of the flow $e^{tA} = Pe^{tB}P^{-1}$ has the form:

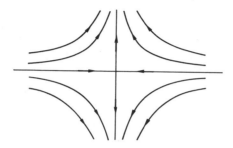

In the original x-coordinates the phase portrait could look, for example, as follows:

In this case the origin is called a *saddle*.

Case 2: All eigenvalues have negative real parts. By the stability criterion (12.11) we then know that

$$\lim_{t \to \infty} u(t) = 0$$

for every solution of (1). In this case the origin is called a *sink* or we say that it is *asymptotically stable*.

We now consider different *subcases*:

(a) *The eigenvalues are real*: $\lambda \leq \mu < 0$. If A is semisimple, then A can be transformed into

$$B = \begin{bmatrix} \lambda & 0 \\ 0 & \mu \end{bmatrix}.$$

Then we obtain the following phase portraits in the transformed coordinates (since $y(t) = (e^{\lambda t} y^1, e^{\mu t} y^2)$):

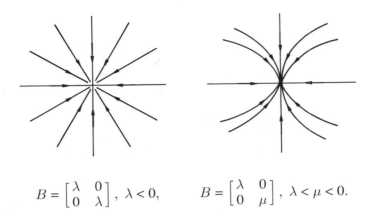

$$B = \begin{bmatrix} \lambda & 0 \\ 0 & \lambda \end{bmatrix}, \ \lambda < 0, \qquad B = \begin{bmatrix} \lambda & 0 \\ 0 & \mu \end{bmatrix}, \ \lambda < \mu < 0.$$

If A is not semisimple (and consequently $\lambda = \mu$), then A can be transformed into the Jordan normal form

$$B = \begin{bmatrix} \lambda & 1 \\ 0 & \lambda \end{bmatrix}, \qquad \lambda < 0$$

(by a "real" transformation $P \in \mathcal{GL}(\mathbb{R}^2)$, since λ is real). Then the transformed equation $\dot{y} = By$ has the solutions

$$\begin{aligned} y^1(t) &= \alpha e^{\lambda t} + \beta t e^{\lambda t}, \\ y^2(t) &= \beta e^{\lambda t}, \end{aligned} \tag{3}$$

where $\alpha, \beta \in \mathbb{R}$. In this case the phase portrait has the form (in the (y^1, y^2) plane):

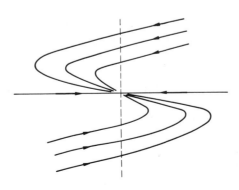

[In order to see this we can, for example (for appropriate constants α and β), express $y^1 =: \xi$ as a function of $y^2 =: \eta$. Then (3) implies that

$$\xi = (\alpha/\beta)\eta + (1/\lambda)\eta \, \ln(\eta/\beta),$$

where $\lambda < 0$.]

In all these cases the origin is called a *(stable) node*. The last case is often referred to as an *improper node*, while the case when $B = \text{diag} \, [\lambda, \lambda]$ is called a *focus*.

(b) *The eigenvalues are complex*, hence complex conjugate, as we already know. Let $A_\mathbb{C} \in \mathcal{L}(\mathbb{C}^2)$ denote the complexification of $A \in \mathcal{L}(\mathbb{R}^2)$. Then conjugating $A_\mathbb{C} z = \lambda z$ implies

$$A_\mathbb{C} \bar{z} = \bar{\lambda} \bar{z},$$

that is,

$$z \in \ker(A_\mathbb{C} - \lambda) \Leftrightarrow \bar{z} \in \ker(A_\mathbb{C} - \bar{\lambda}).$$

Thus, if z is an eigenvector of $A_\mathbb{C}$ corresponding to the eigenvalue λ, then $\mathbb{C}^2$ has the basis

$$\{z, \bar{z}\} = \{x + iy, x - iy\}$$

for some $x, y \in \mathbb{R}^2$. Since $\lambda \notin \mathbb{R}, \{x, y\}$ is a basis for $\mathbb{R}^2$. Moreover, with $\lambda = \alpha + i\omega$ we have:

$$Ax + iAy = A_\mathbb{C}(x + iy) = (\alpha + i\omega)(x + iy) = (\alpha x - \omega y) + i(\alpha y + \omega x),$$

hence

$$Ax = \alpha x - \omega y,$$
$$Ay = \omega x + \alpha y.$$

We see that: If $A \in \mathcal{L}(\mathbb{R}^2)$ *has a nonreal eigenvalue* $\lambda = \alpha + i\omega, \omega \neq 0$, *then also* $\bar{\lambda} = \alpha - i\omega$ *is an eigenvalue and there exists some* $P \in \mathcal{GL}(\mathbb{R}^2)$ *such that*

$$B := P^{-1}AP = \begin{bmatrix} \alpha & -\omega \\ \omega & \alpha \end{bmatrix}, \quad \omega > 0.$$

As usual, we identify $\mathbb{R}^2$ with $\mathbb{C}$ by

$$(\xi, \eta) \leftrightarrow \xi + i\eta, \tag{4}$$

so that e^{tB} can be calculated. It follows from this identification and from

$$\begin{bmatrix} \alpha & -\omega \\ \omega & \alpha \end{bmatrix} \begin{bmatrix} \xi \\ \eta \end{bmatrix} = \begin{bmatrix} \alpha\xi & -\omega\eta \\ \omega\xi & +\alpha\eta \end{bmatrix} \quad \leftrightarrow \quad (\alpha + i\omega)(\xi + i\eta)$$

that B corresponds to multiplication by $\lambda = \alpha + i\omega$. If, as usual, we identify $\mathcal{L}(\mathbb{C})$ with $\mathbb{C}$ (by $M \in \mathcal{L}(\mathbb{C}) \leftrightarrow m := M \cdot 1 \in \mathbb{C}$), it is clear that (4) induces an isomorphism of $\mathbb{R}$-algebras,

$$\mathcal{L}(\mathbb{R}^2) \leftrightarrow \mathcal{L}(\mathbb{C}) = \mathbb{C}.$$

Consequently we have

$$B^n \leftrightarrow \lambda^n, \quad \forall n \in \mathbb{N},$$

and therefore

$$e^{tB} \leftrightarrow e^{\lambda t} = e^{\alpha t}[\cos(\omega t) + i \sin(\omega t)],$$

hence

$$e^{tB} = e^{\alpha t} \begin{bmatrix} \cos \omega t & -\sin \omega t \\ \sin \omega t & \cos \omega t \end{bmatrix}.$$

Geometrically e^{tB} has the effect of a dilation by a factor of $e^{\alpha t}$ and a rotation by the angle ωt in the mathematically positive sense.

Since in our case $\alpha < 0$, we obtain the following phase portrait in the y-coordinates:

$$B = \begin{bmatrix} \alpha & -\omega \\ \omega & \alpha \end{bmatrix}, \quad \alpha < 0 < \omega.$$

In this case the origin is called a *stable spiral*.

Case 3: All eigenvalues have positive real parts. In this case every solution $u \neq 0$ satisfies

$$\lim_{t \to \infty} |u(t)| = \infty \quad \text{and} \quad \lim_{t \to -\infty} u(t) = 0$$

(cf. corollaries (12.12) and (12.13)). The origin is called a *source*.

Since $e^{tA} = e^{-t(-A)}$, the phase portrait is obtained from the phase portrait of case 2 by reversing the arrows. One then speaks of *unstable* foci, nodes and spirals.

Case 4: The eigenvalues are pure imaginary. In this case A can be transformed into the form

$$B = P^{-1}AP = \begin{bmatrix} 0 & -\omega \\ \omega & 0 \end{bmatrix}, \quad \omega > 0.$$

Consequently

$$e^{tB} = \begin{bmatrix} \cos \omega t & -\sin \omega t \\ \sin \omega t & \cos \omega t \end{bmatrix}$$

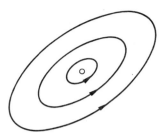

and so *all solutions are periodic with period $2\pi/\omega$.* In the y-coordinates the orbits are circles centered at the origin, and in the x-coordinates they are ellipses.
    In this case the origin is called a *center*.   □

    All the information about the eigenvalues of $A \in \mathcal{L}(\mathbb{R}^2)$ is, of course, contained in the characteristic polynomial

$$\det(A - \lambda) = \lambda^2 - \text{trace}(A)\lambda + \det(A).$$

With the *discriminant D* defined by

$$D := [\text{trace}(A)]^2 - 4 \ \det(A),$$

the eigenvalues are given by

$$\frac{1}{2}(\text{trace}(A) \pm \sqrt{D}).$$

Hence the eigenvalues are real if and only if $D \geq 0$, and they are complex with negative real part if and only if trace $(A) < 0$ and $D < 0$; and so on. The following diagram summarizes the geometric information about the phase portraits of $\dot{x} = Ax$, which can be deduced from the characteristic polynomial.

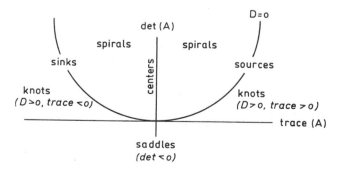

Hyperbolic Linear Flows

We now consider the general case of an arbitrary $\mathbb{K}$-vector space $E = (E, | \cdot |)$ of dimension $m < \infty$. For simplicity we call e^{tA} *the linear flow on E induced by A,* $A \in \mathcal{L}(E)$, (instead of the precise terminology

$$\varphi : \mathbb{R} \times E \to E, \quad (t, x) \mapsto e^{tA} x,$$

as was used before). The origin in E is of course a critical point of e^{tA} (in fact, it is the only one if A is injective). It is called a *sink* or a *source* if for every $x \in E \backslash \{0\}$ we have, respectively,

$$\lim_{t \to \infty} e^{tA} x = 0 \quad \text{or} \quad \lim_{t \to \infty} |e^{tA} x| = \infty.$$

We know from (12.12) and (12.13) that the origin is a sink or a source if and only if

$$Re\,\lambda < 0 \quad \text{or} \quad Re\,\lambda > 0, \quad \forall\,\lambda \in \sigma(A),$$

respectively. If the origin is a sink [resp. source], then the linear flow e^{tA} is called a *contraction* [resp. *expansion*]. We will show now that in the case of a contraction [resp. expansion] every flow line $\varphi_x(t) = e^{tA} x$, $x \neq 0$, converges exponentially to 0 [resp. ∞] as $t \to \infty$. To show this we need the following important lemma (13.1).

If $M \subseteq \mathbb{C}$ is nonempty and if $\beta \in \mathbb{R}$, we write

$$Re\,M < \beta$$

whenever $Re\,m < \beta$ *for all* $m \in M$. Related inequalities should be interpreted similarly. Moreover, a *Hilbert norm* $|| \cdot ||$ denotes a norm which is induced by an inner product, that is, for some appropriate inner product $(\cdot \mid \cdot)$ on E we have $||x||^2 = (x \mid x)$.

(13.1) Lemma. *Let $A \in \mathcal{L}(E)$ and $\alpha \in \mathbb{R}$ and assume that*

$$Re\,\sigma(A) < \alpha.$$

Then there exists a Hilbert norm $|| \cdot ||$ on E such that

$$||e^{tA}|| \leq e^{\alpha t}, \quad \forall\,t \in \mathbb{R}_+.$$

Proof. Assume that $\mathbb{K} = \mathbb{C}$. Then A has the form $A = D + N$ with respect to some appropriate basis, where

$$D = \text{diag}\,[\lambda_1, \ldots, \lambda_1, \lambda_2, \ldots, \lambda_2, \ldots, \lambda_k, \ldots \lambda_k]$$
$$= \text{diag}\,[\mu_1, \ldots, \mu_m]$$

and $N^m = 0$, as well as $DN = ND$. Moreover, we may choose the basis $\{e_1, \ldots, e_m\}$ of E such that $N\,e_j = e_{j-1}$ or $N\,e_j = 0$. If we replace e_j by $a_j :=$

$\delta^j e_j$, where $\delta > 0$, then D remains unchanged and for N we have $N a_j = \delta a_{j-1}$ or $N a_j = 0$.

Hence the matrix of N with respect to the basis $\{a_1, \ldots, a_m\}$ can only have nonzero elements, namely δ, along the upper superdiagonal. If we now apply the Euclidean norm corresponding to this basis, it follows that: For every $\epsilon > 0$ there exists a Hilbert norm $|| \cdot ||$ on E such that $||N|| \leq \epsilon$.

For $D = \mathrm{diag} \, [\mu_1, \ldots, \mu_m]$ we clearly have

$$||D|| = \max_{1 \leq j \leq m} |\mu_j|.$$

Hence

$$e^{tD} = \mathrm{diag} \, [e^{t\mu_1}, \ldots, e^{t\mu_m}]$$

implies the estimate

$$||e^{tD}|| = \max_{1 \leq j \leq m} |e^{t\mu_j}| = \max_{1 \leq j \leq m} e^{tRe\,\mu_j} \leq e^{t(\alpha - \epsilon)},$$

where $\epsilon > 0$ is chosen so small that $Re\,\lambda \leq \alpha - \epsilon$ for all $\lambda \in \sigma(A)$. Therefore proposition (12.4), together with remark (12.3 a), implies the estimate

$$||e^{tA}|| = ||e^{t(D+N)}|| \leq ||e^{tD}|| \, ||e^{tN}|| \leq e^{t(\alpha-\epsilon)} e^{t||N||}$$
$$\leq e^{t(\alpha-\epsilon)} e^{t\epsilon} = e^{\alpha t}$$

for all $t > 0$.

It is well-known (e.g., Yosida [1]) that a norm $|| \cdot ||$ on a vector space over $\mathbb{K}$ is a Hilbert norm if and only if it satisfies the parallelogram law: $||x+y||^2 + ||x-y||^2 = 2(||x||^2 + ||y||^2)$. From this it follows immediately that a Hilbert norm $|| \cdot ||_{E_\mathbb{C}}$, defined on the complexification $E_\mathbb{C}$ of a real vector space E, induces a Hilbert norm $|| \cdot ||_E$ on the real vector subspace E. For $A \in \mathcal{L}(E)$ and $x \in E$ we clearly have $||A x||_E = ||A_\mathbb{C} x||_{E_\mathbb{C}}$, and so it follows that $||A||_{\mathcal{L}(E)} \leq ||A_\mathbb{C}||_{\mathcal{L}(E_\mathbb{C})}$. Hence in the real case ($\mathbb{K} = \mathbb{R}$) we obtain the assertion by applying the above results to the complexification. □

(13.2) Remarks. (a) *If $Re\,\sigma(A) < \alpha$, there exists a constant $\beta \geq 0$ such that*

$$|e^{tA}| \leq \beta e^{\alpha t}, \quad \forall \, t \geq 0.$$

This follows immediately from (13.1) and the fact that all norms on a finite dimensional vector space – so in particular on $\mathcal{L}(E)$ – are equivalent.

(b) *Assume that $A \in \mathcal{L}(E)$ and $\alpha \in \mathbb{R}$ are such that*

$$Re\,\sigma(A) > \alpha.$$

Then there exist a Hilbert norm $|| \cdot ||$ on E and a constant $\gamma > 0$ so that

$$||e^{tA} x|| \geq e^{\alpha t} ||x||$$

and

$$|e^{tA}x| \geq \gamma e^{\alpha t}|x|$$

for all $x \in E$ and $t \geq 0$.

Proof. Using lemma (13.1) and $\sigma(-A) = -\sigma(A)$, we get

$$||e^{-tA}|| = ||e^{t(-A)}|| \leq e^{-\alpha t}, \quad \forall t \geq 0.$$

From this it follows that

$$||x|| = ||e^{-tA}e^{tA}x|| \leq ||e^{-tA}|| \, ||e^{tA}x|| \leq e^{-\alpha t}||e^{tA}x||$$

for all $x \in E$ and $t \geq 0$. This proves the first inequality. The second inequality again follows from the equivalence of the norms. $\square$

After these preliminaries we easily obtain the following theorem about the exponential shrinking or expanding of the flow lines in the case of a sink or a source, respectively.

(13.3) Theorem. *Assume that $A \in \mathcal{L}(E)$. Then the following are equivalent:*

(i)   *The origin is a sink.*
(ii)   *There exist constants $\alpha > 0$ and $\beta > 0$ such that*

$$|e^{tA}x| \leq \beta e^{-\alpha t}|x|, \quad \forall t \geq 0, \quad x \in E.$$

(iii)   *There exist a Hilbert norm $||\cdot||$ on E and a constant $\alpha > 0$ such that*

$$||e^{tA}x|| \leq e^{-\alpha t}||x||, \quad \forall t \geq 0, \quad x \in E.$$

Also the following are equivalent:

(i$'$)   *The origin is a source.*
(ii$'$)   *There exist constants $\alpha > 0$ and $\beta > 0$ such that*

$$|e^{tA}x| \geq \beta e^{\alpha t}|x|, \quad \forall t \geq 0, \quad x \in E.$$

(iii$'$)   *There exist a Hilbert norm $||\cdot||$ on E and a constant $\alpha > 0$ such that*

$$||e^{tA}x|| \geq e^{\alpha t}||x||, \quad \forall t \geq 0, \quad x \in E.$$

Proof. The theorem follows immediately from the stability criterion (12.11), corollary (12.13), lemma (13.1) and remarks (13.2). $\square$

In what follows we denote by

$$m(\lambda)$$

the algebraic multiplicity of the eigenvalue λ of $A \in \mathcal{L}(E)$. Moreover, we partition the spectrum $\sigma(A)$ as follows:

$$\sigma(A) = \sigma_s(A) \cup \sigma_n(A) \cup \sigma_u(A),$$

where

$$\sigma_s(A) := \{\lambda \in \sigma(A) \mid Re\,\lambda < 0\}$$

is the "*stable spectrum*,"

$$\sigma_n(A) := \{\lambda \in \sigma(A) \mid Re\,\lambda = 0\}$$

is the "*neutral spectrum*," and

$$\sigma_u(A) := \{\lambda \in \sigma(A) \mid Re\,\lambda > 0\}$$

is the "*unstable spectrum*." The flow e^{tA}, induced by A, is called *hyperbolic* if $\sigma_n(A) = \emptyset$, that is, if

$$\sigma(A) = \sigma_s(A) \cup \sigma_u(A).$$

The following theorem furnishes the higher dimensional generalization of the two-dimensional saddle.

(13.4) Theorem. *Suppose that e^{tA} is a hyperbolic linear flow. Then E has a direct sum decomposition*

$$E = E_s \oplus E_u$$

which decomposes A, and hence the flow e^{tA},

$$A = A_s \oplus A_u \quad and \quad e^{tA} = e^{tA_s} \oplus e^{tA_u},$$

such that e^{tA_s} is a contraction and e^{tA_u} is an expansion. This decomposition is unique and

$$\dim(E_s) = \sum_{\lambda \in \sigma_s(A)} m(\lambda).$$

Proof. First we consider the complex case $\mathbb{K} = \mathbb{C}$. We set

$$E_s := \bigoplus_{\lambda \in \sigma_s(A)} \ker[(A - \lambda)^{m(\lambda)}]$$

and

$$E_u := \bigoplus_{\lambda \in \sigma_u(A)} \ker[(A - \lambda)^{m(\lambda)}].$$

Then (cf. the proof of theorem (12.7))

$$E = E_s \oplus E_u$$

and this direct sum decomposes A, that is, $A = A_s \oplus A_u$. One easily verifies that

$$\sigma(A_s) = \sigma_s(A) \quad and \quad \sigma(A_u) = \sigma_u(A).$$

It now follows from the stability criterion (12.11) and corollary (12.13) that, respectively, e^{tA_s} is a contraction and e^{tA_u} is an expansion. Also, the formula for $\dim(E_s)$ is clear. It remains to show the uniqueness.

So assume that $E = E_1 \oplus E_2$ is some other direct sum decomposition which decomposes A,

$$A = A_1 \oplus A_2,$$

and such that e^{tA_1} is a contraction and e^{tA_2} is an expansion. Each $x \in E_1$ can be written in the form

$$x = y + z,$$

where $y \in E_s$ and $z \in E_u$. Since $e^{tA}x = e^{tA_1}x \to 0$ as $t \to \infty$, it follows that

$$e^{tA}z = e^{tA}P_u x = P_u e^{tA}x \to 0$$

as $t \to \infty$, where $P_u : E \to E_u$ denotes the canonical projection corresponding to the direct sum $E = E_s \oplus E_u$. Theorem (13.3) implies the existence of constants $\alpha, \beta > 0$ such that

$$|e^{tA}z| = |e^{tA_u}z| \geq \beta e^{\alpha t}|z|, \quad \forall\, t \geq 0.$$

Consequently we must have $z = 0$, hence $E_1 \subseteq E_s$. For symmetry reasons it follows that $E_s \subseteq E_1$ and thus $E_1 = E_s$.

If now $x \in E_2$, then

$$e^{tA}x = e^{tA_2}x \to 0 \quad \text{as} \quad t \to -\infty,$$

(cf. corollaries (12.12) and (12.13)) and consequently

$$e^{tA}y \to 0 \quad \text{as} \quad t \to -\infty.$$

Since $e^{tA_s} = e^{|t|(-A_s)}$ and $\sigma(-A_s) = -\sigma(A_s)$, it follows from theorem (13.3) that

$$|e^{tA}y| = |e^{tA_s}y| = |e^{|t|(-A_s)}y| \geq \beta e^{\alpha|t|}|y|$$

for all $t \leq 0$ and some appropriate constants $\alpha, \beta > 0$. Hence $y = 0$ and so we have $E_2 \subseteq E_u$. Symmetry reasons imply that $E_u \subseteq E_2$, hence $E_2 = E_u$, and therefore the uniqueness of the direct sum decomposition is proved.

Now assume that $\mathbb{K} = \mathbb{R}$. We then can apply the above results to the complexification $E_\mathbb{C} = E + iE$ and $A_\mathbb{C} \in \mathcal{L}(E_\mathbb{C})$. Thus we have

$$E_\mathbb{C} = (E_\mathbb{C})_s \oplus (E_\mathbb{C})_u \quad \text{and} \quad A_\mathbb{C} = (A_\mathbb{C})_s \oplus (A_\mathbb{C})_u,$$

so that $e^{t(A_\mathbb{C})_s}$ is a contraction and $e^{t(A_\mathbb{C})_u}$ is an expansion. We now set

$$E_s := (E_\mathbb{C})_s \cap E \quad \text{and} \quad E_u := (E_\mathbb{C})_u \cap E$$

and show that

$$(E_\mathbb{C})_s = (E_s)_\mathbb{C} \quad \text{and} \quad (E_\mathbb{C})_u = (E_u)_\mathbb{C}. \tag{5}$$

To show this we first consider the case when $\sigma(A) = \{\lambda\} \subseteq \mathbb{R}$. Then it follows from section 12 that $E_\mathbb{C}$ has a direct sum decomposition $X_1 \oplus \cdots \oplus X_n$ which reduces $A_\mathbb{C}$, and such that each $X = X_j$ has a basis $\{e_1, \ldots, e_m\}$ with the property that $(A_\mathbb{C} - \lambda)e_k = e_{k-1}$, where we have set $e_0 := 0$. Conjugation implies $(A_\mathbb{C} - \lambda)\bar{e}_k = \bar{e}_{k-1}$ and so $\bar{e}_1, \ldots, \bar{e}_m$ also belong to X.

Let $x \in X \cap E$. Then $e^{tA}x = e^{tA_\mathbb{C}}x$ shows that x has, as an element of X, the same asymptotic behavior (with respect to the flow) as in $X \cap E$. Let $z = x + iy \in X$. Then it follows from the representation $z = \sum_{j=1}^m \alpha_j e_j$ and the considerations above that also $\bar{z} = \sum_{j=1}^m \bar{\alpha}_j \bar{e}_j$ belongs to X. Therefore $x = (z + \bar{z})/2$ and $y = (z - \bar{z})/2i$ are elements of $X \cap E$. If $e^{tA_\mathbb{C}}$ is a contraction (resp. expansion), then $e^{tA}x = e^{tA_\mathbb{C}}x = e^{tA_\mathbb{C}}(z + \bar{z})/2$ implies that e^{tA} is also a contraction (resp. expansion) on $X \cap E$. From this, assertion (5) follows in this case.

Next, we consider the case when $\sigma(A) = \{\lambda, \bar{\lambda}\}$, where $Im\,(\lambda) \neq 0$. Then $E_\mathbb{C}$ has the direct sum decomposition $X_1 \oplus Y_1 \oplus \cdots \oplus X_n \oplus Y_n$ such that $X_j \oplus Y_j$ has a basis $\{e_1, \ldots, e_m, \bar{e}_1, \ldots, \bar{e}_m\}$ with the properties that $(A_\mathbb{C} - \lambda)e_k = e_{k-1}$ and $(A_\mathbb{C} - \bar{\lambda})\bar{e}_k = \bar{e}_{k-1}$ for all $k = 1, \ldots, m$, where $e_0 := 0$ (cf. case 2. (b)). As above, from this we obtain assertion (5). Assertion (5) now follows also in the general case, because the general case is comprised of such subcases (by direct sums). The theorem can now be deduced from (5) and the applicability of the theorem in the complex case. $\square$

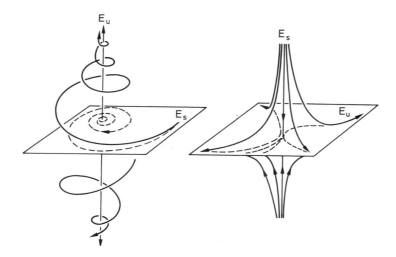

The invariant subspaces E_s and E_u of the hyperbolic linear flow e^{tA} are called, respectively, *stable* and *unstable subspaces* of the flow. A hyperbolic linear flow may be a contraction ($E_u = \{0\}$) or an expansion ($E_s = \{0\}$). If $E = \mathbb{R}^3$ and if neither the stable nor the unstable subspace is trivial, then typical trajectories may look like the ones shown in the figures.

Flow Equivalence

Now the question arises what the characteristic properties of the phase portraits of this section are. Is it possible to introduce appropriate nonlinear coordinates such that a saddle is transformed into a node, or a stable node into an unstable spiral? We will show that this is not the case, but that it is possible to transform a stable node into a stable spiral. For this we must first define the concept of equivalent flows.

Assume that M and N are metric spaces and let φ and ψ be flows on M and N with domains Ω_φ and Ω_ψ, respectively. Then φ and ψ are called (*topologically*) *equivalent* if there exists some orientation preserving automorphism $\alpha : \mathbb{R} \to \mathbb{R}$ and a homeomorphism $h : M \to N$ such that

$$h(\varphi(t, x)) = \psi(\alpha(t), h(x)), \quad \forall\, (t, x) \in \Omega_\varphi.$$

Every pair (α, h) with these properties is called a (*topological*) *flow equivalence*. Hence (α, h) is a topological flow equivalence if and only if the following diagram commutes:

$$
\begin{array}{ccc}
\mathbb{R} \times M \supset \Omega_\varphi & \xrightarrow{\ \varphi\ } & M \\
{\scriptstyle \alpha \times h} \downarrow & & \downarrow {\scriptstyle h} \\
\mathbb{R} \times N \supset \Omega_\psi & \xrightarrow[\ \psi\ ]{} & N
\end{array}\ ,
$$

where, as usual, $\alpha \times h$ is defined by

$$\alpha \times h : \Omega_\varphi \to \mathbb{R} \times N, \quad (t, x) \mapsto (\alpha(t), h(x)).$$

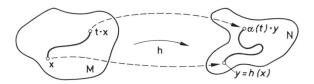

If M and N are differentiable manifolds, if (α, h) is a flow equivalence between φ and ψ, and if h is a C^1-diffeomorphism, we say that φ and ψ are *differentiably equivalent* and (α, h) is called a C^1-*flow equivalence*. If M and N are Banach spaces, if (α, h) is a flow equivalence between the flows φ on M and ψ on N, and if h is a vector space isomorphism, we say that φ and ψ are *linearly equivalent* and (α, h) is called a *linear flow equivalence*.

(13.5) Remarks. (a) Every orientation preserving automorphism α on $\mathbb{R}$ has the form

$$\alpha(t) = \alpha \cdot t, \quad \forall t \in \mathbb{R}, \tag{6}$$

for some unique positive number α. Since also every $\alpha > 0$ defines, by (6), an orientation preserving automorphism, we will, in what follows, identify the automorphism α with the unique positive number it determines.

(b) *If (α, h) is a flow equivalence, then $\alpha \times h$ maps the set Ω_φ homeomorphically onto Ω_ψ.*

Proof. By the definition of flow equivalence we have $\alpha \times h(\Omega_\varphi) \subseteq \Omega_\psi$. Since $\alpha \times h$ is a homeomorphism from $\mathbb{R} \times M$ onto $\mathbb{R} \times N$ and since Ω_φ is open in $\mathbb{R} \times M$, it follows that $\tilde{\Omega} := \alpha \times h(\Omega_\varphi)$ is open in $\mathbb{R} \times N$. In fact $\tilde{\Omega}$ is open in Ω_ψ, because Ω_ψ is open. Using $\psi \circ (\alpha \times h) = h \circ \varphi$, one easily verifies that $\tilde{\psi} := \psi | \tilde{\Omega}$ is a flow on $\tilde{\Omega}$. If now $\tilde{\Omega} \neq \Omega_\psi$, then there exists some $y \in N$ such that $J_{\tilde{\psi}}(y) \underset{\neq}{\subset} J_\psi(y)$ and so either

$$\tau := t^+_{\tilde{\psi}}(y) < t^+_\psi(y) \quad \text{or} \quad \sigma := t^-_{\tilde{\psi}}(y) > t^-_\psi(y).$$

In the first case we have $\tilde{\psi}([0, \tau), y) \subseteq \psi([0, \tau], y)$ and in the second case $\tilde{\psi}((\sigma, 0], y) \subseteq \psi([\sigma, 0], y)$. It now follows from corollary (10.13) that $\tau = \infty$ in the first case and that $\sigma = -\infty$ in the second. This is impossible and hence $\alpha \times h(\Omega_\varphi) = \Omega_\psi$. Therefore $\alpha \times h$ is a homeomorphism from Ω_φ onto Ω_ψ. $\square$

(c) From (b) we immediately deduce: *(Topological) flow equivalence, C^1-flow equivalence and linear flow equivalence are all equivalence relations.*

(d) *If (α, h) is a flow equivalence between φ and ψ, then the homeomorphism $h : M \rightarrow N$ maps the orbits of φ onto the orbits of ψ and preserves the orientations.* $\square$

It is easy now to determine the equivalence classes of the linear flows with respect to linear flow equivalence.

(13.6) Proposition. *Assume that $A, B \in \mathcal{L}(E)$. Then e^{tA} and e^{tB} are linear flow equivalent if and only if there exists some $\alpha > 0$ with*

$$\sigma(A) = \sigma(\alpha B)$$

and such that their eigenvalues have the same geometric and algebraic multiplicities.

Proof. If (α, h) is a linear flow equivalence between e^{tA} and e^{tB}, then

$$he^{tA} = e^{\alpha tB}h, \quad \forall t \in \mathbb{R},$$

and hence

$$e^{tA} = h^{-1}e^{\alpha tB}h = e^{th^{-1}(\alpha B)h}, \quad \forall t \in \mathbb{R},$$

by proposition (12.4). Since the infinitesimal generator of a flow is unique, it follows that $A = h^{-1}(\alpha B)h$. Now $\alpha > 0$ and $h \in \mathcal{GL}(E)$ and hence $\sigma(A) = \sigma(\alpha B)$ – for instance, because $\det(h^{-1}(\alpha B)h - \lambda) = \det(\alpha B - \lambda)$.

Conversely, if $\sigma(A) = \sigma(\alpha B)$ for some $\alpha > 0$, we know from linear algebra (as an immediate consequence of the Jordan normal form) that $A = h^{-1}(\alpha B)h$ for some $h \in \mathcal{GL}(E)$. Hence e^{tA} and e^{tB} are linearly flow equivalent because

$$e^{tA} = e^{th^{-1}(\alpha B)h} = h^{-1}e^{\alpha tB}h, \quad \forall t \in \mathbb{R}.$$

$\square$

The next proposition shows that we do not get anything new from the differentiable classification.

(13.7) Proposition. *Let $A, B \in \mathcal{L}(E)$. Then e^{tA} and e^{tB} are C^1-flow equivalent if and only if they are linearly flow equivalent.*

Proof. Let (α, h) be a C^1-flow equivalence between e^{tA} and e^{tB}. The diffeomorphism $h \in C^1(E, E)$ maps the critical point $x = 0$ of the flow e^{tA} into a critical point y of the flow e^{tB}, that is, into a $y \in E$ such that $e^{sB}y = y$ holds for all $s \in \mathbb{R}$. If $T : E \to E$ denotes the translation $x \mapsto x - y$, then

$$T \circ h \circ e^{tA}x = h \circ e^{tA}x - y = e^{\alpha tB}h(x) - y$$
$$= e^{\alpha tB}h(x) - e^{\alpha tB}y = e^{\alpha tB}(T \circ h)(x), \quad \forall x \in E, t \in \mathbb{R},$$

show that $(\alpha, T \circ h)$ is a flow equivalence between e^{tA} and e^{tB}. Moreover, $T \circ h(0) = 0$ and $C := D(T \circ h)(0) \in \mathcal{GL}(E)$. Differentiating

$$(T \circ h) \circ e^{tA}x = e^{\alpha tB}(T \circ h)(x)$$

at $x = 0$, we get

$$Ce^{tA} = e^{\alpha tB}C, \quad \forall t \in \mathbb{R}.$$

Therefore (α, C) is a linear flow equivalence between e^{tA} and e^{tB}. The converse is trivial.

$\square$

For the substantially more difficult *topological classification* of linear flows we need the following lemma.

(13.8) Lemma. *Suppose $A \in \mathcal{L}(E)$ is such that*

$$Re\, \sigma(A) < 0$$

and let φ denote the linear flow on E induced by A, that is, $\varphi^t = e^{tA}$ for all $t \in \mathbb{R}$. Then there exists a Hilbert norm $\|\cdot\|$ on E such that

$$\bar{\varphi} : \mathbb{R} \times \mathbb{S} \to E \backslash \{0\}, \quad (t, x) \mapsto \varphi(t, x),$$

is a homeomorphism, where $\mathbb{S} := \{x \in E \mid \|x\| = 1\}$.

Proof. Choose some $\alpha > 0$ so that $Re\,\sigma(A) < -\alpha < 0$. Lemma (13.1) implies the existence of a Hilbert norm $\|\cdot\|$ on E such that

$$\|e^{tA}\| \leq e^{-\alpha t}, \quad \forall\, t \geq 0.$$

Now let $y \in E \backslash \{0\}$ be arbitrary. Then $\varphi^t(y) = e^{tA} y \neq 0$ for all $t \in \mathbb{R}$ and

$$\|\varphi^t(y)\| \leq e^{-\alpha t} \|y\| \quad \text{for all} \ \ t \geq 0. \tag{7}$$

This implies

$$\|y\| = \|\varphi^t \circ \varphi^{-t}(y)\| \leq e^{-\alpha t} \|\varphi^{-t}(y)\|,$$

and thus

$$\|\varphi^{-t}(y)\| \geq e^{\alpha t} \|y\|, \quad \forall\, t \geq 0. \tag{8}$$

From (7) and (8) we immediately deduce the fact that every noncritical trajectory intersects the sphere $\mathbb{S}$ in exactly one point. Therefore

$$\bar{\varphi} : \mathbb{R} \times \mathbb{S} \to E \backslash \{0\}$$

is a continuous bijection. It remains to show that the inverse function is continuous.

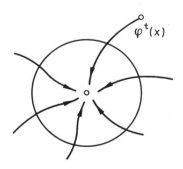

So assume then that (y_k) is a sequence in $E \backslash \{0\}$ converging to $y \in E \backslash \{0\}$. Then there exist a sequence (t_k) in $\mathbb{R}$ and a sequence (x_k) in $\mathbb{S}$ such that $y_k = \varphi(t_k, x_k)$. Now $\mathbb{S}$ is compact, and so by going to a subsequence we may, in fact, assume that $x_k \to x \in \mathbb{S}$. By taking a further subsequence we may also assume that (t_k) converges to some $t \in \bar{\mathbb{R}}$. If $t \in (0, \infty]$, then (7) implies that

$$\|y_k\| = \|\varphi^{t_k}(x_k)\| \leq e^{-\alpha t_k} \|x_k\| = e^{-\alpha t_k}$$

for large k. Hence $||y|| \leq e^{-\alpha t}$ and therefore t is finite, because $y \neq 0$. If $t \in [-\infty, 0)$, then (8) implies that

$$||y_k|| = ||\varphi^{t_k}(x_k)|| \geq e^{-\alpha t_k}||x_k|| = e^{-\alpha t_k}$$

for large k. Hence $||y|| \geq e^{-\alpha t} = e^{\alpha|t|}$, from which we deduce that $t > -\infty$. Therefore $t \in \mathbb{R}$ and so the continuity of φ implies that $y = \varphi(t, x) = \varphi^t(x)$. Since this holds for every convergent subsequence, we see that the inverse function $\bar{\varphi}^{-1}$ (i.e., the "projection" of $y \in E\backslash\{0\}$ onto $\mathbb{S}$ "along the orbit" $\gamma(y)$) is continuous. □

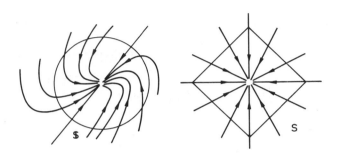

Geometrically the above lemma says that for the unit sphere $\mathbb{S}$ with respect to some appropriate Hilbert norm we have: Every noncritical trajectory crosses $\mathbb{S}$ *"transversely."* The following lemma graphically shows that the orbits of a contraction can be "straightened."

(13.9) Lemma. *Let $A \in \mathcal{L}(E)$ be such that $\operatorname{Re}\sigma(A) < 0$. Then there exists a flow equivalence of the form $(1, h)$ between e^{tA} and $e^{-t}I$.*

Proof. Lemma (13.8) shows that there exists a Hilbert norm $||\cdot||$ on E so that for the corresponding unit sphere $\mathbb{S}$ we have:

$$\mathbb{R} \times \mathbb{S} \to E\backslash\{0\}, \quad (t, x) \mapsto e^{tA}x =: \varphi^t(x),$$

is a homeomorphism. Let S denote the unit sphere corresponding to the original norm $|\cdot|$ on E. We then define a bijection $h : E \to E$ by $h(0) := 0$ and

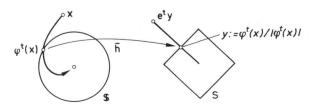

$$h(x) := e^t \frac{\varphi^t(x)}{|\varphi^t(x)|}, \quad x \in E \backslash \{0\},$$

where t is the unique real number, as guaranteed by lemma (13.8), such that $\varphi^t(x) \in \mathbb{S}$. Since the mapping

$$\bar{h} : \mathbb{S} \rightarrow S, \quad y \mapsto y/|y|$$

is clearly a homeomorphism and since $h(x) = e^t \bar{h} \circ \varphi^t(x)$, it follows from lemma (13.8) that h is a homeomorphism from $E \backslash \{0\}$ onto itself. In order to show that h is continuous at $0 \in E$, we let V denote any neighborhood of 0 with $V \subseteq \mathbb{B}$. Then there exists some $t_0 > 0$ so that $e^{-t} S \subseteq V$ for all $t \geq t_0$. Set

$$U := \{x \in E \mid ||x|| < 1/||e^{-t_0 A}||\}.$$

Then for $x \in U$ we have

$$||\varphi^{-t_0}(x)|| = ||e^{-t_0 A} x|| \leq ||e^{-t_0 A}|| \, ||x|| < 1.$$

It follows from (8) that for all $t \geq 0$ the inequality

$$||\varphi^{t-t_0}(x)|| = ||\varphi^t(\varphi^{-t_0}(x))|| \leq e^{-\alpha t}||\varphi^{-t_0}(x)|| < e^{-\alpha t} \leq 1$$

holds for some appropriate $\alpha > 0$. Hence we see that the unique $t = t(x)$, with $\varphi^t(x) \in \mathbb{S}$, satisfies $t < -t_0$. And so $h(U) \subseteq V$ follows from the definition of h. Therefore h is continuous at 0. Similarly one shows that h^{-1} is continuous at 0. So consequently h is a homeomorphism from E onto itself.

For $(1, h)$ to be a flow equivalence we must show that

$$h \circ \varphi^t(x) = e^{-t} h(x), \quad \forall x \in E, \quad \forall t \in \mathbb{R}.$$

For $x = 0$ this is trivially true. If $x \neq 0$, then $x = \varphi^s(y)$ for some appropriate $(s, y) \in \mathbb{R} \times \mathbb{S}$. Hence it follows from the definition of h that

$$h \circ \varphi^t(x) = h \circ \varphi^t \circ \varphi^s(y) = h \circ \varphi^{t+s}(y)$$
$$= e^{-(t+s)} \frac{y}{|y|} = e^{-t} \left(e^{-s} \frac{y}{|y|} \right)$$
$$= e^{-t} \left(e^{-s} \frac{\varphi^{-s}(x)}{|\varphi^{-s}(x)|} \right) = e^{-t} h(x)$$

for all $t \in \mathbb{R}$. $\qquad \qquad \square$

After these preliminaries we can prove the central classification theorem for hyperbolic flows. For $A \in \mathcal{L}(E)$ we now set

$$m_-(A) := \sum_{\lambda \in \sigma_s(A)} m(\lambda).$$

(13.10) Theorem. *Two hyperbolic linear flows e^{tA} and e^{tB} are flow equivalent if and only if $m_-(A) = m_-(B)$. That is, the only invariant of flow equivalence for hyperbolic linear flows is the dimension of the stable subspace.*

Proof. "$\Leftarrow$" By theorem (13.4) there exists a direct sum decomposition

$$E = E_s \oplus E_u, \quad e^{tA} = e^{tA_s} \oplus e^{tA_u},$$

with $\dim(E_s) = m_-(A)$ and such that e^{tA_s} is a contraction and e^{tA_u} is an expansion. From the stability criterion (12.11) and from lemma (13.9) we deduce the existence of a flow equivalence $(1, h_s)$ between e^{tA_s} and $e^{-t}id_{E_s}$. Because $e^{tA_u} = e^{-t(-A_u)}$, we similarly obtain from lemma (13.9) a flow equivalence $(1, h_u)$ between e^{tA_u} and $e^{t}id_{E_u}$. One immediately verifies that $(1, h_s \oplus h_u)$ is a flow equivalence between $e^{tA} = e^{tA_s} \oplus e^{tA_u}$ and $e^{-t}id_{E_s} \oplus e^{t}id_{E_u}$, where

$$h_s \oplus h_u : E_s \oplus E_u \to E_s \oplus E_u, \quad x + y \mapsto h_s(x) + h_u(y).$$

Likewise, there exist direct sum decompositions

$$E = \bar{E}_s \oplus \bar{E}_u, \quad e^{tB} = e^{tB_s} \oplus e^{tB_u}$$

and a flow equivalence $(1, \bar{h}_s \oplus \bar{h}_u)$ between e^{tB} and $e^{-t}id_{\bar{E}_s} \oplus e^{t}id_{\bar{E}_u}$. Because $\dim(E_s) = \dim(\bar{E}_s)$, there exist isomorphisms $T_s : E_s \to \bar{E}_s$ and $T_u : E_u \to \bar{E}_u$. Then clearly $(1, T_s \oplus T_u)$ is a flow equivalence between the flows $e^{-t}id_{E_s} \oplus e^{t}id_{E_u}$ and $e^{-t}id_{\bar{E}_s} \oplus e^{t}id_{\bar{E}_u}$. Hence transitivity implies that e^{tA} and e^{tB} are flow equivalent.

"$\Rightarrow$" If (α, h) is a flow equivalence between e^{tA} and e^{tB}, then

$$h(e^{tA}x) = e^{\alpha tB}h(x), \quad \forall (t, x) \in \mathbb{R} \times E.$$

Thus $x \in E_s$ implies $\lim_{t \to \infty} e^{\alpha tB}h(x) = \lim_{t \to \infty} h(e^{tA}x) = h(\lim_{t \to \infty} e^{tA}x) = h(0) = 0$, which shows that $h(x) \in \bar{E}_s$. Hence $h(E_s) \subseteq h(\bar{E}_s)$ and similarly $h(\bar{E}_s) \subseteq h(E_s)$. Therefore h is a homeomorphism from the vector space E_s onto the vector space $\bar{E}_s$. It follows from the theorem on the invariance of domain (cf. Dugundji [1]) that $\dim(\bar{E}_s) = \dim(E_s)$. Hence we obtain from theorem (13.4) that $m_-(A) = m_-(B)$. $\qquad \square$

(13.11) Remarks. (a)   The topological classification of linear flows e^{tA} with $\sigma(A) \subseteq i\mathbb{R}$, that is, $\sigma(A) = \sigma_n(A)$, is an unsolved problem.

(b) It is not difficult to see that *the set of all $A \in \mathcal{L}(E)$, such that $\sigma(A) = \sigma_s(A) \cup \sigma_u(A)$, is open and dense in $\mathcal{L}(E)$.* In other words, the property of inducing a hyperbolic flow is a *generic* property; it is shared by "almost all" $A \in \mathcal{L}(E)$. With the aid of theorem (13.10) we can therefore classify "almost all" linear flows (which, however, is useless if we are especially interested in nonhyperbolic flows).

(c) A hyperbolic flow e^{tA} is, in particular, flow equivalent to the simple "higher dimensional" saddle

$$\dot{x} = -x, \quad x \in \mathbb{K}^{m_-},$$
$$\dot{y} = y, \quad y \in \mathbb{K}^{m_+},$$

where $m_{\pm} := m_{\pm}(A)$.

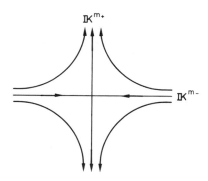

$\square$

Problems

1. Describe the phase portraits of planar linear flows for all those cases that are not treated in the text, that is, if at least one eigenvalue is zero.

2. Describe the phase portraits of the linear flow e^{tA}, where $A \in \mathcal{L}(\mathbb{R}^3)$, i.e, of the three-dimensional linear flows under all possible distributions of the eigenvalues of A in $\mathbb{C}$.

3. Illustrate the phase portrait of the linear flow e^{tA}, where $A = \text{diag } [\omega_1, -\omega_1, \omega_2, -\omega_2] \in \mathcal{L}(\mathbb{R}^4)$.

4. Prove that $\{A \in \mathcal{L}(E) \mid \sigma(A) \cap i\mathbb{R} = \emptyset\}$ is open and dense in $\mathcal{L}(E)$.

14. Higher Order Linear Differential Equations

In this section J denotes an open interval in $\mathbb{R}$ and

$$a_0, \ldots, a_m, b \in C(J, \mathbb{K})$$

are such that

$$a_m(t) \neq 0, \quad \forall t \in J.$$

We define the mth *order linear ordinary differential operator*

$$A(t, D) : C^m(J, \mathbb{K}) \to C(J, \mathbb{K})$$

by

$$A(t, D) := \sum_{j=0}^{m} a_j D^j,$$

that is,

$$[A(t, D)u](t) = a_m(t)u^{(m)}(t) + a_{m-1}(t)u^{(m-1)}(t) + \cdots + a_1(t)\dot{u}(t) + a_0(t)u(t),$$

where $u \in C^m(J, \mathbb{K})$ and $t \in J$.

(14.1) Remarks. (a) The variable t in $A(t, D)$ merely symbolizes the fact that the *coefficients* $a_0, \ldots, a_m$ are functions of $t \in J$. If all $a_0, \ldots, a_m$ are constant, we simply write

$$A(D) = \sum_{j=0}^{m} a_j D^j$$

and call $A(D)$ an mth order linear differential operator *with constant coefficients.*

(b) It is clear that

$$A(t, D) : C^m(J, \mathbb{K}) \to C(J, \mathbb{K})$$

is a linear map.

(c) If $b \neq 0$, we call the equation

$$A(t, D)u = b$$

nonhomogeneous, otherwise we call it *homogeneous*.   □

We already know from section 5 that the equation $A(t, D)u = b$ is equivalent to the system (with $x^0 := u$)

$$\dot{x}^0 = x^1$$
$$\dot{x}^1 = x^2$$
$$\vdots$$
$$\dot{x}^{m-2} = x^{m-1}$$
$$\dot{x}^{m-1} = -\frac{a_0}{a_m}x^0 - \frac{a_1}{a_m}x^1 - \cdots - \frac{a_{m-1}}{a_m}x^{m-1} + \frac{b}{a_m}.$$

This system can be written as

$$\dot{x} = A(t)x + \vec{b}(t),$$

where

$$A := \begin{bmatrix} 0 & 1 & 0 & \cdots & 0 & 0 \\ 0 & 0 & 1 & \cdots & 0 & 0 \\ 0 & 0 & 0 & \cdots & 0 & 0 \\ \vdots & \vdots & \vdots & \ddots & \vdots & \vdots \\ 0 & 0 & 0 & \cdots & 0 & 1 \\ -\frac{a_0}{a_m} & -\frac{a_1}{a_m} & \frac{a_2}{a_m} & \cdots & -\frac{a_{m-2}}{a_m} & -\frac{a_{m-1}}{a_m} \end{bmatrix}, \quad \vec{b} := \begin{bmatrix} 0 \\ 0 \\ 0 \\ \vdots \\ 0 \\ \frac{b}{a_m} \end{bmatrix}.$$

Therefore the IVP

$$\dot{x} = A(t)x + \vec{b}(t), \quad x(\tau) = \xi \in \mathbb{K}^m,$$

corresponds to the IVP

$$\begin{cases} A(t, D)u = b \\ u(\tau) = \xi^0, \quad \dot{u}(\tau) = \xi^1, \quad \ldots, \quad u^{(m-1)}(\tau) = \xi^{m-1}, \end{cases} \tag{1}$$

where $\xi^0, \ldots, \xi^{m-1} \in \mathbb{K}$. Hence we can apply the results about first order linear systems immediately to (1) and we thus obtain the following fundamental theorem.

(14.2) Existence Theorem. *The IVP (1) has a unique solution $u(\cdot, \tau, \xi_0, \ldots, \xi_{m-1})$ for every "initial condition" $(\tau, \xi_0, \ldots, \xi_{m-1}) \in J \times \mathbb{K}^m$, and*

$$u \in C(J \times J \times \mathbb{K}^m, \mathbb{K}).$$

Moreover, if also $a_0, \ldots, a_m, b \in C^n(J, \mathbb{K}), n \geq 1$, then

$$u \in C^n(J \times J \times \mathbb{K}^m, \mathbb{K})$$

(that is, the general solution u has as many derivatives as the data [i.e., the coefficient functions] of the problem). The totality of all solutions of the homogeneous differential equation $A(t, D)u = 0$ forms an m-dimensional vector subspace V of $C^m(J, \mathbb{K})$. The set of all solutions of the nonhomogeneous equation $A(t, D)u = b$ forms the affine subspace

$$V + v \subseteq C^m(J, \mathbb{K}),$$

where v is an arbitrary "particular" solution of $A(t, D)u = b$.

We call every set of m linearly independent solutions of the homogeneous equation $A(t, D)u = 0$ a *fundamental set of solutions*. The *Wronskian* of m arbitrary functions $u_1, \ldots, u_m \in C^m(J, \mathbb{K})$ is defined by

$$W(u_1, \ldots, u_m) := \det \begin{bmatrix} u_1 & \cdots & u_m \\ \dot{u}_1 & \cdots & \dot{u}_1 \\ \vdots & \ddots & \vdots \\ u_1^{(m-1)} & \cdots & u_m^{(m-1)} \end{bmatrix}.$$

It now follows from corollary (11.5) that *the solutions $u_1, \ldots, u_m$ form a fundamental set of solutions if and only if the Wronskian does not vanish at $t \in J$ (and therefore nowhere)*.

Assume that $\{u_1, \ldots, u_m\}$ is a fundamental set of solutions and let

$$X := \begin{bmatrix} u_1 & \cdots & u_m \\ \dot{u}_1 & \cdots & \dot{u}_m \\ \vdots & \ddots & \vdots \\ u_1^{(m-1)} & \cdots & u_m^{(m-1)} \end{bmatrix}$$

denote the corresponding fundamental matrix. We then know from the variation of constants formula (theorem (11.13)) that a particular solution y of the system

$$\dot{x} = A(t)x + \vec{b}(t)$$

is given by

$$y(t) = X(t) \int_\tau^t X^{-1}(s)\vec{b}(s)ds, \quad t \in J.$$

Set

$$\vec{a}(s) := X^{-1}(s)\vec{b}(s),$$

then

$$X\vec{a} = \vec{b}. \tag{2}$$

It follows from Cramer's rule that the solution of this system of equations is given by

$$a^j = \frac{V_j}{W}, \quad j = 1, \ldots, m,$$

where $W := W(u_1, \ldots, u_m)$ and

$$V_j := \det \begin{bmatrix} u_1 & \cdots & u_{j-1} & 0 & u_{j+1} & \cdots & u_m \\ \dot{u}_1 & \cdots & \dot{u}_{j-1} & 0 & \dot{u}_{j+1} & \cdots & \dot{u}_m \\ \vdots & \vdots & \vdots & \vdots & \vdots & \vdots & \vdots \\ u_1^{(m-2)} & \cdots & u_{j-1}^{(m-2)} & 0 & u_{j+1}^{(m-2)} & \cdots & u_m^{(m-2)} \\ u_1^{(m-1)} & \cdots & u_{j-1}^{(m-1)} & \frac{b}{a_m} & u_{j+1}^{(m-1)} & \cdots & u_m^{(m-1)} \end{bmatrix}.$$

Therefore

$$V_j = (-1)^{m+j} \frac{b}{a_m} W(u_1, \ldots, u_{j-1}, u_{j+1}, \ldots, u_m),$$

where of course the last expression denotes the Wronskian of $m - 1$ functions, an $(m - 1) \times (m - 1)$ determinant.

Since the first component of the vector y represents a particular solution of the nonhomogeneous equation $A(t, D)u = b$, we have proved the following proposition.

(14.3) Proposition. *Assume that $\{u_1, \ldots, u_m\}$ is a fundamental set of solutions of the homogeneous equation $A(t, D)u = 0$. Then the function*

$$v(t) := \sum_{j=1}^{m} (-1)^{m+j} \int_{\tau}^{t} \frac{W(u_1, \ldots, u_{j-1}, u_{j+1}, \ldots, u_m)}{W(u_1, \ldots, u_m)} \frac{b}{a_m} \, ds \, u_j(t) \qquad (3)$$

is a particular solution of the nonhomogeneous equation $A(t, D)u = b$.

(14.4) Remarks. (a) The particular solution (3) can also be obtained by finding a solution of the form

$$v(t) = \sum_{j=1}^{m} c_j(t) u_j(t), \qquad t \in J,$$

with the method of *variation of constants*.

(b) In the *special case* when $m = 2$, a particular solution is given by

$$v(t) = - \int_{\tau}^{t} \frac{u_2 b}{a_2 W(u_1, u_2)} \, ds \, u_1(t) + \int_{\tau}^{t} \frac{u_1 b}{a_2 W(u_1, u_2)} \, ds \, u_2(t), \qquad t \in J,$$

where $\{u_1, u_2\}$ is a fundamental set of solutions and $W(u_1, u_2) = u_1 \dot{u}_2 - \dot{u}_1 u_2$.   □

We now consider the *constant coefficient case*

$$A(D) = \sum_{j=0}^{m} a_j D^j, \qquad a_j \in \mathbb{K}, \; a_m \neq 0.$$

Then the characteristic polynomial corresponding to the matrix $A \in \mathbb{M}^m(\mathbb{K})$ is given by

$$\det(A - \lambda) = \det \begin{bmatrix} -\lambda & 1 & \cdots & 0 & 0 \\ 0 & -\lambda & \cdots & 0 & 0 \\ \vdots & \vdots & \ddots & \vdots & \vdots \\ 0 & 0 & \cdots & -\lambda & 1 \\ -\frac{a_0}{a_m} & -\frac{a_1}{a_m} & \cdots & -\frac{a_{m-2}}{a_m} & -\frac{a_{m-1}}{a_m} - \lambda \end{bmatrix}.$$

By expanding the determinant according to the last row, we get

$$\det(A - \lambda) = (-1)^{m+1} \left(-\frac{a_0}{a_m} \right) + (-1)^{m+2} \left(-\frac{a_1}{a_m} \right) (-\lambda)$$

$$+ (-1)^{m+3} \left(-\frac{a_2}{a_m} \right) (-\lambda)^2 + \cdots + (-1)^{2m-1} \left(-\frac{a_{m-2}}{a_m} \right) (-\lambda)^{m-2}$$

$$+ (-1)^{2m} \left(-\frac{a_{m-1}}{a_m} - \lambda \right) (-\lambda)^{m-1}$$

$$= \frac{(-1)^m}{a_m} [a_0 + a_1 \lambda + \cdots + a_{m-1} \lambda^{m-1} + a_m \lambda^m].$$

Hence λ *is an eigenvalue of the matrix* A *with multiplicity* $m(\lambda)$ *if and only if* λ *is a zero of the polynomial*

$$A(\lambda) := \sum_{j=0}^{m} a_j \lambda^j$$

with multiplicity $m(\lambda)$. *This polynomial is called the* characteristic polynomial of *the differential operator* $A(D)$. *It can be obtained from the* "operator polynomial" $A(D)$ *by replacing the indeterminate* D *by* λ.

It is easy now to determine a fundamental set of solutions for the homogeneous equation $A(D)u = 0$.

(14.5) Theorem. *Assume that*

$$A(D) = \sum_{j=0}^{m} a_j D^j, \quad a_j \in \mathbb{K}, \ a_m \neq 0,$$

is a linear differential operator with constant coefficients, and let $\lambda_1, \ldots, \lambda_k$ *denote the distinct roots of the characteristic polynomial*

$$A(\lambda) := \sum_{j=0}^{m} a_j \lambda^j$$

with multiplicities $m(\lambda_j)$. *Then the functions*

$$e^{\lambda_l t}, te^{\lambda_l t}, \ldots, t^{m(\lambda_l)-1} e^{\lambda_l t}, \quad 1 \leq l \leq k, \quad t \in \mathbb{R}, \tag{4}$$

form a fundamental set of solutions for the homogeneous equation $A(D)u = 0$.

If all the coefficients are real, $a_j \in \mathbb{R}$, *we obtain a real fundamental set of solutions by decomposing* (4) *into real and imaginary parts. Thus, if*

$$\lambda_1, \ldots, \lambda_r, \alpha_1 \pm i\omega_1, \ldots, \alpha_s \pm i\omega_s,$$

are all the distinct zeros of the characteristic polynomial, where $\lambda_1, \ldots, \lambda_r, \alpha_1, \ldots, \alpha_s \in \mathbb{R}$, $\omega_1 > 0, \ldots, \omega_s > 0$, *then the functions*

$$e^{\lambda_\rho t}, \ te^{\lambda_\rho t}, \ \ldots, \ t^{m(\lambda_\rho)-1} e^{\lambda_\rho t}, \quad 1 \leq \rho \leq r$$
$$e^{\alpha_\sigma t} \cos(\omega_\sigma t), \ e^{\alpha_\sigma t} \sin(\omega_\sigma t), \ te^{\alpha_\sigma t} \cos(\omega_\sigma t), \ te^{\alpha_\sigma t} \sin(\omega_\sigma t), \ \ldots$$
$$t^{m_\sigma-1} e^{\alpha_\sigma t} \cos(\omega_\sigma t), \ t^{m_\sigma-1} e^{\alpha_\sigma t} \sin(\omega_\sigma t), \quad 1 \leq \sigma \leq s,$$

with $m_\sigma := m(\alpha_\sigma + i\omega_\sigma)$, *form a real-valued fundamental set of solutions.*

Proof. With the above considerations, the assertion follows immediately from theorem (12.7) and theorem (12.10). □

(14.6) Remarks. (a) One is led to the characteristic polynomial $A(\lambda)$ of the differential operator $A(D)$ with constant coefficients directly by the classical approach of seeking a solution u of $A(D)u = 0$ of the form $u(t) = e^{\lambda t}, t \in \mathbb{R}$. We then evidently have

$$A(D)u = A(\lambda)u$$

and so $A(D)u = 0 \longleftrightarrow A(\lambda) = 0$.

(b) This approach delivers the same number of distinct solutions as the characteristic polynomial has distinct roots. One immediately sees that these solutions are linearly independent (in $C^m(\mathbb{R}, \mathbb{K})$). If $k < m$, that is, if this approach does not furnish a fundamental set of solutions, then the standard (classical) argument to obtain the solutions $t^n e^{\lambda t}$ goes as follows:

Let λ_1 denote a double root of $A(\lambda)$. We then consider a "nearby" differential operator $\tilde{A}(D)$ such that its characteristic polynomial has two simple roots $\tilde{\lambda}_1$ and $\tilde{\lambda}_2$ "near" λ_1. Then $e^{\tilde{\lambda}_1 t}$ and $e^{\tilde{\lambda}_2 t}$ are two linearly independent solutions of $\tilde{A}(D)u = 0$, that is,

$$\tilde{W} := \mathrm{span}\{e^{\tilde{\lambda}_1 t}, e^{\tilde{\lambda}_2 t}\}$$

is a two-dimensional vector subspace of the solution space of $\tilde{A}(D)u = 0$. Now clearly we have

$$\tilde{W} = \mathrm{span}\left\{e^{\tilde{\lambda}_1 t}, \frac{e^{\tilde{\lambda}_2 t} - e^{\tilde{\lambda}_1 t}}{\tilde{\lambda}_2 - \tilde{\lambda}_1}\right\}.$$

If we now "deform" $\tilde{A}(D)$ into $A(D)$ so that $\tilde{\lambda}_1$ and $\tilde{\lambda}_2$ approach λ_1, then "$\tilde{W}$ approaches W in the limit," that is, it "converges" to the two-dimensional subspace

$$W = \mathrm{span}\{e^{\lambda_1 t}, t e^{\lambda_1 t}\}.$$

One verifies directly that $t \mapsto t e^{\lambda_1 t}$ is a solution of $A(D)u = 0$. $\square$

(14.7) Examples. (a) We consider the real, second order linear differential equation with constant coefficients

$$a\ddot{u} + b\dot{u} + cu = 0, \quad a, b, c \in \mathbb{R}, \quad a > 0. \tag{5}$$

The corresponding characteristic polynomial

$$a\lambda^2 + b\lambda + c = 0$$

has the roots

$$\lambda_{1,2} = \frac{-b \pm \sqrt{b^2 - 4ac}}{2a}.$$

Case 1: $b^2 > 4ac$, $c > 0$. In this case $\lambda_1 < \lambda_2$ are real and the general (real) solution is given by

$$u(t) = c_1 e^{\lambda_1 t} + c_2 e^{\lambda_2 t}, \quad \text{where } c_1, c_2 \in \mathbb{R}.$$

If $b > 0$, then $\lambda_1 < \lambda_2 < 0$ and all solutions decay exponentially, that is, we have the case of *damping*.

In the *phase plane* $(x = u, y = \dot{u})$ we have a stable node.

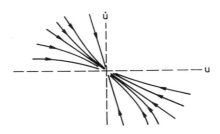

If, however, $b < 0$, then $0 < \lambda_1 < \lambda_2$ and we have *excitation*. In the phase plane we have an unstable node.

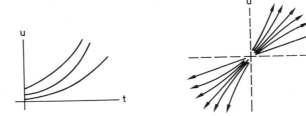

Case 2: $b^2 = 4ac$, $b \neq 0$. In this case the general solution of (5) is given by

$$u(t) = c_1 e^{\lambda t} + c_2 t e^{\lambda t}, \quad t \in \mathbb{R},$$

where $\lambda = -b/2a \neq 0$, and we call this the *aperiodic critical case*. If $b > 0$ (the case with damping), then $\lambda < 0$ and the phase plane is a stable improper node.

If, on the other hand, $b < 0$ (the case with excitation), then $\lambda > 0$ and the phase plane is an unstable improper node.

Case 3: $b^2 < 4ac$, $b \neq 0$. In this case we set

$$\alpha := -\frac{b}{2a} \quad \text{and} \quad \omega := \frac{\sqrt{4ac - b^2}}{2a}.$$

Then the general solution of (5) is given by

$$u(t) = e^{\alpha t}[c_1 \cos(\omega t) + c_2 \sin(\omega t)], \quad t \in \mathbb{R},$$

where $c_1, c_2 \in \mathbb{R}$. If $b > 0$, hence $\alpha < 0$, then we are dealing with a *damped vibration*. The phase plane is then a stable spiral.

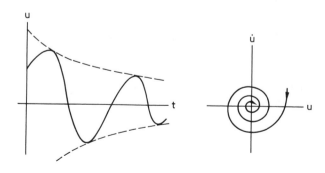

If, however, $b < 0$, hence $\alpha > 0$, then we are dealing with an *excited vibration*. The phase plane is then an unstable spiral.

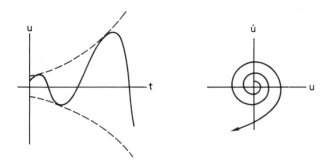

Case 4: $b = 0$, $c > 0$. In this case the general solution of (5) is given by

$$u(t) = c_1 \cos(\omega t) + c_2 \sin(\omega t), \quad t \in \mathbb{R},$$

where $c_1, c_2 \in \mathbb{R}$ and $\omega = \sqrt{c/a}$. Therefore all solutions are ω-periodic and the phase plane is a center.

In this case (5) is equivalent to the differential equation

$$\ddot{u} + \omega^2 u = 0,$$

the equation of the *harmonic oscillator*.

Case 5: $c < 0$. Now the general solution is given by

$$u(t) = c_1 e^{\lambda_1 t} + c_2 e^{\lambda_2 t}, \quad t \in \mathbb{R},$$

where $c_1, c_2 \in \mathbb{R}$ and $\lambda_1 < 0 < \lambda_2$. The phase plane is a saddle and the stability depends on the choice of the constants c_1, c_2, that is, it depends on the initial condition.

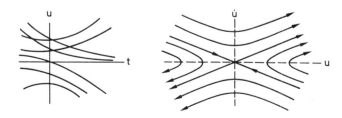

Case 6: $b \neq 0$, $c = 0$. The general solution is given by

$$u(t) = c_1 e^{\lambda t} + c_2, \quad t \in \mathbb{R}, \tag{6}$$

where $c_1, c_2 \in \mathbb{R}$ and $\lambda := -b/a$. If $c_1 \neq 0$, the solutions converge to the constant solutions $u = c_2$, in case $b > 0$, and they grow exponentially in case $b < 0$. From (6) we obtain the flow

$$u(t) = \lambda^{-1}\dot{u}(0)e^{\lambda t} + c, \quad \dot{u}(t) = \dot{u}(0)e^{\lambda t}, \quad c \in \mathbb{R},$$

in the phase plane. Hence every point on the u-axis is a critical point. We find the following qualitative behavior when $b > 0$.

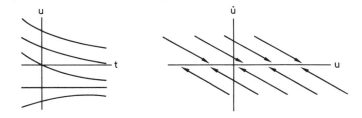

If, however, $b < 0$, hence $\lambda > 0$, then we have instability.

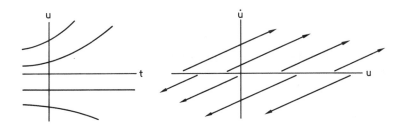

Case 7: $b = 0$, $c = 0$. Of course now the solutions are the straight lines

$$u(t) = c_1 t + c_2, \quad t \in \mathbb{R},$$

where $c_1, c_2 \in \mathbb{R}$ and the flow in the phase plane is given by

$$\begin{bmatrix} u \\ \dot{u} \end{bmatrix}(t) = \begin{bmatrix} c_2 \\ c_1 \end{bmatrix} + \begin{bmatrix} c_1 \\ 0 \end{bmatrix} t, \quad \forall t \in \mathbb{R}.$$

Hence every point on the u-axis is a stationary point and the noncritical orbits are the lines parallel to the u-axis.

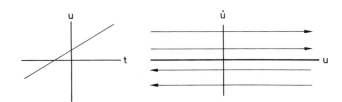

The differential equation

$$\ddot{u} + 2\alpha\dot{u} + \omega^2 u = 0, \quad \alpha, \omega > 0,$$

which is a special case of (5) with $a = 1, b = -2\alpha$, and $c = \omega^2$, plays (for $\alpha^2 \leq \omega^2$) an important role in classical mechanics as the equation of the *damped harmonic oscillator*. In this connection the "damping term" $-2\alpha\dot{u}$ corresponds to a *friction force* which causes the harmonic motion to decay to zero as $t \rightarrow \infty$.

(b) By remark (14.4 b) and using the method of variation of constants, one can easily solve the nonhomogeneous equation

$$a\ddot{u} + b\dot{u} + cu = f(t), \quad t \in \mathbb{R},$$

where $a, b, c \in \mathbb{R}$. However, in many cases – in particular also for higher order equations – it is easier to produce a particular solution directly (e.g. by guessing or by looking for a solution of a special form). This is always possible if $f(t)$ is a sum of *quasipolynomials*, that is, the sum of expressions of the form

$$e^{\lambda t} \sum_{k=0}^{m} \alpha_k t^k.$$

One can in fact show that the following *proposition* holds: *If the right-hand side of the differential equation*

$$\sum_{j=0}^{m} a_j D^j u = f, \quad a_j \in \mathbb{K},$$

is a sum of quasipolynomials, then every solution of this equation also has this form.
For details and examples we refer to section 26 of the book by Arnold [2]. □

Problems

1. Show that the general solution of the equation of the harmonic oscillator $\ddot{u} + \omega_0^2 u = 0$, with $\omega_0 > 0$, can be written in the form

$$u(t) = \beta \sin(\omega_0 t + \gamma), \quad \forall t \in \mathbb{R},$$

for some $\beta, \gamma \in \mathbb{R}$.

2. Show that the differential equation of "forced undamped vibrations"

$$\ddot{u} + \omega_0^2 u = c \sin(\omega t), \quad t \in \mathbb{R},$$

where $\omega_0, \omega > 0$, $\omega_0 \neq \omega$ and $c \in \mathbb{R}$, has the general solution

$$u(t) = A \sin(\omega_0 t + \gamma) + \frac{c}{\omega_0^2 - \omega^2} \sin(\omega t), \quad t \in \mathbb{R},$$

and illustrate the solutions graphically, particularly near the "resonance case" $\omega \approx \omega_0$.

3. Determine the general solution of the differential equation of "forced damped vibrations"

$$\ddot{u} + 2\alpha\dot{u} + \omega_0^2 u = c \sin(\omega t), \quad t \in \mathbb{R},$$

where $\alpha, \omega_0, \omega > 0$, $c \in \mathbb{R}$, and discuss the solution.

Chapter IV
Qualitative Theory

The main goal of this chapter is to obtain a good understanding of the qualitative behavior of flows, induced by ordinary differential equations, near critical points. This problem is closely related to the long-time behavior, the so-called stability theory.

In the first section we will prove the "principle of linearized stability," which makes it possible to deduce information about the Liapunov stability of a critical point from the spectrum of the linearized vector field at this critical point. Following this section, we consider semiflows in the large and obtain criteria for the positive invariance of sets. These concepts are closely related to general stability considerations which are dealt with in the next section.

The concept of Liapunov function is central for Liapunov's stability theory and will be thoroughly discussed. Primarily, we treat the autonomous case since nonautonomous equations can be reduced to the autonomous case by extending the phase space. In order to work out the general ideas on the one hand, and since these concepts are also of great importance for parabolic differential equations on the other, we will develop the relevant theory for general semiflows on metric spaces as much as possible.

In the final section of this chapter we consider, analogous to the classification of linear flows, hyperbolic critical points of differentiable vector fields. We will prove the linearization theorem of Grobman and Hartman, as well as the theorem about the locally stable and unstable manifolds.

15. Liapunov Stability

In this section let $E = (E, |\cdot|)$ denote a finite dimensional Banach space, let $D \subseteq E$ be open, and let $J \subseteq \mathbb{R}$ denote an open interval with $\mathbb{R}_+ \subseteq J$. Moreover, suppose

$$f \in C^{0,1-}(J \times D, E),$$

and let $u \in C^{1-}(\mathcal{D}(f), D)$ denote the solution of the IVP

$$\dot{x} = f(t, x), \quad x(\tau) = \xi, \quad (\tau, \xi) \in J \times D,$$

(cf. theorem (8.3)).

Assume now that $f(\cdot, 0) = 0$ (which of course implies that $0 \in D$), so that the differential equation $\dot{x} = f(t, x)$ has the global *zero solution* $x = 0$. The zero solution is called (*Liapunov*) *stable* if for every neighborhood U of 0 and every $\tau \in J$ there exists a neighborhood V of 0 such that

$$u(t, \tau, \xi) \in U, \quad \forall\ t \in [\tau, t^+(\tau, \xi)), \quad \forall \xi \in V.$$

If the zero solution is not stable, it is called *unstable* (in the sense of Liapunov).

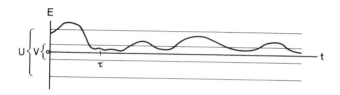

The zero solution is *attractive* if for every $\tau \in J$ there exists a neighborhood W of 0 such that $\xi \in W$ implies

$$t^+(\tau, \xi) = \infty \quad \text{and} \quad \lim_{t \to \infty} u(t, \tau, \xi) = 0. \tag{1}$$

If the zero solution is stable and attractive, it is called *asymptotically stable*.

Finally, we say that the zero solution is *uniformly stable* or *uniformly attractive* if the choice of the neighborhoods V or W, respectively, is independent of $\tau \in J$, and if the limit in (1) is uniform with respect to $(\tau, \xi) \in J \times W$. The last condition means that for every neighborhood $\tilde{U}$ of 0 there exists some $T > 0$ so that $u(t, \tau, \xi) \in \tilde{U}$ for all $t > \tau + T$ and all $(\tau, \xi) \in J \times W$.

The zero solution is *uniformly asymptotically stable* if it is both uniformly stable and uniformly attractive.

(15.1) Remarks. (a) If U is chosen to be a neighborhood of 0 with $\text{dist}(U, \partial D) > 0$, then theorem (7.6) implies that $t^+(\tau, \xi) = \infty$ for all $\xi \in V$. *Consequently, the zero solution is stable if and only if for every neighborhood U of 0 and every $\tau \in J$ there exists a neighborhood V of 0 such that*

$$t^+(\tau, \xi) = \infty \quad \text{and} \quad u(t, \tau, \xi) \in U, \quad \forall (t, \xi) \in [\tau, \infty) \times V.$$

If the zero solution is unstable, then it is possible to find, for each $\tau \in J$, initial values ξ arbitrarily close to 0 and such that $t^+(\tau, \xi) < \infty$.

(b) The concept of stability is clearly an extension of the "continuous dependence on initial values." It means that for every $\tau \in J$ we have:

$$\lim_{\xi \to 0} u(t, \tau, \xi) = 0, \tag{2}$$

uniformly in $t \in [\tau, \infty)$. From theorem (8.3) one can deduce only that the limit in (2) exists uniformly on compact subintervals of J.

(c) *The concepts of stability and attractivity are independent of $\tau \in J$ in the following sense: If $x = 0$ is stable, resp. attractive, with respect to $\tau \in J$, then it has the same property with respect to $\sigma \in J$.*

Indeed, if $x = 0$ is stable with respect to $\tau \in J$, then for every neighborhood U of 0 there exists a neighborhood V of 0 such that $u(t, \tau, \xi) \in U$ for all $(t, \xi) \in [\tau, \infty) \times V$. If $\sigma < \tau$, then there exists a neighborhood $\tilde{V}$ of 0 such that $u(t, \sigma, \eta) \in V$ for all $(t, \eta) \in [\sigma, \tau] \times \tilde{V}$. This follows easily from the continuity of u and the compactness of the interval $[\sigma, \tau]$. Hence we have $u(t, \sigma, \eta) \in U$ for all $(t, \eta) \in [\sigma, \infty) \times \tilde{V}$.

If $\tau < \sigma$, then $\tilde{V} := u(\sigma, \tau, V)$ is a neighborhood of 0 since $u(\sigma, \tau, \cdot)$ is a homeomorphism. So also in this case we have $u(t, \sigma, \eta) \in U$ for all $(t, \eta) \in [\sigma, \infty) \times \tilde{V}$.

(d) *Stability and attractivity are independent concepts.* For instance, a center (cf. section 13) is stable, but not attractive. Conversely, one can show (e.g., Hahn [1], Ch. 40) that there exists an autonomous system in $\mathbb{R}^2$ with a phase portrait of the following form:

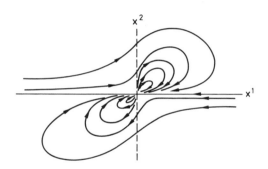

In this case the zero solution is attractive and unstable.

(e) *If f is either independent of t or periodic in t, then the property of being stable or asymptotically stable implies the uniformity of these properties.*

(f) If $\bar{u} := u(\cdot, \tau_0, \xi_0)$ is any global solution of $\dot{x} = f(t, x)$, the differential equation

$$\dot{y} = f(t, y + \bar{u}(t)) - f(t, \bar{u}(t)) \tag{3}$$

has the global zero solution and y "measures the deviation from $\bar{u}$." For this reason we say that *the solution $\bar{u}$ is stable, attractive, etc. if the trivial solution of (3) has the corresponding properties.* In particular, if f is independent of t and if $f(x_0) = 0$, that is, if x_0 is a critical point, then x_0 is called stable, attractive, etc. if the constant solution $\bar{u}(t) := x_0$, $t \in \mathbb{R}$, has the corresponding properties.

(g) Based on remarks (7.10) and (8.5 a), the remarks above also remain valid in the infinite dimensional case. □

First, we now study the stability behavior of autonomous linear differential equations.

(15.2) Theorem. *Let $A \in \mathcal{L}(E)$. Then the zero solution of the linear differential equation $\dot{x} = Ax$ is stable if and only if the following holds:*

(i) $\operatorname{Re} \sigma(A) \leq 0$,
(ii) *every $\lambda \in \sigma(A)$ with $\operatorname{Re} \lambda = 0$ is a semisimple eigenvalue.*

The zero solution is asymptotically stable if and only if

$$\operatorname{Re} \sigma(A) < 0.$$

Proof. Remark (15.1 c) shows that it suffices to consider the case $\tau = 0$, i.e., the flow $e^{tA}\xi$, $\xi \in E$.
    Let $\alpha := \sup\{|e^{tA}| \mid t \in \mathbb{R}_+\} < \infty$. For $\epsilon > 0$ we then have

$$|e^{tA}\xi| \leq |e^{tA}||\xi| < \epsilon, \qquad \forall(t, \xi) \in \mathbb{R}_+ \times \mathbb{B}(0, \epsilon/\alpha),$$

that is, the zero solution is stable. If $\{x_1, \ldots, x_m\}$ is a basis of E, then it follows from $e^{tA}\xi = \sum \xi^i e^{tA} x_i$, where $\xi = \sum \xi^i x_i$, from the equivalence of the norms, and from the definition of the operator norm, that $\alpha < \infty$ holds if and only if every solution $e^{tA} x_i$, $i = 1, \ldots, m$, is bounded. This is the case exactly when every solution of $\dot{x} = Ax$ is bounded and so, by the boundedness criterion (12.14), if and only if (i) and (ii) hold.
    If one of the conditions (i) or (ii) fails to hold, then (12.14) implies the existence of some $x \in E$ such that $|e^{tA}x| \to \infty$ as $t \to \infty$. Then also $|e^{tA}(\epsilon x)|$ is unbounded for every $\epsilon > 0$ and hence the zero solution is unstable.
    The second part of the assertion now follows immediately from the above and the stability criterion (12.11). $\qquad\qquad\qquad\qquad\qquad\qquad\square$

Linearized Stability

Next, we consider "perturbed linear systems" of the form

$$\dot{x} = Ax + g(t, x), \tag{4}$$

where g is a small perturbation in some appropriate sense. More concretely, in what follows we will show that if

$$g(t, x) = o(|x|) \quad \text{as} \quad x \to 0,$$

uniformly in $t \in J$, the perturbed linear system (4) has nearly the same asymptotic stability properties as the unperturbed linear equation $\dot{x} = Ax$ (the "linearization").

We begin with a simple, but crucial remark. If $g \in C^{0,1-}(J \times D, E)$ and if $u(t) := u(t, \tau, \xi)$, $t \in J(\tau, \xi)$, is any solution of the differential equation (4), then the nonhomogeneous linear equation

$$\dot{x} = Ax + g(t, u(t)), \quad t \in J(\tau, \xi),$$

has the unique solution u on $J(\tau, \xi)$ such that $u(\tau) = \xi$. Thus it follows from the variation of constants formula (12.3 c) that u satisfies the (nonlinear) *integral equation*

$$u(t) = e^{(t-\tau)A} \xi + \int_{\tau}^{t} e^{(t-s)A} g(s, u(s)) ds, \quad t \in J(\tau, \xi). \tag{5}$$

This integral equation is the basis for the following stability theorem – which is essentially due to Liapunov. (It is also the basis for numerous existence proofs in the analogous case of infinite dimensional evolution equations, e.g., for parabolic systems.)

(15.3) Theorem *(Asymptotic Stability). Let $A \in \mathcal{L}(E)$ be such that*

$$\operatorname{Re} \sigma(A) < 0.$$

Moreover, let $g \in C^{0,1-}(J \times D, E)$ be such that

$$g(t, x) = o(|x|) \quad as \quad x \to 0, \tag{6}$$

uniformly in $t \in J$. Then the zero solution of the perturbed linear equation

$$\dot{x} = Ax + g(t, x)$$

is uniformly asymptotically stable.

Proof. By (13.2) there exist positive constants α and β so that

$$|e^{tA}| \le \beta e^{-\alpha t}, \quad \forall t \ge 0,$$

where we may assume that $\beta > 1$. Hence from (5) we deduce the estimate

$$|u(t)| \le \beta e^{-\alpha(t-\tau)} |\xi| + \beta \int_{\tau}^{t} e^{-\alpha(t-s)} |g(s, u(s))| ds, \tag{7}$$

for all $\tau \le t < t^+(\tau, \xi)$.

Let now $\epsilon \in (0, \alpha)$ be arbitrary. By (6) there exists some $\delta \in (0, \epsilon)$ such that

$$|g(t, x)| \le (\epsilon/\beta) |x| \quad \text{for all} \quad |x| \le \delta, \quad t \ge \tau. \tag{8}$$

We now claim that $|u(t)| < \delta < \epsilon$ for all $|\xi| < \delta/\beta$ and $t \in [\tau, t^+(\tau, \xi))$, from which the uniform stability of the zero solution follows. Otherwise, there would exist some $\xi \in \mathbb{B}(0, \delta/\beta)$ and some $\bar{t} \in (\tau, t^+(\tau, \xi))$ such that

$$\bar{t} = \inf \{t \in [\tau, t^+(\tau, \xi)) \mid |u(t)| = \delta\}.$$

For $\tau \le t \le \bar{t}$ we then obtain from (7) and (8) that

$$|u(t)| \le \delta e^{-\alpha(t-\tau)} + \epsilon \int_\tau^t e^{-\alpha(t-s)} |u(s)| ds$$

which is equivalent to

$$e^{\alpha t} |u(t)| \le \delta e^{\alpha \tau} + \epsilon \int_\tau^t e^{\alpha s} |u(s)| ds.$$

Gronwall's lemma (6.2) now implies that

$$|u(t)| \le \delta e^{-(\alpha - \epsilon)(t-\tau)} \quad \text{for all} \quad \tau \le t \le \bar{t}, \tag{9}$$

hence

$$\delta = |u(\bar{t})| \le \delta e^{-(\alpha - \epsilon)(\bar{t} - \tau)} < \delta,$$

which is impossible.

Finally, given $\xi \in \mathbb{B}(0, \delta/\beta)$, it follows from the fact that $|u(t)| < \delta$ for all $t \in [\tau, t^+(\tau, \xi))$ that (9) is valid for all $\tau \le t < t^+(\tau, \xi)$. Hence the zero solution is uniformly attractive. $\square$

To prove the corresponding instability theorem, we need the following lemma.

(15.4) Lemma. *Let* $A \in \mathcal{L}(E)$ *be such that*

$$\alpha < Re\, \sigma(A) < \beta.$$

Then there exists a Euclidean norm $\| \cdot \|$ *on* E *such that for the corresponding inner product* $(\cdot \mid \cdot)$ *we have*

$$\alpha \|x\|^2 \le Re(Ax \mid x) \le \beta \|x\|^2, \quad \forall x \in E.$$

Proof. First we consider the case $\mathbb{K} = \mathbb{C}$. From the proof of lemma (13.1) we know that A has the form $A = D + N$, where $D = \text{diag}\,[\mu_1, \dots, \mu_m]$ and $\mu_1, \dots, \mu_m$ are the eigenvalues of A, counted according to their multiplicity, and that for every $\epsilon > 0$ there exists a Euclidean norm $\| \cdot \|$ on E so that $\|N\| \le \epsilon$. We now fix $\epsilon > 0$ (and hence $\| \cdot \|$) so that

$$\epsilon \le \min\{\beta - \max[Re\, \sigma(A)], \min[Re\, \sigma(A)] - \alpha\}.$$

Since $(Dx \mid x) = \sum \mu_j |x^j|^2$, where $x^1, \dots, x^m$ are the coordinates of x with respect to the orthonormal basis used (for the construction of the norm), it follows that

$$\min[Re\, \sigma(A)]\|x\|^2 \le Re(Dx \mid x) \le \max[Re\, \sigma(A)]\|x\|^2.$$

Moreover, since $Re(Ax \mid x) = Re\,(Dx \mid x) + Re(Nx \mid x)$ and $|Re(Nx \mid x)| \le \|N\|\,\|x\|^2 \le \epsilon \|x\|^2$, the assertion follows from

$$Re(Dx \mid x) - \epsilon\|x\|^2 \le Re(Ax \mid x) \le Re(Dx \mid x) + \epsilon\|x\|^2$$

and the choice of ϵ.

Now let $\mathbb{K} = \mathbb{R}$. Then we can apply the above to the complexification $A_{\mathbb{C}}$ in $E_{\mathbb{C}}$. The Hilbert norm $\|\cdot\|_{\mathbb{C}}$ on $E_{\mathbb{C}}$ induces (by restricting onto $E \subseteq E_{\mathbb{C}}$) a Hilbert norm $\|\cdot\|$ on E (cf. the proof of lemma (13.1)). For the corresponding inner products we obtain

$$Re(\xi \mid \eta)_{\mathbb{C}} = (\|\xi + \eta\|_{\mathbb{C}}^2 - \|\xi - \eta\|_{\mathbb{C}}^2)/4, \quad \forall \xi, \eta \in E_{\mathbb{C}},$$

and

$$(x \mid y) = (\|x + y\|^2 - \|x - y\|^2)/4, \quad \forall x, y \in E.$$

From this it follows that

$$\alpha\|x\|^2 = \alpha\|x\|_{\mathbb{C}}^2 \le Re(A_{\mathbb{C}}x \mid x)_{\mathbb{C}} = (Ax \mid x) \le \beta\|x\|_{\mathbb{C}}^2 = \beta\|x\|^2$$

for all $x \in E$. $\qquad\qquad\qquad\qquad\qquad\qquad\qquad\qquad\qquad\qquad\square$

(15.5) Theorem (*Instability*). *Assume the operator $A \in \mathcal{L}(E)$ has at least one eigenvalue with positive real part. Moreover, let $g \in C^{0,1^-}(J \times D, E)$ be such that*

$$g(t, x) = o(|x|) \quad as \quad x \to 0, \tag{10}$$

uniformly in $t \in J$. Then the zero solution of the perturbed linear equation

$$\dot{x} = Ax + g(t, x) \tag{11}$$

is unstable.

Proof. By assumption, the unstable spectrum $\sigma_u(A)$ is nonempty. Consequently there exists some γ so that $0 < \gamma < Re\,\sigma_u(A)$. Since $\sigma(A - \gamma) = \sigma(A) - \gamma$, it follows that the neutral spectrum of $A_\gamma := A - \gamma$, $\sigma_n(A_\gamma)$, is empty and thus A_γ induces a hyperbolic linear flow e^{tA_γ}. Theorem (13.4) implies that there is a direct sum decomposition

$$E = E_- \oplus E_+ \tag{12}$$

which decomposes A_γ, $A_\gamma = (A_\gamma)_- \oplus (A_\gamma)_+$, such that $\sigma_s(A_\gamma) = \sigma((A_\gamma)_-)$ and $\sigma_u(A_\gamma) = \sigma((A_\gamma)_+)$. Evidently (12) also decomposes the operator A,

$$A = A_- \oplus A_+,$$

and we have $\sigma(A_+) = \sigma_u(A)$, as well as $\sigma(A_-) = \sigma_s(A) \cup \sigma_n(A)$. For some appropriate $\alpha > 0$ we therefore have

$$Re\,\sigma(A)_- \le 0 \quad \text{and} \quad Re\,\sigma(A_+) > \alpha > 0.$$

We now choose some fixed $\beta \in (0, \alpha)$. Lemma (15.4) then implies the existence of Hilbert norms, $\|\cdot\|_+$ on E_+ and $\|\cdot\|_-$ on E_-, such that

$$Re(A_-x_- \mid x_-)_- \leq \beta\|x_-\|_-^2, \quad \forall x_- \in E_-, \tag{13}$$

and

$$Re(A_+x_+ \mid x_+)_+ \geq \alpha\|x_+\|_+^2, \quad \forall x_+ \in E_+. \tag{14}$$

Clearly

$$(x_- + x_+ \mid y_- + y_+) := (x_- \mid y_-)_- + (x_+ \mid y_+)_+$$

defines an inner product on $E = E_- \oplus E_+$ and therefore it induces a norm $\|\cdot\|$ given by

$$\|x\|^2 = \|x_- + x_+\|^2 = \|x_-\|_-^2 + \|x_+\|_+^2, \quad \forall x = x_- + x_+ \in E. \tag{15}$$

Finally, we set

$$\Phi(x) := (\|x_+\|^2 - \|x_-\|^2)/2 = (\|Px\|^2 - \|Qx\|^2)/2, \quad \forall x \in E,$$

where $P : E \rightarrow E_+$ and $Q : E \rightarrow E_-$ are the projections corresponding to (12) and $\gamma := (\alpha - \beta)/4$. Using (10), there exists some $\delta > 0$ so that

$$\|g(t, x)\| \leq \gamma\|x\| \quad \text{for all} \quad \|x\| \leq \delta. \tag{16}$$

Let now u denote a solution of (11) satisfying $\|u(0)\| < \delta$ and $\Phi(u(0)) > 0$. Setting $\varphi(t) := \Phi(u(t))$ and using (13), (14) and (16), it then follows that for all $t \geq 0$ with $\|u(t)\| \leq \delta$ we have

$$\begin{aligned}
\dot{\varphi}(t) =& Re(Pu(t) \mid P\dot{u}(t)) - Re(Qu(t) \mid Q\dot{u}(t)) \\
=& Re(A_+u_+(t) \mid u_+(t)) - Re(A_-u_-(t) \mid u_-(t)) \\
& + Re(Pu(t) \mid Pg(t, u(t))) - Re(Qu(t) \mid Qg(t, u(t))) \\
\geq& \alpha\|Pu(t)\|^2 - \beta\|Qu(t)\|^2 - \gamma\|P\| \, \|Pu(t)\| \, \|u(t)\| \\
& - \gamma\|Q\| \, \|Qu(t)\| \, \|u(t)\|.
\end{aligned}$$

By (15) we have

$$\|Px\| \leq \|x\| \quad \text{and} \quad \|Qx\| \leq \|x\|, \quad \forall x \in E,$$

and so it follows that $\|P\| \leq 1$ and $\|Q\| \leq 1$, hence

$$\dot{\varphi}(t) \geq \alpha\|Pu(t)\|^2 - \beta\|Qu(t)\|^2 - \gamma(\|Pu(t)\| + \|Qu(t)\|)\|u(t)\|. \tag{17}$$

From $\varphi(0) > 0$ it follows that for small $t > 0$ we have $\varphi(t) = \Phi(u(t)) \geq 0$ and hence $\|Qu(t)\| \leq \|Pu(t)\|$. So it follows from (15) that

$$\|u(t)\| \leq 2\|Pu(t)\|.$$

We now obtain from (17) that

$$\dot{\varphi}(t) \geq (\alpha - 4\gamma)\|Pu(t)\|^2 - \beta\|Qu(t)\|^2 = 2\beta\varphi(t) \tag{18}$$

for all $t \geq 0$ satisfying $\|u(t)\| \leq \delta$ and $\varphi(t) \geq 0$. Integrating (18) we see that

$$\varphi(t) \geq \varphi(0)e^{2\beta t} \tag{19}$$

for all those t values above. From this it follows in particular that $\varphi(t) > 0$ for all $t \geq 0$ such that $\|u(t)\| \leq \delta$. In other words: No solution of (11) with initial value

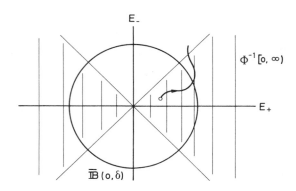

in $\mathbb{B}(0, \delta) \cap \Phi^{-1}[0, \infty)$ leaves the "double cone" $\Phi^{-1}[0, \infty)$ before it leaves the ball $\bar{\mathbb{B}}(0, \delta)$. Moreover, it follows from (19) that every solution with initial value in $\mathbb{B}(0, \delta) \cap \Phi^{-1}[0, \infty)$ and distinct from zero reaches the boundary of $\mathbb{B}(0, \delta)$. This proves that the zero solution is unstable.   □

As a simple corollary of theorems (15.3) and (15.5) we obtain the following *principle of linearized stability* for critical points of autonomous differential equations. This fundamental principle is one of the best-known stability and instability criteria, which has numerous applications, particularly in the applied sciences.

(15.6) Theorem. *Let $f \in C^1(D, E)$ be such that $f(x_0) = 0$. If*

$$Re \, \sigma(Df(x_0)) < 0,$$

then the autonomous differential equation $\dot{x} = f(x)$ has the asymptotically stable critical point x_0. If

$$\sigma(Df(x_0)) \cap \{z \in \mathbb{C} \mid Re \, z > 0\} \neq \emptyset,$$

then x_0 is unstable.

Proof. Let $A := Df(x_0) \in \mathcal{L}(E)$ and $g(y) := f(y+x_0) - Df(x_0)y$, then $g(y) = o(|y|)$ as $y \to 0$ and $\dot{y} = f(y + x_0) = Ay + g(y)$. Therefore the assertion follows immediately from (15.1 f), theorem (15.3) and theorem (15.5).   □

(15.7) Examples. (a) From section 1 we consider the *predator-prey model* with bounded growth:

$$\dot{x} = (\alpha - \beta y - \lambda x)x$$
$$\dot{y} = (\delta x - \gamma - \mu y)y, \tag{20}$$

where $\alpha, \beta, \gamma, \delta, \lambda, \mu$ are positive constants. This system possesses the critical points $(0,0)$, $(0, -\gamma/\mu)$, $(\alpha/\lambda, 0)$ and $\left(\frac{\alpha\mu+\beta\gamma}{\lambda\mu+\beta\delta}, \frac{\alpha\delta-\lambda\gamma}{\lambda\mu+\beta\delta} \right)$, where the last point is the point of intersection z of the straight lines L and M (cf. Figs. 5 and 6 of section 1). Making the obvious identifications, we have

$$Df(x,y) = \begin{bmatrix} \alpha - \beta y - 2\lambda x & -\beta x \\ \delta y & \delta x - \gamma - 2\mu y \end{bmatrix},$$

and hence

$$Df(0,0) = \begin{bmatrix} \alpha & 0 \\ 0 & -\gamma \end{bmatrix}, \quad Df\left(0, -\frac{\gamma}{\mu}\right) = \begin{bmatrix} \alpha + \beta\gamma/\mu & 0 \\ * & \gamma \end{bmatrix},$$

$$Df\left(\frac{\alpha}{\lambda}, 0\right) = \begin{bmatrix} -\alpha & * \\ 0 & \alpha\delta/\lambda - \gamma \end{bmatrix}, \quad Df(\xi,\eta) = \begin{bmatrix} -\lambda\xi & -\beta\xi \\ \delta\eta & -\mu\eta \end{bmatrix},$$

where $z = (\xi, \eta)$. It easily follows from this and from theorem (15.6) that the critical points $(0,0)$ and $(0, -\gamma/\mu)$ are always unstable. The critical point $(\alpha/\lambda, 0)$ is asymptotically stable when $\alpha/\lambda < \gamma/\delta$ and unstable when $\alpha/\lambda > \gamma/\delta$, which agrees with Figs. 5 and 6 of section 1. For the eigenvalues $\lambda_{1,2}$ of $Df(\xi,\eta)$ one easily calculates

$$\lambda_{1,2} = \frac{-(\lambda\xi + \mu\eta) \pm \sqrt{(\lambda\xi + \mu\eta)^2 - 4\xi\eta(\lambda\mu + \delta\beta)}}{2}.$$

In case the straight lines L and M intersect in the first quadrant, i.e., when $\xi > 0$ and $\eta > 0$, which is of interest in applications, one easily obtains from this formula that $\mathrm{Re}\,\sigma(Df(\xi,\eta)) < 0$. From the explicit expression for (ξ, η) it follows that the critical point z is asymptotically stable when $\alpha/\lambda > \gamma/\delta$, which also agrees with Figs. 7 and 8 of section 1.

(b) If we set $\lambda = \mu = 0$ in (20), we obtain the Volterra-Lotka equations of section 1:

$$\dot{x} = (\alpha - \beta y)x$$
$$\dot{y} = (\delta x - \gamma)y. \tag{21}$$

This system possesses the critical points $(0,0)$ and $(\gamma/\delta, \alpha/\beta)$. From the calculations above (with $\lambda = \mu = 0$), it again follows that $(0,0)$ is unstable. Moreover, we have

$$Df\left(\frac{\gamma}{\delta}, \frac{\alpha}{\beta}\right) = \begin{bmatrix} 0 & -\beta\gamma/\delta \\ \alpha\delta/\beta & 0 \end{bmatrix},$$

hence $\lambda_{1,2} = \pm i\sqrt{\alpha\gamma}$. In this case the critical point $(\gamma/\delta, \alpha/\beta)$ is a *center for the linearized equation*, but nothing can be deduced from theorem (15.6) for the nonlinear system (21). □

(15.8) Remarks. (a) Theorem (15.6) makes no stability assertions for the case when $\mathrm{Re}\,\sigma(Df(x_0)) \leq 0$ and $\sigma(Df(x_0)) \cap i\mathbb{R} \neq \emptyset$. In this case the stability behavior is essentially determined by higher order terms. To see this, we consider the system

$$\dot{x} = -y + x^3$$
$$\dot{y} = x + y^3,$$

with unique critical point $(0,0)$, which is a center for the linearized system. With $r^2 :=$ $x^2 + y^2$, we have $r\dot{r} = x\dot{x} + y\dot{y} = -xy + x^4 + xy + y^4 = x^4 + y^4$. Hence

$$\dot{r} = \frac{x^4 + y^4}{r} \quad \text{if} \quad r > 0.$$

Consequently, $\dot{r} > 0$ and so the trajectories recede from $(0,0)$, i.e., $(0,0)$ is unstable.

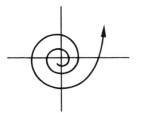

The system

$$\dot{x} = -y - x^3$$
$$\dot{y} = x - y^3$$

has the same linearization at the critical point $(0,0)$. However, now we have $\dot{r} = -(x^4 + y^4)/r < 0$ if $r > 0$. Consequently, all trajectories approach $(0,0)$, i.e., $(0,0)$ is stable.

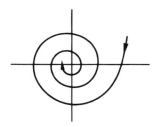

It is now easily seen that higher order perturbations can make the phase portrait much more complicated (cf. problem 1).

(b) The central stability result contained in theorem (15.6) is a *local statement*. It makes no assertions about the *region of attraction* of an asymptotically stable critical point. We will come across results in this direction in the following sections.

(c) Clearly theorem (15.6) remains correct if we only assume that $f \in C^{1-}(D, E)$ and f is differentiable at x_0.

(d) Similar results can also be proved for perturbed nonautonomous equations

$$\dot{x} = A(t)x + g(t, x)$$

under appropriate assumptions on the evolution operator U of the nonautonomous linear equation $\dot{x} = A(t)x$ (cf. theorem (11.13)) (e.g., Hale [1], Ch. III 2, and Daleckii-Krein [1], Ch. VII, 3 and 4). Because such assumptions can almost never be verified in practical cases, we will deal, in what follows, primarily with the particularly important case of autonomous equations.

(e) If we consider in example (15.7 a) only the *linearized equations* at the critical points, then at $(0, 0)$ we have a saddle, at $(0, -\gamma/\mu)$ we have a source, at $(\alpha/\lambda, 0)$ we have either a sink if $\alpha/\lambda < \gamma/\delta$, or a saddle if $\alpha/\lambda > \gamma/\delta$, and at $(\xi, \eta) \in (0, \infty)^2$ we have a sink. The figures of section 1 show that these qualitative structures of the phase portraits are retained also in the nonlinear case near the critical points. For this reason we also say that the critical point x_0 of the autonomous system $\dot{x} = f(x)$ is a *sink*, a *source*, or a *saddle* if the origin is a sink, a source, or a saddle, respectively, of the linearized equation

$$\dot{y} = Df(x_0)y.$$

In section 19 we will justify this terminology by means of an important general "linearization theorem."

(f) In order to actually apply the stability theorems above, we need criteria which allow us to determine whether $Re\,\sigma(A) < 0$ holds. Since the eigenvalues are the roots of the characteristic polynomial

$$\det(\lambda - A) = \lambda^m + a_1\lambda^{m-1} + a_2\lambda^{m-2} + \cdots + a_{m-1}\lambda + a_m$$

(if $\dim(E) = m$), one would like to determine from the coefficients of the polynomial whether all the roots lie in the negative complex half plane.

There are a number of such criteria. The best known is probably the following *Routh-Hurwitz* criterion (e.g., Hahn [1, § 6]).

Let

$$p_m(z) = z^m + a_1 z^{m-1} + \cdots + a_{m-1}z + a_m$$

be a polynomial with real coefficients, and for $k = 1, \ldots, m$ set

$$D_k := \det \begin{bmatrix} a_1 & a_3 & a_5 & a_7 & \cdots & a_{2k-1} \\ 1 & a_2 & a_4 & a_6 & \cdots & a_{2k-2} \\ 0 & a_1 & a_3 & a_5 & \cdots & a_{2k-3} \\ 0 & 1 & a_2 & a_4 & \cdots & a_{2k-4} \\ \vdots & \vdots & \vdots & \vdots & & \vdots \\ 0 & & & & & \end{bmatrix},$$

where $a_j = 0$ for all $j > m$. Then all the roots of p_m have negative real part if and only if the inequalities

$$a_k > 0 \quad \text{and} \quad D_k > 0, \quad k = 1, 2, \ldots, m,$$

are satisfied. □

Problems

1. Show that the phase portrait of the perturbed linear system

$$\dot{x} = -y + xr^2 \sin \frac{\pi}{r}$$

$$\dot{y} = x + yr^2 \sin \frac{\pi}{r},$$

where $r^2 = x^2 + y^2$, has the following qualitative behavior:

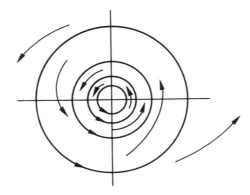

More precisely: There exists a sequence of concentric circles, centered at $(0,0)$ with radii $1/n$, such that "the orbits spiral alternately toward and away from them in the mathematically positive sense."

(*Hint:* Derive a system of differential equations in the polar coordinates (r, φ) (cf. (15.8 a)).)

2. The following system

$$\epsilon \dot{x} = x + y - xy - qx^2$$

$$\dot{y} = 2fz - y - xy \qquad\qquad (F - N)$$

$$p\dot{z} = x - z,$$

set up by *Field* and *Noyes*, is a mathematical model for the description of a chemical oscillation, the so-called *Belousov-Zhabotinsky reaction* (cf. Spektrum der Wissenschaften, 5 (1980), 131–137, for a description of this reaction, in particular, the pictures there!). Here f, p, q and ϵ are positive constants (with $\epsilon \ll 1$) and the equations have already been made dimensionless. The variables x, y and z correspond to chemical concentrations. Consequently, only "nonnegative solutions" are of interest, i.e., solutions in the positive octant $\mathbb{R}^3_+$.

Show that the system (F-N) has, aside from the origin $(0, 0, 0)$, exactly one critical point $(\xi, \eta, \zeta) \in \mathbb{R}^3_+$, and determine the stability properties of these two stationary points. (Apply the Routh-Hurwitz criterion.)

3. For the mathematical description of certain biological models for cell divisions, *Goodwin* proposed the following system:

$$\dot{x}_1 = \frac{1}{1 + x_m^k} - \alpha_1 x_1$$
$$\dot{x}_2 = x_1 - \alpha_2 x_2$$
$$\vdots$$
$$\dot{x}_m = x_{m-1} - \alpha_m x_m.$$

(G)

Here $\alpha_1, \ldots, \alpha_m$ are positive constants and the "Hill coefficient" k is a positive integer. The equations are again dimensionless and $x_1, \ldots, x_m$ correspond to concentrations of certain substances (enzymes, nucleic acids, etc.). Hence only nonnegative solutions in $\mathbb{R}_+^m$ are of interest.

Show that (G) has exactly one critical point in $\mathbb{R}_+^m$ and discuss its stability for the case $m = 3$ and $k = 1$.

4. Discuss the stability of the critical points for the equation of the *damped simple pendulum*

$$\ddot{x} + 2\alpha\dot{x} + \lambda \sin x = 0, \quad x \in \mathbb{R},$$

where $\alpha > 0$ and $\lambda = g/l > 0$ (cf. example (3.4 c)).

16. Invariance

In this section X denotes a metric space and $\varphi : \Omega \to X$ is a *semiflow* on X (cf. (10.2 c)). Here we consider semiflows since we are interested, first of all, in the "future behavior," i.e., as $t \to t^+(x)$, and because the general results about semiflows can also be applied, for example, to parabolic differential equations (reaction–diffusion equations).

(16.1) Remark. If $\varphi : \Omega \to X$ is a flow, then, by "restricting to positive time," we obtain a semiflow φ^+ on X, the *positive semiflow $\varphi^+ : \Omega_+ \to X$ of φ*, that is,

$$\varphi^+ := \varphi|\Omega \cap (\mathbb{R}_+ \times X),$$

or explicitly,

$$\Omega_+ := \{(t, x) \in \mathbb{R}_+ \times X \mid 0 \le t < t^+(x)\}$$

and

$$\varphi^+(t, x) := \varphi(t, x), \quad \forall (t, x) \in \Omega_+.$$

Hence, in particular, φ and φ^+ have the same positive escape times.

Moreover, set $J_-(x) := \{t \in \mathbb{R}_+ \mid -t \in J(x)\}$ and

$$\Omega_- := \{(t, x) \in \mathbb{R}_+ \times X \mid t \in J_-(x)\},$$

as well as

$$\varphi^-(t,x) := \varphi(-t,x), \quad \forall (t,x) \in \Omega_-.$$

Then also $\varphi^- : \Omega_- \to X$ is a semiflow on X, the *negative semiflow of* φ, and $-t^-(x)$ is the positive escape time of φ^-.

Therefore every assertion about semiflows can be applied to the positive, as well as to the negative semiflow of a flow. For this reason one can restrict stability investigations of flows to the study of semiflows. Unless stated otherwise, we will use the following *convention*: *Every statement about semiflows will, in case of a flow* φ, *be applied to the positive semiflow* φ^+. □

If x is a periodic point of φ, e.g., a critical point, then it is intuitively clear that the positive semitrajectory $\gamma^+(x)$ "can also be extended to the left," i.e., it can run through negative times. The concept of a solution of a semiflow makes this situation more precise.

A continuous function $u : J_u \to X$ is called a *solution of the semiflow* $\varphi :$ $\Omega \to X$ (through x) if the following holds:

(i) J_u is an open interval in $\mathbb{R}$ such that $[0, t^+(x)) \subseteq J_u$.

(ii) $u(0) = x$.

(iii)   For all (t,τ) satisfying $(t, u(\tau)) \in \Omega$ and $t + \tau \in J_u$, we have

$$t \cdot u(\tau) = u(t + \tau).$$

(iv)   The function u is maximal in the sense that it has no proper continuous extension satisfying (i)–(iii).

(16.2) Remarks. (a) For a given point $x \in X$, there may not be a solution of φ through x. In this case x is called a *starting point*.

(b) *Let* φ *be a flow and let* φ^+ *denote its positive semiflow. Then* $\varphi_x := \varphi(\cdot, x) : (t^-(x),$ $t^+(x)) \to X$ *is the unique solution of* φ^+ *through* x *for every* $x \in X$. *In particular,* φ^+ *is negatively unique and has no starting points.*

Indeed, φ_x clearly satisfies (i)–(iii). If $u : J_u \to X$ is a proper extension of φ_x, then there exists some $\bar{t} \in J_u$ such that either $\bar{t} < t^-(x)$ or $\bar{t} > t^+(x)$. So either $\gamma^-(x)$ lies in the compact set $u([\bar{t}, 0])$, or we have $\gamma^+(x) \subseteq u([0, \bar{t}])$. In any case, this is impossible by proposition (10.12). Consequently, φ_x is a solution of φ^+ through x.

Finally, if v is some other solution of φ^+ through x, then there exists some $\tau < 0$ such that $(-\tau) \cdot v(\tau) = v(-\tau + \tau) = v(0) = x$ and $(-\tau) \cdot \varphi_x(\tau) = (-\tau) \cdot (\tau \cdot x) = x$. Setting $t := -\tau$, $y := v(\tau)$ and $z := \varphi_x(\tau)$, we have: $t \cdot y = t \cdot z$ and $t > 0$, and, in particular, $y, z \in \Omega_t$. Consequently, $t \cdot y = t \cdot z \in \Omega_{-t}$ and $y = \varphi^{-t}(t \cdot y) = \varphi^{-t}(t \cdot z) = z$ by theorem (10.14). From this it follows that $v = \varphi_x$, that is, the uniqueness follows. □

A subset $M \subseteq X$ is called *positively invariant* if $\gamma^+(M) \subseteq M$, that is, if $m \in M$ always implies $\gamma^+(m) \subseteq M$. The set M is *invariant* if for each $m \in M$

there exists a solution $u_m : J \rightarrow X$ of φ through m such that $u_m(J) \subseteq M$. Finally, if φ is a flow, then M is called *negatively invariant* if $\gamma^-(M) \subseteq M$.

(16.3) Remarks. (a) An invariant set has no starting points.

(b) *Let φ be a flow. Then $M \subseteq X$ is invariant with respect to the positive semiflow φ^+ if and only if M is positively and negatively invariant.*

This follows immediately from (16.2 b).

(c) According to our convention, we say that $M \subseteq X$ is invariant with respect to the flow φ if M is invariant with respect to the positive semiflow φ^+. Hence (b) can be expressed as follows: *Let φ be a flow. Then M is invariant if and only if M is positively and negatively invariant.*

(d) Clearly $\emptyset$ is invariant and X is positively invariant. If φ has no starting points, then X is also invariant. Moreover, arbitrary unions of [positively] invariant sets are [positively] invariant. Also, arbitrary intersections of positively invariant sets are positively invariant. If φ is negatively unique, in particular, if φ is a flow, then arbitrary intersections of invariant sets are invariant.

Hence for every subset $M \subseteq X$ there exists a smallest positively invariant set containing M and a largest [positively] invariant subset of M. If φ is negatively invariant and has no starting points, there also exists a smallest invariant set containing M.

(e) *If M is positively invariant, then so is $\overline{M}$.*

Indeed, for $x \in \overline{M}$ there exists a sequence (x_k) in M such that $x_k \rightarrow x$. By lemma (10.5 i) (and remark (10.16)) we have $t^+(x) \leq \liminf t^+(x_k)$. So for every $t \in [0, t^+(x))$ there exists some $k_t \in \mathbb{N}$ such that $t^+(x_k) > t$ for all $k \geq k_t$. Hence $t \cdot x_k$ is defined for all $k \geq k_t$ and $t \cdot x_k \rightarrow t \cdot x$. Since M is positively invariant and $t \cdot x_k \in M$, it now follows that $t \cdot x \in \overline{M}$.

(f) It is graphically clear that a set M is positively invariant if the trajectory through every boundary point "moves inward." The following simple, but very useful result makes this precise.

A closed set $M \subseteq X$ is positively invariant if and only if for every $x \in \partial M$ there exists some $\epsilon > 0$ such that $[0, \epsilon) \cdot x \subseteq M$.

It is clear that the condition is necessary. If M is not positively invariant, there exist $x \in M$ and $t \in (0, t^+(x))$ so that $t \cdot x \notin M$. Since M is closed, there exists some $s \in [0, t)$ such that $s \cdot x \in M$, but $\tau \cdot x \notin M$ for all $s < \tau \leq t$. Then $y := s \cdot x \in \partial M$ and the condition is not satisfied for y.

(g) *If φ is a flow, then M is positively invariant if and only if M^c is negatively invariant.*

Indeed, if M^c is not negatively invariant, there exist $x \in M^c$ and $t \in (t^-(x), 0)$ such that $y := t \cdot x \in M$. Hence $(-t) \cdot y = (-t) \cdot (t \cdot x) = x \notin M$. Consequently, M is not positively invariant. Analogously, one shows the sufficiency of the condition above.

(h) *Let φ be a flow. If M is [positively] invariant, then so are $\text{int}(M)$ and $\overline{M}$.*

In fact, if $\text{int}(M)$ is not positively invariant, there exist $x \in \text{int}(M)$ and $t \in (0, t^+(x))$ such that $t \cdot x \in \partial M$. Hence there exists a sequence (x_k) in M^c so that $x_k \to t \cdot x$. Because $t \cdot x \in \Omega_{-t}$, and since Ω_{-t} is open, there exists some k_0 so that $x_k \in \Omega_{-t}$ for all $k \geq k_0$. Hence we have $(-t) \cdot x_k \to (-t) \cdot (t \cdot x) = x \in \text{int}(M)$ and therefore $(-t) \cdot x_{k_1} \in \text{int}(M)$ for some $k_1 \geq k_0$. Thus $t \cdot [(-t) \cdot x_{k_1}] = x_{k_1} \in M^c$, which contradicts the fact that M is positively invariant. Therefore, if M is positively invariant, then so is $\text{int}(M)$ and, because of (e), also $\overline{M}$.

If M is negatively invariant, then M^c is positively invariant by (g). Hence also $\overline{M^c}$ and $\text{int}(M^c)$ are positively invariant. Based on (g), it follows that $[\text{int}(M^c)]^c = \overline{M}$ and $[\overline{M^c}]^c = \text{int}(M)$ are negatively invariant. Using (c), it follows that the invariance of M implies that of $\text{int}(M)$ and $\overline{M}$.

(i) *Let φ be a flow. If M is invariant, then so is ∂M. If ∂M is invariant, then so are $\overline{M}$ and $\text{int}(M)$.*

Since $\partial M = \overline{M} \cap (\overline{M^c})$, the first assertion follows from (h), (g) and (d). If ∂M is invariant, then obviously so is $\overline{M}$ and, using (g), $\partial M = \partial M^c$ and $[\overline{M^c}]^c = \text{int}(M)$, also $\text{int}(M)$. $\square$

Let X now be an open subset of a Banach space $(E, |\cdot|)$ and let the flow φ be induced by a vector field $f \in C^{1-}(X, E)$. For every $x \in X$ we then have

$$t \cdot x = x + tf(x) + o(t) \quad \text{as} \quad t \to 0^+$$

(cf. remark (10.4 b)). If $x \in M \subseteq X$ and if M is positively invariant, it follows that

$$\text{dist}(x + tf(x), M) \leq |x + tf(x) - t \cdot x| = o(t)$$

as $t \to 0^+$ (because $t \cdot x \in M$ for all $0 \leq t < \epsilon$). The following theorem (16.5) shows that positively invariant sets can be characterized by this condition. For this we need a few preliminaries about Dini derivatives and *differential inequalities*.

Assume that $-\infty < \alpha < \beta \leq \infty$ and $h : [\alpha, \beta] \to \mathbb{R}$. We then define the *lower right Dini derivative* of h at α by

$$D_+h(\alpha) := \liminf_{\xi \to 0^+} \frac{h(\alpha + \xi) - h(\alpha)}{\xi}.$$

Clearly $D_+h(\alpha)$ is finite whenever h is Lipschitz continuous, and $D_+h(\alpha)$ agrees with the usual (right-hand) derivative if h is differentiable at α.

(16.4) Lemma (*Comparison theorem*). *Assume that J and D are open intervals in $\mathbb{R}$ and let $g \in C^{0,1-}(J \times D, \mathbb{R})$. Moreover, let $u \in C^1(J, D)$ be a solution of the differential equation $\dot{x} = g(t, x)$. Suppose $v \in C(J, D)$ and $\alpha \in J$ are such that*

$$v(\alpha) \leq u(\alpha) \quad and \quad D_+ v(t) \leq g(t, v(t)), \quad \forall\, t \in J \cap [\alpha, \infty).$$

Then $v \leq u$ on $J \cap [\alpha, \infty)$.

Proof. Let $\beta \in J$ satisfy $\beta > \alpha$, and let u_λ be the solution of the parameter-dependent IVP

$$\dot{x} = g(t, x) + \lambda, \quad x(\alpha) = u(\alpha),$$

for $0 < \lambda < \lambda_0$, where λ_0 is chosen so small that u_λ exists on $[\alpha, \beta]$. This is possible by the continuity theorem (8.3) (cf. the proof of lemma (10.5 i)).

We now assume that there exist some $\lambda \in (0, \lambda_0)$ and some $t_1 \in (\alpha, \beta)$ with $v(t_1) > u_\lambda(t_1)$. Then there exists some $t_0 \in [\alpha, t_1)$ such that $v(t_0) = u_\lambda(t_0)$ and $v(t) > u_\lambda(t)$ for all $t_0 < t < t_1$. From this it follows that

$$D_+ v(t_0) \geq \dot{u}_\lambda(t_0) = g(t_0, u_\lambda(t_0)) + \lambda > g(t_0, v(t_0)),$$

which contradicts the assumption. Thus we have $v \leq u_\lambda$ on $[\alpha, \beta]$ for every $\lambda \in (0, \lambda_0)$. Now, as $\tau \to 0^+$, we deduce from the continuity theorem (8.3) that $v \leq u$ on $[\alpha, \beta]$. Since $\beta \in J$ with $\beta > \alpha$ was arbitrary, the assertion follows. $\quad \square$

In what follows we let X denote an open subset of a finite dimensional Banach space $(E, |\cdot|)$ and $f \in C^{1-}(X, E)$. The subsequent assertions about positive invariance refer to the flow induced by f.

(16.5) Theorem. *Let $M \subseteq X$ be closed. Then M is positively invariant if and only if for every $x \in M$, the* subtangent condition

$$\liminf_{t \to 0^+} t^{-1} \operatorname{dist}(x + t f(x), M) = 0 \tag{1}$$

is satisfied.

Proof. Based on the motivating considerations above, it is enough to prove the sufficiency of condition (1).

So assume that $x \in M$ and let $r > 0$ be chosen so that $\overline{\mathbb{B}}(x, 4r) \subseteq X$. We set $v(t) := \operatorname{dist}(t \cdot x, M)$ for $0 \leq t < t^+(x)$ and fix some $\epsilon \in (0, t^+(x))$ satisfying $[0, \epsilon] \cdot x \subseteq \overline{\mathbb{B}}(x, r)$. For $t \in [0, \epsilon]$ we certainly have $0 \leq v(t) \leq r$ and, since $M_r := M \cap \overline{\mathbb{B}}(x, 2r)$ is compact, there exists some $p := p_t \in M_r$ such that $v(t) = |t \cdot x - p|$. For $t \in [0, \epsilon)$, let $s \in (0, \epsilon - t)$ be so small that we have $s|f(p)| \leq r$ and hence $p + s f(p) \in \overline{\mathbb{B}}(x, 3r)$. Next, we find some $q \in M \cap \overline{\mathbb{B}}(x, 4r)$ satisfying $|p + s f(p) - q| = \operatorname{dist}(p + s f(p), M)$. From this it follows that

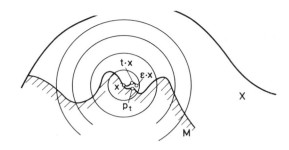

$$v(t+s) \leq |(t+s) \cdot x - q| \leq |t \cdot x - p| + s|f(t \cdot x) - f(p)|$$
$$+ |s \cdot (t \cdot x) - t \cdot x - sf(t \cdot x)| + |p + sf(p) - q|$$
$$= v(t) + s|f(t \cdot x) - f(p)| + o(s) + \mathrm{dist}(p + sf(p), M),$$

and hence

$$D_+ v(t) \leq |f(t \cdot x) - f(p_t)|, \quad \forall\, t \in [0, \epsilon). \tag{2}$$

Since $([0, \epsilon] \cdot x) \cup M_r$ is compact, it follows from proposition (6.4) that there exists some $\lambda > 0$ so that

$$|f(t \cdot x) - f(p_t)| \leq \lambda |t \cdot x - p_t| = \lambda v(t), \quad \forall\, t \in [0, \epsilon).$$

Using (2) we get

$$D_+ v(t) \leq \lambda v(t), \quad \forall\, t \in [0, \epsilon),$$

and, because $v(0) = 0$ and $v \geq 0$, it follows from lemma (16.4) that $v = 0$ on $[0, \epsilon)$. Therefore $[0, \epsilon) \cdot x \subseteq M$ and now the assertion follows from (16.3 f). $\qquad\square$

(16.6) Remarks. (a) Theorem (16.5) also holds when $\dim(E) = \infty$, with a much more complicated proof, however (cf. Deimling [1], Martin [1]). In arbitrary Banach spaces there may be no $p_t \in M$ satisfying $v(t) = |t \cdot x - p_t|$. Consequently, the proof above cannot be carried over.

(b) Condition (1) is obviously satisfied for $x \in \mathrm{int}(M)$. It is therefore only a condition for f on the boundary of M. (It should be noted that $\mathrm{int}(M)$ could be empty, so that $\partial M = M$ is possible.) $\qquad\square$

In the following theorem (16.9) we will show that, if M is a manifold with boundary and if $\dim(M) = \dim(E)$, then condition (1) is equivalent to requiring that the vector field f "points inward" or "lies below the tangent plane" along ∂M. This explains the name "subtangent condition."

To prove this theorem we need the following, useful lemma, which is of independent interest.

(16.7) Approximation Lemma. *Let Y and Z be arbitrary Banach spaces, let $U \subseteq Y$ be open, and let $f \in C(U, Z)$. Then for every $\epsilon > 0$, there exists some $f_\epsilon \in C^{1-}(U, Z)$ such that $f_\epsilon(U) \subseteq \operatorname{co}(f(U))$ and*

$$\sup_{x \in U} |f_\epsilon(x) - f(x)| < \epsilon,$$

where $\operatorname{co}(C)$ denotes the convex hull of C, i.e., the smallest convex set containing C.

Proof. For every $x \in U$ set

$$U_x := \{ y \in U \mid |f(x) - f(y)| < \epsilon/2 \}.$$

Then $\{U_x \mid x \in U\}$ is an open covering of U. Since every metric space is paracompact (e.g., Dugundji [1], Schubert [1]), there exists a locally finite open refinement $\{V_\alpha \mid \alpha \in A\}$ of this covering, that is, $\{V_\alpha \mid \alpha \in A\}$ is an open covering of U such that every $x \in U$ has a neighborhood which intersects only finitely many V_α's, and such that for every $\alpha \in A$ there exists some $x \in U$ with $V_\alpha \subseteq U_x$. Now we set

$$\Psi_\alpha := \operatorname{dist}(\cdot, V_\alpha^c)$$

and

$$\varphi_\alpha := \Psi_\alpha / \sum_{\alpha \in A} \Psi_\alpha.$$

The triangle inequality implies that

$$|x - b| \le |x - y| + |y - b|$$

for all $b \in B \subseteq U$ and all $x, y \in U$. Hence $\operatorname{dist}(x, B) \le |x - y| + |y - b|$ for all $b \in B$ and therefore

$$\operatorname{dist}(x, B) \le |x - y| + \operatorname{dist}(y, B).$$

Exchanging x and y, we obtain

$$|\operatorname{dist}(x, B) - \operatorname{dist}(y, B)| \le |x - y|,$$

that is, $\operatorname{dist}(\cdot, B)$ is uniformly Lipschitz continuous. Since $\{V_\alpha \mid \alpha \in A\}$ is locally finite, one easily verifies that $\varphi_\alpha \in C^{1-}(U, \mathbb{R})$.

For each $\alpha \in A$ we now choose some $y_\alpha \in V_\alpha$ and set

$$f_\epsilon := \sum_{\alpha \in A} \varphi_\alpha f(y_\alpha).$$

Then $f_\epsilon \in C^{1-}(U, Z)$ and, since $\varphi_\alpha \ge 0$ and $\sum \varphi_\alpha = 1$, we get

$$|f_\epsilon(x) - f(x)| = \left| \sum_\alpha \varphi_\alpha(x)[f(y_\alpha) - f(x)] \right| \le \sum_\alpha \varphi_\alpha(x) |f(y_\alpha) - f(x)|.$$

If now $\varphi_\alpha(x) \neq 0$, then $x \in V_\alpha \subseteq U_{x_\lambda}$ for some $x_\lambda \in U$. Using this and $y_\alpha \in V_\alpha$, it follows that $|f(y_\alpha) - f(x)| \leq \epsilon$ and therefore $|f_\epsilon(x) - f(x)| \leq \epsilon \sum_\alpha \varphi_\alpha(x) = \epsilon$. Finally, since $f_\epsilon(x)$ is a convex combination of elements in $f(U)$, it follows that $f_\epsilon(U) \subseteq \mathrm{co}(f(U))$. $\qquad\square$

(16.8) Remark. In the proof above we have shown that *every locally finite family* $\{V_\alpha \mid \alpha \in A\}$ *of open subsets of a Banach space* Y *has a subordinate* $\underline{C^{1-}\text{-partition of unity}}$, i.e., a collection $\{\varphi_\alpha \mid \alpha \in A\}$ such that $\varphi_\alpha \in C^{1-}(Y, \mathbb{R}_+)$,

$$\mathrm{supp}\ \varphi_\alpha := \overline{\{y \in Y \mid \varphi_\alpha(y) \neq 0\}} \subseteq \overline{V_\alpha},$$

and $\sum_\alpha \varphi_\alpha = 1$. $\qquad\square$

After these preliminaries we now can prove the following, important and intuitively obvious theorem.

(16.9) Theorem. *Let* $E = \mathbb{R}^m$ *and let* $\Phi \in C^1(X, \mathbb{R})$ *be such that* $\nabla\Phi(x) \neq 0$ *for all* $x \in \Phi^{-1}(0)$, *that is, assume that* 0 *is a regular value of* Φ. *Then* $M := \Phi^{-1}(-\infty, 0]$ *is positively invariant if and only if*

$$(\nabla\Phi(x) \mid f(x)) \leq 0, \quad \forall\, x \in \partial M = \Phi^{-1}(0). \tag{3}$$

Proof. "$\Longrightarrow$" Suppose that for some $x \in \partial M$ we have $(\nabla\Phi(x) \mid f(x)) > 0$. Then there exists a neighborhood U of x so that $(\nabla\Phi(y) \mid f(y)) > 0$ for all $y \in U$. Choose $\epsilon > 0$ so small that $[0, \epsilon) \cdot x \subseteq U$. Then for all $0 < t < \epsilon$ we have

$$\Phi(t \cdot x) - \Phi(x) = \int_0^t \frac{d}{ds}\Phi(s \cdot x)ds = \int_0^t (\nabla\Phi(s \cdot x) \mid f(s \cdot x))ds > 0$$

and hence $(0, \epsilon) \cdot x \subseteq M^c$. Using (16.3 f), it follows that M is not positively invariant.

"$\Longleftarrow$" Let $x \in \partial M$ be arbitrary, let U be a neighborhood of x, and let α denote a positive number such that $|\nabla\Phi(y)| \geq 2\alpha$ for all $y \in U$. Lemma (16.7) implies the existence of some $g \in C^{1-}(X, \mathbb{R}^m)$ such that $|\nabla\Phi(y) - g(y))| \leq \alpha$ for all $y \in X$. From this we get that

$$\begin{aligned}(\nabla\Phi(y) \mid g(y)) &= |\Phi(y)|^2 - (\nabla\Phi(y) \mid \nabla\Phi(y) - g(y)) \\ &\geq |\nabla\Phi(y)|^2 - |\nabla\Phi(y)|\alpha \geq |\nabla\Phi(y)|^2/2 \geq \alpha^2\end{aligned} \tag{4}$$

for all $y \in U$.

For each $\lambda \geq 0$, let φ_λ denote the vector field on U induced by $f - \lambda g$, and let $y \in \partial M \cap U$ be arbitrary. It follows from (3) and (4) that

$$(\nabla\Phi(y) \mid f(y) - \lambda g(y)) \leq -\lambda\alpha^2.$$

Hence for each $\lambda > 0$ there exists a neighborhood V of y so that $(\nabla\Phi \mid f - \lambda g) \leq 0$ on V. We now choose $\epsilon \in (0, t_{\varphi_\lambda}^+(y))$ so small that $\varphi_\lambda([0, \epsilon], y) \subseteq V$. Then we get

$$\Phi(\varphi_\lambda^t(y)) - \Phi(y) = \int_0^t (\nabla\Phi(\varphi_\lambda^s(y)) \mid f(\varphi_\lambda^s(y)) - \lambda g(\varphi_\lambda^s(y)))ds \le 0$$

for all $0 \le t \le \epsilon$, i.e., $\varphi_\lambda([0, \epsilon), y) \subseteq M$. Therefore, using (16.3 f), $M \cap U$ is positively invariant with respect to the flow induced by $f - \lambda g$.

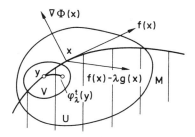

By the continuity theorem (8.3), there exist positive numbers $\bar\epsilon$ and $\bar\lambda$ such that $\varphi_\lambda(\cdot, x)$ is defined on $[0, \bar\epsilon]$ for all $\lambda \in [0, \bar\lambda]$, and so that

$$\varphi_\lambda(t, x) \subseteq U \quad \text{for all} \quad (\lambda, t) \in [0, \bar\lambda] \times [0, \bar\epsilon].$$

Since by the above discussion we have $\varphi_\lambda(t, x) \in M$ for all $(\lambda, t) \in (0, \bar\lambda] \times [0, \bar\epsilon]$, and since by theorem (8.3), $\varphi_\lambda(t, x) \to \varphi_0(t, x) = t \cdot x$ for all $t \in [0, \bar\epsilon]$ as $\lambda \to 0$, it follows that $[0, \bar\epsilon] \cdot x \subseteq M$. By (16.3 f), M is therefore positively invariant. $\square$

(16.10) Corollary. *Assume that $E = \mathbb{R}^m$ and let $\Phi_1, \dots, \Phi_k \in C^1(X, \mathbb{R})$. Moreover, assume that 0 is a regular value of each $\Phi_j, j = 1, \dots, k$, and let*

$$M := \bigcap_{j=1}^k \Phi_j^{-1}(-\infty, 0].$$

If

$$(\nabla\Phi_j(x) \mid f(x)) \le 0, \quad \forall\, x \in \Phi_j^{-1}(0), \quad j = 1, \dots, k, \tag{5}$$

then M is positively invariant. If, in fact, we have

$$(\nabla\Phi_j(x) \mid f(x)) = 0, \quad \forall\, x \in \Phi_j^{-1}(0), \quad j = 1, \dots, k, \tag{6}$$

then M and all the hypersurfaces $\Phi_j^{-1}(0), j = 1, \dots, k$, are invariant.

Proof. The first assertion follows immediately from theorem (16.9) and remark (16.3 d). Since the negative semiflow φ^- is induced by the vector field $-f$, it follows from (6) and from what we have just proved that M is negatively invariant, in fact, invariant by (16.3 c). Because $\Phi_j^{-1}(0) = \Phi_j^{-1}(-\infty, 0] \cap \Phi_j^{-1}[0, \infty)$, the invariance of $\Phi_j^{-1}(0)$ follows analogously. $\square$

(16.11) Remarks. (a) If 0 is a regular value of $\Phi \in C^1(X, \mathbb{R})$, then we know that $\Phi^{-1}(0)$ is a hypersurface in $\mathbb{R}^m$ which bounds the m-dimensional C^1-manifold $\Phi^{-1}(-\infty, 0]$. Moreover, $\nabla\Phi(x)$ is a vector in the direction of the outward unit normal at $x \in \Phi^{-1}(0)$. Hence corollary (16.10) states, in particular, that M is invariant if at every x, f is a tangent vector on every hypersurface $\Phi_j^{-1}(0)$ through that point.

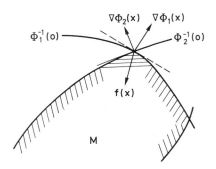

(b) Under the assumptions of theorem (16.9), it follows from (16.5) and (16.9) that the subtangent condition (1) is equivalent to (3).

(c) If $(E, (\cdot \mid \cdot))$ is a general real Hilbert space and if $f \in C^1(X, \mathbb{R})$, then the Riesz representation theorem uniquely defines the gradient $\nabla f(x)$ of f at x by the formula

$$(\nabla f(x) \mid y) = Df(x)y, \quad \forall\, y \in E.$$

With this definition and based on remark (8.5), one easily verifies that theorem (16.9) and corollary (16.10) remain true in an arbitrary (infinite dimensional) real Hilbert space.   □

(16.12) Examples. *Ecological Models.* In what follows we consider *two-species models* of the form

$$\dot{x} = a(x, y)x$$
$$\dot{y} = b(x, y)y, \tag{7}$$

where x and y denote the populations of the two species, and a and b denote the corresponding growth rates. We assume that

$$a, b \in C^1(\mathbb{R}^2, \mathbb{R})$$

and, of course, we are only interested in nonnegative solutions, that is, solutions in $\mathbb{R}_+^2$.

It follows immediately from corollary (16.10) that the positive quadrant $\mathbb{R}_+^2$ and each of the two semiaxes, $\mathbb{R}_+ \times \{0\}$ and $\{0\} \times \mathbb{R}_+$, are invariant with respect to the flow induced by (7). We may therefore completely restrict our analysis to the flow on $\mathbb{R}_+^2$.

(a) *Predator-Prey Models.* Suppose that y denotes the predator and x the prey population. Then the following assumptions are meaningful:

(i) If there is not enough prey, the predator population declines, i.e., there exists some $B > 0$ so that

$$b(x, y) < 0 \quad \text{whenever} \quad x < B, \ y \in \mathbb{R}_+.$$

(ii) An increase in the prey population increases the growth rate of the predator, i.e.,

$$D_1 b > 0.$$

(iii) If no predators are present, a small prey population increases. Hence we have

$$a(0, 0) > 0.$$

(iv) If the prey population goes beyond a certain size, it must decrease, i.e., there exists some $A > 0$ so that

$$a(x, y) < 0 \quad \text{whenever} \quad x > A, \ y \in \mathbb{R}_+.$$

(v) If the predator population increases, the growth rate of the prey declines, i.e.,

$$D_2 a < 0.$$

These assumption are clearly satisfied by the simple model

$$a(x, y) = \alpha - \beta y - \lambda x, \quad b(x, y) = \delta x - \gamma - \mu y \tag{8}$$

of section 1. Figures 1 and 2 show typical situations of the general case. In each one of these cases, the indicated rectangles are positively invariant by corollary (16.10), and the hollow points are critical points. In addition, the vertically and horizontally hatched regions in Fig. 2 are also positively invariant. For the special case (8), Fig. 1 and 2 correspond to Fig. 6 and 5, respectively, of section 1.

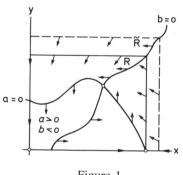

Figure 1

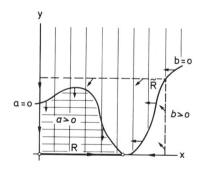

Figure 2

(b) *Competing Species Models.* We consider now two species which compete for a common food supply. In this connection the following assumptions are reasonable:

(i) If one population increases, the growth rate of the other decreases, i.e.

$$D_2 a < 0 \quad \text{and} \quad D_1 b < 0.$$

(ii) If a population is very large, it can no longer grow ("limited growth"), i.e., there exist positive constants A and B such that

$$a(x,y) \le 0 \quad \text{and} \quad b(x,y) \le 0 \quad \text{if} \quad x \ge A \quad \text{or} \quad y \ge B.$$

(iii) If a species is absent, the other has first a positive, then a negative growth rate, i.e., there exist positive constants α and β such that

$$a(x,0)(\alpha - x) > 0, \quad \forall\, x \in \mathbb{R}_+\backslash\{\alpha\},$$

and

$$b(0,y)(\beta - y) > 0, \quad \forall\, y \in \mathbb{R}_+\backslash\{\beta\}.$$

Typical situations are depicted in Figures 3 and 4.

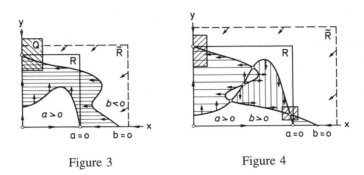

Figure 3   Figure 4

In any case, the horizontally and vertically hatched regions between the curves $a = 0$ and $b = 0$, as well as the indicated rectangles, are positively invariant. Because the obliquely hatched rectangles can be chosen arbitrarily small, it follows that the critical points contained in them are (asymptotically) stable. In general we have: If the curves $a = 0$ and $b = 0$ intersect transversely in a point $P \in (0,\infty)^2$, if both slopes at P are negative and if the slope of the curve $a = 0$ is smaller than that of $b = 0$, then P is (asymptotically) stable. In fact, in this case we can choose arbitrarily small positively invariant rectangles around P so that their corners lie in four different "quadrants." (The fact that we have asymptotic stability is of course intuitively clear, but its exact proof will be deferred to section 18.)

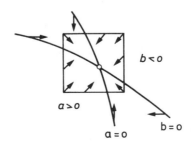

(c) *Symbiosis Models.* We now assume that two species are living in symbiosis. More precisely, we make the following assumption:

(i) If one population increases, the growth rate of the other also increases, i.e.,

$$D_2 a > 0 \quad \text{and} \quad D_1 b > 0.$$

(ii) Neither population can grow beyond a certain size, i.e., there exist positive constants A and B such that

$$a(x, y) < 0 \quad \text{if} \quad x > A, \; y \in \mathbb{R}_+$$

and

$$b(x, y) < 0 \quad \text{if} \quad y > B, \; x \in \mathbb{R}_+.$$

(iii) If both populations are very small, they both increase, i.e.,

$$a(0, 0) > 0 \quad \text{and} \quad b(0, 0) > 0.$$

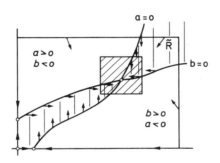

Figure 5

A typical situation is again depicted in Figure 5. The rectangles, as well as the hatched regions between the curves are positively invariant. Since the obliquely hatched rectangle can be chosen arbitrarily small, the critical point contained in it is (asymptotically) stable.

The positively invariant rectangles in Figures 1–4 are minimal. (R has degenerated to an interval $[0, x_0]$ on the x-axis in Fig. 2.) Every rectangle $\tilde{R}$, which is parallel to the axes and contains R, is also positively invariant. From this and corollary (10.13) it follows that (7) induces a global flow on $\mathbb{R}_+^2$. Graphically it is also clear that

$$t \cdot (\xi, \eta) \to R \quad \text{as} \quad t \to \infty$$

for every $(\xi, \eta) \in \mathbb{R}_+^2 \backslash R$, i.e., all trajectories "run into R," or "R attracts (ξ, η)." The techniques that will be developed in the next section allow us to correctly prove these assertions. □

Problems

1. Let J and D be open intervals in $\mathbb{R}$ and let $f \in C^{0,1-}(J \times D, \mathbb{R})$. A function $v \in C^1(J, D)$ is called a *subsolution* of the IVP

$$\dot{x} = f(t, x), \quad x(\tau) = \xi, \tag{9}$$

if

$$\dot{v}(t) \le f(t, v(t)), \quad v(\tau) \le \xi, \tag{10}$$

holds for all $t \in J$. If the inequalities in (10) are reversed, v is called a *supersolution* of the IVP (9). Prove the following: *If $\bar{v}$ is a subsolution and $\hat{v}$ is a supersolution of (9) satisfying $\bar{v} \le \hat{v}$, then (9) has a unique solution u defined on $J \cap [\tau, \infty)$, and on $J \cap [\tau, \infty)$ we have $\bar{v} \le u \le \hat{v}$.*

2. Show that $\mathbb{R}_+^m$ is positively invariant with respect to the flow induced by the Goodwin equations (G). Also show that the positive semiflow on $\mathbb{R}_+^m$ is global. Determine a compact, positively invariant subset $M \subseteq \mathbb{R}_+^m$.

3. Show that $\mathbb{R}_+^3$ is positively invariant with respect to the Field-Noyes equations (F-G), and for $q < 1$ determine a compact, positively invariant parallelepiped as small as possible. Show that the positive semiflow on $\mathbb{R}_+^3$ is global.

4. In their investigations about the transmission of impulses in nerves, for which they received the 1959 Nobel prize, *Hodgkin* and *Huxley* set up the following system of reaction-diffusion equations:

$$
\begin{aligned}
cu_t - \frac{1}{R}u_{xx} &= g(u, v, w, z) \\
v_t - \epsilon_1 v_{xx} &= g_1(u)(h_1(u) - v) \\
w_t - \epsilon_2 w_{xx} &= g_2(u)(h_2(u) - w) \\
z_t - \epsilon_3 z_{xx} &= g_3(u)(h_3(u) - z).
\end{aligned}
\tag{H$-$H}
$$

Here

$$g(u, v, w, z) := k_1 v^3 w(c_1 - u) + k_2 z^4(c_2 - u) + k_3(c_3 - u),$$

c, R, k_i are positive constants, $c_1 > c_3 > 0 > c_2$ and $\epsilon_i \ge 0$, $i = 1, 2, 3$. g_i and h_i are C^1-functions and satisfy the inequalities $g_i > 0$ and $0 < h_i < 1$ pointwise, $i = 1, 2, 3$. With $U := (u, v, w, z)$ and using the (one-dimensional) Laplace operator $\Delta U := (\Delta u, \Delta v, \Delta w, \Delta z)$, (H-H) can be written in the form

$$\frac{\partial U}{\partial t} - D\Delta U = f(U), \tag{11}$$

where the diffusion matrix D is a diagonal matrix with nonnegative entries. Positively invariant axes-parallel parallelepipeds with respect to the corresponding ordinary differential equation – the so-called *kinetic equation* corresponding to (11) –

$$\dot{y} = f(y), \quad y \in \mathbb{R}^m \quad (m = 4 \text{ in (11))},$$

play an important role in the theory of such reaction-diffusion equations.

For the kinetic equation corresponding to (H-H) determine a positively invariant, axes-parallel parallelepiped in $\mathbb{R}^4$ as small as possible.

5. Since the Hodgkin-Huxley equations lead to serious mathematical problems, simpler models were developed which are able to reproduce the essential features of the full H-H model. A particularly well-known model is the *FitzHugh-Nagumo* system:

$$u_t - \epsilon u_{xx} = \sigma v - \gamma u$$
$$v_t - v_{xx} = g(v) - u. \tag{FH - N}$$

Here σ and γ are positive constants, $\epsilon \geq 0$, and $g \in C^1(\mathbb{R}, \mathbb{R})$ has the qualitative form of a cubic polynomial, i.e.,

$$g(v) = -v(v - a)(v - b),$$

where $0 < a < b$.

For the kinetic equation corresponding to (FH-N) determine a positively invariant rectangle as small as possible.

17. Limit Sets and Attractors

In this section we let X denote a metric space and let $\varphi : \Omega \to X$ be a *semiflow* on X.

For each $x \in X$, we define the *positive limit set* (the ω-*limit set*) of x by

$$\omega(x) := \bigcap_{t>0} \overline{\gamma^+(t \cdot x)}. \tag{1}$$

If φ is a *flow*, we define the *negative limit set* (the α-*limit set*) of x by

$$\omega^-(x) := \bigcap_{t<0} \overline{\gamma^-(t \cdot x)}.$$

(17.1) Remarks. (a) If φ is a flow, then clearly $\omega^-(x)$ is the positive limit set of x with respect to the corresponding negative semiflow φ^- (cf. remark (16.1)). In general it therefore suffices to consider only positive limit sets.

(b) If $t^+(x) < \infty$, then $\omega(x)$ is empty.

(c) *For every* $x \in X$ *with* $t^+(x) = \infty$ *we have*:

$$\omega(x) = \{y \in X \mid \exists t_k \to \infty \text{ such that } t_k \cdot x \to y\}.$$

In fact, if (t_k) is a sequence in $\mathbb{R}_+$ such that $t_k \to \infty$ and $t_k \cdot x \to y$, then $y \in \overline{\gamma^+(x)}$. Because $y = \lim t_k \cdot x = \lim(t_k - t) \cdot (t \cdot x)$, it follows that $y \in \overline{\gamma^+(t \cdot x)}$ for all $t \geq 0$. Hence $y \in \omega(x)$.

Conversely, assume that $y \in \omega(x)$, hence $y \in \overline{\gamma^+(t \cdot x)}$ for all $t \geq 0$. Then for each $k \in \mathbb{N}$, there exists some $n(k) > k$ so that $n(k) \cdot x \in \mathbb{B}(y, 1/k)$. Setting $t_k := n(k)$, we then have $t_k \to \infty$ and $t_k \cdot x \to y$.

(d) $\overline{\gamma^+(x)} = \gamma^+(x) \cup \omega(x)$ *for every* $x \in X$ *satisfying* $t^+(x) = \infty$.

(e) $\overline{\gamma^+(x)}$ *and* $\omega(x)$ *are closed and positively invariant for all* $x \in X$.

The closure is clear. Since $\gamma^+(x)$ is positively invariant, the positive invariance of $\overline{\gamma^+(x)}$ follows from (16.3 e). We now obtain the positive invariance of $\omega(x)$ from (1) and (16.3 d).

(f) $\omega(x) = \omega(t \cdot x)$ *for all* $0 \le t < t^+(x)$. □

Under an additional compactness condition, we obtain the following important generalization of (17.1 e). Here and in what follows, we say that $t \cdot x$ *converges to the set* $M \subseteq X$ *as* $t \to t^+$,

$$t \cdot x \to M \quad \text{as} \quad t \to t^+,$$

if for every neighborhood U of M there exists some $t_U < t^+$ such that $t \cdot x \in U$ for all $t \ge t_U$. We define the convergence of a point sequence to a set with the obvious modifications.

(17.2) Theorem. *Assume that* $\gamma^+(x)$ *is relatively compact. Then* $\omega(x)$ *is nonempty, compact, connected, invariant and*

$$t \cdot x \to \omega(x) \quad \text{as} \quad t \to \infty.$$

Proof. Corollary (10.13) implies that $t^+(x) = \infty$. Consequently $\gamma^+(t \cdot x) \ne \emptyset$ for all $t \in \mathbb{R}_+$. From $0 \le t \le s$ it always follows that $\gamma^+(t \cdot x) \supseteq \gamma^+(s \cdot x)$, and therefore $\overline{\gamma^+(t \cdot x)}$ is nonempty and compact for all $t \in \mathbb{R}_+$. It also follows that the sets $\gamma^+(t \cdot x)$, $t \ge 0$, have the "finite intersection property" (i.e., the intersection of finitely many sets is nonempty). Therefore $\omega(x) = \bigcap_{t \ge 0} \overline{\gamma^+(t \cdot x)}$ is nonempty and compact (e.g., Dugundji [1], Schubert [1]). By (17.1 c), $\omega(x)$ is positively invariant.

Let U now be an open neighborhood of $\omega(x)$ and let $t_k \to \infty$ be a sequence so that $t_k \cdot x \notin U$. By the compactness of $U^c \cap \overline{\gamma^+(x)}$, there exists a subsequence $(t_{k'})$ and some $y \in U^c$ such that $t_{k'} \cdot x \to y$. Then (17.1 c) implies that $\omega(x) \cap U^c \ne \emptyset$, which is impossible. Therefore $t \cdot x$ converges to $\omega(x)$ as $t \to \infty$.

Next, assume that $\omega(x)$ is not connected. Then $\omega(x) = \omega_1 \cup \omega_2$ for some closed, nonempty and disjoint sets ω_1 and ω_2. Because $\omega(x)$ is compact, it follows that also the ω_i's are compact. And since a metric space is normal, there exist disjoint open neighborhoods U_i of ω_i, $i = 1, 2$. Because $t \cdot x \to \omega(x)$, there exists some $t \in \mathbb{R}_+$ so that $\gamma^+(t \cdot x) \subseteq U_1 \cup U_2$, and $\gamma^+(t \cdot x) \cap U_i \ne \emptyset$ follows from $\omega_i \ne \emptyset$, $i = 1, 2$, and by (17.1 c). Therefore $\gamma^+(t \cdot x)$ is disconnected, which is false since $\gamma^+(t \cdot x)$ is the image of the interval $[t, \infty)$ under a continuous function. Therefore $\omega(x)$ is connected.

Let $y_0 \in \omega(x)$. It follows from (17.1 c) that there exists a sequence $t_k \to \infty$ such that $t_k \cdot x \to y_0$. By the compactness of $\overline{\gamma^+(x)}$, there exists a subsequence $t_{k_1} \to \infty$ so that $\lim_{k_1 \to \infty}(t_{k_1} - 1) \cdot x = y_1$ for some appropriate $y_1 \in \omega(x)$. Inductively, for each $j \in \mathbb{N}$ we obtain a subsequence (t_{k_j}) of the preceding subsequence and some $y_j \in \omega(x)$ such that $\lim_{k_j \to \infty}(t_{k_j} - j) \cdot x = y_j$. For the diagonal sequence $t_i' := t_{i_i}$, $i \in \mathbb{N}$, we have

$$(t_i' - j) \cdot x \to y_j \quad \text{as} \quad i \to \infty$$

for all $j \in \mathbb{N}$.

Let now $0 \le k \le j$. Then it follows from

$$y_k = \lim_{i \to \infty} (t_i' - k) \cdot x = \lim_{i \to \infty} (t_i' - j + j - k) \cdot x$$
$$= \lim_{i \to \infty} (j - k) \cdot (t_i' - j) \cdot x = (j - k) \cdot \lim_{i \to \infty} (t_i' - j) \cdot x = (j - k) \cdot y_j$$

that for some $t \in \mathbb{R}$ with $-t \le k \le j$ we have

$$(t + j) \cdot y_j = (t + k + j - k) \cdot y_j = (t + k) \cdot (j - k) \cdot y_j = (t + k) \cdot y_k.$$

Therefore, we may define $u : \mathbb{R} \to X$ by

$$u(t) := (t + j) \cdot y_j \quad \text{if} \quad t + j \ge 0, \quad j \in \mathbb{N}.$$

Since $\omega(x)$ is positively invariant, it follows that $u(\mathbb{R}) \subseteq \omega(x)$ and it is clear that u is a solution of the semiflow through y_0. Since $y_0 \in \omega(x)$ was arbitrary, $\omega(x)$ is invariant.   □

(17.3) Remark. The proof above shows that if $\gamma^+(x)$ is relatively compact, there exists a solution u through every point $y \in \omega(x)$ and defined on all of $\mathbb{R}$ – a *global solution* – such that $u(\mathbb{R}) \subseteq \omega(x)$.   □

(17.4) Examples. (a) *x is a periodic point if and only if* $\gamma^+(x) = \omega(x)$ *and* $\gamma^+(x)$ *is compact.*

The necessity of the condition is clear. So assume that $\gamma^+(x) = \omega(x)$ and $\gamma^+(x)$ is compact. Then by (17.3), there exists a global solution u through x. It follows that $(-\tau) \cdot u(0) = u(-\tau)$ for every $\tau > 0$. Because $u(\mathbb{R}) \subseteq \omega(x) = \gamma^+(x)$, we have that for each fixed $\tau > 0$ there exists some $\tau' \ge 0$ such that $u(-\tau) = \tau' \cdot x$. Hence for every $t \ge 0$ we have

$$(t + \tau + \tau') \cdot x = t \cdot \tau \cdot \tau' \cdot x = t \cdot \tau \cdot u(-\tau) = t \cdot u(0) = t \cdot x.$$

Therefore x is periodic with period $\tau + \tau' > 0$.

If the semiflow is not negatively unique and if x is a periodic point, then, of course, there may be solutions through x which do not lie in $\omega(x)$.

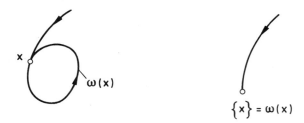

Such a situation cannot occur with flows. In fact here we have:

(b) *If φ is a flow, then x is periodic if and only if $\gamma(x) = \omega(x)$ and $\gamma(x)$ is compact.*

This follows immediately by applying (a) to the positive and negative semiflow of φ.

(c) For the planar flow induced by the differential equations $\dot{r} = r(1 - r), \dot{\vartheta} = 1$ (polar coordinates) we have : $\omega(0) = \{0\}$, $\omega(x) = \mathbb{S}^1$ if $x \in \mathbb{R}^2 \setminus \{0\}$, $\omega^-(x) = \emptyset$ if $|x| > 1$, $\omega^-(x) = \mathbb{S}^1$ if $|x| = 1$, and $\omega^-(x) = \{0\}$ if $|x| < 1$.

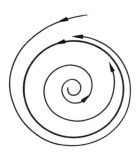

(d) Let φ be a hyperbolic linear flow e^{tA}.

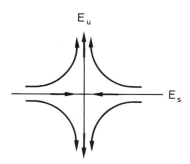

Then we have:

$$\begin{aligned}
\omega(x) &= \{0\}, & \forall x \in E_s, \\
\omega(x) &= \emptyset, & \forall x \in E \setminus E_s, \\
\omega^-(x) &= \{0\}, & \forall x \in E_u, \\
\omega^-(x) &= \emptyset, & \forall x \in E \setminus E_u.
\end{aligned}$$

(e) Let $t \cdot (\xi, \eta) := (e^{2\pi i t}\xi, e^{2\pi i \alpha t}\eta)$ be a "linear" flow on the torus $T^2 = \mathbb{S}^1 \times \mathbb{S}^1$. If $\alpha \in \mathbb{Q}$, every point is periodic (cf. problem 2(c) of section 10). Therefore $\gamma(x) = \omega(x)$ for all $x \in T^2$. If, however, α is irrational, then every orbit is dense in T^2. It follows that in this case $\omega(x) = T^2$ for all $x \in T^2$.

(f) If for $\alpha \in \mathbb{R} \setminus \mathbb{Q}$ we restrict the above linear flow on T^2 to a single orbit, i.e., $X = \gamma(x)$ for some fixed $x \in T^2$, then we obtain a flow on X for which we clearly have $\gamma(x) = \omega(x)$, but x is not periodic. This shows that we cannot omit the compactness assumptions in (a) and (b).
□

As a generalization of the concepts introduced in section 15, we will say that a point $x \in X$ is *attracted* by a set $M \subseteq X$ if

$$t^+(x) = \infty \quad \text{and} \quad t \cdot x \to M \quad \text{as} \quad t \to \infty.$$

The set

$$\mathcal{A}(M) := \{x \in X \mid x \text{ is attracted by } M\}$$

is called the *region of attraction* of M, and we say that M is an *attractor* (or that M *attracts*) if $\mathcal{A}(M)$ is a neighborhood of M. If $\mathcal{A}(M) = X$, we say that M is a *global attractor*.

(17.5) Remarks. (a) Evidently $\mathcal{A}(M)$ *is positively invariant.* Moreover, it is clear that $u(J) \subseteq \mathcal{A}(M)$ for every $x \in \mathcal{A}(M)$ and every solution $u : J \to X$ through x. *Hence if φ is a flow, then $\mathcal{A}(M)$ is invariant.*

(b) *If M is an attractor, then $\mathcal{A}(M)$ is open.*

Indeed, since $U := \text{int}(\mathcal{A}(M))$ is an open neighborhood of M, then for every $x \in \mathcal{A}(M)$ there exists some $\tau \geq 0$ such that $t \cdot x \in U$ for all $t \geq \tau$. Since $\varphi^\tau : \Omega_\tau \to X$ is continuous and since Ω_τ is open in X, it follows that $W := (\varphi^\tau)^{-1}(U)$ is an open neighborhood of x in X. Therefore $\mathcal{A}(M)$ is open.

(c) $\omega(x) \subseteq \overline{M}$ *for every* $x \in \mathcal{A}(M)$.

This follows immediately from (17.1 c). □

(17.6) Examples. (a) If $X \subseteq E$ is open and if $f \in C^{1-}(X, E)$ is such that $f(x_0) = 0$, then $\{x_0\}$ is an attractor of the flow induced by f if and only if x_0 is an attractive critical point in the sense of section 15.

(b) For the flow on $\mathbb{R}^2$, induced by

$$\dot{x} = -y + x \sin^2(\pi/r)$$
$$\dot{y} = x + y \sin^2(\pi/r),$$

where $r^2 := x^2 + y^2$, we have: The origin is a critical point and the circles $(1/n)\mathbb{S}^1$, centered at 0 with radii $1/n$, are periodic orbits. All other trajectories are spirals which

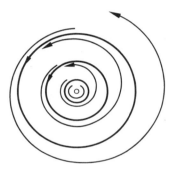

tend to the outer circles (or "toward ∞") as $t \to \infty$. Therefore $(1/n)\mathbb{S}^1$ attracts every point $(x, y) \in \overline{\mathbb{B}}(0, 1/n) \setminus \overline{\mathbb{B}}(0, 1/(n+1))$, but none of the orbits $(1/n)\mathbb{S}^1$ is an attractor.

(c) For the flow described in problem 1 of section 15, every circle $(1/2n)\mathbb{S}^1$ is an attractor with $\mathcal{A}((1/2n)\mathbb{S}^1) = \mathbb{B}(0, 1/(2n-1)) \setminus \overline{\mathbb{B}}(0, 1/(2n+1))$ for all $n \in \mathbb{N}^*$.

(d) If γ^+ is relatively compact, then theorem (17.2) and remarks (17.1 d) and (17.1 f) imply that $\gamma^+(x) \subseteq \mathcal{A}(\omega(x))$. □

A set $M \subseteq X$ is called *stable* if for every $x \in M$ and every neighborhood U of M there exists a neighborhood V of x such that $t^+(y) = \infty$ for all $y \in V$ and $t \cdot V \subseteq U$ for all $t \geq 0$. A set is called *asymptotically stable* if it is a stable attractor. If M is not stable, it is called *unstable*.

(17.7) Remarks. (a) *Assume that X is locally compact and M is compact. Then M is stable if and only if for every $x \in M$ and every neighborhood U of M there exists a neighborhood V of x such that*

$$t \cdot y \in U, \quad \forall t \in [0, t^+(y)), \quad \forall y \in V.$$

The necessity of the condition is clear. Conversely, if the condition above is satisfied, then, by the compactness of M and the local compactness of X, we may choose compact neighborhoods U of M. From corollary (10.13) it follows that $t^+(y) = \infty$ for all $y \in V$.

(b) If x_0 is a critical point of the flow induced by the vector field $f \in C^{1-}(X, E)$, $X \subseteq E$ open, then $\{x_0\}$ is (asymptotically) stable if and only if x_0 is (asymptotically) stable in the sense of Liapunov.

(c) If $\gamma^+(x)$ is stable (is an attractor or is asymptotically stable), we say that x (φ_x or $\gamma^+(x)$) is *orbitally stable* (is an *orbital attractor* or is *orbitally asymptotically stable*), respectively.

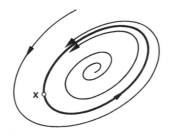

orbitally asymptotically stable x

(d) *If M is stable and closed, then M is positively invariant.*

This follows immediately from the obvious relation: $M = \cap\{U \mid U$ is a neighborhood of $M\}$.

$\square$

Problems

1. Draw the phase portrait of the flow in $\mathbb{R}^2$, induced by

$$\dot{r} = r(1 - r), \qquad \dot{\vartheta} = \sin^2(\vartheta/2)$$

(polar coordinates), and determine the positive and negative limit sets for each $x \in \mathbb{R}^2$.

2. Let φ be a semiflow on a metric space X. Prove that if $\{x_0\}$ is stable, then x_0 is a critical point.

3. Let φ be a semiflow on a metric space X. A flow line ("motion") φ_x is called *Liapunov stable* if for every $\epsilon > 0$ there exists some $\delta > 0$ such that

$$t^+(y) = \infty \quad \text{and} \quad d(t \cdot y, t \cdot x) < \epsilon, \qquad \forall t \geq 0,$$

holds for every $y \in \mathbb{B}(x, \delta)$.

Show that:

(i) If x is a critical point, then φ_x is Liapunov stable if and only if x is stable (in the old sense!).

(ii) If φ_x is Liapunov stable, then $\gamma^+(x)$ is orbitally stable.

(iii) The sets $X_b := \{x \in X \mid \gamma^+(x) \text{ is bounded}\}$ and $X_c := \{x \in X \mid \gamma^+(x) \text{ is relatively compact}\}$ are positively invariant. If all flow lines are Liapunov stable, then X_b and X_c are closed.

18. Liapunov Functions

In the discussion of Hamiltonian systems in section 3 we have made (e.g., to draw the phase portraits of the simple pendulum) considerable use of the fact that the energy is conserved, i.e., that a first integral exists. When, in fact, a first integral exists, the trajectories must lie on level sets. This describes – e.g., for bounded level sets – some kind of stability statement. We now want to generalize this idea considerably, which leads to important and extremely useful methods in stability analysis.

Again, let X denote a metric space and let $\varphi : \Omega \to X$, $(t, x) \mapsto t \cdot x$, be a semiflow on X. Also, assume that $M \subseteq X$ and that $V : M \to \mathbb{R}$ is continuous. For every $x \in M$, we now define the *orbital derivative* $\dot{V}(x)$ at x by

$$\dot{V}(x) := \liminf_{t \to 0^+} t^{-1}(V(t \cdot x) - V(x)),$$

whenever x is a limit point of $M \cap [0, \epsilon) \cdot x$ for some $\epsilon \in (0, t^+(x))$, and $\dot{V}(x) := -\infty$ otherwise. If $x \in M$ is such that there exists some $\epsilon \in (0, t^+(x))$ satisfying $[0, \epsilon) \cdot x \subseteq M$, then $\dot{V}(x)$ is exactly the lower right Dini derivative of the function $[0, \epsilon) \to \mathbb{R}$, $t \mapsto V(t \cdot x)$ at the point 0. The continuous function $V : M \to \mathbb{R}$ is called a *Liapunov function for φ on M* (or *for M*) if

$$\dot{V}(x) \le 0, \quad \forall x \in M.$$

(18.1) Remarks. (a)   Let X be an open subset of a real Banach space E and assume that φ is a flow induced by $f \in C^{1-}(X, E)$. If $V : X \to \mathbb{R}$ is differentiable, it immediately follows from the chain rule that

$$\dot{V}(x) = \langle DV(x), f(x) \rangle, \quad \forall x \in X. \tag{1}$$

Here $\langle \cdot, \cdot \rangle : E' \times E \to \mathbb{K}$ denotes the "duality pairing" between the dual space of E, E', and E, i.e., $\langle x', x \rangle$ is the value of the continuous linear functional $x' \in E'$ at x. If $(E, (\cdot \mid \cdot))$ is a Hilbert space (in particular $\mathbb{R}^m$) and if the gradient of V, ∇V, is defined by

$$(\nabla V(x) \mid y) = \langle DV(x), y \rangle, \quad \forall x \in X, \ \forall y \in E,$$

(cf. 16.11 c), then

$$\dot{V}(x) = (\nabla V(x) \mid f(x)), \quad \forall\, x \in X. \tag{2}$$

Formula (1) shows, in particular, that *under the assumptions above, the "orbital derivative"* $\dot{V}(x)$, *i.e., the derivative of* V *along the trajectory, can be calculated directly from the vector field without prior knowledge of the flow.*

(b) The most important property of a Liapunov function is the fact that *it is decreasing along orbits.* More precisely we have:

Let V *be a Liapunov function for* φ *on* M *and suppose that for some* $T \in (0, t^+(x))$ *we have* $t \cdot x \in M$ *for all* $0 \le t \le T$. *Then the function* $t \mapsto V(t \cdot x)$ *is decreasing on* $[0, T]$ *and*

$$V(t \cdot x) \le V(x) + \int_0^t \dot{V}(\tau \cdot x)d\tau \quad \text{for all} \quad 0 \le t < T. \tag{3}$$

Conversely, if M *is positively invariant and if the continuous function* $V : M \to \mathbb{R}$ *is decreasing along orbits, then* V *is a Liapunov function.*

To see this, we set $f(t) := V(t \cdot x)$ for all $0 \le t < T$. For the lower right Dini derivative we have

$$D_+ f(t) = \dot{V}(t \cdot x) \le 0, \quad \forall\, t \in [0, T).$$

Let $\epsilon > 0$ be arbitrary and set $f_\epsilon(t) := f(t) - \epsilon t$. Then $D_+ f_\epsilon(t) \le -\epsilon$ for all $t \in [0, T)$. If f_ϵ is nondecreasing, there exist numbers $0 \le t_0 < t_1 < T$ such that $f_\epsilon(t_1) > f_\epsilon(t_0)$. By the continuity of f_ϵ, we may assume that $f_\epsilon(t) > f_\epsilon(t_0)$ for all $t \in (t_0, t_1)$ (otherwise we increase t_0). It follows that $D_+ f_\epsilon(t_0) \ge 0$, which is impossible. Therefore f_ϵ is decreasing. Taking the limit $\epsilon \to 0$, it follows from this that f is also decreasing and is differentiable a.e. Moreover, we have: $f' = D_+ f$ a.e., f' is integrable on $[0, T]$ and

$$f(t) - f(0) = \varphi(t) + \int_0^t f'(s)ds, \quad 0 \le t < T,$$

for some continuous, decreasing function φ whose derivative vanishes a.e. (see, for example, theorem 8.18 in Rudin [1]). This proves the first part of the assertion. The converse is trivial.

(c) *Suppose* V *is a Liapunov function on* M *and assume that there exists some* $\alpha > 0$ *so that* $\dot{V}(y) \le -\alpha V(y)$ *for all* $y \in M$. *Then for each* $x \in M$ *and* $T \in [0, t^+(x))$ *satisfying* $[0, T] \cdot x \subseteq M$ *we have*

$$V(t \cdot x) \le e^{-\alpha t} V(x), \quad \forall\, t \in [0, T).$$

Indeed, since $t \mapsto V(t \cdot x)$ is decreasing, it follows from (3) that

$$V(x) - V(t \cdot x) \ge \alpha \int_0^t V(\tau \cdot x)d\tau \ge \alpha t V(t \cdot x)$$

for all $t \in [0, T)$. We thus obtain

$$V(x) \geq \left(1 + \frac{\alpha t}{n}\right) V\left(\frac{t}{n} \cdot x\right) \geq \left(1 + \frac{\alpha t}{n}\right)^2 V\left(\frac{2t}{n} \cdot x\right) \geq \cdots$$

$$\geq \left(1 + \frac{\alpha t}{n}\right)^n V(t \cdot x)$$

for all $t \in [0, T)$ and $n = 2, 3, \ldots$ As $n \to \infty$, we get that $V(x) \geq e^{\alpha t} V(t \cdot x)$, which is the assertion.

(d) For some investigations it is advantageous to admit discontinuous Liapunov functions as well (e.g., Bhatia and Szegö [1]). To simplify matters, we will not consider any of these generalizations.   □

The following theorem shows that Liapunov functions can be utilized to determine positively invariant sets.

(18.2) Theorem. *Let* $-\infty \leq \gamma < \beta < \infty$ *and assume that* $V \in C(X, \mathbb{R})$ *is a Liapunov function on*

$$\{x \in X \mid \gamma < V(x) < \beta\}.$$

Then

$$M_\alpha := \{x \in X \mid V(x) \leq \alpha\}$$

is positively invariant for each $\alpha \in [\gamma, \beta)$.

Proof. Assume that $\gamma < \alpha < \beta$ and let $x \in \partial M_\alpha \subseteq V^{-1}(\alpha)$ be arbitrary. Since $U := V^{-1}(\gamma, \beta)$ is a neighborhood of x, there exists some $\epsilon > 0$ such that $[0, \epsilon) \cdot x \subseteq U$. Because V is a Liapunov function on U, it follows from (3) that $V(t \cdot x) \leq V(x) = \alpha$ for all $t \in [0, \epsilon)$. Hence $[0, \epsilon) \cdot x \subseteq M_\alpha$ and now (16.3 f) implies that M_α is positively invariant. Finally, since $M_\gamma = \cap\{M_\alpha \mid \gamma < \alpha < \beta\}$, we deduce the positive invariance of M_γ from (16.3 d).   □

Evidently, theorem (18.2) is closely related to theorem (16.9). However, the latter theorem is a much sharper result (in those special situations in which it applies), since it involves a condition which must be satisfied only on the boundary of M_α, while in theorem (18.2) V must be a Liapunov function (i.e., decreasing along orbits) in a whole neighborhood of ∂M_α.

Liapunov functions can be employed to localize positive limit sets, as the following *LaSalle invariance principle* shows.

(18.3) Theorem. *Let* $M \subseteq X$ *be closed and assume that* $V \in C(M, \mathbb{R})$ *is a Liapunov function on* $M \subseteq X$. *If* $\gamma^+(x) \subseteq M$, *there exists some* $\alpha \in \mathbb{R}$ *so that* $\omega(x) \subseteq V^{-1}(\alpha)$. *In particular,*

$$\omega(x) \subseteq \{y \in M \mid \dot{V}(y) = 0\}.$$

Proof. Without loss of generality we may assume that $\omega(x) \neq \emptyset$. Then $t^+(x) = \infty$ by (17.1 b). Using (17.2 d), we get $\omega(x) \subseteq \overline{\gamma^+(x)}$, hence $\omega(x) \subseteq M$. And since V is a Liapunov function on M, it follows from (18.1 b) that the function $t \mapsto V(t \cdot x)$ is decreasing on $\mathbb{R}_+$. Therefore $V(t \cdot x)$ converges to

$$\alpha := \inf\{V(t \cdot x) \mid t \in \mathbb{R}_+\}$$

as $t \to \infty$. Based on the continuity of V, it follows from (17.1 c) that $V(y) = \alpha$ for all $y \in \omega(x)$. Consequently α is finite. The last assertion now follows immediately from the positive invariance of $\omega(x)$ (cf. (17.1 e)). □

(18.4) Corollary. *Let* $V \in C(M, \mathbb{R})$ *be a Liapunov function on the closed set* $M \subseteq X$ *and assume that* $\gamma^+(x)$ *is relatively compact with* $\gamma^+(x) \subseteq M$. *Moreover, let* M_V *denote the largest invariant subset of* $\{x \in M \mid \dot{V}(x) = 0\}$. *Then we have*

$$t \cdot x \to M_V \quad \text{as } t \to \infty.$$

Proof. This follows immediately from theorems (17.2) and (18.3). □

(18.5) Corollary. *Let* $M \subseteq X$ *be closed and positively invariant, and assume that* $V \in C(M, \mathbb{R})$ *is a Liapunov function on* M. *If every positive semitrajectory in* M *is relatively compact, then* M_V *attracts every point in* M, *i.e.,* $M \subseteq \mathcal{A}(M_V)$.

Of course every semitrajectory $\gamma^+(x) \subseteq M$ is relatively compact if M is compact. This is the prevailing situation for flows induced by autonomous ordinary differential equations. In the case of parabolic differential equations (e.g., reaction-diffusion equations), M is in general not compact, although it is often possible to establish the relative compactness of semitrajectories in M.

For stability investigations we will, in what follows, employ Liapunov functions. For this we need the following notation: If $M \subseteq X$ is nonempty, let

$$d(x, M) := \text{dist}(x, M) \quad \text{and} \quad \mathbb{B}(M, r) := \{x \in X \mid d(x, M) < r\}.$$

It is clear that $\mathbb{B}(M, r)$ is an open neighborhood – *the r-neighborhood* – of M.

Finally, we say that the set M is *exponentially stable* (with exponent α), if there exist positive constants α, γ and δ such that for each $x \in \mathbb{B}(M, \delta)$ we have:

$$t^+(x) = \infty \quad \text{and} \quad d(t \cdot x, M) \leq \gamma e^{-\alpha t} d(x, M) \tag{4}$$

for all $t \geq 0$. If there exist positive constants α and γ so that (4) holds for all $x \in X$, then M is called *globally exponentially stable* (with exponent α).

(18.6) Remarks. (a) *If* M *is compact and globally exponentially stable, then* M *is asymptotically stable.*

Indeed, if U is an arbitrary neighborhood of M, then the compactness of M and the continuity of $d(\cdot, M)$ imply that $r := d(M, U^c) := \inf_{m \in M} d(m, U^c)$ is positive. Hence

$\mathbb{B}(M, r) \subseteq U$, and from (4) we deduce the existence of some $T > 0$ such that $t \cdot x \in U$ for all $t \geq T$. (If M is closed, but not compact, and if U is a neighborhood of M, there may not be an r-neighborhood satisfying $\mathbb{B}(M, r) \subseteq U$.

Example: $X = \mathbb{R}_+^2$, $M = \{(x, y) \in X \mid xy \geq 1\}$, $U = (0, \infty)^2$.)

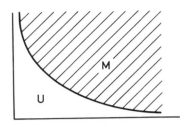

(b) *Let X be an open subset of a finite dimensional Banach space E and assume that $f \in C^{1-}(X, E)$ satisfies $f(x_0) = 0$. If*

$$Re\, \sigma(Df(x_0)) < \alpha < 0,$$

then x_0 is an exponentially stable critical point (i.e., $\{x_0\}$ is exponentially stable) with exponent α.

This follows immediately from formula (9) in the proof of theorem (15.3).   □

In the following theorem we let $\mathcal{W}_r$ denote the set of all increasing functions $g : [0, r] \rightarrow \mathbb{R}$ satisfying $g(0) = 0$ and $g(\xi) > 0$ if $\xi > 0$.

(18.7) Theorem (*Stability*). *Let $V \in C(X, \mathbb{R})$ and assume that $M := V^{-1}(-\infty, 0]$ is compact and nonempty. Moreover, let $r > 0$ and assume that V is a Liapunov function on $\mathbb{B}(M, r) \backslash M$. Finally, suppose every positive semitrajectory in $\mathbb{B}(M, r)$ is relatively compact.*

(i) *If for some $g \in \mathcal{W}_r$ we have*

$$V(x) \geq g(d(x, M)), \quad \forall x \in \mathbb{B}(M, r) \backslash M, \tag{5}$$

then M is stable.

(ii) *If, in addition, there exists some $h \in \mathcal{W}_r$ such that*

$$\dot{V}(x) \leq -h(d(x, M)), \quad \forall x \in \mathbb{B}(M, r) \backslash M, \tag{6}$$

then M is asymptotically stable.

(iii) *Finally, if there exist positive constants $\alpha, \beta, \overline{\gamma}$ and $\hat{\gamma}$ satisfying*

$$\dot{V}(x) \le -\alpha V(x) \tag{7}$$

and

$$\bar{\gamma}[d(x, M)]^\beta \le V(x) \le \hat{\gamma}[d(x, M)]^\beta \tag{8}$$

for all $x \in \mathbb{B}(M, r)$, then M is exponentially stable with exponent α/β.

Moreover, in cases (ii) *and* (iii) *we also have*

$$\mathcal{A}(M) \supseteq V^{-1}(-\infty, \bar{\alpha}),$$

where

$$\bar{\alpha} := \sup\{\alpha \ge 0 \mid V^{-1}(-\infty, \alpha] \subseteq \mathbb{B}(M, r)\}.$$

Proof. First we deduce from corollary (10.13) that $t^+(x) = \infty$ for every x with $\gamma^+(x) \subseteq \mathbb{B}(M, r)$. If U is an arbitrary neighborhood of M, it follows from the compactness of M that for some $\epsilon \in (0, r)$ we have $\mathbb{B}(M, \epsilon) \subseteq U$.

(i) Let now $0 < \beta < g(\epsilon)$. Since g is increasing, it follows that

$$M_\beta := \{x \in X \mid V(x) \le \beta\} \subseteq \mathbb{B}(M, \epsilon) \subseteq U.$$

The continuity of V implies that for each $\alpha \in (0, \beta)$, M_α is a neighborhood of M. Since V is a Liapunov function on $V^{-1}(0, \beta)$, it follows from theorem (18.2) that M_α is positively invariant. Therefore M is stable.

(ii) If $0 < \alpha < \beta < g(r)$, then V is a Liapunov function on the closed set $M_\beta \setminus \text{int}(M_\alpha) \subseteq \mathbb{B}(M, r) \setminus M$. Using (6), it follows from the invariance theorem (18.3) that for no $x \in M_\beta \setminus \text{int}(M_\alpha)$ can the semiorbit $\gamma^+(x)$ lie completely in $M_\beta \setminus \text{int}(M_\alpha)$ (for otherwise we would have $\omega(x) \subseteq M_\beta \setminus \text{int}(M_\alpha)$). Hence for each $x \in M_\beta \setminus M_\alpha$ there exists some $t > 0$ such that $t \cdot x \in M_\alpha$. Since, by theorem (18.2), M_α is positively invariant, we have $\gamma^+(t \cdot x) \subseteq M_\alpha$. Because for every neighborhood U of M there exists some $\alpha \in (0, \beta)$ so that $M_\alpha \subseteq U$, this consideration shows that M is an attractor and that $M_\beta \subseteq \mathcal{A}(M)$.

(iii) Let $x \in M_\gamma \subseteq \mathbb{B}(M, r)$. Since (5) follows from (8), we have that M_γ is positively invariant. Therefore $t \cdot x \in \mathbb{B}(M, r)$ for all $t \ge 0$. From (7), (8) and remark (18.1 c) we now obtain

$$\bar{\gamma}[d(t \cdot x, M)]^\beta \le V(t \cdot x) \le e^{-\alpha t} V(x) \le e^{-\alpha t} \hat{\gamma}[d(x, M)]^\beta,$$

and hence

$$d(t \cdot x, M) \le \delta e^{-(\alpha/\beta)t} d(x, M)$$

for all $x \in M_\gamma$ and $t \ge 0$, where we have set $\delta := (\hat{\gamma}/\bar{\gamma})^{1/\beta}$. Therefore M is exponentially stable with exponent α/β and $\mathcal{A}(M) \supseteq M_\gamma$.

The last assertion is now evident. $\square$

This stability criterion can be significantly simplified if X is assumed to be a locally compact metric space. For reasons of simplicity, the next corollary will be confined to the following situation: *E is a finite dimensional real Banach space, X is open in E, and the flow is induced by a vector field $f \in C^{1-}(X, E)$.*

(18.8) Corollary. *Let $V \in C(X, \mathbb{R})$ and assume that $M := V^{-1}(-\infty, 0]$ is compact and nonempty. Moreover, assume that for some $r > 0$ we have $\overline{\mathbb{B}}_E(M, r) \subseteq X$ and that V is continuously differentiable on $\mathbb{B}(M, r)\backslash M$. If*

$$\langle DV(x), f(x)\rangle \le 0, \quad \forall\, x \in \mathbb{B}(M, r)\backslash M, \tag{9}$$

then M is stable. If, in fact,

$$\langle DV(x), f(x)\rangle < 0, \quad \forall\, x \in \mathbb{B}(M, r)\backslash M, \tag{10}$$

then M is asymptotically stable and $\mathcal{A}(M) \supseteq V^{-1}(-\infty, \bar{\alpha}]$, where $\bar{\alpha} := \sup\{\alpha \ge 0 \mid V^{-1}(-\infty, \alpha] \subseteq \mathbb{B}(M, r)\}$.

Proof. Since M is compact and hence bounded, it follows that also $\mathbb{B}(M, r)$ is bounded and therefore relatively compact (because $\dim(E) < \infty$). Hence every semitrajectory $\gamma^+(x) \subseteq \mathbb{B}(M, r)$ is relatively compact. If follows from (9) and (1) that V is a Liapunov function on $\mathbb{B}(M, r)\backslash M$.

We now set $g(0) := 0$ and for each $\xi \in [0, r)$ we set

$$g(\xi) := \min\{V(x) \mid \xi \le d(x, M) \le r\}.$$

For reasons of compactness we have $g \in \mathcal{W}_r$ and V satisfies (5). Analogously, it follows from (10) that $h \in \mathcal{W}_r$, where h is defined by

$$h(\xi) := \min\{-\langle DV(x), f(x)\rangle \mid \xi \le d(x, M) \le r\}, \quad \text{if } 0 < \xi < r$$

and $h(0) := 0$. Therefore (6) holds, and so the assertions follow from theorem (18.7) (with $X = \mathbb{B}(M, r)$). $\square$

(18.9) Remarks. (a) If – under the assumptions of corollary (18.8) – $M = \{x_0\}$, that is, if x_0 is an isolated minimum of V, then x_0 is a critical point of f (cf. problem 4 of section 10). Hence, in this case, corollary (18.8) furnishes sufficient conditions for the (asymptotic) stability of a critical point. Because for this method one does not need to know the solutions of the differential equation $\dot{x} = f(x)$, it is also called *Liapunov's direct method* (or *Liapunov's second method*). There are, however, no general criteria which make possible the construction of appropriate Liapunov functions. In many practical cases one must depend on luck and ingenuity, and it is often useful to consider the origin of the equations (i.e., the physical, ecological, etc. theories which are behind the mathematical model). For examples in this regard we refer to the literature (e.g., Hahn [1], Rouche-Habets-Laloy [1]).

So-called *converse theorems*, which assert when the existence of Liapunov functions follows from the (asymptotic) stability, are also of interest in this connection (cf. the literature cited above).

(b) Let $J \subseteq \mathbb{R}$ be an open interval so that $\mathbb{R}_+ \subseteq J$ and assume that $g \in C^{1-}(J \times X, E)$ satisfies $g(\cdot, 0) = 0$. Then, as is well-known, the differential equation

$$\dot{x} = g(t, x) \tag{11}$$

can be transformed into an autonomous equation by "extending the phase space." More precisely, we set: $\hat{X} := J \times X \subseteq \mathbb{R} \times E =: \hat{E}$, $\hat{x} := (t, x) \in \hat{X}$ and $\hat{f} := (1, g) \in C^{1-}(\hat{X}, \hat{E})$. Then (11) is equivalent to

$$\dot{\hat{x}} = \hat{f}(\hat{x}), \quad \text{that is,} \quad \dot{t} = 1, \quad \dot{x} = g(t, x). \tag{12}$$

A function $\hat{V} \in C^1(J \times X, \mathbb{R})$ is called a *Liapunov function for the nonautonomous differential equation* (11) *on* $M \subseteq X$ if the following holds:

$$D_1 \hat{V}(t, x) + \langle D_2 \hat{V}(t, x), g(t, x) \rangle_E \le 0 \tag{13}$$

for all $(t, x) \in J \times M$, i.e., if $\hat{V}$ is a Liapunov function for (12) on $\hat{M} := J \times M$. Then the zero solution of (11) corresponds to the set $\hat{M}_0 := J \times \{0\} \subseteq \hat{X}$, which is certainly not compact. Consequently, corollary (18.8) cannot be applied directly to this case.

In the proof of theorem (18.7), however, we have used the compactness of M only twice; in order to show that $t^+(x) = \infty$ for all $x \in \mathbb{B}(M, r)$ and to show that for every neighborhood U of M there exists an ϵ-neighborhood $\mathbb{B}(M, \epsilon) \subseteq U$. In the case of the nonautonomous problem (11), the definition of (asymptotic) stability of the zero solution (cf. section 15) does not agree with the definition of (asymptotic) stability of the set $\hat{M}_0$ of section 17. In this case one need not consider general neighborhoods $\hat{U}$ of $\hat{M}_0$, but rather just "cylindrical neighborhoods" of the form $\hat{U} = J \times U$, where U is a neighborhood of 0 in X. However, it is directly clear that for every such cylindrical neighborhood $J \times U$ of $\hat{M}_0$ there exists some "cylindrical ϵ-neighborhood" $\hat{\mathbb{B}}(\hat{M}_0, \epsilon) := J \times \mathbb{B}_E(0, \epsilon)$ of $\hat{M}_0$ such that $\hat{\mathbb{B}}(\hat{M}_0, \epsilon) \subseteq J \times U$. It then follows from remark (15.1 a) that $\hat{t}^+(\hat{x}) := t^+(\tau, x) = \infty$ for all $\hat{x} = (\tau, x) \in \hat{\mathbb{B}}(\hat{M}_0, r)$. As a result of these considerations we see that parts (i) and (ii) of theorem (18.7) can be applied to the stability problem of the zero solution of (11) if conditions (5) and (6) are replaced by, respectively,

$$\hat{V}(t, x) \ge g(|x|) \tag{5'}$$

and

$$\dot{\hat{V}}(t, x) \le -h(|x|), \tag{6'}$$

for all $(t, x) \in J \times \mathbb{B}(0, r)$, where $\dot{\hat{V}}$ is defined by the left-hand side of (13). Since the following conditions, (14) and (16), imply the inequalities (5') and (6'), respectively, as in the proof of corollary (18.8), we finally obtain the following analogue of corollary (18.8) for the nonautonomous case:

Let $J \subseteq \mathbb{R}$ be an open interval such that $\mathbb{R}_+ \subseteq J$ and assume that $g \in C^{1-}(J \times X, E)$ satisfies $g(\cdot, 0) = 0$. Moreover, let $r > 0$ be such that $\bar{\mathbb{B}}_E(0, r) \subseteq X$ and assume that $\hat{V} \in C^1(J \times \mathbb{B}_E(0, r), \mathbb{R}_+)$ satisfies the following conditions:

$$D_1\hat{V}(t,x) + \langle D_2\hat{V}(t,x), g(t,x)\rangle \le 0 \qquad (\alpha)$$

for all $(t,x) \in J \times \mathbb{B}_E(0,r)$.

$$\hat{V}(t,0) = 0, \qquad \forall\, t \in J. \qquad (\beta)$$

If there exists a function $W \in C(\mathbb{B}_E(0,r), \mathbb{R})$ *such that* $W(x) > 0$ *for all* $|x| > 0$ *and*

$$\hat{V}(t,x) \ge W(x), \qquad \forall\, (t,x) \in J \times \mathbb{B}_E(0,r), \qquad (14)$$

then the zero solution of the nonautonomous differential equation

$$\dot{x} = g(t,x) \qquad (15)$$

is stable. If, in addition, there exists a function $W_1 \in C(\mathbb{B}_E(0,r), \mathbb{R})$ *such that* $W_1(x) > 0$ *for all* $|x| > 0$ *and*

$$D_1\hat{V}(t,x) + \langle D_2\hat{V}(t,x), g(t,x)\rangle \le -W_1(x) \qquad (16)$$

for all $(t,x) \in J \times \mathbb{B}_E(0,r)$, *then the zero solution of (15) is asymptotically stable.* □

Our next theorem shows that Liapunov functions can also be employed to prove instability. For reasons of simplicity, we only consider the case of a critical point, and we leave it to the reader to transfer this to the general case of sets.

(18.10) Theorem. *Let* $V \in C(X, \mathbb{R})$ *and assume that* $x_0 \in X$ *is a critical point satisfying* $V(x_0) = 0$. *Moreover, suppose there exists some* $r > 0$ *and some* $h \in \mathcal{W}_\infty$ *such that*

$$\dot{V}(x) \le -h(-V(x)) \qquad (17)$$

for all

$$x \in \{y \in X \mid V(y) < 0\} \cap \mathbb{B}(x_0, r) =: A_r.$$

If for each $\epsilon > 0$, A_ϵ *is nonempty, and if* $t^+(x) = \infty$ *for all* x *such that* $\gamma^+(x) \subseteq A_r$, *then* x_0 *is unstable.*

Proof. Choose $\delta \in (0,r)$ so small that $V(x) \ge -1$ for all $x \in A_\delta$. If x_0 is stable then there exists some $\epsilon > 0$ such that $\gamma^+(y) \subseteq \mathbb{B}(x_0, \delta)$ for all $y \in A_\epsilon$. Because of (17), V is a Liapunov function on A_r, and so it follows from (18.1 b) that $\gamma^+(y) \subseteq A_\delta$ for all $y \in A_\epsilon$. Moreover, from (3) and from the fact that V is decreasing along the trajectories in A_δ, we obtain

$$V(t \cdot y) \le V(y) + \int_0^t \dot{V}(\tau \cdot y)d\tau \le V(y) - \int_0^t h(-V(\tau \cdot y))d\tau$$
$$\le V(y) - th(-V(y)).$$

This implies that $V(t \cdot y) \to -\infty$, which contradicts $V(t \cdot y) \ge -1$. □

We leave it to the reader to formulate a simplified version of this instability criterion for the finite dimensional case. It should also be pointed out that the proof of theorem (15.5) is based on the same idea as the proof of the instability theorem above.

(18.11) Examples. (a) *Gradient Systems.* Let $X \subseteq \mathbb{R}^m$ be open and $V \in C^2(X, \mathbb{R})$. A system of differential equations of the form

$$\dot{x} = -\operatorname{grad} V(x) \tag{18}$$

is called a *gradient system.* Using (2), we get

$$\dot{V}(x) = -|\operatorname{grad} V(x)|^2, \quad \forall x \in X. \tag{19}$$

Therefore V is a Liapunov function on X and $\dot{V}(x) = 0$ if and only if x is a critical point. If x_0 is a regular point of V, that is, if $\nabla V(x_0) \neq 0$, then we know that $\nabla V(x_0)$ is a vector orthogonal to the level set $V(x) = V(x_0)$ at x_0. We therefore have:

At regular points the trajectories intersect the level sets orthogonally. Nonregular points are stationary points of the gradient flow, i.e., the flow induced by (18).

If we apply the invariance theorem (18.3) to the positive and negative semiflow, we obtain:

The α- and ω-limit points of a gradient flow correspond exactly to the critical points of V.

If x_0 is an isolated minimum of V, there exists an open neighborhood U of x_0 such that

$$V(x) - V(x_0) > 0, \quad \forall x \in U \setminus \{x_0\}. \tag{20}$$

If, in addition, x_0 is also an isolated critical point, then U can be chosen so that we also have

$$[V - V(x_0)]^\cdot(x) < 0, \quad \forall x \in U \setminus \{x_0\} \tag{21}$$

(cf. (19)). Hence we may apply corollary (18.8) to U and the Liapunov function $V - V(x_0)$ and we obtain:

Isolated minima of V, which are also isolated critical points, are asymptotically stable stationary points of the gradient flow.

A good description of a (two-dimensional) gradient flow is obtained if one imagines water flowing down the surface, described by the graph of V in $\mathbb{R}^3$, under the influence of constant "gravity" and no friction. Then the individual trajectories correspond to the lines of steepest descend, that is, they are the gradient lines in the direction of the "valleys" and sinks, where they come to rest.

In particular, we see that a gradient flow possesses no noncritical periodic points. Systems of differential equations which are supposed to describe periodic phenomena (e.g., in biology or chemistry) therefore cannot be gradient systems.

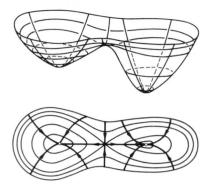

(b) *Hamiltonian Systems.* Let X be an open subset of $\mathbb{R}^n \times \mathbb{R}^n$ and let $H \in C^2(X, \mathbb{R})$. Then it follows from the general theorem about conservation of energy (corollary (3.12)) that the Hamiltonian H is a first integral of the Hamiltonian system

$$\dot{x} = H_y, \quad \dot{y} = -H_x. \tag{22}$$

Consequently H is a Liapunov function for (22) on X and $\dot{H}(x, y) = 0$ for all $(x, y) \in X$ (the trajectories of (22) lie on the level sets $H^{-1}(\alpha), \alpha \in \mathbb{R}$). From corollary (18.8) we deduce:

If (x_0, y_0) is an isolated minimum of the Hamiltonian, then (x_0, y_0) is a stable critical point.

If now H has the special form

$$H(x, y) = \frac{1}{2}(A(x)y \mid y) + U(x) = E_{\text{kin}} + E_{\text{pot}}$$

for some uniformly positive definite symmetric matrix $A(x) \in \mathsf{M}^n(\mathbb{R})$, and if x_0 is an isolated local minimum of the potential energy U, then the point $(x_0, 0)$ is a stable stationary point (see, for example, the phase portrait of the simple pendulum (Fig. 2 in section 3)).

As a simple, but important, consequence of Liouville's theorem about the conservation of volume (example (11.10)) we want to note:

No critical point or orbit of a Hamiltonian system can be asymptotically stable or asymptotically orbitally stable, respectively.

(c) *Norms as Liapunov Functions.* Let $(E, (\cdot \mid \cdot))$ be a real Hilbert space with corresponding norm $|\cdot|$. Then the function

$$V(x) := \frac{1}{2}|x|^2, \quad x \in E, \tag{23}$$

is continuously differentiable and we have

$$\langle DV(x), y \rangle = (\nabla V(x) \mid y) = (x \mid y), \quad \forall\, x, y \in E, \tag{24}$$

i.e., $\nabla V(x) = x$. The level set $V^{-1}(\alpha)$, $\alpha > 0$, is the sphere centered at 0 with radius $\sqrt{2\alpha}$. If $f \in C^{1-}(E, E)$ is a vector field satisfying

$$(x \mid f(x)) \le 0, \quad \forall\, x \in V^{-1}(\alpha, \beta), \tag{25}$$

then V is a Liapunov function for the flow induced by f on $V^{-1}(\alpha, \beta)$, $0 < \alpha < \beta < \infty$. Since the vector x has the direction of the outward normal, (25) says that the vector field f "points inward." This geometric interpretation is of course also possible for arbitrary norms

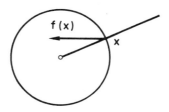

in Banach spaces, even if the norm is not differentiable (as, for instance, the maximum-norm $|\cdot|_\infty$ in $\mathbb{R}^m$). It is then necessary – graphically speaking – to replace the radius vector, i.e., the outward normal, by the "cone of outward normals," $\vartheta(x)$, i.e., the set of all outward normals to the support hyperplanes of the ball $\overline{\mathbb{B}}(0, \sqrt{2\alpha})$ (in the corresponding norm). In the general case of an *arbitrary real Banach space* $(E, |\cdot|)$, $\vartheta(x)$ is a subset of

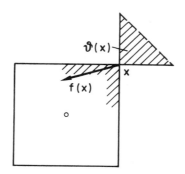

the dual space E'. By definition we have

$$\vartheta(x) := \{x' \in E' \mid \langle x', x \rangle = |x|^2,\ |x'| = |x|\}$$

for every $x \in E$. By the Hahn-Banach theorem (e.g., Yosida [1]) it follows that for every $x \in E$ satisfying $|x| = 1$ there exists some $x' \in E'$ such that $|x'| = 1$ and $\langle x', x \rangle = 1$. Hence for each $x \in E$, $\vartheta(x)$ is nonempty. The map

$$E \to 2^{E'}, \quad x \mapsto \vartheta(x),$$

is called the *duality mapping on E*, where $2^{E'}$ denotes the power set, i.e., the set of all subsets of E'. We clearly have

$$\vartheta(x) = \{x' \in E' \mid \langle x', x \rangle \ge |x|^2\} \cap \{x' \in E' \mid |x'| \le |x|\}, \tag{26}$$

i.e., $\vartheta(x)$ is the intersection of a closed half space with a closed ball $\overline{\mathbb{B}}_{E'}(0, |x|)$. Therefore $\vartheta(x)$ is convex, closed, bounded, and hence by Alaoglu's theorem (e.g., Yosida [1]), it is also w^*-compact. Thus for every pair $x, y \in E$ there exists some

$$y'(y) \in \vartheta(y) \quad \text{such that} \quad \langle y'(y), x \rangle = \max\{\langle y', x \rangle \mid y' \in \vartheta(y)\}.$$

For each pair $x, y \in E$ we now define a *semi inner product* $[y \mid x]$ by

$$[y \mid x] := \max\{\langle y', x \rangle \mid y' \in \vartheta(y)\}.$$

It clearly follows that

$$[y \mid x_1 + x_2] \leq [y \mid x_1] + [y \mid x_2],$$

$$|[y \mid x]| \leq |y||x| \quad \text{and} \quad [\alpha x \mid \beta y] = \alpha\beta[x \mid y]$$

for all $x, y, x_1, x_2 \in E$ and all $\alpha, \beta \in \mathbb{R}_+$. But, in general, $[\cdot \mid \cdot]$ is neither bilinear nor continuous.

If E' is *strictly convex*, i.e., if the unit sphere in E' does not contain line segments with more than one point, it follows directly from (26) that $\vartheta(x)$ is a one-point set. This is the case, for example, in every Hilbert space and also in every L^p-space, $1 < p < \infty$. From this it follows, in particular, that in a Hilbert space the semi inner product agrees with the inner product (cf. remark (11.16 a)).

Assume now that $C \subseteq E$ is a closed convex set with nonempty interior. By the separation theorem for convex sets (i.e., the geometric form of the Hahn-Banach theorem, e.g., Schäfer [1]), for each $x \in \partial C$ there exists some $x' \in E' \backslash \{0\}$ such that

$$\langle x', y \rangle \leq \langle x', x \rangle, \quad \forall y \in C, \tag{27}$$

a so-called *support functional to C (at x)*. In this case we call

$$H_{x'} := \{y \in E \mid \langle x', y \rangle \leq \langle x', x \rangle\}$$

a *support half space to C at x*, and the intersection of all such support half spaces to C at x is the *support cone $K_x(C)$ to C at $x \in \partial C$*. Clearly $K_x(C)$ is a closed convex cone with vertex x (i.e., $y \in K_x(C)$ implies that $x + t(y - x) \in K_x(C)$ for all $t \geq 0$) and we have $C \subseteq K_x(C)$ for every $x \in \partial C$. In particular then, $K_x(C)$ has interior points.

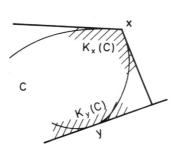

More precisely:

y is an interior point of $K_x(C)$ if and only if $\langle x', y \rangle < \langle x', x \rangle$ for every support functional
x' to C at $x \in \partial C$.

Indeed, the necessity of this condition is clear. Conversely, if $y \notin \text{int}(K_x(C))$, then by the separation theorem quoted above, there exists some $y' \in E' \setminus \{0\}$ such that $\langle y', y \rangle \geq \langle y', z \rangle$ for all $z \in K_x(C)$. Since $u \in K_x(C)$ always implies that $x + t(u - x) \in K_x(C)$ for all $t \geq 0$, it follows from

$$\langle y', y \rangle \geq \langle y', x + t(u - x) \rangle = \langle y', x \rangle + t\langle y', u - x \rangle$$

that $\langle y', u - x \rangle \leq 0$, i.e., $\langle y', u \rangle \leq \langle y, x \rangle$ for all $u \in K_x(C)$. Because $C \subseteq K_x(C)$, y' is a support functional to C at x and we have $\langle y', y \rangle \geq \langle y', x \rangle$. Therefore the condition above is also sufficient.

For some $x \neq 0$, let $C := \overline{\mathbb{B}}(0, |x|)$. Assume that $x' \in E'$ is a support functional to C at x. Since $|x'| = \sup\{\langle x', y \rangle \mid |y| \leq 1\}$, we obtain from (27) that $|x'| = \langle x', x \rangle / |x|$. For $u := (|x|/|x'|)x'$ we therefore have $|u| = |x|$ and $\langle u, x \rangle = |x|^2$, i.e., $u \in \vartheta(x)$. Conversely, since

$$\langle x', x \rangle = |x|^2 = |x'||x| \geq |x'||y| \geq \langle x', y \rangle, \quad \forall y \in \overline{\mathbb{B}}(0, |x|),$$

every $x' \in \vartheta(x)$ is a support functional to C at x. Therefore $\{tx' \mid t > 0, x' \in \vartheta(x)\}$ is the set of all support functionals to $\overline{\mathbb{B}}(0, |x|)$ at x. From this we obtain the important relation

$$K_x(\overline{\mathbb{B}}(0, |x|)) = x + \{y \in E \mid [x \mid y] \leq 0\}. \tag{28}$$

Let $T_x(E)$ denote the tangent space of E at x, i.e., $T_x(E) = \{(x, y) \mid y \in E\}$. Then (28) expresses the following:

The vector $(x, y) \in T_x(E)$ lies in the support cone $K_x(\overline{\mathbb{B}}(0, |x|))$
if and only if $[x \mid y] \leq 0$. $\tag{29}$

We also deduce from the above considerations that:

$(x, y) \in T_x(E)$ lies in the interior of the support cone
$K_x(\overline{\mathbb{B}}(0, |x|))$ if and only if $[x \mid y] < 0$. $\tag{30}$

With this we have found a mathematically precise formulation of the expression "the vector y at $x \in \overline{\mathbb{B}}(0, |x|)$ points inward."

The following lemma now gives us a generalization of formula (24).

(18.12) Lemma. *For every $x, y \in E$ we have*

$$\lim_{t \to 0^+} \frac{|x + ty|^2 - |x|^2}{2t} = [x \mid y].$$

Proof. Since $|x + ty|^2 - |x|^2 = (|x + ty| + |x|)(|x + ty| - |x|)$, it suffices to show (for fixed $x \in E$) that

$$\varphi(y) := \lim_{t \to 0^+} (|x + ty| - |x|)/t$$

exists and that $|x|\varphi(y) = [x \mid y]$.

Since $t \mapsto |x + ty|$ is convex, it easily follows that $t \mapsto (|x + ty| - |x|)/t$ is increasing for $t > 0$. Since also $|x + ty| - |x| \geq -t|y|$, $\varphi : E \to \mathbb{R}$ is well-defined.

From the definition of φ it immediately follows that

$$\varphi(y) \leq |y|, \quad \varphi(sy) = s\varphi(y), \quad \forall s > 0, \quad \forall y \in E, \tag{31}$$

and

$$\varphi(\lambda x) = \lambda\varphi(x), \quad \forall \lambda \in \mathbb{R}. \tag{32}$$

Because $2|x + t(y_1 + y_2)/2| \leq |x + ty_1| + |x + ty_2|$, the subadditivity of φ now follows from (31):

$$\varphi(y_1 + y_2) \leq \varphi(y_1) + \varphi(y_2), \quad \forall y_1, y_2 \in E. \tag{33}$$

From this, in turn, it follows that $0 = \varphi(0) = \varphi(y - y) \leq \varphi(y) + \varphi(-y)$ and hence $-\varphi(-y) \leq \varphi(y)$. Together with (31) this implies

$$s\varphi(y) \leq \varphi(sy), \quad \forall s \in \mathbb{R}, \quad \forall y \in E. \tag{34}$$

For $x' \in \vartheta(x)$ and $t > 0$ we immediately obtain $\langle x', y \rangle = ((\langle x', x + ty \rangle - \langle x', x \rangle))/t \leq |x|(|x + ty| - |x|)/t$, hence

$$[x \mid y] \leq |x|\varphi(y), \quad \forall y \in E. \tag{35}$$

We now define a linear functional u on $E_0 := \text{span}\{x, y\}$ by

$$u(\xi x + \eta y) := \xi|x| + \eta\varphi(y), \quad \forall \xi, \eta \in \mathbb{R}.$$

If we have $\xi x + \eta y = 0$ and $\eta \neq 0$, that is, if $y = \zeta x$ with $\zeta = -\xi/\eta$, then (32) implies that $\varphi(y) = \zeta|x|$, hence $\xi|x| + \eta\varphi(y) = 0$. Therefore u is well-defined.

From

$$|x + t(\xi x + \eta y)| - |x| = (1 + t\xi)\{|x + t(1 + t\xi)^{-1}\eta y| - |x|\} + t\xi|x|$$

we obtain $\varphi(\xi x + \eta y) = \xi|x| + \varphi(\eta y)$, and, using (34), we get

$$u(z) \leq \varphi(z), \quad \forall z \in E_0.$$

Because of (33), the Hahn-Banach theorem guarantees the existence of a linear extension of u, $\bar{u} : E \to \mathbb{R}$, such that $\bar{u}(z) \leq \varphi(z) \leq |z|$ for all $z \in E$. Therefore $\bar{u} \in E'$ and $|\bar{u}| \leq 1$. From $\bar{u}(x) = u(x) = |x|$ we deduce $|\bar{u}| = 1$. It follows that $x' := |x|\bar{u} \in \vartheta(x)$ and since $\bar{u}(y) = u(y) = \varphi(y)$, we have $\langle x', y \rangle = |x|\varphi(y)$. Hence we have equality in (35). $\qquad\square$

(18.13) Corollary. *Let $-\infty < \alpha < \beta \leq \infty$ and assume that $u \in C([\alpha, \beta), E)$ is differentiable from the right (with $\dot{u} = D_+u$). Then*

$$v(t) := |u(t)|^2/2, \quad t \in [\alpha, \beta),$$

is differentiable from the right and we have

$$D_+v(t) = [u(t) \mid \dot{u}(t)], \quad \forall t \in [\alpha, \beta).$$

Proof. Since $u(t + s) = u(t) + s\dot{u}(t) + o(s)$ as $s \to 0^+$, it easily follows that

$$v(t+s) - v(s) = (|u(t) + s\dot{u}(t)|^2 - |u(t)|^2)/2 + o(s)$$

as $s \to 0^+$. Now the assertion follows from lemma (18.12).   □

(18.14) Corollary. *Let $X \subseteq E$ be open and let $f \in C^{1-}(X, E)$. Then for every $\gamma \in \mathbb{R}$,*

$$V(x) := |x|^2/2 + \gamma, \quad \forall\, x \in E,$$

is a Liapunov function on

$$M := \{x \in X \mid [x \mid f(x)] \le 0\}$$

for the flow induced by f.

In order to obtain useful results, we recall that a subset $B \subseteq E$ is called *balanced* if $b \in B$ implies $tb \in B$ for all $-1 \le t \le 1$. If, moreover, B is a closed, bounded, balanced and convex subset of E with interior points, then the *Minkowski functional* of B,

$$|x|_B := \inf\{t > 0 \mid x \in tB\},$$

defines a norm on E which is equivalent to $|\cdot|$ and B is precisely the closed unit ball in $(E, |\cdot|_B)$ (e.g., Yosida [1]).

After these considerations we are ready to prove the following, intuitively obvious, *geometric stability criterion.*

(18.15) Theorem. *Assume that $U \subseteq \mathbb{R}^m$ is open and let $f \in C^{1-}(U, \mathbb{R}^m)$. Moreover, suppose $X \subseteq U$ is a closed (in $\mathbb{R}^m$), convex and positively invariant subset for the flow induced by f. Let B be a compact, convex and balanced subset of $\mathbb{R}^m$ such that $0 \in int(B)$, and let*

$$M_\alpha := (x_0 + \alpha B) \cap X$$

for some $x_0 \in \mathbb{R}^m$ and $\alpha \ge 0$. Finally, assume that $0 \le \beta < \gamma < \infty$ and let $\partial_X M_\alpha$ denote the boundary of M_α in (the subspace topology of) X.

(i) *If for all $\alpha \in (\beta, \gamma]$ and all $x \in \partial_X M_\alpha$ we have that*

$$(x, f(x)) \in T_x(\mathbb{R}^m) \quad \text{lies in the support cone} \quad K_x(x_0 + \alpha B),$$

then M_β is stable.

(ii) *If for all $\alpha \in (\beta, \gamma]$ and all $x \in \partial_X M_\alpha$ we have that*

(a) *$(x, f(x)) \in T_x(\mathbb{R}^m)$ lies in the interior of $K_x(x_0 + \alpha B)$,*
(b) *$M_\beta \ne \emptyset$,*

then M_β is asymptotically stable and $\mathcal{A}(M_\beta) \supseteq M_\gamma$.

Proof. (i) Let $|\cdot|_B$ denote the Minkowski functional of B and set

$$V(x) := (|x - x_0|_B^2 - \beta^2)^2/2, \quad \forall\, x \in X.$$

Then

$$M_\alpha = V^{-1}(-\infty, (\alpha^2 - \beta^2)/2) = \mathbb{B}(M_\beta, \alpha - \beta),$$

where $\mathbb{B}$ denotes the unit ball with respect to $|\cdot|_B$. Setting $r := \gamma - \beta$, it follows from (29) and corollary (18.14) that V is a Liapunov function on $\mathbb{B}(M_\beta, r)\backslash M_\beta$. Also every semitrajectory in $\mathbb{B}(M_\beta, r)$ is relatively compact.

If $x \in \mathbb{B}(M_\beta, r)\backslash M_\beta$, we clearly have

$$V(x) \geq [d(x, M_\beta)]^2/2.$$

Hence the assertion follows from theorem (18.7 i).

(ii) Using assumption (a) and (30), it follows from corollary (18.13) that $\dot{V}(x) < 0$ for all $x \in M_\gamma \backslash M_\beta$. Now, as in the proof of theorem (18.7 ii), the invariance principle (corollary (18.4)) implies that M_β is an attractor and $\mathcal{A}(M_\beta) \supseteq M_\gamma$.   □

We want to employ theorem (18.15) now to show that the rectangles R, depicted in Figures 1–4 of section 16, are *globally asymptotically stable*. We only consider the case of Fig. 2. The analogous, but simpler, proofs for the other cases are left to the reader.

Since we already know from section 16 that every rectangle $\tilde{R}$ is positively invariant, it directly follows that R is stable. Therefore, it remains to show that R is an attractor and that $\mathcal{A}(R) = \mathbb{R}_+^2$.

So choose some arbitrary $x \in \mathbb{R}_+^2$. We then find a closed rectangle of the form S (cf. the figure below) with $x \in S$ and such that the line connecting 0 with the "upper right-hand corner" y lies completely in the set $\{x \in \mathbb{R}_+^2 \mid b(x) < 0\}$. Since S is positively invariant, it follows, in particular, that $t^+(x) = \infty$. Now let T denote the smallest closed rectangle of the form τS, $\tau > 0$, and such that it completely contains R, and let B be the smallest closed rectangle in $\mathbb{R}^2$ which is symmetric with respect to the axes and such that it contains T. We now set $X := \mathbb{R}_+^2$, $\gamma := 1/\tau$, $x_0 := 0$ and $M_\alpha := (\alpha B) \cap X$ for all $\alpha \geq 0$.   Then

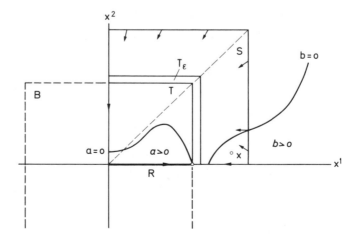

$T = M_1$ and $S = M_\gamma$, and the assumptions of theorem (18.15 ii) are satisfied for $\beta := 1$. Therefore T is asymptotically stable and $\mathcal{A}(T) \supseteq S$.

Choose now $\epsilon > 0$ so small that $T_\epsilon := M_{1+\epsilon}$ lies in $\{x \in X \mid b(x) < 0\}$. Then T_ϵ is a positively invariant neighborhood of T in X. Hence there exists some $t_\epsilon \geq 0$ such that $t \cdot x \in T_\epsilon$ for all $t \geq t_\epsilon$. In a second step we now show that R is an attractor and $\mathcal{A}(R) \supseteq T_\epsilon$. Then we clearly have $t \cdot x \to R$ as $t \to \infty$ and, since $x \in X$ was arbitrary, it follows that R is a global attractor.

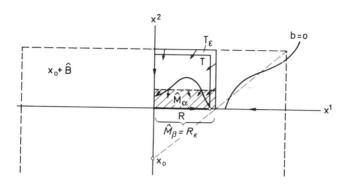

For this we set $X := T_\epsilon$ and choose some x_0 of the form $x_0 = (0, \eta)$, where $\eta < 0$. We let $x_0 + \hat{B}$ denote the smallest closed rectangle in $\mathbb{R}^2$ with sides parallel to the axes and center x_0 such that it contains T_ϵ. Then there exists some $\beta \in (0, 1)$ so that $\hat{M}_\beta := (x_0 + \beta\hat{B}) \cap X = R_\epsilon$. Moreover, the assumptions of theorem (18.15 ii) are satisfied with $\gamma = 1$. Hence $\hat{M}_\beta$ is asymptotically stable. For each $\delta \in (0, 1-\beta)$, $\hat{M}_{\beta+\delta}$ is a neighborhood of R in X. It follows that there exists some $t_\delta \geq 0$ such that $t \cdot x \in \hat{M}_{\beta+\delta}$ for all $t \geq t_\delta$. Finally, since ϵ and δ can be chosen to be arbitrarily small, we deduce the asymptotic stability of R.

It is now easy to show that the critical point z_0, distinct from 0, is asymptotically stable and that $\mathcal{A}(z_0) = \mathbb{R}_+^2 \setminus (\{0\} \times \mathbb{R}_+)$. For this we again consider an arbitrary point $x \in \mathbb{R}_+^2$ with $x^1 \neq 0$. We now choose a closed rectangle Q in $\mathbb{R}_+^2$ with sides parallel to the axes and such that it satisfies $R \subseteq \text{int}(Q) \subseteq Q \subseteq \{x \in \mathbb{R}_+^2 \mid b(x) < 0\}$. In addition, we choose Q such that its side in the x^2-direction is half as small as the distance from 0 to the intersection of $\{x \in \mathbb{R}_+^2 \mid a(x) = 0\}$ with the x^2-axis. Then Q is a positively invariant neighborhood of

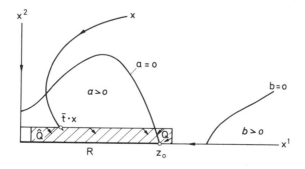

R. Hence there exists some $\bar{t} > 0$ such that $t \cdot x \in Q$ for all $t \geq \bar{t}$. If we choose $\bar{t}$ to be the first time for which $t \cdot x \in Q$, then certainly $\bar{t} \cdot x$ cannot lie on the x^2-axis (because this positive semiaxis is itself a trajectory of the flow (restricted to $\mathbb{R}_+^2$) and distinct trajectories cannot cross). Hence we may reduce Q to $\hat{Q}$ by "moving the left-hand side a little bit to the right" so that the new, closed rectangle $\hat{Q}$ is still positively invariant and contains the point $\bar{t} \cdot x$, but no point from the x^2-axis.

We now set $X := \hat{Q}$, $x_0 := z_0$ and $B := \{x \in \mathbb{R}^2 \mid |x^i| \leq 1, i = 1, 2\}$. Then there exists some $\gamma > 0$ such that $M_\gamma := (z_0 + \gamma B) \cap X = \hat{Q}$. If $\beta := 0$, the assumptions of theorem (18.15 ii) are satisfied. Therefore $M_0 = \{z_0\}$ is asymptotically stable and $\mathcal{A}(z_0) \supseteq \hat{Q}$.

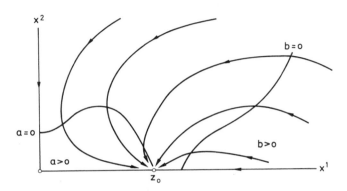

In particular, we have $t \cdot x \to z_0$ as $t \to \infty$. Since $x \in \mathbb{R}_+^2 \setminus (\{0\} \times \mathbb{R}_+)$ was arbitrary, it follows that $\mathcal{A}(z_0) = \mathbb{R}_+^2 \setminus (\{0\} \times \mathbb{R}_+)$.

We have therefore shown that: If in the predator-prey problem of example (16.12 a) it is the case that the zero sets, $a^{-1}(0)$ and $b^{-1}(0)$, have no points of $\mathbb{R}_+^2$ in common, then we have $t \cdot x \to z_0 := a^{-1}(0) \cap (\mathbb{R}_+ \times \{0\})$ as $t \to \infty$ for every $x \in \mathbb{R}_+^2$ satisfying $x^1 \neq 0$. In this case, the predator becomes extinct and the prey population approaches (as $t \to \infty$) an equilibrium state z_0 (if initially prey was present).

Problems

1. Let E be a finite dimensional Banach space and assume that $A \in \mathcal{L}(E)$ satisfies $\operatorname{Re} \sigma(A) < -\alpha < 0$. Show – using some appropriate Hilbert norm $\| \cdot \|$ – that the function

$$V(x) := \int_0^\infty \|e^{\tau A} x\|^2 d\tau, \quad x \in E,$$

is a Liapunov function on E for the linear flow e^{tA}. Use this to prove that the origin is globally exponentially stable.

2. With the aid of the function V in problem 1, prove the stability theorem (15.3) under the assumption that $g \in C^{1-}(J \times D, E)$.

3. Determine a Liapunov function of the form $V(x, y) = F(x) + G(y)$ on $(0, \infty)^2$ for the *Volterra-Lotka equations*

$$\dot{x} = (\alpha - \beta y)x, \quad \dot{y} = (\delta x - \gamma)y, \quad \alpha, \beta, \gamma, \delta > 0,$$

and show that the critical point $(\gamma/\delta, \alpha/\beta)$ is stable.

4. The following (special) *Liénard equation*

$$\dot{x} = y - f(x), \quad \dot{y} = -x, \quad f \in C^1(\mathbb{R}, \mathbb{R}), \tag{L}$$

(or written as a second order equation: $\ddot{x} + f'(x)\dot{x} + x = 0$) plays an important role in the theory of electrical circuits. As a special case one obtains for $f(x) = x^3 - x$ the *Van der Pol equation*.

(i) Determine the stability of the stationary point of (L) as a function of $f'(0)$.

(ii) If $xf(x) > 0$ for $x \neq 0$, show that the origin is globally asymptotically stable. (In circuit theory, f is called the "characteristic" of the resistor, and a resistor for which $xf(x) > 0$ if $x \neq 0$, is called "passive.")

(*Hint for* (ii): It is useful to know that $x^2 + y^2$ can be interpreted as energy.)

5. Let φ denote the flow on $\mathbb{R}_+^2$ induced by the two-species models. Show that:

(i) The closed rectangles R in Figures 1, 3 and 4 of section 16 are asymptotically stable.

(ii) If the zero sets $a^{-1}(0)$ and $b^{-1}(0)$ do not intersect in $\mathbb{R}_+^2$, then, in the case of the competing species model, the nontrivial critical point z on the x^2-axis is asymptotically stable and $\mathcal{A}(z) = \mathbb{R}_+^2 \backslash (\mathbb{R}_+ \times \{0\})$ (cf. Fig. 3 of section 16).

(iii) In the case of the competing species model the following situation holds: If the curves $a^{-1}(0)$ and $b^{-1}(a)$ have only finitely many transversal intersections, then $t \cdot x \rightarrow \{y \in \mathbb{R}_+^2 \mid y$ is a critical point$\}$. Hence in these cases there are no periodic solutions.

6. Assume that $-\infty < \alpha < \beta < \infty$ and let E be a Banach space. Then $g \in C^1([\alpha, \beta], E)$ can be thought of as a parametrization of a curve Γ in E. As in the finite dimensional case, one defines the *length of* Γ, $l(\Gamma)$, as the supremum over the lengths of all inscribed finite polygonal arcs, and, as in the finite dimensional case, one shows that

$$l(\Gamma) = \int_\alpha^\beta \|Dg(t)\| dt, \tag{36}$$

where g is an arbitrary C^1-parametrization of Γ. (Verify these assertions by convincing yourself that proofs for curves in $\mathbb{R}^m$ make no use of the dimension.) Finally, if Γ is a "half-open curve," i.e., if there exists a parametrization $g \in C^1([\alpha, \beta), E)$ of Γ, then one defines the length of Γ by

$$l(\Gamma) := \lim_{\gamma \to \beta^-} \int_\alpha^\gamma \|Dg(t)\| \, dt.$$

Prove that: If Γ is a half-open curve in E with C^1-parametrization $g \in C^1([\alpha, \beta), E)$ and finite length, then Γ is relatively compact.

Application: If φ is a semiflow on E and if $\gamma^+(x)$ has finite length, i.e.,

$$l(\gamma^+(x)) := \lim_{t \to t^+(x)^-} \int_0^t \left\| \frac{d}{d\tau}(\tau \cdot x) \right\| d\tau < \infty, \tag{37}$$

then $\gamma^+(x)$ is relatively compact.

(*Hint*: Since E is complete, it suffices to show that Γ is *totally bounded*, i.e., for every $\epsilon > 0$ there exist finitely many points $x_1, \ldots, x_m \in \Gamma$ such that $\Gamma \subseteq \bigcup_{j=1}^m \mathbb{B}(x_j, \epsilon)$. This follows easily from (36) and (37) by considering sufficiently small parametrization intervals and using the fundamental theorem of calculus.)

7. Let E be a real Banach space, $g \in C^1(E, \mathbb{R})$ and $v \in C^{1-}(E, E)$. Then v is called a *pseudo-gradient vector field* (PGVF) *for g* if

$$\|v(x)\| \le 2\|Dg(x)\| \quad \text{and} \quad \langle Dg(x), v(x) \rangle \ge \|Dg(x)\|^2/2$$

for all $x \in E$. If $E = (E, (\cdot \mid \cdot))$ is a Hilbert space, then clearly $v := \operatorname{grad} g$ is a PGVF for g. PGVF's play an important role in the calculus of variations in the large and nonlinear functional analysis, where one tries to obtain statements about the existence and multiplicity of critical points of g.

Show that: If $\gamma^+(x)$ is a positive semitrajectory of the flow induced by $-v$ and if $\gamma^+(x)$ has finite length, then g has a critical point at the level $\alpha := \inf g(\gamma^+(x))$, i.e., $g^{-1}(\alpha) \cap \{x \in E \mid Dg(x) = 0\} \ne \emptyset$.

19. Linearizations

In this section we let $(E, |\cdot|)$ denote a finite dimensional Banach space, M an open subset of E and $f \in C^1(M, E)$.

The main goal is to study the flow φ, induced by f, in the neighborhood of a critical point x_0, in particular, in situations in which the principle of linearized stability (theorem (15.6)) is not applicable. For reasons of simplicity, we only consider the case when $Df(x_0)$ induces a hyperbolic linear flow. We will show that locally, i.e., near x_0, the flow φ is flow equivalent to the linear flow $e^{tDf(x_0)}$, that is to say, the structure of a saddle is preserved. In addition, we will derive precise statements about the "stable and unstable manifolds" W_s and W_u.

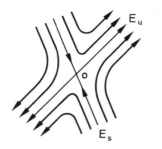

Let X and Y be either metric spaces or differentiable manifolds and assume that $\varphi : \Omega_\varphi \to X$ and $\psi : \Omega_\psi \to Y$ are flows on X and Y, respectively. We say that φ at $x_0 \in X$ is (locally) C^k-flow equivalent to ψ at $y_0 \in Y$, or for short: $\varphi|x_0$ is C^k-flow equivalent to $\psi|y_0$, $0 \le k \le \infty$, if there exist neighborhoods U and V of x_0 and y_0, respectively, such that the flow φ restricted to U is C^k-flow equivalent to the flow ψ restricted to V, for short: $\varphi|U$ and $\psi|V$ are C^k-flow equivalent.

In what follows, we always let $\varphi : \Omega \to M$ denote the flow induced by f on M and, again, we write $t \cdot x$ for $\varphi(t, x)$.

The first proposition shows that one can "straighten out the trajectories" in a neighborhood of a *regular point*, i.e., a noncritical point. Here we have $\dim_\mathbb{R}(E) = 2\dim_\mathbb{C}(E)$ whenever $\mathbb{K} = \mathbb{C}$ ("decomposition into real and imaginary parts").

(19.1) Proposition. *Let $x_0 \in M$ be a regular point of φ and let ψ denote the flow induced by the constant vector field $y \mapsto e_1 = (1, 0, \ldots, 0)$ on $\mathbb{R}^m$, $m := \dim_\mathbb{R}(E)$, that is to say,*

$$\psi(t, y) = y + te_1, \quad \forall (t, y) \in \mathbb{R} \times \mathbb{R}^m.$$

Then $\varphi|x_0$ and $\psi|0$ are C^1-flow equivalent.

Proof. By decomposing f and x into real and imaginary parts, we may assume that $\mathbb{K} = \mathbb{R}$. By means of a translation we can obtain $x_0 = 0$, and by introducing a basis with first basis vector $f(0)$, we may identify E with $\mathbb{R}^m$ and $f(0)$ with e_1. Then there exists some appropriate neighborhood U of $0 \in \mathbb{R}^m = \mathbb{R} \times \mathbb{R}^{m-1}$ such that the function $h : U \to \mathbb{R}^m$, defined by

$$h(t, \eta) := \varphi(t, (0, \eta)),$$

is well-defined and of class C^1.

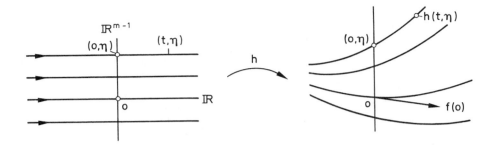

For all $(s, \eta), (t, \eta) \in U$ such that $(s + t, \eta) \in U$ we then have

$$h \circ \psi(t, (s, \eta)) = h((s, \eta) + te_1) = h(s + t, \eta)$$
$$= \varphi(s + t, (0, \eta)) = \varphi(t, \varphi(s, (0, \eta))) = \varphi(t, h(s, \eta)),$$

that is, near $0 \in \mathbb{R}$ we have

$$h \circ \psi = \varphi \circ (id \times h). \tag{1}$$

Since $h|\{0\} \times \mathbb{R}^{m-1} = (0, id_{\mathbb{R}^{m-1}})$ and $D_1 h(0) = D_1 \varphi(0, 0) = f(0) = e_1$, it follows that

$$Dh(0) = \begin{bmatrix} 1 & 0 \\ 0 & id_{\mathbb{R}^{m-1}} \end{bmatrix} = id_{\mathbb{R}^m}.$$

Hence, by the inverse function theorem, h is a local C^1-diffeomorphism near 0 and therefore – because of (1) – a local C^1-flow equivalence. $\square$

(19.2) Remarks. (a) It follows from the differentiability theorem (9.5) and the proof above that $\varphi|x_0$ and $\psi|0$ are C^k-flow equivalent whenever $f \in C^k(M, E), 1 \le k \le \infty$, and $\mathbb{K} = \mathbb{R}$.

(b) The proof above shows that $\varphi|x_0$ and $\psi|0$ are *isochron flow equivalent*, i.e., the time variable is unchanged. $\square$

The Linearization Theorem

We now turn to the main task of this section, the study of a flow in a neighborhood of a *hyperbolic critical point*. Here the critical point x_0 of the flow induced by f is called *hyperbolic* if $\sigma_n(Df(x_0)) = \emptyset$, that is, if the linear flow $e^{tDf(x_0)}$ is hyperbolic.

There exists a close connection between flows and homeomorphisms. In fact, according to theorem (10.14), φ^t is a homeomorphism from Ω_t onto Ω_{-t} for all $t \in \mathbb{R}$. In particular, for every linear flow e^{tA} it follows that $e^{tA} \in \mathcal{GL}(E)$ for every $t \in \mathbb{R}$, that is, e^{tA} is an automorphism on E. It is therefore reasonable (and useful for applications) to consider the case of homeomorphisms first.

To motivate the next definition, we first prove the following special case of the *spectral mapping theorem* (see, for example, Yosida [1]).

(19.3) Lemma. *If $A \in \mathcal{L}(E)$, then*

$$\sigma(e^A) = e^{\sigma(A)} := \{e^\lambda \mid \lambda \in \sigma(A)\}.$$

Proof. Since $\sigma(A) := \sigma(A_\mathbb{C})$, we may assume – by passing to the complexification – that E is a complex Banach space. If $\lambda_1, \ldots, \lambda_k$ denote the distinct eigenvalues of A, we know from section 12 that E has the direct sum decomposition $E = E_1 \oplus \cdots \oplus E_k$ which reduces A as well as e^A. Hence it suffices to prove that $\sigma(e^{A_j}) = e^{\sigma(A_j)}$, where $A_j := A|E_j$ for $j = 1, \ldots, k$. We may therefore assume (cf. section 12) that the following holds: $\sigma(A) = \{\lambda\}$ and $A = \lambda + N$ for some nilpotent operator $N \in \mathcal{L}(E)$. It follows that there exists some $x \in E \setminus \{0\}$ such that $Ax = \lambda x$, i.e., $Nx = 0$. From this we deduce that

$$e^A x = e^\lambda e^N x = e^\lambda x,$$

that is, $\sigma(e^A) \supseteq e^{\sigma(A)}$. Conversely, if we have $e^A y = \mu y$ for some $\mu \in \mathbb{C}$ and $y \in E \setminus \{0\}$, then

$$\mu y = e^\lambda e^N y = e^\lambda \sum_{k=0}^{m} \frac{1}{k!} N^k y. \tag{2}$$

Then there exists a smallest index l such that $0 \le l \le m$ and $N^{l+1} y = 0$. Applying N^l to (2), we obtain

$$\mu N^l y = e^\lambda N^l y.$$

Since $N^l y \ne 0$, we have $\mu = e^\lambda$, which implies that $\sigma(e^A) \subseteq e^{\sigma(A)}$. □

Assume now that $A \in \mathcal{L}(E)$ and let A induce the hyperbolic linear flow e^{tA}, that is, $\sigma_n(A) = \sigma(A) \cap i\mathbb{R} = \emptyset$. It then follows from lemma (19.3) that $\sigma(e^A) \cap \mathbb{S}_\mathbb{C} = \emptyset$, where $\mathbb{S}_\mathbb{C} := \{z \in \mathbb{C} \mid |z| = 1\}$ denotes the unit circle in the complex plane. In other words, $e^A \in \mathcal{GL}(E)$ has no eigenvalues of norm 1. In general, if T has no eigenvalues of norm 1, i.e., if $\sigma(T) \cap \mathbb{S}_\mathbb{C} = \emptyset$, *the automorphism $T \in \mathcal{GL}(E)$ is called hyperbolic.*

If $T \in \mathcal{GL}(E)$ is hyperbolic, then

$$\sigma(T) = \sigma_0(T) \cup \sigma_\infty(T),$$

where

$$\sigma_0(T) := \{\lambda \in \sigma(T) \mid |\lambda| < 1\}$$

and

$$\sigma_\infty(T) := \{\lambda \in \sigma(T) \mid |\lambda| > 1\}.$$

If, again, we let $m(\lambda)$ denote the algebraic multiplicity of the eigenvalue $\lambda \in \sigma(T)$, it follows – for the case $\mathbb{K} = \mathbb{C}$ – that

$$E_0 := \bigoplus_{\lambda \in \sigma_0(T)} \ker[(\lambda - T)^{m(\lambda)}]$$

and

$$E_\infty := \bigoplus_{\lambda \in \sigma_\infty(T)} \ker[(\lambda - T)^{m(\lambda)}]$$

are invariant subspaces of E which reduce T, that is,

$$E = E_0 \oplus E_\infty \quad \text{and} \quad T = T_0 \oplus T_\infty, \tag{3}$$

and it also follows that

$$\sigma(T_0) = \sigma_0(T) \quad \text{and} \quad \sigma(T_\infty) = \sigma_\infty(T). \tag{4}$$

If $\mathbb{K} = \mathbb{R}$, we apply this decomposition to the complexification and subsequently restrict to the real subspaces, i.e.,

$$E_0 := (E_\mathbb{C})_0 \cap E \quad \text{and} \quad E_\infty := (E_\mathbb{C})_\infty \cap E,$$

as well as

$$T_0 := (T_\mathbb{C})_0 | E_0 \quad \text{and} \quad T_\infty := (T_\mathbb{C})_\infty | E_\infty.$$

Then one easily verifies that relations (3) and (4) also hold for the real case (cf. the proof of theorem (13.4)).

The following lemma represents an analogue of lemma (13.1). To simplify the formulation, we make use of the descriptive notation

$$|\sigma(A)| < \alpha \quad \Leftrightarrow \quad |\lambda| < \alpha, \quad \forall \lambda \in \sigma(A).$$

Other inequalities are to be interpreted analogously.

(19.4) Lemma. *Let* $T \in \mathcal{GL}(E)$ *be hyperbolic and assume that for some* $\alpha \in \mathbb{R}_+$ *we have*

$$|\sigma(T_0)| < \alpha \quad \text{and} \quad |\sigma((T_\infty)^{-1})| < \alpha.$$

Then there exists a Hilbert norm $\| \cdot \|$ *on* E *such that*

$$\max\{\|T_0\|, \|(T_\infty)^{-1}\|\} \le \alpha,$$

and so that E_0 *and* E_∞ *are orthogonal.*

Proof. Since $\|A_\mathbb{C}\| = \|A\|$ (cf. the proof of lemma (13.1)), we may, without loss of generality, assume that $\mathbb{K} = \mathbb{C}$. From the proof of lemma (13.1), we know that $T_0 = D + N$ for some nilpotent operator $N \in \mathcal{L}(E_0)$ and some diagonal operator $D = \text{diag}[\mu_1, \ldots, \mu_k]$ (with respect to some appropriate basis), where $\mu_1, \ldots, \mu_k$ are the eigenvalues of T_0, counted according to their multiplicity. In addition, we know that we can choose the basis such that the corresponding Euclidean norm $\| \cdot \|_0$ on E_0 satisfies

$$\|N\|_0 \le \alpha - \max\{|\mu_j| \mid j = 1, \ldots, k\}.$$

From this it immediately follows that

$$\|T_0\|_0 \le \|D\|_0 + \|N\|_0 \le \max\{|\mu_j| \mid 1 \le j \le k\} + \|N\|_0 \le \alpha.$$

Similarly, we find a Hilbert norm $\| \cdot \|_\infty$ for E_∞ such that for the corresponding operator norm we have $\|T_\infty^{-1}\|_\infty \le \alpha$. Then

$$\|x\|^2 := \|x_0\|_0^2 + \|x_\infty\|_\infty^2, \quad \forall x = x_0 + x_\infty \in E_0 \oplus E_\infty = E,$$

defines the desired Hilbert norm on E. $\qquad\square$

(19.5) Remark. If $T \in \mathcal{GL}(E)$ is hyperbolic, then

$$|\sigma_0(T)| < 1 < |\sigma_\infty(T)|. \tag{5}$$

Since for each $B \in \mathcal{GL}(E)$ we trivially have

$$\sigma(B^{-1}) = [\sigma(B)]^{-1} := \left\{ \frac{1}{\lambda} \mid \lambda \in \sigma(B) \right\},$$

it follows from (4) and (5) that

$$|\sigma(T_\infty^{-1})| < 1.$$

Then lemma (19.4) implies the existence of some $\alpha < 1$ and a norm $\| \cdot \|$ on E such that

$$\|T_0\| \le \alpha < 1 \quad \text{and} \quad \|T_\infty^{-1}\| \le \alpha < 1.$$

From this it follows that for every $x \in E_0$ we have

$$T^k x = (T_0)^k x \to 0 \quad \text{as} \quad k \to \infty$$

and for every $y \in E_\infty$ we have

$$T^{-k} y = (T_\infty)^{-k} y \to 0 \quad \text{as} \quad k \to \infty.$$

Analogous to the situation for linear flows, E_0 is therefore called the *stable* and E_∞ the *unstable* subspace corresponding to T (or, more precisely: corresponding to the *discrete flow* induced by T). $\qquad\square$

For a topological space X we set

$$BC(X, E) := B(X, E) \cap C(X, E),$$

where $B(X, E)$ denotes the Banach space of all bounded maps $u : X \to E$ with the *sup-norm*

$$\|u\|_\infty := \sup_{x \in X} |u(x)|_E.$$

The theorem on the continuity of the limit function of a uniformly convergent sequence of continuous functions implies that $BC(X, E)$ is a closed subspace of $B(X, E)$, and therefore $BC(X, E)$ is itself a Banach space with respect to the sup-norm, the *space of bounded continuous functions* (on X with values in E). If X is compact, then, of course, $BC(X, E) = C(X, E)$ and

$$\|u\|_\infty = \max_{x \in X} |u(x)| =: \|u\|_C.$$

Moreover, it is also clear that if we replace the norm on E by an equivalent norm, we obtain an equivalent norm on $BC(X, E)$.

Let now $E = E_1 \oplus E_2$ be a direct sum decomposition of E and assume that the corresponding projections $P_i : E \to E_i$, $i = 1, 2$, satisfy $|P_i| \leq 1$, $i = 1, 2$. Then every element $u \in B := BC(X, E)$ can be written uniquely in the form

$$u = P_1 u + P_2 u$$

and

$$P_i u \in BC(X, E_i) =: B_i, \quad i = 1, 2.$$

Moreover, we trivially have

$$\|P_i u\|_{B_i} = \sup_{x \in X} |P_i u(x)| \leq |P_i| \|u\|_\infty \leq \|u\|_\infty$$

for $i = 1, 2$. Consequently

$$(P_i u)(x) := P_i u(x), \quad \forall x \in X,$$

defines continuous projections $P_i : B \to B_i$, $i = 1, 2$, satisfying $P_1 + P_2 = \mathrm{id}_B$, i.e., we have:

$$B = B_1 \oplus B_2$$

and $P_i : B \to B_i$ are the corresponding projections. Finally, we set

$$\|u\|_B := \max\{\|P_1 u\|_{B_1}, \|P_2 u\|_{B_2}\}.$$

It follows from

$$(1/2)\|u\|_\infty = (1/2)\|P_1 u + P_2 u\|_\infty \leq (1/2)[\|P_1 u\|_\infty + \|P_2 u\|_\infty] \tag{6}$$
$$= (1/2)[\|P_1 u\|_{B_1} + \|P_2 u\|_{B_2}] \leq \|u\|_B \leq \|u\|_\infty$$

that $\|\cdot\|_B$ is an equivalent norm on B.

What is still needed is an analogue of the concept of "flow equivalence" for the case of homeomorphisms. So let X and Y be topological spaces and let $f : X \to X$ and $g : Y \to Y$ be homeomorphisms. Then a homeomorphism $h : X \to Y$ is called a *topological conjugacy from f to g* if $h \circ f = g \circ h$, that is to say, if the diagram

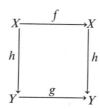

commutes. If X and Y are differentiable manifolds (e.g., open subsets of Banach spaces) and if f, g and h are C^k-diffeomorphisms, $1 \le k \le \infty$, then h is called a C^k-conjugacy. Lastly, f and g are said to be *topologically* (resp. C^k-)conjugate if there exists a topological (resp. C^k-)conjugacy from f to g. This trivially defines an equivalence relation in the class of all homeomorphisms (resp. C^k-diffeomorphisms).

After these preliminaries we now can prove the *global Hartman linearization theorem*.

(19.6) Proposition. *If $T \in \mathcal{GL}(E)$ is hyperbolic and if $g \in BC(E, E)$ is a uniformly Lipschitz continuous function with a sufficiently small Lipschitz constant, then the maps T and $T + g$ are topologically conjugate.*

Proof. By lemma (19.4) and remark (19.5), there exists a Hilbert norm $\| \cdot \|$ on E such that

$$\max\{\|T_0\|, \|T_\infty^{-1}\|\} \le \alpha < 1. \tag{7}$$

Since the passage to an equivalent norm on E implies that the Lipschitz constant of g will be multiplied by a positive factor, we may assume that the norm $\| \cdot \|$ on E satisfies

$$\|g(x) - g(y)\| \le \lambda \|x - y\|, \quad \forall x, y \in E, \tag{8}$$

where $2\lambda < \min\{1 - \alpha, \|T^{-1}\|^{-1}\}$.

(i) First we show that $T + g \in C(E, E)$ is a homeomorphism. Since for every $z \in E$ the equation $Tx + g(x) = z$ is equivalent to the fixed point equation

$$x = T^{-1}(z - g(x)) =: f_z(x),$$

it follows that $T + g$ is bijective whenever $f_z : E \to E$ has a unique fixed point $x(z)$. That f_z has a unique fixed point follows, however, from

$$\begin{aligned}
\|f_z(x) - f_z(y)\| &\le \|T^{-1}\| \, \|g(y) - g(x)\| \le \lambda \|T^{-1}\| \, \|x - y\| \\
&\le (1/2)\|x - y\|
\end{aligned} \tag{9}$$

for all $x, y \in E$ and the Banach fixed point theorem. From (9) we obtain

$$\|x(z) - x(\tilde{z})\| = \|f_z(x(z)) - f_{\tilde{z}}(x(\tilde{z}))\|$$
$$\le \|f_z(x(z)) - f_z(x(\tilde{z}))\| + \|f_z(x(\tilde{z})) - f_{\tilde{z}}(x(\tilde{z}))\|$$
$$\le (1/2)\|x(z) - x(\tilde{z})\| + \|T^{-1}\| \|z - \tilde{z}\|.$$

for all $z, \tilde{z} \in E$ (cf. problem 2 in section 7), hence $\|x(z) - x(\tilde{z})\| \le 2\|T^{-1}\| \|z - \tilde{z}\|$. Consequently $x(\cdot) = (T + g)^{-1} : E \to E$ is (uniformly) Lipschitz continuous.

(ii) Let now $h \in BC(E, E) =: B$ be a second function which is uniformly Lipschitz continuous with Lipschitz constant λ. Assume further that corresponding to every pair (g, h) of such functions there exists a unique $H := H(g, h) \in C(E, E)$ such that

$$H - id \in B \tag{10}$$

and

$$(T + g) \circ H = H \circ (T + h). \tag{11}$$

Then for $a := H(g, 0)$ we have

$$(T + g) \circ a = a \circ T, \tag{12}$$

and for $b := H(0, g)$ we have

$$T \circ b = b \circ (T + g). \tag{13}$$

It follows from (12) and (13) that

$$(T + g) \circ a \circ b = a \circ T \circ b = a \circ b \circ (T + g). \tag{14}$$

Since $a = id + u$ and $b = id + v$ for some $u, v \in B$, it follows that $a \circ b = id + w$, where $w = v + u \circ b \in B$. Based on the uniqueness of H, we thus obtain from (14) that $a \circ b = H(g, g) = id$. Similarly it follows that $b \circ a = id$. Therefore a is a homeomorphism from E onto E, hence – by (12) – a topological conjugacy from $T + g$ to T.

(iii) With $H = id + u$ it remains to be shown that there exists a unique $u \in B$ such that

$$(T + g) \circ (id + u) = (id + u) \circ (T + h). \tag{15}$$

Since, according to (i), $T + h$ is a homeomorphism, (15) is equivalent to

$$id + u = (T + g) \circ (id + u) \circ (T + h)^{-1}$$
$$= g \circ (id + u) \circ (T + h)^{-1} + T(T + h)^{-1} + Tu \circ (T + h)^{-1}.$$

Because $id = (T + h) \circ (T + h)^{-1}$, the last equation is equivalent to

$$u = Tu \circ (T + h)^{-1} + G(u) =: \tilde{F}(u), \tag{16}$$

where

$$G(u) := g \circ (id + u) \circ (T + h)^{-1} - h \circ (T + h)^{-1}. \tag{17}$$

Clearly $\tilde{F}$ maps the Banach space B into itself. It remains to show that $\tilde{F}$ has a unique fixed point in B.

Because $E = E_0 \oplus E_\infty$ and since E_0 and E_∞ are orthogonal, it follows that for the corresponding projections we have $\|P_0\|, \|P_\infty\| \leq 1$ (cf. the proof of theorem (15.5)). The fixed point equation (16) is therefore equivalent to the system of equations

$$P_0 u = T_0 P_0 u \circ (T + h)^{-1} + P_0 G(u) =: F_0(u) \tag{18}$$

$$P_\infty u = T_\infty P_\infty u \circ (T + h)^{-1} + P_\infty G(u). \tag{19}$$

Since equation (19) is transformed into the equivalent equation

$$P_\infty u = T_\infty^{-1} P_\infty u \circ (T + h) - T_\infty^{-1} P_\infty G(u) \circ (T + h) =: F_\infty(u) \tag{20}$$

by multiplying from the left by T_∞^{-1} and from the right by $T + h$, (16) is equivalent to the system (18) and (20).

By the considerations preceding this proposition, the decomposition $E = E_0 \oplus E_\infty$ induces the decomposition $B = B_0 \oplus B_\infty$ and for the norm

$$\|u\|_B := \max\{\|P_0 u\|_\infty, \|P_\infty u\|_\infty\}$$

we have

$$(1/2)\|u\|_\infty \leq \|u\|_B \leq \|u\|_\infty, \quad \forall u \in B.$$

Hence

$$F := F_0 + F_\infty$$

defines a map from B into itself such that the fixed point problem $u = F(u)$ is equivalent to the fixed point problem $u = \tilde{F}(u)$.

For $u, v \in B$ and $x \in E$ we set $y := (T + h)^{-1}(x)$ and $z := (T + h)(x)$, and so by making use of (7) we obtain the estimates

$$\|F_0(u)(x) - F_0(v)(x)\| \leq \alpha\|P_0 u(y) - P_0 v(y)\| + \|g(y + u(y)) - g(y + v(y))\|$$
$$\leq \alpha\|P_0(u - v)\|_\infty + \lambda\|u - v\|_\infty$$

and

$$\|F_\infty(u)(x) - F_\infty(v)(x)\| \leq \alpha\|P_\infty u(z) - P_\infty v(z)\|$$
$$+ \|g(x + u(x)) - g(x + v(x))\|$$
$$\leq \alpha\|P_\infty(u - v)\|_\infty + \lambda\|u - v\|_\infty.$$

Therefore

$$\|F_0(u) - F_0(v)\|_\infty \leq \alpha\|P_0(u - v)\|_\infty + 2\lambda\|u - v\|_B$$
$$\leq (\alpha + 2\lambda)\|u - v\|_B$$

and

$$\|F_\infty(u) - F_\infty(v)\|_\infty \leq (\alpha + 2\lambda)\|u - v\|_B,$$

and so, since $F_0(B) \subseteq B_0$ and $F_\infty(B) \subseteq B_\infty$, we deduce that

$$\|F(u) - F(v)\|_B \le (\alpha + 2\lambda)\|u - v\|_B, \quad \forall u, v \in B.$$

The existence of a unique fixed point of F now follows from the Banach fixed point theorem and the fact that $\alpha + 2\lambda < 1$. □

(19.7) Remarks. (a) The proof above shows that there exists a unique topological conjugacy h from T to $T + g$ which satisfies $h - \mathrm{id} \in BC(E, E)$ (of course, if the Lipschitz constant of g is sufficiently small).

(b) If $g \in C^k(E, E)$, $1 \le k \le \infty$, one would naturally expect the topological conjugacy from $T + g$ to T to have the corresponding differentiability properties, i.e., that $T + g$ and T are C^k-conjugate. This, however, is in general not true. For further investigations along these lines we refer to Hartman [1]. □

We need the following simple lemma for the local version of the above linearization theorem.

(19.8) Lemma. *Let F be an arbitrary NVS and let $r_\alpha : F \to \overline{\mathbb{B}}(0, \alpha)$ denote the radial retraction, that is,*

$$r_\alpha(x) := \begin{cases} x, & \text{if } |x| \le \alpha \\ \alpha x/|x|, & \text{if } |x| \ge \alpha. \end{cases}$$

Then r_α is uniformly Lipschitz continuous with Lipschitz constant 2.

Proof. For $|x| > \alpha \ge |y|$ we have

$$|r_\alpha(x) - r_\alpha(y)| = |\alpha|x|^{-1}x - y| \le \alpha|x|^{-1}|x - y| + \left|\alpha|x|^{-1}y - y\right|$$

$$\le |x - y| + |x|^{-1}|y|(|x| - \alpha)$$

$$\le |x - y| + |x| - |y| \le 2|x - y|.$$

If $|x| > \alpha$ and $|y| > \alpha$, we obtain

$$|r_\alpha(x) - r_\alpha(y)| = \left|\alpha|x|^{-1}x - \alpha|y|^{-1}y\right|$$

$$\le \alpha|x|^{-1}|x - y| + \alpha|y|\left||x|^{-1} - |y|^{-1}\right|$$

$$\le |x - y| + ||x| - |y|| \le 2|x - y|.$$

This proves the lemma. □

After these preliminaries we can now easily prove the main result of this section. To do this, we recall that E is a finite dimensional Banach space, that M is open in E, and that $f \in C^1(M, E)$.

(19.9) Theorem *(Grobman, Hartman). Let x_0 be a hyperbolic critical point of φ. Then $\varphi|x_0$ and $e^{tDf(x_0)}|0$ are isochronally flow equivalent.*

Proof. (i) Since a translation is evidently an isochron flow equivalence, we may assume, without loss of generality, that $x_0 = 0$. If $\lambda > 0$ is arbitrary, there exists some $\alpha > 0$ so that $|Df(x) - Df(0)| \le \lambda/2$ for all $x \in \overline{\mathbb{B}}(0, \alpha)$. It follows from the mean value theorem that the function $x \mapsto f(x) - Df(0)x$ is uniformly Lipschitz continuous on $\overline{\mathbb{B}}(0, \alpha)$ with Lipschitz constant $\lambda/2$. Employing the radial retraction $r_\alpha : E \to \overline{\mathbb{B}}(0, \alpha)$, we now define $g \in BC(E, E)$ by

$$g := [f - Df(0)] \circ r_\alpha.$$

It follows from lemma (19.8) (and the proof of lemma (8.1 iii) that g is globally Lipschitz continuous with Lipschitz constant λ. Setting $A := Df(0) \in \mathcal{L}(E)$, we have

$$(A + g)|\mathbb{B}(0, \alpha) = f|\mathbb{B}(0, \alpha). \tag{21}$$

If ψ denotes the flow on E, induced by $A + g$, then (21) shows that ψ coincides with φ on $\mathbb{B}(0, \alpha)$ (cf. theorem (10.3) and remark (10.4 b)). It therefore suffices to show that ψ and e^{tA} are isochronally flow equivalent.

(ii) Because g is globally Lipschitz continuous, g – and therefore $A + g$ – is linearly bounded. Hence by proposition (7.8), ψ is a global flow. Since $\dot{x} = Ax + g(x)$, we deduce from the variation of constants formula (cf. formula (5) in section 15) that

$$\psi^t(x) = e^{tA}x + \int_0^t e^{(t-\tau)A}g(\psi^\tau(x))d\tau, \quad \forall t \in \mathbb{R}. \tag{22}$$

This implies that

$$|\psi^t(x) - \psi^t(y)| \le e^{t|A|}|x - y| + \int_0^t e^{(t-\tau)|A|}\lambda|\psi^\tau(x) - \psi^\tau(y)|d\tau$$

for all $t \ge 0$ and all $x, y \in E$. After multiplying this inequality by $e^{-t|A|}$, we may apply Gronwall's inequality (corollary (6.2)) and obtain

$$|\psi^t(x) - \psi^t(y)| \le |x - y|e^{(\lambda + |A|)t}, \quad \forall x, y \in E, \quad t \ge 0. \tag{23}$$

Since $g \in BC(E, E)$, it follows from (22) that

$$|\psi^t(x) - e^{tA}x| \le \|g\|_\infty \left|\int_0^t e^{|t-\tau||A|}d\tau\right|$$

for all $t \in \mathbb{R}$ and and $x \in E$, hence

$$\psi^t - e^{tA} \in BC(E, E), \quad \forall t \in \mathbb{R}. \tag{24}$$

Finally, it follows from (22) and (23) that

$$|(\psi^t - e^{tA})(x) - (\psi^t - e^{tA})(y)| \le \int_0^t e^{(t-\tau)|A|}\lambda|\psi^\tau(x) - \psi^\tau(y)|d\tau$$

$$\le \lambda|x - y|e^{t|A|}\int_0^t e^{\lambda\tau}\,d\tau \qquad (25)$$

$$= |x - y|e^{t|A|}(e^{\lambda t} - 1)$$

for all $t \ge 0$ and $x, y \in E$.

(iii) By assumption, 0 is a hyperbolic critical point and thus $\sigma(A) \cap i\mathbb{R} = \emptyset$. It therefore follows from lemma (19.3) that $T := e^A$ is a hyperbolic automorphism on E. Since λ can be chosen arbitrarily small, it follows from (25) that the Lipschitz constant of $\tilde{g} := \psi^1 - T$ can be made arbitrarily small. Because $\tilde{g} \in BC(E, E)$ (cf. (24)), we may assume, by proposition (19.6), that T and $\psi^1 = T + \tilde{g}$ are topologically conjugate. From remark (19.7 a) we also know that there exists a unique topological conjugacy h from T to ψ^1 satisfying $h - id \in BC(E, E)$.

From $h \circ T = \psi^1 \circ h$ we deduce that for each $t \in \mathbb{R}$ we have

$$\psi^1 \circ (\psi^t \circ h \circ T^{-t}) = \psi^t \circ \psi^1 \circ h \circ T^{-t} = \psi^t \circ h \circ T \circ T^{-t}$$

$$= (\psi^t \circ h \circ T^{-t}) \circ T,$$

where $T^t := e^{tA}$. Therefore $\psi^t \circ h \circ T^{-t}$ is also a topological conjugacy from T to ψ^1. Because we have

$$\psi^t \circ h \circ T^{-t} - id = (\psi^t - T^t) \circ h \circ T^{-t} + T^t \circ (h - id) \circ T^{-t},$$

it follows from (24) that $\psi^t \circ h \circ T^{-t} - id \in BC(E, E)$. Therefore $\psi^t \circ h \circ T^{-t} = h$ and hence $\psi^t \circ h = h \circ e^{tA}$ for all $t \in \mathbb{R}$, that is to say, ψ and e^{tA} are isochronally flow equivalent. $\square$

Stable Manifolds

The preceding theorem states that in a neighborhood of a hyperbolic critical point x_0, the phase portrait of φ has the same topological structure as the phase portrait of the linearization in a neighborhood of 0. Analogous to the stable subspace E_s and unstable subspace E_u of a linear flow, one defines the *stable manifold* $W_s(x_0)$ *of* φ *at* x_0 and the *unstable manifold* $W_u(x_0)$ *of* φ *at* x_0 by

$$W_s(x_0) := \{x \in E \mid t^+(x) = \infty \text{ and } t \cdot x \to x_0 \text{ as } t \to \infty\}$$

and

$$W_u(x_0) := \{x \in E \mid t^-(x) = -\infty \text{ and } t \cdot x \to x_0 \text{ as } t \to -\infty\},$$

respectively. Clearly $x_0 \in W_s(x_0) \cap W_u(x_0)$. We will now show that in a neighborhood of x_0 the sets $W_s(x_0)$ and $W_u(x_0)$ are indeed differentiable submanifolds of E which intersect transversely at x_0, and that the tangent spaces of $W_s(x_0)$ and $W_u(x_0)$ are translates of E_s and E_u, respectively, that is to say, $T_{x_0}W_s(x_0) = x_0 + E_s$ and $T_{x_0}W_u(x_0) = x_0 + E_u$.

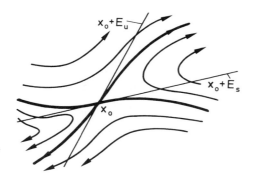

In order to prove these facts, we again consider first homeomorphisms from E onto itself. If $h : E \to E$ is a homeomorphism such that $h(0) = 0$, then the set

$$W_0 := \{x \in E \mid h^n(x) \to 0 \ \text{as} \ n \to \infty\}$$

is called the *stable set* of h at 0, where h^n denotes the nth iterate of h. Similarly, we define the *unstable set* of h at 0 by

$$W_\infty := \{x \in E \mid h^{-n}(x) \to 0 \ \text{as} \ n \to \infty\},$$

where we have set $h^{-n} := (h^{-1})^n$. Evidently W_0 and W_∞ change into each other if we replace h by h^{-1}. Hence it suffices to consider W_0.

Now let $T \in \mathcal{GL}(E)$ be hyperbolic and let

$$E = E_0 \oplus E_\infty, \quad T = T_0 \oplus T_\infty$$

be the decomposition into the stable and unstable subspace, introduced a few pages back.

(19.10) Proposition. *Let* $g : E \to E$ *be uniformly Lipschitz continuous and assume that* $g(0) = 0$. *If the Lipschitz constant of* g *is sufficiently small, there exists a unique uniformly Lipschitz continuous function* $h : E_0 \to E_\infty$ *such that the graph of* h *is the stable set* W_0 *of* $T + h$ *at 0. If in some neighborhood of 0,* g *is of class* $C^k, 1 \le k \le \infty$, *then so is the function* h. *In this case there exists a neighborhood* V *of 0 in* E_0 *such that*

$$W_0^V := \{(x, h(x)) \mid x \in V\}$$

is a C^k-*manifold. If, in addition,* $Dg(0) = 0$, *then* $T_0 W_0^V = E_0$.

Proof. Set

$$B_0 := \{u : \mathbb{N} \to E \mid u(k) \to 0 \text{ as } k \to \infty\}.$$

Then it is easily seen that B_0 is a closed subspace of the Banach space of all bounded sequences in $E, (B(\mathbb{N}, E), \|\cdot\|_\infty)$. Hence B_0 itself is a Banach space with respect to the sup-norm.

Let

$$\mathbb{W}_0 := \{u \in B_0 \mid u(k) = (T + g)^k(x), k \in \mathbb{N}, x \in E\}.$$

Then clearly

$$W_0 = \{u(0) \mid u \in \mathbb{W}_0\}.$$

Moreover, we have

$$u \in \mathbb{W}_0 \quad \Leftrightarrow \quad u(k + 1) = (T + g)(u(k)), \quad \forall k \in \mathbb{N}.$$

Using the canonical projections $P_0 : E \to E_0$ and $P_\infty : E \to E_\infty$, the last equation is equivalent to the system

$$P_0 u(k + 1) = T_0 P_0 u(k) + P_0 g(u(k))$$
$$P_\infty u(k + 1) = T_\infty P_\infty u(k) + P_\infty g(u(k)),$$

and hence to the system

$$P_0 u(k + 1) = T_0 P_0 u(k) + P_0 g(u(k))$$
$$P_\infty u(k) = T_\infty^{-1} P_\infty u(k + 1) - T_\infty^{-1} P_\infty g(u(k))$$

for all $k \in \mathbb{N}$. If for $x \in E$ and $u \in B_0$ we now set

$$F_x(u)(k) := \begin{cases} T_0 P_0 u(k - 1) + T_\infty^{-1} P_\infty u(k + 1) + P_0 g(u(k - 1)) \\ \qquad\qquad\qquad\qquad\qquad\qquad\qquad -T_\infty^{-1} P_\infty g(u(k)) \quad (26) \\ P_0 x + T_\infty^{-1}(P_\infty u(1) - P_\infty g(u(0))), \quad \text{if } k = 0, \end{cases}$$

we see that $x \in E_0$ belongs to W_0 if and only if u is a fixed point of F_x in B_0.

As in the proof of proposition (19.6), we may assume that

$$|T_0|, \ |T_\infty^{-1}| \leq \alpha < 1, \quad |P_0|, \ |P_\infty| \leq 1$$

and that

$$|g(x) - g(y)| \leq \lambda |x - y|, \quad \forall x, y \in E, \tag{27}$$

where $2\lambda < 1 - \alpha$. In addition, we may employ the equivalent norm $\|u\|_B := \max\{\|P_0 u\|_\infty, \|P_\infty u\|_\infty\}$ in B_0, for which we have

$$(1/2)\|u\|_\infty \leq \|u\|_B \leq \|u\|_\infty, \quad \forall u \in B_0.$$

For $x \in E$ and $u, v \in B_0$ we then easily obtain the estimates

$$\|F_x(u) - F_x(v)\|_B \leq (\alpha + 2\lambda)\|u - v\|_B \tag{28}$$

and

$$|F_x(u)(k)| \leq (\alpha + \lambda)|u(k - 1)| + \alpha\lambda|u(k)| + \alpha|u(k + 1)| \tag{29}$$

for all $k \geq 1$. It follows from (29) that F_x maps the Banach space B_0 into itself, and (28) shows that $F_x : B_0 \to B_0$ is a contraction with contraction constant $\alpha + 2\lambda < 1$, which is independent of $x \in E$. Therefore the Banach fixed point theorem implies the existence of a unique fixed point u_x of F_x in B_0. From the estimate

$$\|u_x - u_y\|_B = \|F_x(u_x) - F_y(u_y)\|_B$$
$$\leq \|F_x(u_x) - F_x(u_y)\|_B + \|F_x(u_y) - F_y(u_y)\|_B$$
$$\leq (\alpha + 2\lambda)\|u_x - u_y\|_B + |P_0 x - P_0 y|$$

we deduce that

$$\|u_x - u_y\|_B \leq |P_0 x - P_0 y|/(1 - \alpha - 2\lambda), \quad \forall x, y \in E. \tag{30}$$

We now set

$$h(x) := P_\infty u_x(0), \quad \forall x \in E_0.$$

Because of (20), h is locally uniformly Lipschitz continuous and maps the space E_0 into E_∞, and from (26) we read off the relation

$$W_0 = \{(x, h(x)) \mid x \in E_0\} = \operatorname{graph}(h).$$

We let $S_1, S_{-1} : B_0 \to B_0$ denote the "shift operators"

$$(S_{-1} u)(k) := u(k + 1), \quad \forall k \in \mathbb{N},$$

and

$$(S_1 u)(k) := \begin{cases} u(k - 1), & \text{if } k \in \mathbb{N}^* \\ 0, & \text{if } k = 0. \end{cases}$$

Clearly S_{-1} and S_1 are continuous and

$$\|S_{-1}\|_B, \quad \|S_1\|_B \quad \leq 1. \tag{31}$$

Finally, we define $G : B_0 \to B_0$ by

$$G(u)(k) := g(u(k)), \quad \forall k \in \mathbb{N}.$$

If for some $\beta > 0$ and some $k \in \mathbb{N}^* \cup \{\infty\}$ we have

$$g \in C^k(\mathbb{B}_E(0, \beta), E),$$

then one easily verifies that

$$G \in C^k(\mathbb{B}_{B_0}(0, \beta), B_0) \tag{32}$$

and (since $Dg(0) = 0$)

$$DG(0) = 0.$$

Using the "unit vector" $e_0 := (1, 0, \ldots) \in B_0$, F_x can be written in the form

$$F_x = (P_0 x)e_0 + T_0 P_0(S_1 + G \circ S_1) + T_\infty^{-1} P_\infty(S_{-1} - G).$$

Setting

$$H(x, u) := u - F_x(u)$$

it then follows from (32) that

$$H \in C^k(E \times \mathbb{B}_{B_0}(0, \beta), B_0)$$

and

$$D_2 H(0, 0) = id_{B_0} - T_0 P_0 S_1 - T_\infty^{-1} P_\infty S_{-1} =: id_{B_0} - K. \tag{33}$$

With the aid of (31) we obtain the estimate

$$\|K\|_{\mathcal{L}(B_0)} \leq \alpha < 1,$$

from which it easily follows (by means of the Neumann series (e.g., Yosida [1])) that $D_2 H(0, 0) \in \mathcal{GL}(B_0)$ (cf. remark (25.6 a)). Because $H(0, 0) = 0$ and since for each $x \in E_0$, the element $u_x \in B_0$ is the unique solution of the equation

$$H(x, u) = 0,$$

it follows from the implicit function theorem (cf. Dieudonné [1]) that in some neighborhood of $x = 0$, the function

$$E_0 \to B_0, \quad x \mapsto u_x$$

is of class C^k. Therefore the function $h : E_0 \to E_\infty$ is – because the "evaluation map" $B_0 \to E$, $u \mapsto u(0)$, is clearly linear and continuous – a C^k-function in some neighborhood V of 0. Therefore W_0^V is, as a graph of a C^k-function, a C^k-manifold of dimension $\dim_{\mathbb{R}}(E_0)$. Since

$$V \ni x \mapsto (x, h(x)) \in E$$

defines a parametrization on W_0^V, we know that

$$T_0 W_0^V = im(id_{E_0}, Dh(0))$$

(where we may assume, without loss of generality, that $\mathbb{K} = \mathbb{R}$). By differentiating the identity $H(x, u_x) = 0$ at the point $x = 0$, it follows that

$$D_1 H(0, 0) + D_2 H(0, 0) Du_0 = 0, \tag{34}$$

where $Du_0 \in \mathcal{L}(E_0, B_0)$ and

$$D_1 H(0, 0)\xi = -(P_0\xi)e_0, \quad \forall \xi \in E.$$

We thus obtain $P_\infty D_1 H(0, 0)\xi = 0$, and from (33) and (34) we deduce (by applying P_∞ to (34)) that

$$P_\infty Du_0 - T_\infty^{-1} P_\infty S_{-1} Du_0 = 0,$$

that is,

$$P_\infty[(Du_0)x](k) = T_\infty^{-1} P_\infty[(Du_0)x](k + 1)$$

for all $k \in \mathbb{N}$ and all $x \in E_0$. From this it follows that

$$|P_\infty[(Du_0)x](k)| \leq \alpha \|P_\infty Du_0\|_{\mathcal{L}(E_0, B_0)} \|x\|_E, \quad \forall k \in \mathbb{N},$$

and therefore

$$\|P_\infty Du_0\|_{\mathcal{L}(E_0, B_0)} \leq \alpha \|P_\infty Du_0\|_{\mathcal{L}(E_0, B_0)},$$

i.e., $P_\infty Du_0 = 0$, since $\alpha < 1$. Because

$$Dh(0)\xi = P_\infty[(Du_0)\xi](0)$$

for all $\xi \in E_0$, we obtain $Dh(0) = 0$ and hence the assertion. $\square$

Let x_0 be a critical point of the flow φ. If V is a neighborhood of x_0, we define the *local stable* and *unstable manifolds of φ at x_0 with respect to V* by, respectively,

$$W_s^V(x_0) := \{x \in W_s(x_0) \mid t \cdot x \in V \text{ for all } t \geq 0\}$$

and

$$W_u^V(x_0) := \{x \in W_u(x_0) \mid t \cdot x \in V \text{ for all } t \leq 0\}.$$

As the adjoining figure shows, it need not be true, in general, that $W_s^V(x_0) = W_s(x_0) \cap V$ or $W_u^V(x_0) = W_u(x_0) \cap V$.

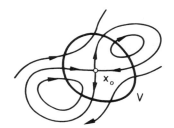

After these preliminaries we can prove the announced *theorem about the local stable and unstable manifolds*, which essentially goes back to Hadamard and Perron.

(19.11) Theorem. *Let M be open in the finite dimensional real Banach space E and let $f \in C^k(M, E)$ for some $k \in \mathbb{N}^* \cup \{\infty\}$. Moreover, let x_0 be a hyperbolic critical point of the flow φ induced by f. Then there exists a neighborhood V of x_0 such that $W_s^V(x_0)$ and $W_u^V(x_0)$ are C^k-manifolds. In addition, we have*

$$T_{x_0} W_s^V(x_0) = x_0 + E_s \quad \text{and} \quad T_{x_0} W_u^V(x_0) = x_0 + E_u,$$

where E_s and E_u denote the stable and unstable subspaces of the linear flow $e^{tDf(x_0)}$.

Proof. Without loss of generality we may assume that $x_0 = 0$. As in the proof of theorem (19.9), we set

$$g := [f - Df(0)] \circ r_\alpha,$$

where $r_\alpha : E \to \overline{\mathbb{B}}(0, \alpha)$ denotes the radical retraction. Then g is uniformly Lipschitz continuous, $g(0) = 0$, $Dg(0) = 0$ and $g \in C^k(\mathbb{B}(0, \alpha), E)$. Moreover, by a judicious choice of $\alpha > 0$, the Lipschitz constant of g can be made arbitrarily small. With $A := Df(0) \in \mathcal{L}(E)$ we have

$$(A + g)|\mathbb{B}(0, \alpha) = f|\mathbb{B}(0, \alpha).$$

Hence the global flow ψ induced by $A + g$ agrees with the flow φ induced by f on $\mathbb{B}(0, \alpha)$.

We now set $T := e^A$. Then T is a hyperbolic automorphism and, as in the proof of theorem (19.9), it follows that $\tilde{g} := \psi^1 - T$ is globally Lipschitz continuous, where, by an appropriate choice of α, the Lipschitz constant of $\tilde{g}$ can be made arbitrarily small (cf. (25)). It also follows from theorem (10.3) that in some neighborhood of 0, $\tilde{g}$ is a C^k-function. Evidently $\tilde{g}(0) = 0$, and since theorem (9.2) implies that $D_2\psi(\cdot, 0)$ is the solution of the linearized IVP

$$\dot{z} = [A + Dg(\psi(t, 0))]z, \quad z(0) = id_E,$$

it follows from $\psi(t, 0) = 0$ and $Dg(0) = 0$ that $D_2\psi(t, 0) = e^{tA}$, therefore $D\tilde{g}(0) = 0$. It follows that T and $\tilde{g}$ satisfy the assumptions of proposition (19.10). Therefore the stable set W_0 of $\psi^1 = T + \tilde{g}$ can be represented as the graph of a globally Lipschitz continuous function $h : E_0 \to E_\infty$. Moreover, there exists a neighborhood V_0 of 0 in E_0 such that $h \in C^k(V_0, E_\infty)$ and so that

$$W_0^{V_0} := \{(x, h(x)) \in E \mid x \in V_0\}$$

is a C^k-manifold and $T_0 W_0^{V_0} = E_0 = E_s$.

We now assert that $W_0 = \tilde{W}_s(0)$, where $\tilde{W}_s(0)$ is the stable manifold of Ψ at the point 0. Since $\lim_{t \to \infty} \psi^t(x) = 0$ implies $\psi^k(x) \to 0$ as $k \to \infty$, it follows that $\tilde{W}_s(0) \subseteq W_0$. To prove the reverse inclusion, we first note that for every $\epsilon > 0$ there exists some $\delta > 0$ such that

$$|\psi^t(x)| \leq \epsilon \quad \text{for all} \quad |t| \leq 1 \text{ and } |x| \leq \delta. \tag{35}$$

For otherwise there would exist some $\epsilon > 0$ and a sequence (t_k, x_k) in $[-1, 1] \times E$ such that $x_k \to 0$ and $|\psi^{t_k}(x_k)| \geq \epsilon$. By taking an appropriate subsequence, we may assume that $t_k \to \bar{t} \in [-1, 1]$, from which we deduce $|\psi^{\bar{t}}(0)| \geq \epsilon$, which contradicts $\psi(\cdot, 0) = 0$.

Let $\epsilon > 0$ now be arbitrary and assume that

$$\psi^k(x) \to 0 \quad \text{as} \quad k \to \infty.$$

Then there exists some $k(\epsilon) \in \mathbb{N}$ so that $|\psi^k(x)| \leq \delta$ for all $k \geq k(\epsilon)$, where $\delta > 0$ is chosen as in (35). For $t \geq k(\epsilon)$ and $k := [t]$ it follows from (35) that

$$|\psi^t(x)| = |\psi^{t-k}(\psi^k(x))| \leq \epsilon,$$

i.e., $\psi^t(x) \to 0$ as $t \to \infty$ and therefore $W_0 \subseteq \tilde{W}_s(0)$.

It follows from the proof of proposition (19.10) that $h(x) = P_\infty u_x(0)$ for all $x \in E_0$, where the function

$$E \to B_0, \quad y \mapsto u_y,$$

is continuous, vanishes at $y = 0$, and satisfies $u_y(k) = (T + \tilde{g})^k(y) = \psi^k(y)$ for all $k \in \mathbb{N}$. Hence for every $\epsilon > 0$ there exists some $\delta > 0$ such that

$$\{(x, h(x)) \mid |x| \leq \delta\} \subseteq \{y \in W_0 \mid |\psi^k(y)| \leq \epsilon, \ \forall k \in \mathbb{N}\}.$$

As in the proof of the inclusion $W_0 \subseteq \tilde{W}_s(0)$, it follows from (35) that – for a given $\epsilon > 0$ – the number $\delta > 0$ can be chosen such that

$$\{(x, h(x)) \mid |x| \leq \delta\} \subseteq \{y \in W_0 \mid |\psi^t(y)| \leq \epsilon, \ \forall t \geq 0\}.$$

Because $W_0 = \tilde{W}_s(0)$, there exist neighborhoods of 0, V in E and $\hat{V} \subseteq V_0$ in E_0, such that $W_0^{\hat{V}} \subseteq \tilde{W}_s^V(0)$. Since the flows agree in a neighborhood of 0, we can choose V so small that $\tilde{W}_s^V(0) = W_s^V(0)$. In particular, we have $W_s^V(0) \subseteq W_0$. And since W_0 is the graph of a function defined on E_0, we have $W_s^V(0) \subseteq W_0^{V_0}$ for some sufficiently small neighborhood V of 0. Therefore $W_s^V(0)$ is a C^k-

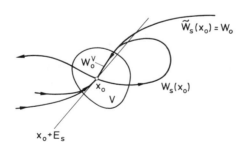

manifold and $T_{x_0} W_s^V(0) = E_s$. The assertion about $W_u^V(0)$ now follows by "reversing the time." $\square$

(19.12) Remarks. (a) The fact that E is finite dimensional was only used in the proofs of lemmas (19.3) and (19.4). By Dunford's operational calculus (e.g., Dunford-Schwartz [1], Yosida [1]) lemma (19.3) also holds for an arbitrary Banach space E. And in this case $T \in \mathcal{GL}(E)$ is also called hyperbolic if $\sigma(T) \cap \mathbb{S}^1_\mathbb{C} = \emptyset$ (that is to say, if the whole spectrum of T is a positive distance away from the unit circle in the complex plane). With the aid of the Dunford calculus, one then, again, shows that we have the decomposition $E = E_0 \oplus E_\infty$ and $T = T_0 \oplus T_\infty$, and that the following holds: $\sigma(T_0) = \sigma_0(T) := \sigma(T) \cap \mathbb{B}_\mathbb{C}(0, 1)$ and

$\sigma(T_\infty) = \sigma_\infty(T) := \sigma(T) \setminus \sigma(T_0)$. Using spectral theory, one can also show that there exists an equivalent norm $\|\cdot\|$ on E (but in general no Hilbert norm) satisfying

$$\max\{\|T_0\|, \|T_\infty^{-1}\|\} \le \alpha < 1.$$

Then proposition (19.6), theorem (19.9), proposition (19.10) and theorem (19.11) remain correct with these modifications also when $\dim(E) = \infty$ (we merely have to replace the inequality $\|u\|_B \le \|u\|_\infty$ by $\|u\|_B \le \beta \|u\|_\infty$, where $\beta := \max\{\|P_0\|, \|P_\infty\|\}$).

(b) The theorem of Grobman and Hartman can also be expressed in the following manner: If x_0 is a hyperbolic critical point of the vector field $f \in C^1(M, E)$, then there exists a homeomorphism $h : U \to V$, from some neighborhood U of x_0 onto some neighborhood V of 0, satisfying $h(x_0) = 0$ and such that the solutions of the differential equation

$$\dot{x} = f(x)$$

in U are mapped homeomorphically by $y = h(x)$ onto the solutions of the linear differential equation

$$\dot{y} = Ay$$

in V, where $A := Df(x_0)$. Naturally, the inevitable question about what happens if x_0 is a nonhyperbolic critical point now arises. In this case there exists a decomposition

$$E = E_n \oplus E_h, \quad A = A_n \oplus A_h$$

such that

$$\sigma(A_n) = \sigma_n(A) = \sigma(A) \cap i\mathbb{R}$$

and

$$\sigma(A_h) = \sigma(A) \setminus \sigma_n(A).$$

In other words: A_h induces a hyperbolic linear flow on E_h. In this case one can "partially linearize," i.e., there exists a homeomorphism h, from a neighborhood U of x_0 onto a neighborhood V of 0 satisfying $h(x_0) = 0$, and a function $g \in C^1(V, E^n)$ such that the solutions of the differential equation

$$\dot{x} = f(x)$$

in U are mapped homeomorphically by $y = h(x)$ onto the solutions of the differential equation

$$\dot{y}_n = A_n y_n + g(y_n, y_h) \tag{36}$$
$$\dot{y}_h = A_h y_h,$$

where $y = (y_n, y_h) \in V$. One can show, furthermore, that there exists a neighborhood V_n of 0 in E_n and a function $G \in C^1(V_n, E_h)$ satisfying $G(0) = 0$ and $DG(0) = 0$, and such that for every solution v in V_0 of the equation

$$\dot{v} = A_n v + g(v, G(v)),$$

the function $t \mapsto (v(t), G(v(t)))$ is a solution of the "full" system (36). Therefore the graph of G is a C^1-manifold which has the space E_n as tangent space at the point 0. This manifold Z – the *center manifold* – is characterized by requiring that it contain all solutions of (36) with bounded projections y_h in E_h. For proofs and further investigations in this direction we refer to the literature (e.g., Palmer [1 – 3], Abraham-Robbin [1], Marsden-McCracken [1], Knobloch-Kappel [1]).

(c) If x_0 is a hyperbolic critical point of the flow induced by $f \in C^k(M, E)$, one can show that $W_s(x_0)$ and $W_u(x_0)$ are *immersed C^k-manifolds*, that is to say, there exist C^k-atlases for $W_s(x_0)$ and $W_u(x_0)$ such that the inclusion maps $W_s(x_0) \hookrightarrow E$ and $W_u(x_0) \hookrightarrow E$ are immersions (i.e., at every point the tangent maps are injective) (see, for example, Irwin [1]). In general, however, $W_s(x_0)$ and $W_u(x_0)$ are *not* embedded submanifolds of E, as the following figure shows.

(d) A point $x \in M$ is called *heteroclinic* if $x \in W_s(x_0) \cap W_u(x_1)$, where x_0 and x_1 are distinct hyperbolic critical points. A point $y \in M$ is called *homoclinic* if $y \in W_s(x_0) \cap W_u(x_0)$. In these cases we also say that $\gamma(x)$ is a *heteroclinic* and that $\gamma(y)$ is a *homoclinic orbit*. The phase portrait of a flow is in general extremely complicated (see, for example, the figures in Abraham-Marsden [1]) and even in relatively easy cases, one is very remote from a complete description.   □

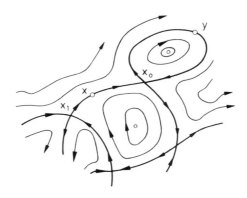

Problems

In what follows we always assume that $f \in C^1(\mathbb{R}, \mathbb{R})$ and

$$f(0) = f(1) = 0,$$

and we consider the simple reaction-diffusion equation

$$\frac{\partial v}{\partial t} - \frac{\partial^2 v}{\partial x^2} = f(v). \tag{37}$$

In many cases (e.g., chemistry or biology), so-called *traveling waves* are of interest, that is, solutions of (37) which have the form

$$v(t, x) = u(x + ct), \quad (t, x) \in \mathbb{R} \times \mathbb{R},$$

so that $u \in C^2(\mathbb{R}, \mathbb{R})$ and $c \neq 0$, and such that the limits

$$u(\pm\infty) := \lim_{\xi \to \pm\infty} u(\xi)$$

exist. Here a traveling wave is said to be of *wave front type* if

$$0 \leq u \leq 1 \quad \text{and} \quad u(-\infty) = 0, \quad u(\infty) = 1,$$

and it is called a *soliton* if

$$u \neq 0, \quad u \geq 0 \quad \text{and} \quad u(-\infty) = u(\infty) = 0.$$

1. Clarify these concepts geometrically.

2. Show that traveling waves of wave front type can be characterized as the heteroclinic orbits of the system

$$\begin{aligned} \dot{x} &= y \\ \dot{y} &= cy - f(x), \end{aligned} \tag{38}$$

which lie in $[0, 1] \times \mathbb{R}$ and connect the critical points $(0, 0)$ and $(1, 0)$. The solitons correspond to the homoclinic orbits in $\mathbb{R}_+ \times \mathbb{R}$ of the critical point $(0, 0)$.

3. Show that the heteroclinic orbits, which correspond to solutions of the wave front type, lie completely in $(0, 1) \times (0, \infty)$. In the special case when

$$f(u) = ku(1 - u),$$

where $k > 0$ is some constant, (37) is called *Fisher's equation*, named after R.A. Fisher who suggested and analyzed it in 1937 as a model in genetics.

4. Show that for every $c > 2\sqrt{k}$, Fisher's equation has a unique traveling wave of wave front type.

5. Show (in the general case) that if $f'(0) > 0$, then (37) has no solitons.

6. Let M be open in $\mathbb{R}^2$ and let $g \in C^1(M, \mathbb{R}^2)$. Furthermore, let $x_0 \in M$ be a critical point of g such that 0 is a spiral point of the linearized equation $\dot{y} = Dg(x_0)y$. By introducing polar coordinates, show that x_0 is also a "spiral" for the nonlinear equation $\dot{x} = g(x)$, i.e., show that, in a neighborhood of x_0, the orbits have the same structure as the orbits of $e^{tDg(x_0)}$: They either spiral toward the point x_0 or away from it.

7. Make use of problem 6 to show that if $c^2 < 4f'(0)$, then (37) has no solutions of wave front type.

Chapter V
Periodic Solutions

This chapter is devoted to existence and stability questions concerning periodic solutions of ordinary differential equations.

In the first section we study the existence of periodic solutions for linear equations with periodic coefficients. As a basis for the stability theory treated later, we develop the Floquet theory.

The existence problem of periodic solutions of nonlinear, nonautonomous T-periodic differential equations is equivalent to a fixed point problem by means of the shift operator. In order to treat such fixed point problems (which are also of great importance for nonlinear boundary value problems and, in general, in nonlinear functional analysis), we devote a whole section to an essentially complete development of the Brouwer degree. We prove, among others, Borsuk's theorem. In subsequent sections this theory will be applied to show the existence of periodic solutions of nonautonomous equations. Here the knowledge about invariant sets and Liapunov functions, gained from the last chapter, is especially useful.

Using Floquet theory, we reduce the stability problem for periodic solutions of nonautonomous differential equations to the case of the Liapunov stability of critical points. With the aid of the Poincaré map, we study the orbital stability of periodic orbits of an autonomous equation. Moreover we will prove (so to speak, as a view to the stability theory of Hamiltonian systems) the Poincaré recurrence theorem.

In a separate section we study plane flows, where, of course, the Poincaré-Bendixson theorem is of central interest. Then we prove, in a general setting (i.e., in $\mathbb{R}^m$), that the winding number of vector fields, which is defined by the Kronecker integral on the boundary of an open set, agrees with the Brouwer degree. This fact will be used to deduce the existence or nonexistence of critical points and periodic orbits.

20. Linear Periodic Differential Equations

In this section we again let $E = (E, |\cdot|)$ denote a finite dimensional Banach space over $\mathbb{K}$.

We begin with some simple general remarks. Let $M \subseteq E$ be open and $f \in C^{0,1^-}(\mathbb{R} \times M, E)$, and let $u(\cdot, \tau, \xi)$ denote the (maximal) solution of the IVP

$$\dot{x} = f(t, x), \quad x(\tau) = \xi.$$

Now let $T \in \mathbb{R}$. Then we define the *shift operator* (or *time-T-map,* or the *T-translation operator*)

$$u_T : \mathrm{dom}(u_T) \subseteq M \to M$$

by

$$\mathrm{dom}(u_T) := \{\xi \in M \mid t^-(0, \xi) < T < t^+(0, \xi)\}$$

and

$$u_T(\xi) := u(T, 0, \xi).$$

From theorem (8.3) it follows that

$$\mathcal{D}(f) = \{(t, \tau, \xi) \in \mathbb{R} \times \mathbb{R} \times M \mid t^-(\tau, \xi) < t < t^+(\tau, \xi)\}$$

is open in $\mathbb{R} \times \mathbb{R} \times M$ and, since $\mathrm{dom}(u_T)$ is the projection of the slice

$$\mathcal{D}(f) \cap (T \times \{0\} \times M)$$

in M, $\mathrm{dom}(u_T)$ is open in M and – again based on theorem (8.3) –

$$u_T \in C^{1^-}(\mathrm{dom}(u_T), M).$$

The following simple, but important theorem, which is essentially due to Poincaré, reduces the existence problem of T-periodic solutions $(T > 0)$ of the equation $\dot{x} = f(t, x)$ to a fixed point problem for the shift operator.

(20.1) Theorem. *Let $f \in C^{0,1^-}(\mathbb{R} \times M, E)$ be T-periodic in t, i.e.,*

$$f(t + T, x) = f(t, x), \quad \forall t \in \mathbb{R}, \ x \in M.$$

Then the differential equation $\dot{x} = f(t, x)$ has a T-periodic solution if and only if the shift operator u_T has a fixed point.

Proof. "$\Rightarrow$" Assume that $u(\cdot, \tau, \xi)$ is a T-periodic solution of $\dot{x} = f(t, x)$. Because of periodicity, $(t^-(0, \xi), t^+(0, \xi)) = J(\tau, \xi) = \mathbb{R}$. Therefore, without loss of generality, we may assume that $\tau \leq 0$. Now set $\xi_0 := u(0, \tau, \xi)$ and note that

$$u(t, 0, \xi_0) = u(t, \tau, \xi).$$

It follows from the T-periodicity of $u(\cdot, \tau, \xi)$ that

$$u_T(\xi_0) = u(T, 0, \xi_0) = u(T, \tau, \xi) = u(0, \tau, \xi) = \xi_0.$$

"$\Leftarrow$" If $\xi \in M$ is a fixed point of u_T, then set

$$x(t) := u(t + T, 0, \xi) \text{ for } t \in J(0, \xi) - T.$$

Then we have $x(0) = u(T, 0, \xi) = u_T(\xi) = \xi$ and

$$\dot{x}(t) = \dot{u}(t + T, 0, \xi) = f(t + T, x(t)) = f(t, x(t)).$$

Hence x solves the IVP

$$\dot{x} = f(t, x), \quad x(0) = \xi,$$

and from the uniqueness it follows that

$$x(t) = u(t + T, 0, \xi) = u(t, 0, \xi) \quad \text{for all} \quad t \in J(0, \xi) - T.$$

Inductively we obtain that $u(\cdot, 0, \xi)$ is defined on all of $\mathbb{R}$ and is a T-periodic solution of $\dot{x} = f(t, x)$.   □

(20.2) Remarks. (a) The proof above shows that $\xi \in M$ is a fixed point of u_T if and only if $u(\cdot, 0, \xi)$ is a T-periodic solution of $\dot{x} = f(t, x)$.

(b) If we have the linear differential equation

$$\dot{x} = A(t)x + a(t), \quad A \in C(\mathbb{R}, \mathcal{L}(E)), \quad a \in C(\mathbb{R}, E),$$

then, by theorem (7.9) and theorem (11.13), the shift operator is given for every $T \in \mathbb{R}$ by $\text{dom}(u_T) = E$ and

$$u_T(\xi) := U(T, 0)\xi + \int_0^T U(T, \tau)a(\tau) \, d\tau, \quad \xi \in E.$$

(c) The one-dimensional problem $\dot{x} = 1$ shows that the existence problem for T-periodic solutions is nontrivial. This equation is evidently T-periodic for every $T > 0$ (i.e., the right-hand side of the equation is T-periodic in t), but it has no T-periodic solutions.   □

In what follows we consider *T-periodic linear differential equations*

$$\dot{x} = A(t)x + a(t) \tag{1}$$

with $A \in C(\mathbb{R}, \mathcal{L}(E))$, $a \in C(\mathbb{R}, E)$, as well as

$$A(t + T) = A(t), \quad a(t + T) = a(t), \quad \forall t \in \mathbb{R},$$

where $T > 0$.

(20.3) Theorem. *The T-periodic linear differential equation* (1) *has a T-periodic solution if and only if it has a uniformly bounded solution.*

Proof. "$\Rightarrow$" trivial.

"$\Leftarrow$" We know from theorem (20.1) and remark (20.2 b) that (1) has a T-periodic solution if and only if there exists some $\xi \in E$ such that

$$\xi = U(T)\xi + \eta, \tag{2}$$

where

$$\eta := \int_0^T U(T, \tau) a(\tau) \, d\tau$$

and

$$U(T) := U(T, 0).$$

We therefore have to show that whenever (2) cannot be solved, (1) has only unbounded solutions.

From linear algebra we know that (2) has no solution if and only if there exists some $\zeta \in E'$ such that

$$\zeta = [U(T)]' \zeta \quad \text{and} \quad \langle \zeta, \eta \rangle \neq 0 \tag{3}$$

(where, as usual, the prime denotes the dual space or dual operator and $\langle \cdot, \cdot \rangle :$ $E' \times E \to \mathbb{K}$ denotes the duality pairing).

If now x is an arbitrary solution of (1), then

$$x(t) = U(t, 0)\xi + \int_0^t U(t, \tau) a(\tau) \, d\tau, \quad \forall t \in \mathbb{R},$$

for some appropriate $\xi \in E$ (cf., theorem (11.13)). Thus it follows that

$$x(T) = U(T)\xi + \eta \tag{4}$$

and

$$\dot{x}(t + kT) = A(t + kT)x(t + kT) + a(t + kT)$$
$$= A(t)x(t + kT) + a(t)$$

for all $k \in \mathbb{N}$ and $t \in \mathbb{R}$. Hence by uniqueness, $x_k(t) := x(t + kT)$ is the solution of the IVP

$$\dot{y} = A(t)y + a(t), \quad y(0) = x(kT).$$

Thus from (4) it follows that

$$x_k(T) = x((k + 1)T) = U(T)x(kT) + \eta$$

and therefore by induction

$$x_k(T) = [U(T)]^{k+1}\xi + \sum_{j=0}^{k} [U(T)]^j \eta.$$

Now, from (3) we obtain

$$\langle \zeta, x_k(T) \rangle = \langle [U(T)']^{k+1}\zeta, \xi \rangle + \sum_{j=0}^{k} \langle [U(T)']^j \zeta, \eta \rangle$$
$$= \langle \zeta, \xi \rangle + (k + 1)\langle \zeta, \eta \rangle.$$

Because $\langle \zeta, \eta \rangle \neq 0$, it follows that $|\langle \zeta, x_k(T) \rangle| = |\langle \zeta, x((k+1)T) \rangle| \rightarrow \infty$ as $k \rightarrow \infty$. Therefore x is unbounded.   □

The following theorem gives a necessary and sufficient condition for (1) to have a unique uniformly bounded – and therefore a unique T-periodic – solution if A is constant.

(20.4) Proposition. *Let $A \in \mathcal{L}(E)$ and $g \in BC(\mathbb{R}, E)$. Then the equation*

$$\dot{x} = Ax + g(t) \tag{5}$$

has a unique solution $u \in BC(\mathbb{R}, E)$ if and only if e^{tA} is hyperbolic. If this is the case, the solution is given by

$$u(t) := \int_{-\infty}^{t} e^{(t-\tau)A} P_s g(\tau) \, d\tau - \int_{t}^{\infty} e^{(t-\tau)A} P_u g(\tau) \, d\tau, \tag{6}$$

where $P_s : E \rightarrow E_s$ and $P_u : E \rightarrow E_u$ are the projections corresponding to $E = E_s \oplus E_u$.

Proof. "$\Rightarrow$" If (5) has a unique solution in $BC(\mathbb{R}, E)$, then the homogeneous equation $\dot{x} = Ax$ has only the trivial solution in $BC(\mathbb{R}, E)$. If e^{tA} were not hyperbolic, there would exist some $\lambda = i\omega \in \sigma(A) \cap i\mathbb{R}$. It now follows from theorem (12.7) that the function $t \mapsto ce^{i\omega t}$, $c \in E \setminus \{0\}$, is a nontrivial solution of $\dot{x} = Ax$ in $BC(\mathbb{R}, E)$. Hence e^{tA} must be hyperbolic.

"$\Leftarrow$" If e^{tA} is hyperbolic, theorem (13.4) shows that every nontrivial solution of $\dot{x} = Ax$ is unbounded. Hence $\dot{x} = Ax$ has only the trivial solution in $BC(\mathbb{R}, E)$. Subsequently, (5) can have at most one solution in $BC(\mathbb{R}, E)$. It therefore suffices to show that (6) is a solution of (5) in $BC(\mathbb{R}, E)$.

As a consequence of remarks (13.2) it follows that there exist constants $\alpha, \beta > 0$ so that

$$\max\{|e^{tA} P_s x|, |e^{-tA} P_u x|\} \leq \beta e^{-t\alpha} |x|, \quad \forall t \geq 0, \; x \in E.$$

With this we obtain, for the function defined by (6), that

$$|u(t)| \leq \beta \left\{ \int_{-\infty}^{t} e^{-(t-\tau)\alpha} \, d\tau + \int_{t}^{\infty} e^{-(\tau-t)\alpha} \, d\tau \right\} \|g\|_\infty \leq (2\beta/\alpha) \|g\|_\infty$$

for all $t \in \mathbb{R}$. Therefore u is in $B(\mathbb{R}, E)$, and from the theorem on the differentiability of integrals with parameters it follows that u is continuously differentiable and that

$$\dot{u}(t) = P_s g(t) + \int_{-\infty}^{t} A e^{(t-\tau)A} P_s g(\tau)\, d\tau + P_u g(t)$$
$$- \int_{t}^{\infty} A e^{(t-\tau)A} P_u g(\tau)\, d\tau = g(t) + A u(t),$$

for all $t \in \mathbb{R}$. $\square$

(20.5) Remark. If e^{tA} is hyperbolic, then equation (5) is equivalent to the system

$$\dot{x}_s = A_s x_s + P_s g(t), \quad \dot{x}_u = A_u x_u + P_u g(t), \tag{7}$$

where $x_s := P_s x$ and $x_u := P_u x$. So for every $\xi = \xi_s + \xi_u$, the unique solution of (7) with initial condition (t_0, ξ) is given by

$$x_s(t) = e^{(t-t_0)A_s}\xi_s + \int_{t_0}^{t} e^{(t-\tau)A_s} P_s g(\tau)\, d\tau$$

$$x_u(t) = e^{(t-t_0)A_u}\xi_u + \int_{t_0}^{t} e^{(t-\tau)A_u} P_u g(\tau)\, d\tau.$$

Since e^{tA_s} is a contraction on E_s, it easily follows that $x_s \in B([t_0, \infty), E)$. Analogously we obtain that $x_u \in B((-\infty, t_0], E)$, because e^{tA_u} is an expansion on E_u (cf., theorem (13.4)). If now $x(t) = x_s(t) + x_u(t)$ is assumed to be bounded on all of $\mathbb{R}$, then in the first equation we let t_0 go to $-\infty$ and in the second we let t_0 go to ∞. We then obtain (formally)

$$x_s(t) = \int_{-\infty}^{t} e^{(t-\tau)A} P_s g(\tau)\, d\tau$$

and

$$x_u(t) = \int_{\infty}^{t} e^{(t-\tau)A} P_u g(\tau)\, d\tau = - \int_{t}^{\infty} e^{(t-\tau)A} P_u g(\tau)\, d\tau$$

and so by setting $u := x_s + x_u$ we obtain equation(6). That this expression is indeed a solution of (5) in $BC(\mathbb{R}, E)$ was shown in the proof of proposition (20.4). $\square$

(20.6) Corollary. *Assume that A is constant and e^{tA} hyperbolic. Then the T-periodic linear equation*

$$\dot{x} = Ax + a(t)$$

has exactly one T-periodic solution. This solution is given by

$$u(t) := \int_{-\infty}^{t} e^{(t-\tau)A} P_s a(\tau)\, d\tau - \int_{t}^{\infty} e^{(t-\tau)A} P_u a(\tau)\, d\tau.$$

Proof. This follows immediately from proposition (20.4) and theorem (20.3). $\square$

The Floquet Theory

In what follows we will show that the T-periodic equation (1) can be transformed into a T-periodic equation with constant main part. For this we need the following lemma.

(20.7) Lemma. *Let* $\mathbb{K} = \mathbb{C}$ *and* $C \in \mathcal{GL}(E)$. *Then there exists some* $B \in \mathcal{L}(E)$ *such that* $C = e^B$.

Proof. By using the decomposition

$$E = E_1 \oplus \cdots \oplus E_k \quad \text{and} \quad C = C_1 \oplus \cdots \oplus C_k$$

where $E_j := \ker[(\lambda_j - C)^{m(\lambda_j)}]$ and $C_j := C|E_j$, we may assume that C has the form

$$C = \lambda + N \quad \text{with} \quad \lambda \neq 0 \quad \text{and} \quad N^m = 0$$

for some $m \in \mathbb{N}$ (cf. section 12). Because $\lambda \neq 0$, there exists some $\beta \in \mathbb{C}$ so that $\lambda = e^\beta$ (e.g., we can choose β to be the principal value of the complex logarithm $\beta = \ln \lambda$). Now set

$$L := \ln(1 + \lambda^{-1} N) := \sum_{j=1}^{m-1} \frac{(-1)^{j+1}}{j}(\lambda^{-1} N)^j,$$

i.e., we define L by the power series (in $\mathcal{L}(E)$) for $\ln(1 + x)$ near $x = 0$. This power series terminates because N is nilpotent. One easily verifies that $e^L = 1 + \lambda^{-1} N$ by substituting L in the power series for e^L. Hence we get

$$C = \lambda(1 + \lambda^{-1} N) = e^\beta e^L = e^{\beta + L}$$

and so the claim follows with $B := \beta + L$. $\qquad\qquad\square$

(20.8) Remarks. (a) Because the complex logarithm is not single valued, B is not uniquely determined by C.

(b) If $\mathbb{K} = \mathbb{R}$ and $C \in \mathcal{GL}(E)$, the complexification $C_\mathbb{C}$ is in $\mathcal{GL}(E_\mathbb{C})$. Hence we can apply lemma (20.7) to $C_\mathbb{C}$ and find some $B \in \mathcal{L}(E_\mathbb{C})$ such that $C_\mathbb{C} = e^B$. Then of course $C = e^B|E$, i.e., $e^B(E) \subseteq E$, but, in general, B is not real, that is, B is not necessarily in $\mathcal{L}(E)$. This is clearly seen from the example: $E = \mathcal{L}(E) = \mathbb{R}$ and $C = -1$. $\qquad\square$

(20.9) Theorem *(Floquet). Let* $\mathbb{K} = \mathbb{C}$. *Then there exists a* T-periodic function $Q \in C^1(\mathbb{R}, \mathcal{GL}(E))$ *and some* $B \in \mathcal{L}(E)$ *such that the* <u>Floquet representation</u>

$$U(t, 0) = Q(t)e^{tB}, \quad \forall t \in \mathbb{R},$$

holds.

Proof. For brevity we set $U(t) := U(t,0)$ and $V(t) := U(t+T)U^{-1}(T) = U(t+T)U(0,T)$ (cf. theorem (11.13)). Then

$$\dot{V}(t) = \dot{U}(t+T)U^{-1}(T) = A(t+T)V(t) = A(t)V(t), \quad \forall t \in \mathbb{R},$$

and $V(0) = I := id_E$. From this and the uniqueness of the solution of the IVP

$$\dot{X} = A(t)X, \quad X(0) = I \quad \text{in } \mathcal{L}(E),$$

it follows that $V(t) = U(t)$ and hence

$$U(t+T) = U(t)U(T), \quad \forall t \in \mathbb{R}. \tag{8}$$

Because $U(T) \in \mathcal{GL}(E)$, there exists by lemma (20.7) some $B \in \mathcal{L}(E)$ such that

$$U(T) = e^{TB}. \tag{9}$$

We now define $Q \in C^1(\mathbb{R}, \mathcal{GL}(E))$ by

$$Q(t) := U(t)e^{-tB}. \tag{10}$$

It then follows from (8) and (10) that

$$Q(t+T) = U(t+T)e^{-(t+T)B} = U(t)U(T)e^{-TB}e^{-tB}$$
$$= U(t)e^{-tB} = Q(t)$$

for all $t \in \mathbb{R}$. Therefore Q is T-periodic. $\qquad\square$

(20.10) Corollary. *If* $\mathbb{K} = \mathbb{C}$, *the transformation*

$$x = Q(t)y$$

converts the T-periodic linear equation

$$\dot{x} = A(t)x + a(t)$$

into the T-periodic linear equation with constant main part

$$\dot{y} = By + b(t),$$

where

$$b(t) := Q^{-1}(t)a(t).$$

Proof. From $\dot{x} = \dot{Q}y + Q\dot{y}$ it follows that

$$\dot{y} = Q^{-1}(Ax + a - \dot{Q}y) = Q^{-1}(AQ - \dot{Q})y + b.$$

Because $Q(t) = U(t,0)e^{-tB}$, we have

$$\dot{Q}(t) = \dot{U}(t,0)e^{-tB} - U(t,0)e^{-tB}B = A(t)Q(t) - Q(t)B,$$

from which the assertion follows. $\qquad\square$

(20.11) Remarks. (a) The operator

$$U(T) := U(T, 0) \in \mathcal{GL}(E),$$

that is, the shift operator of $\dot{x} = A(t)x$, is called the *monodromy operator* (even if $\mathbb{K} = \mathbb{R}$) of the T-periodic homogeneous linear equation

$$\dot{x} = A(t)x. \tag{11}$$

The eigenvalues of the monodromy operator $U(T)$ are called *Floquet* (or *characteristic*) *multipliers* of equation (11). If $\lambda \in \sigma(U(T))$ is a Floquet multiplier of (11), then every $\beta \in \mathbb{C}$ with $\lambda = e^{T\beta}$ is called a *Floquet* (or *characteristic*) *exponent* of equation (11). The Floquet exponents are evidently only determined up to integer multiples of $2\pi i/T$.

(b) *The monodromy operator of* (11) *is the unique operator* $C \in \mathcal{L}(E)$ *satisfying*

$$U(t + T, 0) = U(t, 0)C, \quad \forall t \in \mathbb{R}. \tag{12}$$

Indeed, because of (8), $U(T)$ satisfies equation (12). On the other hand, it follows from (8) and (12) that

$$C = U^{-1}(t, 0)U(t + T, 0) = U(T),$$

hence the uniqueness. (Note that in the proof of (8) the assumption $\mathbb{K} = \mathbb{C}$ was not used.)

(c) *If* $\mathbb{K} = \mathbb{C}$*, the monodromy operator of* (11) *has the form*

$$U(T) = e^{TB}$$

for some appropriate $B \in \mathcal{L}(E)$*, and so the eigenvalues of B can be taken as the Floquet exponents.*

This follows from (9) and the spectral theorem (lemma (19.3)).

(d) The Floquet representation in theorem (20.9) contains important information about the solutions of the T-periodic homogeneous linear equation $\dot{x} = A(t)x$ in the case when $\mathbb{K} = \mathbb{C}$. By theorem (12.7), we do in fact know that every function $e^{tB}\xi$, $\xi \in E$, is a linear combination of functions of the form

$$p(t)e^{\beta t}, \tag{13}$$

where β is a characteristic exponent (an eigenvalue of B) and $p(t)$ is a polynomial in t of degree less than $m(\beta)$ (= algebraic multiplicity of the eigenvalue β of B). Hence it follows from theorem (20.9) *that every solution of* $\dot{x} = A(t)x$ *is a linear combination of functions of the form* (13), *where the coefficients of the polynomials $p(t)$ are T-periodic functions.* If B is semisimple, then all polynomials $p(t)$ have degree 0, that is, in this case every solution of $\dot{x} = A(t)x$ is a linear combination of functions of the form

$$c(t)e^{\beta t},$$

where the β's are the characteristic exponents and the c's are T-periodic functions. $\quad\square$

(20.12) Proposition. *The T-periodic homogeneous linear equation $\dot{x} = A(t)x$ has a nontrivial T-periodic solution if and only if 1 is a Floquet multiplier.*

Proof. By theorem (20.1) and remark (20.2) we know that $\xi \in E$ is a fixed point of the monodromy operator $U(T)$ if and only if $U(\cdot, 0)\xi$ is a T-periodic solution of $\dot{x} = A(t)x$. Hence there exists a nontrivial T-periodic solution if and only if there exists some $\xi \in E \setminus \{0\}$ such that $\xi - U(T)\xi = 0$, i.e., if and only if $\ker(1 - U(T)) \neq \{0\}$. $\qquad\square$

It is very difficult to determine the Floquet multipliers of a periodic equation $\dot{x} = A(t)x$. This, of course, is due to the fact that in principle one has to know the solutions of $\dot{x} = A(t)x$ completely in order to determine the monodromy operator $U(T)$. Yet the Floquet theory is of fundamental theoretical importance, in particular, for stability questions to be treated in section 23.

Problems

1. Assume that $\mathbb{K} = \mathbb{C}$ and let $\mu_j = e^{\beta_j T}$, $j = 1, \ldots, m := \dim E$, be the characteristic multipliers of the T-periodic equation $\dot{x} = A(t)x$, counted according to their multiplicities.

Prove that:

$$\prod_{j=1}^{m} \mu_j = \exp\left(\int_0^T \operatorname{trace} A(s)\, ds \right) \tag{14}$$

and

$$\sum_{j=1}^{m} \beta_j \equiv \frac{1}{T} \int_0^T \operatorname{trace} A(s)\, ds \quad (\bmod\ 2\pi i/T).$$

(*Hint:* Liouville's theorem.)

Remark. If $m - 1$ Floquet multipliers are known, one can use formula (14) to find another one.

2. Prove that the zero solution of the T-periodic linear equation $\dot{x} = A(t)x$ is stable if and only if:

(a) $|\mu| \leq 1$ for every Floquet multiplier μ.
(b) If the Floquet multiplier μ satisfies $|\mu| = 1$, then μ is a semisimple eigenvalue of the monodromy operator $U(T)$.

The zero solution is asymptotically stable if and only if $|\mu| < 1$ for every Floquet multiplier μ.

(*Hint:* Corollary (20.10)).

3. Prove the following: Let $\mathbb{K} = \mathbb{C}$. Then $\beta \in \mathbb{C}$ is a Floquet exponent of the T-periodic equation $\dot{x} = A(t)x$ if and only if the equation

$$\dot{y} = (A(t) - \beta)y$$

has a nontrivial T-periodic solution.

(*Hint:* Obtain a relation between the monodromy operators of both equations.)

21. Brouwer Degree

Based on theorem (20.1), the existence problem of T-periodic solutions of the T-periodic differential equation $\dot{x} = f(t, x)$ is equivalent to the fixed point problem of the shift operator u_T. In this section we therefore introduce some general considerations about fixed point problems. But because of

$$g(x) = x \quad \Leftrightarrow \quad f(x) := x - g(x) = 0,$$

this is equivalent to considerations about the existence and number of zeros of continuous functions.

In what follows let $E = (E, | \cdot |)$ denote a finite dimensional Banach space over $\mathbb{K}$, and let Ω be a bounded open subset of E.

We consider a continuous function $f : \overline{\Omega} \to E$ and pose the following problem: Find an easily calculable quantity which allows us to deduce the existence or the number of zeros of f in $\overline{\Omega}$. In simple cases this quantity must be the exact number of zeros of f and it must be invariant under perturbations (i.e., deformations) of f. Simple one-dimensional examples show that we must certainly assume $0 \notin f(\partial\Omega)$. Moreover it is clear that the

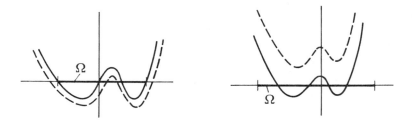

number of zeros, even in the case of finitely many zeros, is not invariant under continuous deformations of f which let "no zeros wander across the boundary $\partial\Omega$." However, if one assigns an "orientation" to every zero, that is, if one assigns the value 1 to the zero

$x_0 \in \Omega \subseteq \mathbb{R}$ when $f'(x_0) > 0$, and the value -1 when $f'(x_0) < 0$, then one can define the "algebraic number of zeros" by

$$a(f, \Omega) := \sum_{x \in f^{-1}(0)} \mathrm{sign} f'(x). \tag{1}$$

One can see, at least in the simple one-dimensional pictures, and when f has only finitely many zeros, that $a(f, \Omega)$ has the desired continuity property for $f \in C^1(\Omega, \mathbb{R}) \cap C(\overline{\Omega}, \mathbb{R})$ and $0 \notin f(\partial\Omega)$.

Formula (1) can be generalized to the m-dimensional case, that is, the case when $E = \mathbb{R}^m$, because then the orientation of the zero is given by the Jacobian determinant. Therefore, if $f \in C^1(\Omega, \mathbb{R}^m) \cap C(\overline{\Omega}, \mathbb{R}^m)$ with $0 \notin f(\partial\Omega)$, and if f has only finitely many regular (i.e., we have $\det Df(x) \neq 0$) zeros in Ω, then the generalization of (1) is given by:

$$a(f, \Omega) := \sum_{x \in f^{-1}(0)} \mathrm{sign}(\det Df(x)). \tag{2}$$

In order to make the above ideas more precise, we let $E = \mathbb{R}^m$. A mapping $f : \overline{\Omega} \to \mathbb{R}^n$ is called *smooth* if there exists an open neighborhood U of $\overline{\Omega}$ and a function $\overline{f} \in C^\infty(U, \mathbb{R}^n)$ such that $\overline{f}|\overline{\Omega} = f$ (for this we use the notation: $f \in \overline{C}^\infty(\overline{\Omega})$). If $f : \overline{\Omega} \to \mathbb{R}^n$ is a smooth function, then $y \in \mathbb{R}^n$ is called a *regular value* of f if there exists an open neighborhood U of $\overline{\Omega}$ and a function $\overline{f} \in C^\infty(U, \mathbb{R}^n)$ with $\overline{f}|\overline{\Omega} = f$ and such that y is a regular value of $\overline{f}$, i.e., such that $D\overline{f}(x) \in \mathcal{L}(\mathbb{R}^m, \mathbb{R}^n)$ is surjective for all $x \in \overline{f}^{-1}(y)$. By definition, it therefore follows that every $y \in \mathbb{R}^n \setminus \overline{f}(U)$ is a regular value of $\overline{f}$. An important theorem of differential topology, the so-called *Sard's lemma*, (cf., Hirsch [1] or Milnor [1]) says that for every mapping $g \in C^\infty(U, \mathbb{R}^n)$, with $U \subseteq \mathbb{R}^m$ open, the set of regular values is dense in $\mathbb{R}^n$.

In what follows let

$$\overline{C}_r^\infty(\overline{\Omega}, \mathbb{R}^n) := \{f : \overline{\Omega} \to \mathbb{R}^n \mid f \text{ is smooth and } 0 \text{ is a regular value of } f\}.$$

The following approximation result is then a simple consequence of Sard's lemma.

(21.1) Lemma. *Let $\Omega \subseteq \mathbb{R}^m$ be open and bounded. Then $\overline{C}_r^\infty(\overline{\Omega}, \mathbb{R}^n)$ is dense in the Banach space $C(\overline{\Omega}, \mathbb{R}^n)$.*

Proof. Let any $\epsilon > 0$ and $f \in C(\overline{\Omega}, \mathbb{R}^n)$ be given. By remark (2.12) there exists a function $f_\epsilon \in \overline{C}^\infty(\overline{\Omega}, \mathbb{R}^n)$ such that $\|f_\epsilon - f\|_\infty < \epsilon/2$. By Sard's lemma there exists some $y \in \mathbb{R}^n$ with $|y| < \epsilon/2$ and such that y is a regular value of f_ϵ. It follows that $g_\epsilon := f_\epsilon - y \in \overline{C}_r^\infty(\overline{\Omega}, \mathbb{R}^n)$ and

$$\|f - g_\epsilon\|_\infty \leq \|f - f_\epsilon\|_\infty + |y| < \epsilon,$$

which proves the lemma. $\qquad \square$

Assume now that $E = \mathbb{R}^m$, $f \in \overline{C}_r^\infty(\overline{\Omega}, \mathbb{R}^m)$ and $0 \notin f(\partial\Omega)$. Then we define the *algebraic number of zeros of f in Ω* by

$$a(f, \Omega) := \sum_{x \in f^{-1}(0)} \text{sign} \det Df(x) \tag{3}$$

with the usual convention that $\sum_{\emptyset} = 0$. The inverse function theorem implies that every

point $x \in f^{-1}(0)$ has an open neighborhood U_x which is mapped diffeomorphically onto an open neighborhood V_x of 0 by the function f. Since $\overline{\Omega}$ is compact, it follows, in particular, that f has only finitely many zeros, hence the sum in (3) is finite and therefore $a(f, \Omega)$ is well-defined. If $\det Df(x) > 0$, then $f|U_x$ represents an orientation preserving diffeomorphism onto V_x, and if $\det Df(x) < 0$, then $f|U_x$ reverses the orientation. This fact explains the term "algebraic number."

(21.2) Lemma. *Let $f \in \overline{C}_r^\infty(\overline{\Omega}, \mathbb{R}^m)$. Then there exists an open neighborhood V of 0 so that every $y \in V$ is a regular value of f and such that*

$$a(f - y, \Omega) = a(f, \Omega), \quad \forall y \in V.$$

Proof. The continuity of f and the compactness of $\overline{\Omega}$ imply that $f(\overline{\Omega})$ is also compact and therefore closed. So trivially the statement is true when $0 \notin f(\overline{\Omega})$.

Assume then that $f^{-1}(0) = \{x_1, \ldots, x_k\}$ and that $U_1, \ldots, U_k$ are mutually disjoint neighborhoods of $x_1, \ldots, x_k$, diffeomorphic to some neighborhoods $V_1, \ldots, V_k$ of 0 under f. Then the set

$$V := \bigcap_{j=1}^k V_j \setminus f\left(\overline{\Omega} \setminus \bigcup_{j=1}^k U_j\right)$$

has the desired properties. □

The following lemma contains a fundamental statement about the continuity of the function $f \to a(f, \Omega)$, which allows us to extend this function to a larger class of mappings.

(21.3) Lemma. *The mapping*

$$a(\cdot, \Omega) : \{f \in \overline{C}_r^\infty(\overline{\Omega}, \mathbb{R}^m) \mid 0 \notin f(\partial\Omega)\} \to \mathbb{Z}$$

is locally constant, that is, for every $f_0 \in \overline{C}_r^\infty(\overline{\Omega}, \mathbb{R}^m)$ satisfying $0 \notin f_0(\partial\Omega)$ there exists a neighborhood U of f_0 in $C(\overline{\Omega}, \mathbb{R}^m)$ such that $0 \notin f(\partial\Omega)$ and $a(f, \Omega) = a(f_0, \Omega)$ for all

$$f \in U \cap \overline{C}_r^\infty(\overline{\Omega}, \mathbb{R}^m).$$

Proof. Assume that $f_0 \in \overline{C}_r^\infty(\overline{\Omega}, \mathbb{R}^m)$ satisfies $0 \notin f_0(\partial\Omega)$. Since $\partial\Omega$ is compact and since $f_0 : \overline{\Omega} \to \mathbb{R}^m$ is continuous, the set $f_0(\partial\Omega)$ is compact and hence $\text{dist}(0, f_0(\partial\Omega)) > 0$.

Now take any $f_1 \in \overline{C}_r^\infty(\overline{\Omega}, \mathbb{R}^m)$ with

$$\|f_0 - f_1\|_\infty < \text{dist}(0, f_0(\partial\Omega)).$$

For $\lambda \in [0, 1]$ let

$$h : \overline{\Omega} \times [0, 1] \to \mathbb{R}^m$$

be defined by

$$h(x, \lambda) := (1 - \lambda)f_0(x) + \lambda f_1(x).$$

Then evidently h is smooth and because of

$$|h(x, \lambda)| \geq |f_0(x)| - \lambda|f_0(x) - f_1(x)|$$
$$\geq \text{dist}(0, f_0(\partial\Omega)) - \|f_0 - f_1\|_\infty > 0, \quad \forall x \in \partial\Omega, \ \forall \lambda \in [0, 1],$$

it satisfies

$$0 \notin h(\partial\Omega \times [0, 1]).$$

Based on Sard's lemma and because of lemma (21.2), there exists some $y \in \mathbb{R}^m$ with $|y| < \text{dist}(0, h(\partial\Omega \times [0, 1]))$, such that y is a regular value of h, f_0 and f_1, and so that

$$a(f_j, \Omega) = a(f_j - y, \Omega), \quad j = 0, 1.$$

Consequently, if we replace f_j by $f_j - y$, we may assume that 0 is a regular value of h, f_0 and f_1, and that

$$0 \notin h(\partial\Omega \times [0, 1])$$

holds.

By assumption, there exist an open neighborhood W of $\overline{\Omega} \times [0, 1]$ in $\mathbb{R}^m \times \mathbb{R}$ and a function $\overline{h} \in C^\infty(W, \mathbb{R}^m)$ so that 0 is a regular value of $\overline{h}$ and $\overline{h}|\overline{\Omega} \times [0, 1] = h$. But then $\overline{h}^{-1}(0)$ is a one-dimensional C^∞-manifold. Since by assumption $\overline{h}^{-1}(0) \cap (\partial\Omega \times [0, 1]) = \emptyset$, it follows that

$$N := h^{-1}(0) \cap (\Omega \times [0, 1]) = \overline{h}^{-1}(0) \cap (\Omega \times [0, 1])$$

is a one-dimensional submanifold of the C^∞-manifold with boundary $Z := \Omega \times [0, 1]$. From the compactness of $\overline{\Omega} \times [0, 1]$ and because $N = \overline{h}^{-1}(0) \cap (\overline{\Omega} \times [0, 1])$, it follows that N is compact and hence it is a finite union of pairwise disjoint, connected, one-dimensional C^∞-submanifolds of Z.

Observe that

$$\partial Z = (\Omega \times \{0\}) \cup (\Omega \times \{1\})$$

and

$$h|(\overline{\Omega} \times \{j\}) = f_j, \quad j = 0, 1.$$

Consequently 0 is a regular value of h and $h|\partial Z$. It then follows easily from the implicit

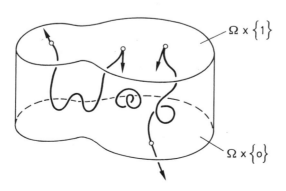

function theorem that N is a C^∞-manifold with boundary and $\partial N = \overline{h}^{-1}(0) \cap \partial Z$ (cf. theorem I.4.1 in Hirsch [1] or lemma (2.4) in Milnor [1]).

In differential topology it is shown that every connected compact one-dimensional C^∞-manifold is either diffeomorphic to $[0, 1]$ or to the circle $\mathbb{S}^1$ (cf. Guillemin-Pollack [1] or Milnor [1]). Hence N consists of finitely many smooth curves. Such a curve either is closed and lies in the interior of Z, that is, in $\Omega \times (0, 1)$, or it is diffeomorphic to the interval $[0,1]$ with its end points in ∂Z.

Let C now be one of the curves above. We then give C an orientation according to the following rule: for $z \in C$ let $[v_1, \dots, v_{m+1}]$ be a positively oriented basis for $T_z Z = T_z \mathbb{R}^{m+1}$ such that $v_{m+1} \in T_z C$. $T_z C$ is positively oriented if and only if $Dh(z)$ transforms the basis $[v_1, \dots, v_m]$ into a positively oriented basis of $T_0 \mathbb{R}^m$. Let now $t(z) \in T_z C$ denote the positively oriented unit tangent vector on C at z and assume $\partial C = \{a, b\} \neq \emptyset$. Then, if t points away from Z at one end point, say at a, t points into Z at the other. If $a \in \Omega \times \{j\}$, it then follows easily from $Dh(a, j)|(\mathbb{R}^m \times \{0\}) = Df_j(a)$ that sign $\det Df_j(a) = 1$ for $j = 1$, and sign $\det Df_j(a) = -1$ for $j = -1$. Analogously we obtain sign $\det Df_j(b) = -1$ for $j = 1$, and sign $\det Df_j = 1$ for $j = 0$.

For each one of the finitely many oriented curves C, which make up N, there are four possibilities:

(i) C is a closed curve in $\Omega \times (0, 1)$.
(ii) C has both end points in the same component $\Omega \times \{j\}$ of Z.
(iii) C goes from $\Omega \times \{0\}$ to $\Omega \times \{1\}$.
(iv) C goes from $\Omega \times \{1\}$ to $\Omega \times \{0\}$.

In both cases (i) and (ii) there is no contribution to $a(f_j, \Omega)$, while, based on the considerations above, it follows directly that in cases (iii) and (iv) we have $a(f_0, \Omega) = a(f_1, \Omega)$. From this and

$$U := \{ f \in C(\overline{\Omega}, \mathbb{R}^m) \mid \|f - f_0\| < \text{dist}(0, f_0(\partial \Omega)) \}$$

the assertion follows.   □

Finally, we need the following simple technical lemma, which we formulate in more generality than presently necessary, for later purposes.

(21.4) Lemma. *Let F be a Banach space, X a nonempty set and $Y \subseteq X$. Then the set*

$$\{ f \in B(X, F) \mid 0 \notin \overline{f(Y)} \}$$

is open in $B(X, F)$.

Proof. For $f \in B(X, F)$ satisfying $0 \notin \overline{f(Y)}$ define $\epsilon := \text{dist}(0, f(Y))$. We then have for all $g \in B(X, F)$ with $\|f - g\|_\infty < \epsilon/2$ and all $y \in Y$ that

$$\|g(y)\| \geq \|f(y)\| - \|f(y) - g(y)\| > \epsilon/2.$$

Hence the lemma follows.   □

After these preparations we are ready to prove the central result of this section. For this we equip $\mathbb{Z}$ with the discrete topology (i.e., every one-point set is open) and for every

$y \in E$ we set

$$\mathcal{D}_y(\Omega, E) := \{f \in C(\overline{\Omega}, E) \mid y \notin f(\partial\Omega)\}.$$

(21.5) Theorem. *Let E be a finite dimensional Banach space. Then for every bounded open subset Ω of E and every $y \in E$, there exists a mapping*

$$\deg(\cdot, \Omega, y) : \mathcal{D}_y(\Omega, E) \to \mathbb{Z},$$

called the <u>*Brouwer degree*</u>, *with the following properties:*

(i) *(Normalization): If $y \in \Omega$, then $\deg(id, \Omega, y) = 1$.*

(ii) *(Additivity): For every two disjoint, open subsets Ω_1 and Ω_2 of Ω and for every $f \in C(\overline{\Omega}, E)$ with $y \notin f(\overline{\Omega} \setminus \Omega_1 \cup \Omega_2)$ we have*

$$\deg(f, \Omega, y) = \deg(f|\overline{\Omega}_1, \Omega_1, y) + \deg(f|\overline{\Omega}_2, \Omega_2, y).$$

(iii) *(Continuity): $\deg(\cdot, \Omega, y)$ is continuous.*

(iv) *(Translation invariance): For all $f \in \mathcal{D}_y(\Omega, E)$ we have $\deg(f, \Omega, y) = \deg(f-y, \Omega, 0)$*

Proof. Evidently it suffices to prove the existence of a mapping

$$\deg(\cdot, \Omega, 0) : \mathcal{D}_0(\Omega, E) \to \mathbb{Z}$$

having properties (i)–(iii). The general result then follows by means of the definition

$$\deg(f, \Omega, y) := \deg(f - y, \Omega, 0).$$

(a) Let $E = \mathbb{R}^m$. By lemma (21.4), $\mathcal{D}_0(\Omega, \mathbb{R}^m)$ is open in $C(\overline{\Omega}, \mathbb{R}^m)$. Hence it follows from lemma (21.1) that

$$\mathcal{D} := \mathcal{D}_0(\Omega, \mathbb{R}^m) \cap \overline{C_r^\infty}(\Omega, \mathbb{R}^m)$$

is dense in $\mathcal{D}_0(\Omega, \mathbb{R}^m)$. The mapping defined in (3),

$$a(\cdot, \Omega) : \mathcal{D} \to \mathbb{Z},$$

clearly satisfies properties (i) and (ii), and, by lemma (21.3), the mapping $a(\cdot, \Omega)$ is also continuous if $\mathcal{D}$ carries the induced topology from $C(\overline{\Omega}, \mathbb{R}^m)$. Because $\mathbb{Z}$ has the discrete topology, it follows immediately that $a(\cdot, \Omega)$ can be extended to a continuous function $\deg(\cdot, \Omega, 0)$ on all of $\mathcal{D}_0(\Omega, \mathbb{R}^m)$. One also verifies easily that $\deg(\cdot, \Omega, \mathbb{R}^m)$ satisfies property (ii).

(b) Now let $\mathbb{K} = \mathbb{R}$ and $m := \dim E$. Then there exists a topological isomorphism $T \in \mathcal{L}(E, \mathbb{R}^m)$ and so we set

$$\deg(f, \Omega, 0) := \deg(T \circ f \circ T^{-1}, T(\Omega), 0)$$

for every $f \in \mathcal{D}_0(\Omega, E)$. One immediately verifies that this definition makes sense and that properties (i)–(iii) are satisfied.

In order to show that $\deg(\cdot, \Omega, 0)$ is independent of T, let $S \in \mathcal{L}(E, \mathbb{R}^m)$ be some other topological isomorphism. Then

$$T \circ f \circ T^{-1} = (T \circ S^{-1}) \circ S \circ f \circ S^{-1} \circ (S \circ T^{-1})$$

and hence $T \circ f \circ T^{-1} = R \circ (S \circ f \circ S^{-1}) \circ R^{-1}$, where $R := T \circ S^{-1} \in \mathcal{GL}(\mathbb{R}^m)$. Therefore

$$\deg(T \circ f \circ T^{-1}, T(\Omega), 0) = \deg(S \circ f \circ S^{-1}, S(\Omega), 0)$$

follows if we can show that

$$\deg(f, \Omega, 0) = \deg(R \circ f \circ R^{-1}, R(\Omega), 0)$$

holds for $E = \mathbb{R}^m$, $f \in \mathcal{D}_0(\Omega, \mathbb{R}^m)$ and $R \in \mathcal{GL}(\mathbb{R}^m)$.

By the definition of $\deg(\cdot, \Omega, 0)$ it suffices to consider the case $f \in \mathcal{D}$. In this case it follows from the chain rule that

$$\deg(R \circ f \circ R^{-1}, R(\Omega), 0) = a(R \circ f \circ R^{-1}, R(\Omega), 0)$$
$$= \sum_{x \in (R \circ f \circ R^{-1})^{-1}(0)} \operatorname{sign} \det D(R \circ f \circ R^{-1})(x)$$
$$= \sum_{y \in f^{-1}(0)} \operatorname{sign} \det(R Df(y) R^{-1}) = a(f, \Omega) = \deg(f, \Omega, 0),$$

and therefore the assertion is proved.

(c) Finally, let $\mathbb{K} = \mathbb{C}$ and $m = \dim E$. Then E is topologically isomorphic to $\mathbb{C}^m$. Therefore it suffices to show (as in (b)) that $\deg(\cdot, \Omega, 0)$ can be defined when $E = \mathbb{C}^m$, and that for every $f \in \mathcal{D}_0(\Omega, \mathbb{C}^m)$ and every $T \in \mathcal{GL}(\mathbb{C}^m)$ we have

$$\deg(T \circ f \circ T^{-1}, T(\Omega), 0) = \deg(f, \Omega, 0). \tag{4}$$

From $\mathbb{C}^m = \mathbb{R}^m + i\mathbb{R}^m$ it follows that $\mathbb{C}^m$ is homeomorphic to $\mathbb{R}^{2m}$ by means of the canonical homeomorphism

$$h : \mathbb{C}^m \to \mathbb{R}^{2m}, \quad x + iy \mapsto (x, y).$$

One easily verifies that h induces a ring isomorphism

$$\mathcal{L}(\mathbb{C}^m) \to \mathcal{L}(\mathbb{R}^{2m}), \quad T \mapsto \begin{bmatrix} R & -S \\ S & R \end{bmatrix},$$

where $R, S \in \mathcal{L}(\mathbb{R}^m)$ are defined by

$$Tx = Rx + iSx \in \mathbb{R}^m + i\mathbb{R}^m, \quad \forall x \in \mathbb{R}^m.$$

If now we define

$$\deg(f, \Omega, 0) := \deg(h \circ f \circ h^{-1}, h(\Omega), 0)$$

for all $f \in \mathcal{D}_0(\Omega, \mathbb{C}^m)$, then one can easily see that $\deg(\cdot, \Omega, 0)$ satisfies properties (i)–(iii), as well as (4). $\qquad \square$

To simplify the notation, we now adopt the following *convention*: Assume that $\Omega \subseteq E$ is a bounded open subset and let Δ be an arbitrary subset of E such that $\overline{\Omega} \subseteq \Delta$. For every function $f : \Delta \to E$ with $f|\overline{\Omega} \in \mathcal{D}_y(\Omega, E)$, we set

$$\deg(f, \Omega, y) := \deg(f|\overline{\Omega}, \Omega, y).$$

With this convention we obtain the following simple, but very important corollary.

(21.6) Corollary. *The Brouwer degree has the following properties:*

(v) *(Excision property): If Ω_1 is an open subset of Ω and if $f \in C(\overline{\Omega}, E)$ satisfies $0 \notin f(\overline{\Omega} \setminus \Omega_1)$, then*

$$\deg(f, \Omega, 0) = \deg(f, \Omega_1, 0).$$

(vi) *(Solution property): If $\deg(f, \Omega, y) \neq 0$, then $f(\Omega)$ is a neighborhood of y in E.*

(vii) *(Homotopy invariance): Let $I \subseteq \mathbb{R}$ be a nonempty compact interval. Moreover, assume, that $h \in C(\overline{\Omega} \times I, E)$ and $y \in C(I, E)$ satisfy*

$$y(\lambda) \notin h(\partial\Omega \times \{\lambda\}), \quad \forall \lambda \in I.$$

Then

$$\deg(h(\cdot, \lambda), \Omega, y(\lambda))$$

is well-defined and independent of $\lambda \in I$.

(viii) *(Dependence on boundary values): For every two functions $f, g \in \mathcal{D}_y(\Omega, E)$ with $f|\partial\Omega = g|\partial\Omega$ we have*

$$\deg(f, \Omega, y) = \deg(g, \Omega, y).$$

(ix) *(Component dependence): If $f \in C(\overline{\Omega}, E)$, then $\deg(f, \Omega, \cdot)$ is constant on the connected components of $E \setminus f(\partial\Omega)$.*

Proof. Because $C(\emptyset, E) = \{\emptyset\}$, where $\emptyset : \emptyset \to E$ denotes the "empty map," one immediately verifies that for $\Omega = \emptyset$ there exists exactly one mapping satisfying properties (i)–(iv), namely $\deg(f, \emptyset, y) = 0$.

(v) This follows immediately from (ii) with $\Omega_2 = \emptyset$.

(vi) Assume $y \notin f(\Omega)$. Then it follows from (v), with $\Omega_1 = \emptyset$, that $\deg(f, \Omega, y) = 0$. This contradicts our assumption.

Since $\mathcal{D}_y(\Omega, E)$ is open in $C(\overline{\Omega}, E)$, there exists some $\epsilon > 0$ such that $f + z \in \mathcal{D}_y(\Omega, E)$ for all $z \in \epsilon \mathbb{B}_E$. Consequently, it follows from (iii) that

$$\deg(f + z, \Omega, 0) \neq 0$$

for all $z \in \epsilon \mathbb{B}_E$ and a sufficiently small $\epsilon > 0$. Therefore $y + \epsilon \mathbb{B}_E \subseteq f(\Omega)$.

(vii) The function $h : \overline{\Omega} \times I \to E$ is uniformly continuous since $\overline{\Omega} \times I$ is compact. From this it easily follows that the mapping

$$I \to C(\overline{\Omega}, E), \quad \lambda \mapsto h(\cdot, \lambda) - y(\lambda),$$

is continuous. As a result of this and using (iii) and (iv), we get that the mapping

$$I \to \mathbb{Z}, \quad \lambda \mapsto \deg(h(\cdot, \lambda), \Omega, y(\lambda)) = \deg(h(\cdot, \lambda) - y(\lambda), \Omega, 0)$$

is well-defined and continuous. But this mapping must be constant because I is connected and $\mathbb{Z}$ is discrete.

(viii) This follows immediately from (vii) with $I = [0, 1]$, $y(\lambda) := y$ and

$$h(\cdot, \lambda) := (1 + \lambda)f + \lambda g.$$

(ix) From (iii) and (iv) it follows that for every $f \in C(\overline{\Omega}, E)$, the mapping

$$E \setminus f(\partial\Omega) \to \mathbb{Z}, \quad y \mapsto \deg(f - y, \Omega, 0) = \deg(f, \Omega, y)$$

is well-defined and continuous. Therefore this mapping is necessarily constant on the connected components of $E \setminus f(\partial\Omega)$. $\square$

(21.7) Remarks. (a) There are many possibilities to define the Brouwer degree. For a historical overview we refer to Siegberg [1]. All definitions of the degree are equivalent because one can show that there is exactly one mapping $\mathcal{D}_y(\Omega, E) \to \mathbb{Z}$ satisfying properties (i)–(iv). In other words: *The Brouwer degree is uniquely determined by the axioms* (i)–(iv) (cf. Amann-Weiss [1], Eisenack-Fenske [1], Lloyd [1], Zeidler [1]).

(b) Let $f \in C(\overline{\Omega}, E)$ and assume that $x_0 \in \Omega$ is an isolated zero of f, i.e., there exists some $\epsilon_0 > 0$ such that $f(x) \neq 0$ for all $x \in x_0 + \epsilon_0 \mathbb{B}$ with $x \neq x_0$. Then for all $\epsilon \in (0, \epsilon_0)$,

$$\deg(f, x_0 + \epsilon\mathbb{B}, 0) \tag{5}$$

is defined and, based upon property (v), independent of ϵ. Hence (5) only depends on f and x_0. For this reason we set

$$i_0(f, x_0) := \deg(f, x_0 + \epsilon\mathbb{B}, 0), \quad 0 < \epsilon < \epsilon_0,$$

and call $i_0(f, x_0)$ the *local index of the isolated zero x_0 of f*.

(c) *Let $f \in C(\overline{\Omega}, E)$ be such that $0 \notin f(\partial\Omega)$ and assume that f has only finitely many zeros* $x_1, \ldots, x_k$. *Then*

$$\deg(f, \Omega, 0) = \sum_{j=1}^{k} i_0(f, x_j).$$

This follows immediately from the additivity property of the degree and the definition of the index. $\square$

(21.8) Proposition. *Let* $\mathbb{K} = \mathbb{R}$ *and* $f \in C(\overline{\Omega}, E)$, *and for some* $x_0 \in \Omega$, *assume that* $f(x_0) = 0$. *Moreover, assume that* f *is differentiable at* x_0 *with* $Df(x_0) \in \mathcal{GL}(E)$. *Then* x_0 *is an isolated zero of* f *and*

$$i_0(f, x_0) = i_0(Df(x_0), 0) = \operatorname{sign} \det Df(x_0).$$

Proof. Since $Df(x_0) \in \mathcal{GL}(E)$, there exists some $\alpha > 0$ such that $|Df(x_0)y| \geq 2\alpha|y|$ for all $y \in E$. Next, we can find some $\epsilon > 0$ so that

$$|f(x) - f(x_0) - Df(x_0)(x - x_0)| \leq \alpha|x - x_0|, \quad \forall x \in x_0 + \epsilon\mathbb{B}.$$

Since $f(x_0) = 0$, it follows that for $x \in x_0 + \epsilon\mathbb{B}$ and $\lambda \in [0, 1]$ we have

$$|(1 - \lambda)f(x) + \lambda Df(x_0)(x - x_0)|$$
$$\geq |Df(x_0)(x - x_0)| - (1 - \lambda)|f(x) - Df(x_0)(x - x_0)| \geq \alpha|x - x_0|.$$

In particular (for $\lambda = 0$), we have $|f(x)| \geq \alpha|x - x_0|$ for all $x \in x_0 + \epsilon\mathbb{B}$. This proves that x_0 is an isolated zero of f. From the homotopy invariance we now obtain

$$i_0(f, x_0) = \deg(f, x_0 + \epsilon\mathbb{B}, 0) = \deg(Df(x_0) - y_0, x_0 + \epsilon\mathbb{B}, 0),$$

where $y_0 := Df(x_0)x_0$. Since the mapping

$$E \to E, \quad x \mapsto Df(x_0)x - y_0$$

has exactly one zero, namely x_0, it follows from the excision property that

$$\deg(Df(x_0) - y_0, x_0 + \epsilon\mathbb{B}, 0) = \deg(Df(x_0) - y_0, \varrho\mathbb{B}, 0)$$

for every $\varrho > |x_0|$. Since evidently

$$0 \neq Df(x_0)x - \lambda y_0$$

for every $\lambda \in [0, 1]$ and every $x \in \varrho\partial\mathbb{B}$ with

$$\varrho > \max\{|x_0|, |[Df(x_0)]^{-1}y_0|\},$$

it follows from the homotopy invariance that

$$\deg(Df(x_0) - y_0, \varrho\mathbb{B}, 0) = \deg(Df(x_0), \varrho\mathbb{B}, 0).$$

Hence, again based on the excision property, one gets

$$\deg(Df(x_0), \varrho\mathbb{B}, 0) = i_0(Df(x_0), 0).$$

Finally, to evaluate the last term we let $T \in \mathcal{L}(E, \mathbb{R}^m)$ be an isomorphism. We then know that

$$\deg(Df(x_0), \varrho\mathbb{B}, 0) = \deg(T \circ Df(x_0) \circ T^{-1}, T(\epsilon\mathbb{B}), 0)$$

holds. The mapping

$$y \mapsto T \circ Df(x_0) \circ T^{-1}y =: g(y)$$

evidently belongs to $\mathcal{D}_0(\tilde{\Omega}, \mathbb{R}^m) \cap C_r^\infty(\tilde{\Omega}, \mathbb{R}^m)$, where $\tilde{\Omega} := T(\epsilon\mathbb{B}) \subseteq \mathbb{R}^m$. Because it has exactly one zero, namely 0, and since

$$\det(Dg(0)) = \det(T \circ Df(x_0) \circ T^{-1}) = \det(Df(x_0))$$

holds, it follows from the definition of the Brouwer degree that

$$i_0(Df(x_0), 0) = \deg(g, \Omega, 0) = a(g, \Omega)$$
$$= \text{sign} \det Dg(0) = \text{sign} \det Df(x_0).$$

This proves the assertion.   □

As a corollary of this proposition we obtain a useful formula to calculate the Brouwer degree. It shows that also for C^1-mappings the degree agrees with the algebraic number of zeros.

(21.9) Corollary. *Let $f \in C(\overline{\Omega}, E) \cap C^1(\Omega, E)$ be such that $0 \notin f(\partial\Omega)$ and suppose $\mathbb{K} = \mathbb{R}$. Moreover, assume that 0 is a regular value of $f|\Omega$. Then*

$$\deg(f, \Omega, 0) = a(f, \Omega) = \sum_{x \in f^{-1}(0)} \text{sign} \det Df(x).$$

Proof. This follows immediately from remark (21.7 c) and proposition (21.8).   □

Suppose $f \in C^1(\Omega, E)$ satisfies $f(x_0) = 0$ for some $x_0 \in \Omega$. Then f determines a flow φ on Ω with x_0 a stationary point of φ. If x_0 is a hyperbolic critical point of φ, that is, if the linearized flow $e^{tDf(x_0)}$ is hyperbolic, then, in particular, $Df(x_0) \in \mathcal{GL}(E)$ and x_0 is an isolated critical point of φ. The following proposition establishes a relation between the local index of f at x_0 and the stability properties of the flow φ near x_0.

(21.10) Proposition. *Assume that $f \in C^1(\Omega, E)$ and $\mathbb{K} = \mathbb{R}$, and let $x_0 \in \Omega$ be a hyperbolic critical point of the flow induced by f. Then*

$$i_0(f, x_0) = (-1)^{\dim(E_S)},$$

where E_S denotes the stable manifold of the linearized flow $e^{tDf(x_0)}$.

Proof. By proposition (21.8) we have $i_0(f, x_0) = \text{sign} \det Df(x_0)$. For the purpose of calculating $\det Df(x_0)$ and because $\det Df(x_0) = \det[Df(x_0)]_{\mathbb{C}}$ (see the remark immediately following theorem (12.10)), we may assume without loss of generality that $\mathbb{K} = \mathbb{C}$. Since the determinant is independent of the particular basis, we immediately obtain, using the Jordan canonical form,

$$\det Df(x_0) = \prod_{j=1}^{m} \lambda_j,$$

where $\lambda_1, \ldots, \lambda_m$ denote the eigenvalues of $Df(x_0)$, counted according to their multiplicities. Because in the real case $\overline{\lambda}$ is an eigenvalue of $Df(x_0)$ whenever λ is one, it follows from this equation that

$$\text{sign} \det Df(x_0) = (-1)^n,$$

where n is the number of negative real eigenvalues, counted according to their multiplicities. For the same reason we have

$$\dim(E_S) \equiv n \pmod 2$$

and hence the assertion follows. □

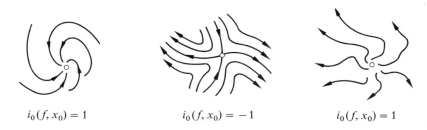

$i_0(f, x_0) = 1$ $i_0(f, x_0) = -1$ $i_0(f, x_0) = 1$

For instance, in the two-dimensional case it follows from the above proposition that a sink or a source has a local index of 1 and a saddle has index -1. In the general m-dimensional case we find that a sink has index $(-1)^m$ and a source always has index $+1$.

Borsuk's Theorem

For the actual calculation of the Brouwer degree it is important to know the degree of as many functions as possible. An important class of mappings, for which one can make a general statement, is the class of odd mappings.

A subset A of a vector space F is called *symmetric* if $-A = A$. Let G be a vector space and $A \subseteq F$ be symmetric. A mapping $g : A \to G$ is called *odd* [resp. *even*] if

$$g(a) = -g(-a) \quad [\text{resp.} \quad g(a) = g(-a)], \quad \forall a \in A.$$

(21.11) Lemma. *Let $E = \mathbb{R}^m$. Assume that Ω is symmetric and that $f \in C(\overline{\Omega}, \mathbb{R}^m)$ is odd. Then for every $\epsilon > 0$ there exists some odd function h in $\overline{C}_r^\infty(\Omega, \mathbb{R}^m)$ with $\|h - f\|_\infty < \epsilon$.*

Proof. (a) For $u \in \overline{C}^\infty(\overline{\Omega}, \mathbb{R}^m)$ and $v \in \overline{C}^\infty(\overline{\Omega}, \mathbb{R})$ we set

$$h_y := u - vy, \quad \forall y \in \mathbb{R}^m.$$

If now $v(x) \neq 0$ for all x in an open subset U of Ω, it follows that: The set of all $y \in \mathbb{R}^m$ such that 0 is a regular value of $h_y|U$ is dense in $\mathbb{R}^m$.

Indeed, 0 is a regular value of $h_y|U$ if and only if y is a regular value of $(u/v)|U$. If $x \in U$ and $u(x)/v(x) = y$, then $h_y(x) = 0$, and from the quotient rule we get $D(u/v)(x) = Dh_y(x)/v(x)$. So the assertion follows from Sard's lemma.

(b) By remark (2.12) there exists a smooth function $\tilde{f} : \overline{\Omega} \to \mathbb{R}^m$ with $\|f - \tilde{f}\|_\infty < \epsilon/2$. Then the function

$$g : \overline{\Omega} \to \mathbb{R}^m, \quad x \mapsto [\tilde{f}(x) - \tilde{f}(-x)]/2$$

is odd, smooth and satisfies (because $f(x) = -f(-x)$)

$$\|f - g\|_\infty < \epsilon/2.$$

(c) Let $\mu := \max\{|x| \mid x \in \overline{\Omega}\}$ and $0 < \delta < \epsilon/4\mu$, where we also assume that $\delta \notin \sigma(Dg(0))$ if $0 \in \Omega$. Hence

$$g_\delta : \overline{\Omega} \to \mathbb{R}^m, \quad x \mapsto g(x) - \delta x$$

is smooth, odd and satisfies $\|f - g_\delta\|_\infty < 3\epsilon/4$. Moreover, if $0 \in \Omega$, then $Dg_\delta(0)$ is invertible.

(d) We now define inductively a collection of smooth, odd functions $h_1, \ldots, h_m : \overline{\Omega} \to \mathbb{R}^m$ with $\|f - h_k\|_\infty < \epsilon$ so that 0 is a regular value of each $h_k|\Omega_k$, where

$$H_k := \{x \in \mathbb{R}^m \mid x^k = 0\} \quad \text{and} \quad \Omega_k := \Omega \setminus (H_1 \cap \cdots \cap H_k).$$

Choose $y_1 \in \mathbb{R}^m$ such that $|y_1| < \epsilon/4m\mu^3$ and 0 is a regular value of $h_1|\Omega_1$, where $h_1 := g_\delta(x) - (x^1)^3 y_1$. By (a) this is possible.

If h_k is already defined, we set

$$h_{k+1}(x) := h_k(x) - (x^{k+1})^3 y_{k+1},$$

where $y_{k+1} \in \mathbb{R}^m$ is chosen in such a way that: $|y_{k+1}| < \epsilon/4m\mu^3$ and 0 is a regular value of $h_{k+1}|(\Omega \setminus H_{k+1})$.

For $x \in \Omega \cap H_{k+1}$ we evidently have $h_{k+1}(x) = h_k(x)$ and $Dh_{k+1}(x) = Dh_k(x)$. This and the induction hypothesis imply that $Dh_{k+1}(x)$ is invertible for all $x \in \Omega_k \cap H_{k+1}$ with $h_{k+1}(x) = 0$. From $(\Omega_k \cap H_{k+1}) \cup (\Omega \setminus H_{k+1}) = \Omega_{k+1}$ it now follows that 0 is a regular value of $h_{k+1}|\Omega_{k+1}$.

Set $h := h_m$, then $\|h - f\|_\infty < \epsilon$, h is smooth, and 0 is a regular value of $h|(\Omega \setminus \{0\})$. Clearly $Dh(0) = Dg_\delta(0)$ in case $0 \in \Omega$, and therefore the assertion follows from (c). $\qquad \square$

(21.12) Theorem (*Borsuk's Theorem*). *Let $\Omega \subset E$ be a symmetric neighborhood of 0 and $f \in C(\overline{\Omega}, E)$. If $f|\partial\Omega$ is odd and $0 \notin f(\partial\Omega)$, then*

$$\deg(f, \Omega, 0) \equiv 1 \pmod 2.$$

Proof. Evidently it suffices to consider the case $E = \mathbb{R}^m$ (cf. the proof of theorem (21.5)). The mapping

$$g : \overline{\Omega} \to \mathbb{R}^m, \quad x \mapsto (f(x) - f(-x))/2$$

is odd and satisfies $f|\partial\Omega = g|\partial\Omega$. Corollary (21.6 viii) implies that $\deg(f, \Omega, 0) = \deg(g, \Omega, 0)$. Using the continuity of $\deg(\cdot, \Omega, 0)$ and lemma (21.11), there exists an odd function $h \in \overline{C_r^\infty}(\overline{\Omega}, \mathbb{R}^m)$ such that

$$\deg(g, \Omega, 0) = \deg(h, \Omega, 0) = \sum_{x \in h^{-1}(0)} \text{sign} \det Dh(x).$$

Now $h(0) = 0$ since h is odd, and $h(x) = 0$ implies that $h(-x) = 0$. Therefore the number of terms in this sum is odd. $\qquad \square$

(21.13) Corollary (*Borsuk-Ulam*). *Let $\Omega \subseteq E$ be a symmetric neighborhood of 0 and let $g \in C(\partial\Omega, E)$. If $g(\partial\Omega)$ is contained in a proper vector subspace E_0 of E, then there exists at least one $x \in \partial\Omega$ such that $g(x) = g(-x)$, i.e., at least one pair of antipodal points has the same image.*

Proof. By introducing an appropriate basis and decomposition into real and imaginary parts, we may assume that $E = \mathbb{R}^m$ and $E_0 = \mathbb{R}^n$ with $n < m$. Applying the Tietze extension theorem to every component function of g, we get an extension $\overline{g} \in C(\overline{\Omega}, E_0)$ of g (cf. Dugundji [1], Schubert [1]). Then

$$f : \overline{\Omega} \to E, \quad x \mapsto [\overline{g}(x) - \overline{g}(-x)]/2$$

is continuous and odd with $f(\overline{\Omega}) \subseteq E_0$. If it were true that $g(x) \neq g(-x)$ for all $x \in \partial\Omega$, then theorem (21.12) would imply that $\deg(f, \Omega, 0) \neq 0$. Hence by corollary (21.6 vi), $f(\Omega)$ would be a neighborhood of 0 in E, which contradicts $f(\overline{\Omega}) \subseteq E_0$. $\qquad\square$

The Brouwer Fixed Point Theorem

As a further, simple application of the Brouwer degree, we will prove the celebrated Brouwer fixed point theorem. For this we need a few preliminaries.

A subspace A of a topological space X is called a *retract of* X if there exists a continuous mapping $r : X \to A$ such that $r|A = id_A$. Each such mapping is a *retraction of* X *onto* A.

(21.14) Remarks. (a) The closed α-ball $\overline{\mathbb{B}}(0, \alpha)$, $\alpha > 0$, is a retract of F for every normed vector space F. A retraction is given by the radial retraction (cf. lemma (19.8)).

(b) A topological space X is said to have the *fixed point property* (f.p.p.) if every continuous mapping from X into X has a fixed point.

If X has the f.p.p. and A is a retract of X, then also A has the f.p.p.

Indeed, if $g \in C(A, A)$ and if $r : X \to A$ is a retraction, then $i \circ g \circ r \in C(X, X)$, where $i : A \hookrightarrow X$ denotes the inclusion map. Hence there exists some $x \in X$ so that $i \circ g \circ r(x) = x$. Therefore $x \in A$ and so $r(x) = x$, which implies that $g(x) = x$.

(c) It is clear that the f.p.p. is a topological property, i.e., *if the topological spaces X and Y are homeomorphic, then X has the f.p.p. if and only if Y has the f.p.p.* $\qquad\square$

(21.15) Lemma. *Every nonempty, compact, convex subset of E is a retract of E.*

Proof. Since E is finite dimensional and since we want to prove a topological property, we may assume that E is a real Hilbert space with norm $|\cdot|$ and scalar product $(\cdot \mid \cdot)$.

So let $C \neq \emptyset$ be compact and convex. Then for every $x \in E$, there exists some $p(x) \in C$ such that

$$|y - p(x)| \leq |x - y|, \quad \forall y \in C,$$

that is, $p(x)$ realizes the shortest distance (with respect to the norm used) from the point x to C. Since for every $y \in C$ the function

$$\varphi_y : [0, 1] \to \mathbb{R}, \quad t \mapsto |x - p(x) - t(y - p(x))|^2$$

assumes its minimum at $t = 0$, we obtain from $D\varphi_y(0) = -2(x - p(x) \mid y - p(x)) \geq 0$ the inequality

$$(x - p(x) \mid y - p(x)) \leq 0, \quad \forall y \in C. \tag{6}$$

If $\tilde{x} \in E$ and if $p(\tilde{x})$ is a point in C with $|\tilde{x} - p(\tilde{x})| = \min\{|\tilde{x} - y| \mid y \in C\}$, then it follows analogously that

$$(\tilde{x} - p(\tilde{x}) \mid z - p(\tilde{x})) \leq 0, \quad \forall z \in C. \tag{7}$$

Setting $y = p(\tilde{x})$ in (6) and $z = p(x)$ in (7) and adding both inequalities we get

$$(x - p(x) - \tilde{x} + p(\tilde{x}) \mid p(\tilde{x}) - p(x)) \leq 0,$$

hence

$$|p(\tilde{x}) - p(x)|^2 \leq (x - \tilde{x} \mid p(x) - p(\tilde{x})) \leq |x - \tilde{x}||p(x) - p(\tilde{x})|$$

and therefore

$$|p(x) - p(\tilde{x})| \leq |x - \tilde{x}|, \quad \forall x, \tilde{x} \in E.$$

In particular, it follows from this that for every $x \in E$ the point $p(x) \in C$, with $|x - p(x)| = \text{dist}(x, C)$, is uniquely determined and $p : E \to C$ is continuous. Because $p|C = id_C$, p is a retraction. $\qquad\square$

(21.16) Brouwer Fixed Point Theorem. *Let $C \subseteq E$ be nonempty, compact and convex. Then every continuous function $f : C \to C$ has a fixed point.*

Proof. Since C is compact there exists some $\alpha > 0$ so that $C \subseteq \alpha\mathbb{B}$. By lemma (21.15) C is a retract of $\alpha\overline{\mathbb{B}}$ and so by remark (21.14 b) it suffices to consider the case $C = \alpha\overline{\mathbb{B}}$. Changing the norm, we may assume without loss of generality that $\alpha = 1$ (see also (21.14 c)).

For $x \in \overline{\mathbb{B}}$ we set $g(x) := x - f(x)$. Then $g \in C(\overline{\mathbb{B}}, E)$ and we may assume that $0 \notin g(\partial\mathbb{B})$, for otherwise f would have a fixed point on $\partial\mathbb{B}$. Now consider the homotopy

$$h : \overline{\mathbb{B}} \times [0, 1] \to E, \quad (x, t) \mapsto x - tf(x).$$

Because $h(\cdot, 1) = g$ and $tf(x) \in \mathbb{B}$ for $0 \leq t < 1$ and $x \in \overline{\mathbb{B}}$, it follows that $h(x, t) \neq 0$ for all $(x, t) \in \partial\mathbb{B} \times [0, 1]$. Using the homotopy invariance and the normalization property of the degree, we get

$$\deg(g, \mathbb{B}, 0) = \deg(h(\cdot, 0), \mathbb{B}, 0) = \deg(id, \mathbb{B}, 0) = 1.$$

Consequently, due to the solution property, the function g has a zero in $\mathbb{B}$, that is, f has a fixed point in $\mathbb{B}$. $\qquad\square$

In the proof above we only used the convexity and compactness of $\overline{\mathbb{B}}$, the fact that 0 was in the interior, as well as $f(\partial\mathbb{B}) \subseteq \overline{\mathbb{B}}$. We therefore have already proved the following useful result.

(21.17) Corollary. *Let $C \subseteq E$ be compact and convex with $0 \in \text{int}(C)$, and for $f \in C(C, E)$ assume that $f(\partial C) \subseteq C$. Then f has a fixed point. If $f(x) \neq x$ for all $x \in \partial C$, then*

$$\deg(id - f, \text{int}(C), 0) = 1.$$

For further results on the Brouwer degree see theorem (24.19) and proposition (25.7).

Problems

1. Prove the *Poincaré-Brouwer theorem:* If $\Omega \subseteq \mathbb{R}^{2m+1}$ is a bounded open neighborhood of zero and if $f \in C(\partial\Omega, \mathbb{R}^{2m+1})$ has no zero, then there exists some $x \in \partial\Omega$ and some $\lambda \in \mathbb{R} \setminus \{0\}$ such that

$$f(x) = \lambda x.$$

(*Hint:* Use the Tietze extension theorem to extend f to $\overline{\Omega}$, consider homotopies of the form

$$\lambda \mapsto (1 - \lambda)f \pm \lambda id, \quad 0 \le \lambda \le 1,$$

and show that $\deg(-id, \Omega, 0) = -1$.)

2. Prove the so-called *"hairy ball theorem"*: Every continuous vector field on $\mathbb{S}^{2m}$ (i.e., every continuous field of tangent vectors on $\mathbb{S}^{2m}$) has at least one zero. ("A hairy ball cannot be combed.") Give an example of a continuous vector field on $\mathbb{S}^{2m+1}$ which is nowhere zero.

3. Let E be a finite dimensional Hilbert space and let $\Omega \subseteq E$ be open and bounded. Assume that for $f \in C(\overline{\Omega}, E)$ there exists some $x_0 \in \Omega$ such that

$$(f(x) \mid x - x_0) \ge 0, \quad \forall x \in \partial\Omega.$$

Use the Brouwer degree to show that f has a zero in $\overline{\Omega}$.

4. Let E be a finite dimensional Banach space and let $\Omega \subseteq E$ be open and bounded. Prove that $\partial\Omega$ *is not a retract of* $\overline{\Omega}$.

5. Let $\Omega_1 \subseteq \mathbb{R}^m$ and $\Omega_2 \subseteq \mathbb{R}^n$ be open and bounded. Suppose that $f_1 \in C(\overline{\Omega}_1, \mathbb{R}^m)$ and $f_2 \in C(\overline{\Omega}_2, \mathbb{R}^n)$ are such that $0 \notin f_j(\partial\Omega_j)$, $j = 1, 2$. Prove that

$$\deg(f_1 \times f_2, \Omega_1 \times \Omega_2, 0) = \deg(f_1, \Omega_1, 0)\deg(f_2, \Omega_2, 0).$$

6. Let Ω be open and bounded in a real, finite dimensional Banach space E of dimension m, and for $f \in C(\overline{\Omega}, E)$ assume that $0 \notin f(\partial\Omega)$. Show that

$$\deg(-f, \Omega, 0) = (-1)^m \deg(f, \Omega, 0).$$

22. Existence of Periodic Solutions

In this section we assume that *E is a finite dimensional Banach space, $X \subseteq E$ is open and*

$$f \in C^{0,1-}(\mathbb{R} \times X, E)$$

is T-periodic with respect to $t \in \mathbb{R}$ for some $T > 0$. Our goal is to develop some general results about the existence of T-periodic solutions of the differential equation

$$\dot{x} = f(t, x). \tag{1}$$

Based upon theorem (20.1), we will, of course, try to show that the shift operator u_T of (1) possesses fixed points.

Our first theorem deals with the case of *asymptotically linear equations.*

(22.1) Theorem. *Let $A \in C(\mathbb{R}, \mathcal{L}(E))$ and $g \in C^{0,1-}(\mathbb{R} \times E, E)$ be T-periodic with respect to $t \in \mathbb{R}$ such that*

$$g(t, x) = o(|x|) \ \ as \ \ |x| \to \infty \tag{2}$$

holds uniformly in $t \in [0, T]$. Then the T-periodic asymptotically linear equation

$$\dot{x} = A(t)x + g(t, x) \tag{3}$$

has a T-periodic solution if 1 is not a Floquet multiplier of $\dot{x} = A(t)x$.

Proof. (a) Because of (2), the right-hand side of (3) is linearly bounded, and hence all solutions of (3) exist globally, i.e., on all of $\mathbb{R}$ (cf. proposition (7.8)). If u is the general solution of (3), then it follows from the variation of constants formula that $u(\cdot, 0, \xi)$ satisfies for all $\xi \in E$ the integral equation

$$u(t, 0, \xi) = U(t, 0)\xi + \int_0^t U(t, \tau)g(\tau, u(\tau, 0, \xi)) \, d\tau, \quad t \in \mathbb{R}, \tag{4}$$

where U denotes the evolution operator of the linear equation $\dot{x} = A(t)x$ (see the discussion immediately preceding theorem (15.3)).

In what follows we set

$$\alpha := \max\{|U(t, \tau)| \mid 0 \le t, \ \tau \le T\}$$

and let $\epsilon \in (0, 1)$ be arbitrary. Because of (2), there exists some $\beta_\epsilon > 0$ so that

$$|g(t, \xi)| \le \beta_\epsilon + \epsilon|\xi|, \quad \forall t \in [0, T], \ \xi \in E. \tag{5}$$

Thus from (4) it follows that

$$|u(t, 0, \xi)| \le \alpha|\xi| + \alpha\beta_\epsilon T + \epsilon\alpha \int_0^t |u(\tau, 0, \xi)| \, d\tau, \quad 0 \le t \le T,$$

and hence by Gronwall's inequality (corollary (6.2)) we get

$$|u(t, 0, \xi)| \le \gamma |\xi| + \delta(\epsilon), \quad \forall t \in [0, T], \ \xi \in E,$$

where

$$\gamma := \alpha e^{\alpha T} \quad \text{and} \quad \delta(\epsilon) := \alpha \beta_\epsilon T e^{\alpha T}.$$

Therefore we obtain, again based on (4) and (5), that

$$|u(t, 0, \xi) - U(t, 0)\xi| \le \alpha T (\beta_\epsilon + \epsilon \delta(\epsilon) + \epsilon \gamma |\xi|)$$

and hence

$$\limsup_{|\xi| \to \infty} \frac{|u(t, 0, \xi) - U(t, 0)\xi|}{|\xi|} \le \epsilon \alpha \gamma T,$$

uniformly in $t \in [0, T]$. It follows from the arbitrary choice of $\epsilon \in (0, 1)$ that

$$u(t, 0, \xi) - U(t, 0)\xi = o(|\xi|) \quad \text{as} \quad |\xi| \to \infty \qquad (6)$$

uniformly in $t \in [0, T]$.

(b) The mapping

$$F : E \to E, \quad \xi \mapsto [I - U(T)]^{-1}(u_T(\xi) - U(T)\xi)$$

is continuous, since 1 is not an eigenvalue of the monodromy operator $U(T)$. From (6) we get that

$$F(\xi) = o(|\xi|) \quad \text{as} \quad |\xi| \to \infty.$$

Consequently there exists some $\varrho > 0$ such that

$$|F(\xi)| \le \varrho + |\xi|/2, \quad \forall \xi \in E.$$

Therefore F maps the ball $\overline{\mathbb{B}}(0, 2\varrho)$ into itself and so by the Brouwer fixed point theorem it has a fixed point. Evidently $F(\xi) = \xi$ if and only if $u_T(\xi) = \xi$ and hence the assertion follows from theorem (20.1). □

As a corollary we obtain the following generalization of corollary (20.6).

(22.2) Corollary. *Assume that $A \in \mathcal{L}(E)$ is such that*

$$\sigma(A) \cap \frac{2\pi i}{T} \mathbb{Z} = \emptyset, \qquad (7)$$

and let $g \in C^{0,1^-}(\mathbb{R} \times E, E)$ be T-periodic and satisfy

$$g(t, \xi) = o(|\xi|) \quad \text{as} \quad |\xi| \to \infty,$$

uniformly in $t \in [0, T]$. Then the equation

$$\dot{x} = Ax + g(t, x)$$

has at least one T-periodic solution.

Proof. Condition (7) and the spectral theorem (19.3) imply that 1 is not an eigenvalue of $U(T) = e^{TA}$, that is, 1 is not a Floquet multiplier of $\dot{x} = Ax$.   □

(22.3) Remark. Let F and G be Banach spaces. A mapping $B : F \to G$ is called *asymptotically linear* if there exists some $B_\infty \in \mathcal{L}(F, G)$ such that

$$\lim_{\|x\| \to \infty} \frac{B(x) - B_\infty(x)}{\|x\|} = 0.$$

The mapping B_∞ is uniquely determined by B and is called the *"derivative of B at infinity."* In the proof of theorem (22.1) we have shown that:

If $f \in C^{0,1}(\mathbb{R} \times E, E)$ is asymptotically linear with respect to $x \in E$ and uniformly in $t \in [0, T]$, and if $A(t) \in \mathcal{L}(E), 0 \le t \le T$, is the derivative of $f(t, \cdot)$ at infinity, then the shift operator is defined on all of E, is asymptotically linear, and has $U(T, 0)$ as its derivative at infinity, where U denotes the evolution operator of the equation $\dot{x} = A(t)x$. □

The Method of Guiding Functions

In order to make statements about the Brouwer degree of the shift operator we need the following technical lemma.

(22.4) Lemma. *Let $f \in C^{0,1^-}(\mathbb{R} \times X, E)$ and assume that $K \subseteq \operatorname{dom} u_T$ is compact. Then the function*

$$h : K \times [0, T] \to E,$$

defined by

$$h(\xi, t) := \begin{cases} (u(t, 0, \xi) - \xi)t^{-1}, & \text{if } t > 0 \\ f(0, \xi), & \text{if } t = 0 \end{cases}$$

is continuous.

Proof. For $t > 0$ the assertion follows immediately from theorem (8.3). So assume that $(\xi_j, t_j) \to (\xi, 0)$ in $K \times [0, T]$. In order to show that $h(\xi_j, t_j) \to h(\xi, 0) = f(0, \xi)$ it suffices to consider the case when all $t_j > 0$, as is easily seen. Then we have

$$h(\xi_j, t_j) - f(0, \xi) = \frac{1}{t_j} \int_0^{t_j} [f(\tau, u(\tau, 0, \xi_j)) - f(0, \xi)]d\tau. \tag{8}$$

Since $K \times [0, T]$ is compact, there exists a compact set M in E so that $\xi, u(\tau, 0, \xi_j) \in M$ for all $\tau \in [0, T]$ and $j \in \mathbb{N}$. Let $\epsilon > 0$ be given. Then the continuity of f implies the existence of some $\delta > 0$ such that

$$|f(t, \eta) - f(0, \xi)| < \epsilon \tag{9}$$

for $t \in [0, \delta]$ and $\eta \in M \cap \mathbb{B}(\xi, \delta)$. Moreover, due to the continuity of $u(\cdot, 0, \cdot)$: $[0, \tau] \times K \to E$, there exists some $N \in \mathbb{N}$ such that

$$|u(\tau, 0, \xi_j) - \xi| = |u(\tau, 0, \xi_j) - u(0, 0, \xi)| < \delta$$

for $0 < \tau < t_j$ and $j \geq N$. Hence we see from (8) and (9) that

$$|h(\xi_j, t_j) - f(0, \xi)| \leq \epsilon, \quad \forall j \geq N,$$

which proves the assertion.   □

It is now easy to give general existence criteria for T-periodic solutions of $\dot{x} = f(t, x)$. For that purpose, a point $\xi \in X$ is called T-irreversible with respect to the equation $\dot{x} = f(t, x)$ if $u(t, 0, \xi) \neq \xi$ for all $0 < t \leq T$.

(22.5) Proposition. *Let $f \in C^{0,1-}(\mathbb{R} \times X, E)$ and assume that $\Omega \subseteq X$ is open and bounded with $\overline{\Omega} \subseteq \operatorname{dom} u_T$, and such that every $x \in \partial\Omega$ is T-irreversible with respect to*

$$\dot{x} = f(t, x). \tag{10}$$

Moreover, assume that $f(0, \xi) \neq 0$ for all $\xi \in \partial\Omega$ and

$$\deg(f(0, \cdot), \Omega, 0) \neq 0.$$

Then equation (10) has at least one T-periodic solution.

Proof. It follows from theorem (20.1) that (10) has a T-periodic solution if and only if the function $h(\xi) := (u_T(\xi) - \xi)/T$ has a zero. The homotopy invariance of the degree, the assumptions above and lemma (22.4) imply that

$$\deg(h, \Omega, 0) = \deg(f(0, \cdot), \Omega, 0) \neq 0.$$

Therefore h has a zero in Ω.   □

One can employ with benefit Liapunov functions to show that every point in $\partial\Omega$ is T-irreversible.

(22.6) Lemma. *Let $V \in C^1(X, \mathbb{R})$ and assume that there exists some $\alpha \in \mathbb{R}$ such that:*

$$\Omega := V^{-1}(-\infty, \alpha) \neq \emptyset$$

and $\overline{\Omega} = V^{-1}(-\infty, \alpha]$ is compact. If $f \in C^{1-}(\mathbb{R} \times X, E)$ satisfies

$$\langle DV(x), f(t, x) \rangle < 0, \quad \forall t \in [0, T], \ x \in \partial\Omega, \tag{11}$$

then every point on $\partial\Omega$ is T-irreversible with respect to $\dot{x} = f(t, x)$ and $\overline{\Omega} \subseteq \operatorname{dom} u_T$.

Proof. Since $[0, T] \times \partial\Omega$ is compact, there exist constants $\beta < \alpha < \gamma$ such that

$$\langle DV(x), f(t,x) \rangle < 0, \quad \forall t \in [0,T], \quad x \in V^{-1}(\beta,\gamma) \cap U, \tag{12}$$

where U is a compact neighborhood of $\partial\Omega$. Now let $\hat\varphi$ denote the flow induced by the vector field $\hat f := (1, f)$ on $\hat X := \mathbb{R} \times X$. It then follows from (12) that

$$\hat V : \hat X \to \mathbb{R}, \quad \hat x := (t,x) \mapsto V(x)$$

is a Liapunov function on

$$\hat M := \{\hat x = (t,x) \in \hat X \mid \beta < \hat V(x) < \gamma, \ x \in U\}.$$

Therefore, by theorem (18.2), $\hat V(-\infty, \alpha] = \mathbb{R} \times \overline\Omega$ is positively invariant for $\hat\varphi$. It follows from corollary (7.7) that $\overline\Omega \subseteq \mathrm{dom}\, u_T$. Because of (12), we have for $\hat x \in \hat M$ that

$$\langle D\hat V(\hat x), \hat f(\hat x) \rangle_{\hat E} < 0,$$

where $\hat E := \mathbb{R} \times E$. It follows easily from

$$\hat V(\hat\varphi^t(\hat x)) - \hat V(\hat x) \le \int_0^t \langle D\hat V(\hat\varphi^\tau(\hat x)), \hat f(\hat\varphi^\tau(\hat x)) \rangle d\tau$$

(cf. remarks (18.1 a and b)) that every point on $\partial\Omega$ is T-irreversible. $\qquad\square$

The following simple lemma establishes a relation between the Brouwer degree of $f(0, \cdot)$ and the gradient of the Liapunov function under the assumption that (12) holds.

(22.7) Lemma. *Assume that $\mathbb{K} = \mathbb{R}$ and let $(\cdot \mid \cdot)$ denote an inner product on E. Moreover, let $\Omega \subseteq E$ be open and bounded. If for $g, h \in C(\overline\Omega, E)$ we have*

$$(g(x) \mid h(x)) < 0, \quad \forall x \in \partial\Omega, \tag{13}$$

then

$$\deg(g, \Omega, 0) = (-1)^m \deg(h, \Omega, 0),$$

where $m := \dim(E)$.

Proof. We consider the convex homotopy

$$k : \overline\Omega \times [0,1] \to E, \quad (x,\lambda) \mapsto \lambda g(x) - (1-\lambda)h(x).$$

Then it follows from (13) that

$$(k(x,\lambda) \mid h(x)) = \lambda(g(x) \mid h(x)) - (1-\lambda)|h(x)|^2 < 0$$

for all $(x,t) \in \partial\Omega \times [0,1]$. Hence the homotopy invariance of the degree gives us

$$\deg(g, \Omega, 0) = \deg(-h, \Omega, 0).$$

If $h \in \overline{C}_r^\infty(\overline\Omega, E)$, then

$$\deg(-h, \Omega, 0) = a(-h, \Omega) = (-1)^m a(h, \Omega) = (-1)^m \deg(h, \Omega, 0)$$

follows immediately from the definition of $a(\cdot, \Omega)$. But this also holds for $h \in C(\overline{\Omega}, E)$ with $0 \notin h(\partial\Omega)$, since it can be approximated by functions in $\overline{C_r^\infty}(\overline{\Omega}, E)$. This proves the lemma. $\square$

Finally, in order to obtain an effective result, we are faced with the problem of determining the Brouwer degree for appropriate Liapunov functions. For this we first prove the following general lemma.

(22.8) Lemma. *Assume that $\mathbb{K} = \mathbb{R}$ and that $|\cdot|$ is a Hilbert norm on E. Moreover, let $U \subseteq E$ be open, and let $g \in C^1(U, \mathbb{R})$ be such that for some $\beta \in \mathbb{R}$ the set $W := g^{-1}(-\infty, \beta)$ is bounded and $\overline{W} \subseteq U$. Finally, suppose that there exist numbers $\alpha < \beta$ and $r > 0$, and a point $x_0 \in U$ such that*

$$g^{-1}(-\infty, \alpha] \subseteq \overline{\mathbb{B}}(x_0, r) \subseteq g^{-1}(-\infty, \beta) = W$$

and

$$\nabla g(x) \neq 0, \quad \forall x \in g^{-1}[\alpha, \beta].$$

Then we have $\deg(\nabla g, W, 0) = 1$.

Proof. Set $\varrho := \min\{|\nabla g(x)| \mid \alpha \leq g(x) \leq \beta\}$. Then $\varrho > 0$ and, by the approximation lemma (16.7), there exists some $h \in C^{1-}(U, E)$ such that

$$|\nabla g(x) - h(x)| \leq \varrho/2, \quad \forall x \in U.$$

Since

$$|(1 - \lambda)\nabla g(x) + \lambda h(x)| \geq |\nabla g(x)| - \lambda|\nabla g(x) - h(x)| \geq \varrho/2$$

for all $\alpha \leq g(x) \leq \beta$ and $\lambda \in [0, 1]$, it follows from the homotopy invariance of the degree that

$$\deg(\nabla g, W, 0) = \deg(h, W, 0). \tag{14}$$

Now let φ denote the flow on U induced by the vector field $-h$. From the estimate

$$(\nabla g(x) \mid h(x)) = |\nabla g(x)|^2 - (\nabla g(x) \mid \nabla g(x) - h(x))$$
$$\geq |\nabla g(x)|^2 - |\nabla g(x)|\varrho/2 \geq |\nabla g(x)|^2/2 \geq \varrho^2/2,$$

which holds for all $x \in g^{-1}[\alpha, \beta]$, and from remark (18.1 a), it follows that g is a Liapunov function for φ on $g^{-1}[\alpha, \beta]$. Theorem (18.2) implies that for $\alpha < \gamma < \beta$, $V := g^{-1}(-\infty, \gamma]$ is positively invariant. V is compact because $V \subseteq \overline{W}$. Therefore, by corollary (10.3), $t^+(x) = \infty$ for all $x \in V$. If $[0, t] \cdot x \subseteq g^{-1}[\alpha, \beta]$, then we have

$$\alpha - \beta \leq g(t \cdot x) - g(x) = \int_0^t \dot{g}(\tau \cdot x) d\tau$$

$$= \int_0^t (\nabla g(\tau \cdot x) \mid -h(\tau \cdot x)) d\tau \leq -\varrho^2 t/2.$$

With $T := 2(\beta - \alpha)/\varrho^2$ we have that

$$T \cdot x \in g^{-1}(-\infty, \alpha] \tag{15}$$

for all $x \in V$. Moreover, for every $x \in \partial V$ and $t > 0$, the relationship

$$t \cdot x \in \text{int}(V) \tag{16}$$

holds. It then follows from (16) and lemma (22.4) that

$$\tilde{h}(x, t) := \begin{cases} (x - t \cdot x)t^{-1}, & \text{if } t > 0 \\ h(x), & \text{if } t = 0 \end{cases}$$

defines a continuous homotopy $\tilde{h} : V \times [0, T] \to E$ satisfying $\tilde{h}(x, t) \neq 0$ for $(x, t) \in \partial V \times [0, T]$. Hence the excision property and the homotopy invariance imply that

$$\deg(h, W, 0) = \deg(h, \text{int}(V), 0) = \deg(\tilde{h}(\cdot, T), \text{int}(V), 0). \tag{17}$$

Next, we consider the homotopy

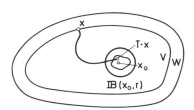

$$k : V \times [0, 1] \to E, \quad (x, \sigma) \mapsto (1 - \sigma)\tilde{h}(x, T) + \sigma(x - x_0)/T.$$

Because

$$k(x, \sigma) = (x - [\sigma x_0 + (1 - \sigma)T \cdot x])T^{-1},$$

and since we may assume without loss of generality that $\overline{\mathbb{B}}(x_0, r) \subseteq \text{int}(V)$, it follows from (15) that

$$k(x, \sigma) \neq 0, \quad \forall (x, \sigma) \in \partial V \times [0, 1].$$

Therefore

$$\deg(\tilde{h}(\cdot, T), \text{int}(V), 0) = \deg(T^{-1}(id - x_0), \text{int}(V), 0).$$

From corollary (21.9) we immediately obtain the value 1 for the last expression. Hence the assertion now follows from (14) and (17).   □

(22.9) Corollary. *Let E be a finite dimensional Hilbert space and let $g \in C^1(E, \mathbb{R})$ be <u>coercive</u>, i.e.,*

$$g(x) \to \infty \quad as \quad |x| \to \infty.$$

If for some $r_0 > 0$

$$\nabla g(x) \neq 0, \quad for \ |x| \geq r_0,$$

then

$$\deg(\nabla g, r\mathbb{B}, 0) = 1, \quad \forall r \geq r_0.$$

Proof. Set $\alpha := \max g(r_0\overline{\mathbb{B}})$ and $r := \max\{|x| \mid g(x) \leq \alpha\}$. Then all the conditions of lemma (22.8) are satisfied with $\beta > \max g(r\overline{\mathbb{B}})$ and $x_0 = 0$. The result now follows from the excision property of the degree. □

After these preparations, we now can prove a general existence theorem for periodic solutions. This result goes back to Krasnosel'skii [1].

(22.10) Theorem. *Let E be a finite dimensional real Hilbert space and assume that*

$$f \in C^{1^-}(\mathbb{R} \times E, E)$$

is T-periodic in $t \in \mathbb{R}$. Moreover, let

$$V \in C^1(E, \mathbb{R})$$

be such that for some $r_0 > 0$ we have

$$(\nabla V(x) \mid f(t, x)) < 0, \quad for \ |x| \geq r_0, \ 0 \leq t \leq T. \tag{18}$$

Now, if

$$either \ V(x) \to \infty \quad or \quad V(x) \to -\infty \quad as \quad |x| \to \infty, \tag{19}$$

then the differential equation

$$\dot{x} = f(t, x) \tag{20}$$

has at least one T-periodic solution.

Proof. (a) We first consider the case

$$V(x) \to \infty \quad as \quad |x| \to \infty.$$

We set $\alpha := \max V(r_0\overline{\mathbb{B}})$ and $\Omega := V^{-1}(-\infty, \alpha)$. Then Ω is relatively compact and, because of lemma (22.6), every point of $\partial\Omega$ is T-irreversible and $\operatorname{dom} u_T \subseteq \overline{\Omega}$. From (18), lemma (22.7) and corollary (22.9) it follows that

$$\deg(f(0, \cdot), \Omega, 0) \neq 0.$$

Now the assertion follows from proposition (22.5).

(b) If now $V(x) \to -\infty$ as $|x| \to \infty$, then, by the periodicity of $f(\cdot, x)$, we can apply case (a) to $-V$ and the equation

$$\dot{y} = -f(-t, y). \tag{21}$$

Therefore (21) has a T-periodic solution v. Setting $u(t) := v(-t)$ we get

$$\dot{u}(t) = -\dot{v}(-t) = f(t, v(-t)) = f(t, u(t)), \quad \forall t \in \mathbb{R}.$$

Hence u is a T-periodic solution of (20).   □

(22.11) Remarks. (a) The proof above also gives us an *a priori estimate* for the T-periodic solutions of the equation $\dot{x} = f(t, x)$. This is an estimate that must be satisfied by every T-periodic solution and can be determined a priori, i.e., without first finding the solutions.

If, for instance, $V(x) \to \infty$ as $|x| \to \infty$, and if $\alpha := \max V(r_0\overline{\mathbb{B}})$, then it follows from lemma (22.6) that every $x \in V^{-1}(\beta)$ with $\beta \geq \alpha$ is T-irreversible with respect to (20). Hence for every T-periodic solution u we have

$$u(t) \in V^{-1}(-\infty, \alpha), \quad \forall t \in \mathbb{R},$$

which implies that

$$|u(t)| \leq r := \max\{|x| \mid V(x) \leq \alpha\}, \quad \forall t \in \mathbb{R}.$$

A similar relationship holds in case $V(x) \to -\infty$ as $|x| \to \infty$.

(b) A function $V \in C^1(E, \mathbb{R})$ satisfying

$$(\nabla V(x) \mid f(t, x)) < 0, \quad \text{for } |x| \geq r_0, \ 0 \leq t \leq T,$$

is called a *guiding function* for the equation

$$\dot{x} = f(t, x).$$

In some cases it is more appropriate to work with several guiding functions. For extensions of the results above, as well as examples, we refer to Krasnosel'skii [1].   □

Stationary Points of Autonomous Equations

The results of this section apply, of course, to autonomous equations as well. In this case $T > 0$ can be chosen arbitrarily. Then the following general proposition shows that the T-periodic solutions are, in general, critical points.

(22.12) Proposition. *Let φ be a semiflow on a metric space M, and for every $t > 0$ let*

$$K_t := \{x \in M \mid t^+(x) > t \ \text{and} \ x = t \cdot x\}$$

be the set of t-periodic points of φ. If for some $t > 0$ we have that

K_t *is compact and*

$K_{t/2^k} \neq \emptyset$ *for all* $k \in \mathbb{N}$,

then the set K of all critical points of φ is nonempty and compact and

$$K = \bigcap_{k \in \mathbb{N}} K_{t/2^k}.$$

Proof. From $x = (t/2^{k+1}) \cdot x$ follows that

$$\frac{t}{2^k} \cdot x = \left(\frac{t}{2^{k+1}} + \frac{t}{2^{k+1}} \right) \cdot x = \frac{t}{2^{k+1}} \cdot \left(\frac{t}{2^{k+1}} \cdot x \right)$$

$$= \frac{t}{2^{k+1}} \cdot x = x,$$

and hence $K_{t/2^{k+1}} \subseteq K_{t/2^k}$ for all $k \in \mathbb{N}$. From the compactness of K_t and the finite intersection property, it therefore follows that

$$K_0 := \bigcap_{k \in \mathbb{N}} K_{t/2^k} \tag{22}$$

is nonempty and compact (because K_s is evidently closed for every $s > 0$). If $y \in K_0$, then (22) implies that $y = t_k \cdot y$, with $t_k := t/2^k$ and $k \in \mathbb{N}$. Hence, by proposition (10.7), $y \in K$, i.e., $K_0 \subseteq K$. Since clearly $K \subseteq K_0$ holds, the assertion follows. $\qquad\square$

As an application of this proposition, we obtain, for instance, that in the autonomous case, the vector field $f \in C^{1-}(E, E)$ has at least one critical point (in the ball $r_0 \mathbb{B}$) if assumptions (18) and (19) of theorem (22.10) are satisfied. This result can, however, also be derived from lemma (22.7) and corollary (22.9). For this reason we give the following application of proposition (22.12), the proof of which is less evident with other aids.

(22.13) Proposition. *Let $f \in C^{1-}(X, E)$ and assume that $M \subseteq X$ is compact, convex, nonempty and positive invariant with respect to the flow induced by f. Then f has at least one zero in M.*

Proof. Corollary (10.13) implies that the time-T-operator of $\dot{x} = f(x)$ is defined on M for every T and maps M continuously into itself. Hence, by the Brouwer fixed point theorem, K_t is for every $t > 0$ nonempty and (as zero set of the continuous function

$$M \to E, \quad x \mapsto t \cdot x - x)$$

closed, hence compact. Now the assertion follows from proposition (22.12). $\qquad\square$

(22.14) Corollary. *Let $E = \mathbb{R}^m$ and $\Phi_1, \ldots, \Phi_k \in C^1(X, \mathbb{R})$. Moreover, let 0 be a regular value for each Φ_j, $j = 1, \ldots, k$. If now*

$$M := \bigcap_{j=1}^{m} \Phi_j^{-1}(-\infty, 0]$$

is compact, convex and nonempty, and if for some $f \in C^{1-}(X, E)$ we have

$$(\nabla \Phi_j(x) \mid f(x)) \le 0, \quad \forall x \in \Phi_j^{-1}(0), \; j = 1, \ldots, k,$$

then f has at least one zero in M.

Proof. This follows immediately from proposition (22.13) and corollary (16.10).□

(22.15) Remark. Here we have given only existence proofs based on the use of the time-T-operator. For nonautonomous equations one can also derive existence results from appropriate integral relations. For a detailed treatment of this method we refer to Rouche-Mawhin [1]. □

Problems

1. Let X be open in $\mathbb{R}^m$ and let $f \in C^{1-}(\mathbb{R} \times X, \mathbb{R}^m)$ be T-periodic in $t \in \mathbb{R}$. Also, let $\Phi_1, \ldots, \Phi_k \in C^1(X, \mathbb{R})$ be such that 0 is a regular value of each Φ_j, $j = 1, \ldots, k$. Moreover, assume that

$$(\nabla \Phi_j(x) \mid f(t, x)) \le 0, \quad \forall t \in [0, T], \; \forall x \in \Phi_j^{-1}(0), \; j = 1, \ldots, k,$$

and let

$$M := \bigcap_{j=1}^{m} \Phi_j^{-1}(-\infty, 0]$$

be nonempty, compact and convex. Show that the equation $\dot{x} = f(t, x)$ has at least one T-periodic solution.

2. (a) Let E be a finite dimensional real Hilbert space. Assume that $V \in C^1(E, \mathbb{R})$ satisfies

$$\alpha |x|^k \le |\nabla V(x)| \le \beta |x|^k, \quad \forall |x| \ge R, \tag{23}$$

for some suitable positive constants α, β, k, R. Moreover, assume that either $V(x) \to \infty$ or $V(x) \to -\infty$ as $|x| \to \infty$. Show the following: If $g \in C^{1-}(\mathbb{R} \times E, E)$ is T-periodic in $t \in \mathbb{R}$ and if

$$g(t, x) = o(|x|^k) \quad \text{as } |x| \to \infty$$

holds uniformly in $t \in [0, T]$, then the equation

$$\dot{x} = \operatorname{grad} V(x) + g(t, x)$$

has at least one T-periodic solution.

(b) Show that (23) is satisfied whenever V is positive homogeneous of degree $k+1$ (i.e., if

$$V(tx) = t^{k+1} V(x), \quad \forall x \in E, \ t > 0)$$

and grad $V(x) \neq 0$ for $x \neq 0$.

(c) Show that the system

$$\dot{x} = ax^3 + g^1(t, x, y)$$
$$\dot{y} = by^3 + g^2(t, x, y)$$

has at least one T-periodic solution if the following holds:

(i) $g^i \in C^{1-}(\mathbb{R} \times \mathbb{R}^2, \mathbb{R})$, $i = 1, 2$, are T-periodic in $t \in \mathbb{R}$,
(ii) $g^i(t, x, y) = o(|x|^3 + |y|^3)$ as $|x| + |y| \to \infty$,
(iii) $a, \ b \in \mathbb{R} \setminus \{0\}$ are such that $\operatorname{sign} a = \operatorname{sign} b$.

(*Hint:* (a) Theorem (22.10). (b) First show that grad V is positive homogeneous of degree k.)

3. Let E be a finite dimensional real Hilbert space, let $U \subseteq E$ be open and let $f \in C^1(U, \mathbb{R})$.

Prove that:
(a) If x_0 is an isolated critical point of f and if $f(x_0)$ is a local minimum, then

$$i_0(\nabla f, x_0) = 1.$$

(b) Assume $U = E$ and f is coercive. Suppose that x_0 is a critical point of f, but that $f(x_0)$ is not the global minimum. If now either f has a local minimum at x_0, or if x_0 is a regular critical point, then f has at least 3 critical points.

(*Hint:* (a) Lemma (22.8). (b) Additivity of the degree.)

4. Let E be a (finite dimensional) Banach space, let $U \subseteq E$ be open and let $J \subseteq \mathbb{R}$ be an open interval with $0 \in J$. Moreover, let $f \in C^{0,1}(J \times U, E)$ and assume that $u^* \in C^1(J, U)$ is a solution of the differential equation

$$\dot{x} = f(t, x). \tag{24}$$

Finally, set

$$A(t) := D_2 f(t, u^*(t)), \quad \forall t \in J,$$

and let W denote the evolution operator of the linear equation

$$\dot{y} = A(t)y.$$

Show that: For each $T \in J$, the time-T-operator u_T of the equation (24) is differentiable at the point $\xi^* := u^*(0)$ and

$$D\, u_T(\xi^*) = W(T, 0).$$

(*Hint:* Theorem (9.2).)

5. Let E be a finite dimensional real Hilbert space, and let $f \in C^{1-}(\mathbb{R} \times E, E)$ be T-periodic in $t \in \mathbb{R}$ with

$$f(t, 0) = 0, \quad \forall t \in \mathbb{R}.$$

Assume also that V is a coercive guiding function for the equation

$$\dot{x} = f(t, x). \tag{25}$$

Prove that: If 1 is not a Floquet multiplier of the T-periodic linear equation

$$\dot{y} = D_2 f(t, 0)y, \tag{26}$$

and if μ is the sum of multiplicities of all real Floquet multipliers of (26) less than 1, then (25) at least one nontrivial T-periodic solution whenever

$$\mu \not\equiv \dim(E) \pmod 2.$$

(*Hint:* For $h(\xi) := (u_T(\xi) - \xi)/T$ it follows from corollary (22.9) and lemmas (22.4) and (22.7) that $\deg(h, \Omega, 0) = (-1)^m$, with $m := \dim(E)$, where we have set $\Omega := V^{-1}(-\infty, \alpha)$ for a sufficiently large α. From problem 4 and proposition (21.8) one derives that $i_0(h, 0) = (-1)^\mu$.)

23. Stability of Periodic Solutions

In what follows, let $(E, |\cdot|)$ denote a finite dimensional Banach space and let $X \subseteq E$ be open. Moreover, let

$$f \in C^{0,1}(\mathbb{R} \times X, E)$$

be T-periodic in $t \in \mathbb{R}$, and let u^* be a T-periodic solution of the equation $\dot{x} = f(t, x)$.

Liapunov Stability

By remark (15.1 f), u^* is (Liapunov) stable, or asymptotically stable, or unstable if the zero solution of the "*equation of perturbed motion*"

$$\dot{y} = f(t, y + u^*(t)) - f(t, u^*(t)) \tag{1}$$

has the corresponding properties. This equation can be written in the form

$$\dot{y} = A(t)y + g(t, y), \tag{2}$$

where

$$A(t) := D_2 f(t, u^*(t))$$

and

$$g(t, y) := f(t, y + u^*(t)) - f(t, u^*(t)) - D_2 f(t, u^*(t))y$$

for all $t \in \mathbb{R}$. If, in order to simplify, we assume that $X = E$, then

$$A \in C(\mathbb{R}, \mathcal{L}(E)) \quad \text{and} \quad g \in C^{0,1}(\mathbb{R} \times E, E),$$

and A, as well as g, are T-periodic in $t \in \mathbb{R}$. Moreover, from the mean value theorem it follows that

$$|g(t, y)| \leq \int_0^1 |D_2 f(t, u^*(t) + \tau y) - D_2 f(t, u^*(t))| \, d\tau |y|,$$

and from the continuity and T-periodicity (in t) of $D_2 f$ and u^* we get that

$$g(t, y) = o(|y|) \quad \text{as } y \to 0,$$

uniformly in $t \in \mathbb{R}$.

Let $\mathbb{K} = \mathbb{C}$. Then, by corollary (20.10), there exists a T-periodic function $Q \in C^1(\mathbb{R}, \mathcal{GL}(E))$ and some $B \in \mathcal{L}(E)$ so that the transformation

$$y = Q(t)z \tag{3}$$

transforms equation (2) into the T-periodic equation with constant main part

$$\dot{z} = Bz + h(t, z), \tag{4}$$

where we have set

$$h(t, z) := Q^{-1}(t)g(t, Q(t)z).$$

Since $Q : \mathbb{R} \to \mathcal{GL}(E)$ is continuous and T-periodic, there exist positive constants γ and δ such that

$$\gamma|z| \leq |Q(t)z| \leq \delta|z|, \quad \forall z \in E, \; t \in \mathbb{R}, \tag{5}$$

and therefore $h \in C^{0,1}(\mathbb{R} \times E, E)$ with

$$h(t, z) = o(|z|) \quad \text{as } z \to 0,$$

uniformly in $t \in \mathbb{R}$. (It suffices to set

$$\gamma := \max_{0 \leq t \leq T} |Q^{-1}(t)|^{-1} \quad \text{and} \quad \delta := \max_{0 \leq t \leq T} |Q(t)|.)$$

After these preparations it is now easy to prove the following "*theorem of linearized stability.*"

(23.1) Theorem. *Let $f \in C^{0,1}(\mathbb{R} \times E, E)$ be T-periodic in $t \in \mathbb{R}$, and let u^* be a T-periodic solution of the equation*

$$\dot{x} = f(t, x).$$

If all the Floquet multipliers μ of the linearized equation

$$\dot{y} = D_2 f(t, u^*(t))y$$

satisfy

$$|\mu| < 1,$$

then u^* is asymptotically stable (in the sense of Liapunov). If, however, there exists a Floquet multiplier μ with $|\mu| > 1$, then u^* in unstable.

Proof. (a) Assume that $\mathbb{K} = \mathbb{C}$. It then follows from estimate (5) that the stability behavior of the zero solutions of equations (2) and (4) is unchanged by transformation (3). Hence we may apply the stability theorem (15.3), respectively, the instability theorem (15.5) to (4) and find that the zero solution of (2) – hence the solution u^* – is asymptotically stable when $Re\,\sigma(B) < 0$ holds, and that u^* is unstable when B has an eigenvalue with strictly positive real part. The monodromy operator $U(T)$ of the linearized equation $\dot{y} = A(t)y$ has the representation

$$U(T) = e^{TB}$$

by remark (20.11 c). Therefore the assertion follows from the spectral mapping theorem (lemma (19.3)).

(b) Assume that $\mathbb{K} = \mathbb{R}$. Then we consider the complexified equation

$$\dot{u} = A_{\mathbb{C}}(t)u + g_{\mathbb{C}}(t, u) \tag{6}$$

in $E_{\mathbb{C}} := E + iE$, where $A_{\mathbb{C}}(t) := [A(t)]_{\mathbb{C}}$ and

$$g_{\mathbb{C}}(t, u) := g(t, \xi) + ig(t, \eta), \quad \forall u = \xi + i\eta \in E_{\mathbb{C}}, \ t \in \mathbb{R}.$$

Then evidently

$$g_{\mathbb{C}}(t, u) = o(|u|) \quad \text{as} \quad u \to 0 \text{ in } E_{\mathbb{C}},$$

uniformly in $t \in \mathbb{R}$. Uniqueness implies that the evolution operator of the complexified equation

$$\dot{v} = A_{\mathbb{C}}(t)v \tag{7}$$

is given by $[U(t, \tau)]_{\mathbb{C}}$, i.e., by the complexification of the evolution operator of $\dot{y} = A(t)y$. Therefore – since $\sigma(U(T)) = \sigma([U(T)]_{\mathbb{C}})$ – the Floquet multipliers of (7) are exactly the Floquet multipliers of $\dot{y} = A(t)y$. Because (6) is clearly equivalent to the system of two real equations

$$\dot{\xi} = A(t)\xi + g(t, \xi)$$
$$\dot{\eta} = A(t)\eta + g(t, \eta),$$

the assertion follows now immediately by applying (a) to (6). $\qquad\square$

(23.2) Remarks. (a) *If u^* is an asymptotically Liapunov stable T-periodic solution of*

$$\dot{x} = f(t, x), \tag{8}$$

then u^* is evidently an isolated periodic solution *of* (8) in the following sense: there exists an $\epsilon > 0$, such that for every periodic solution $u \neq u^*$ of (8) we have

$$|u(t) - u^*(t)| \geq \epsilon, \quad \forall t \in \mathbb{R}.$$

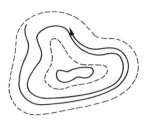

In other words: no other periodic solution intersects the ϵ-neighborhood of the u^*-orbit in E. This follows immediately from the fact that for every $t_0 \in \mathbb{R}$ and every solution u with $|u(t_0) - u^*(t_0)|$ sufficiently small we have:

$$u(t) - u^*(t) \to 0 \quad \text{as} \quad t \to \infty$$

(cf. remark (15.1 c)).

(b) *If u^* is a T-periodic solution of* (8) *and if all Floquet multipliers of the linearized equation*

$$\dot{y} = D_2 f(t, u^*(t))y \tag{9}$$

lie in the open unit disc of the complex plane, then $x_0 := u^*(0)$ is an isolated fixed point of the time-T-map u_T corresponding to (8) *and we have:*

$$i_0(id - u_T, x_0) = 1.$$

Indeed, by problem 4 of section 22, the operator u_T is differentiable at x_0 with

$$D\, u_T(x_0) = U(T),$$

where $U(T)$ is the monodromy operator of equation (9). Since the assumption about the Floquet multipliers implies that $Re\, \sigma(I - U(T)) > 0$, the assertion now follows from proposition (21.10).

(c) *Assume that $f \in C^1(X, E)$ and let u^* be a nonconstant T-periodic solution of the autonomous equation*

$$\dot{x} = f(x). \tag{10}$$

Then 1 is a Floquet multiplier of the linearized equation

$$\dot{y} = Df(u^*(t))y. \tag{11}$$

Indeed, since the right-hand side of the equation $\dot{u}^*(t) = f(u^*(t))$, $t \in \mathbb{R}$, is continuously differentiable, we obtain, by differentiating this equation, that $v := \dot{u}^*$ is a solution of (11). Now $v \neq 0$ since u^* is not constant and so the assertion follows from proposition (20.12).

(d) In particular, from (c) follows that the stability, respectively, the instability criteria of theorem (23.1) cannot be applied to the autonomous equation (10). More is true: If u^* is a nonconstant T-periodic solution, then for every $\tau \in (0, T)$, the map $t \mapsto u^*(t + \tau)$ is also a T-periodic solution of (10). The difference $|u^*(0) - u^*(\tau)|$ can be made arbitrarily small by choosing τ sufficiently close to 0, while evidently $|u^*(t) - u^*(t + \tau)|$ does not converge to 0 as $t \to \infty$. *Hence no nonconstant T-periodic solution of the autonomous equation $\dot{x} = f(x)$ is asymptotically Liapunov stable.* $\qquad\square$

Orbital Stability

The considerations above show that the concept of Liapunov stability is not appropriate for periodic solutions of autonomous differential equations. For this reason we will deal with orbital stability, in what follows, as introduced in remark (17.7 c).

First we will show that, in the case of an asymptotically orbitally stable T-periodic semiorbit, all nearby semiorbits behave after a long time as if they themselves were T-periodic.

Let φ be a semiflow on a metric space M. We say that $x \in M$ has the *asymptotic period* $T \in \mathbb{R}$ if the following holds:

(i) $t^+(x) = \infty$,

(ii) $\lim_{t \to \infty} d((t + T) \cdot x, t \cdot x) = 0$.

With this definition we can make the above heuristic explanations precise in what follows.

(23.3) Theorem. *Let φ be a semiflow on a locally compact metric space M and let $\Gamma := \gamma^+(x)$ be a T-periodic attracting semiorbit. Then there exists a neighborhood U of Γ such that every $x \in U$ has the asymptotic period T.*

Proof. Since Γ is attracting, there exists a neighborhood U of Γ such that for every $x \in U$ we have $t \cdot x \to \Gamma$ as $t \to \infty$. U can be chosen to be compact, since M is locally compact. Hence $\varphi^T | U$ is uniformly continuous.

Choose any $x \in U$ and $\epsilon > 0$. Then there exists some $\delta \in (0, \epsilon)$ such that $d(T \cdot \bar{x}, T \cdot \bar{y}) < \epsilon$ for all $\bar{x}, \bar{y} \in U$ with $d(\bar{x}, \bar{y}) < \delta$. Because $t \cdot x \to \Gamma$, there exists some $t_0 > 0$ with the following property: If $t \geq t_0$, there exists some $y_t \in \Gamma$ with $d(t \cdot x, y_t) < \delta$. Hence $T \cdot y_t = y_t$ implies the estimate

$$d((T+t) \cdot x, t \cdot x) \le d(T \cdot (t \cdot x), T \cdot y_t) + d(T \cdot y_t, t \cdot x)$$
$$= d(T \cdot (t \cdot x), T \cdot y_t) + d(y_t, t \cdot x) \le \epsilon + \delta \le 2\epsilon$$

for all $t \ge t_0$, and therefore the assertion follows. $\square$

In what follows we let $\mathbb{K} = \mathbb{R}$ and $f \in C^1(X, E)$. Moreover, let φ denote the flow on X induced by f. For $x_0 \in X$, let H_{x_0} denote a hyperplane in E through x_0. An open neighborhood V of x_0 in H_{x_0} is called a *local (transversal) section of φ in x_0* (or, of f in x_0) if for all $x \in V$ the vector $(x, f(x)) \in T_x E$ is transversal to H_{x_0}. Since

$$H_{x_0} = x_0 + H$$

for a unique subspace of E of codimension 1, it follows that $(x, f(x)) \in T_x E$ is transversal to H_{x_0} if and only if $f(x) \notin H$. Suppose now that a trajectory through

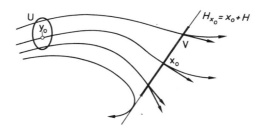

the point y_0 goes through x_0 at time t_0. The following lemma then shows that there exists a neighborhood U of y_0 in E so that every trajectory through a point in U meets the local section and that the corresponding "intersection time" is a continuous function.

(23.4) Lemma. *Let V be a local section of φ in x_0 and let $t_0 \cdot y_0 = x_0$. Then there exists an open neighborhood U of y_0 in E and a unique function $\tau \in C^1(U, \mathbb{R})$ such that $\tau(y_0) = t_0$ and*

$$\tau(y) \cdot y \in V, \quad \forall y \in U.$$

Proof. Let $H_{x_0} = x_0 + H$ be the hyperplane containing V. Then there exists some $h \in E' \setminus \{0\}$ with $H = \ker h$ and consequently $\langle h, f(x_0) \rangle \ne 0$. Theorem (10.3) shows that the function

$$g : (y, t) \mapsto \langle h, t \cdot y \rangle - \langle h, x_0 \rangle = \langle h, t \cdot y - x_0 \rangle$$

is defined in some neighborhood of (y_0, t_0), on which it is also continuously differentiable and satisfies $g(y_0, t_0) = 0$, as well as

$$D_2 g(y_0, t_0) = \langle h, f(t_0 \cdot y_0) \rangle = \langle h, f(x_0) \rangle \ne 0.$$

By the implicit function theorem, there exists a neighborhood U of x_0 in E, a neighborhood I of t_0 in $\mathbb{R}$, and a unique function $\tau \in C^1(U, \mathbb{R})$ such that

$$(y, t) \in U \times I \quad \text{and} \quad g(y, t) = 0 \; \Leftrightarrow \; y \in U \quad \text{and} \quad t = \tau(y).$$

From

$$g(y, \tau(y)) = \langle h, \tau(y) \cdot y - x_0 \rangle = 0$$

we get that $\tau(y) \cdot y \in x_0 + \ker h = H_{x_0}$ for all $y \in U$. $\qquad\qquad\square$

Let γ be a periodic orbit and let $T > 0$ be the minimum period of γ (i.e., for an arbitrary point $x \in \gamma$). Choose any $x_0 \in \gamma$ and let $V = V_{x_0}$ denote a local section of φ in x_0.

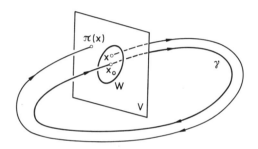

By lemma (23.4) there exists an open neighborhood U of x_0 in E and a unique function $\tau \in C^1(U, \mathbb{R})$ with $\tau(x_0) = T$ and such that $\tau(x) \cdot x \in V$ for all $x \in U$. We now set

$$W := V \cap U$$

and define the *Poincaré map* (with respect to the local section V)

$$\pi \in C^1(W, V)$$

by

$$\pi(x) := \tau(x) \cdot x, \quad \forall x \in W.$$

Evidently $\tau(x)$ is the time of "first return" of the point $x \in W \subseteq V$ into the section V, and $\pi(x)$ is the point in which the semiorbit $\gamma^+(x)$ hits the section V on its first return.

If $H_{x_0} = x_0 + H$ is the uniquely determined hyperplane through x_0 containing V, then $W_0 := W - x_0$ is an open neighborhood of 0 in H and the mapping

$$\pi_0 : W_0 \to H, \quad x \mapsto \pi(x + x_0) - x_0,$$

is a C^1-diffeomorphism from W_0 onto an open neighborhood of 0 in H. In addition, we have $\pi_0(0) = 0$, that is, x_0 is a fixed point of the Poincaré map. In what follows we will use the (slightly incorrect, but convenient) notation

$$D\pi(x_0) := D\pi_0(0) \in \mathcal{L}(H).$$

Now, if the eigenvalues of $D\pi(x_0)$ are in the interior of the unit circle of the complex plane, then, by lemma (19.4), there exists a norm $\|\cdot\|$ on H with

$$\alpha := \|D\pi(x_0)\| < 1.$$

For every $\epsilon \in (0, 1 - \alpha)$ there exists a neighborhood $\tilde{W} \subseteq W$ of x_0 in $x_0 + H$ such that

$$\|\pi(x) - x_0 - D\pi(x_0)(x - x_0)\| \le \epsilon \|x - x_0\|, \quad \forall x \in \tilde{W}.$$

Therefore we get

$$\|\pi(x) - x_0\| \le \beta \|x - x_0\|, \quad \forall x \in \tilde{W}, \tag{12}$$

with $\beta := \alpha + \epsilon < 1$. In other words, all points of a sufficiently small neighborhood of x_0 in V "move" closer to the point x_0 after the "first rotation." It is now intuitively clear that this implies the asymptotic stability of the orbit γ.

(23.5) Proposition. *If*

$$|\sigma(D\pi(x_0))| < 1,$$

then γ is asymptotically stable.

Proof. We must show the following: If U is an arbitrary neighborhood of γ in E, then there exists a neighborhood $U_1 \subseteq U$ of γ in E, such that $t^+(x) = \infty$ for all $x \in U_1$ and $t \cdot U_1 \subseteq U$ for all $t \ge 0$, as well as $t \cdot x \to \gamma$ as $t \to \infty$ and $x \in U_1$.

From the considerations above it follows that there exists a neighborhood $\tilde{W}$ of $x_0 \in V$ such that (12) holds. By shrinking $\tilde{W}$ we may assume that $\tau(x) < 2T$ for all $x \in \tilde{W}$ and that $[0, 2T] \cdot \tilde{W} \subseteq U$. Now (12) implies that $t^+(x) = \infty$ for all $x \in \tilde{W}$, and it easily follows that

$$U_1 := \gamma^+(\tilde{W}) = \{\gamma^+(x) \mid x \in \tilde{W}\}$$

is a positively invariant neighborhood of γ in E and satisfies $U_1 \subseteq U$. Therefore γ is stable.

It follows from (12) that for each $x \in \tilde{W}$ we have $x_k := \pi^k(x) \to x_0$ as $k \to \infty$ and $x_k \in \tilde{W}$ for all $k \in \mathbb{N}^*$. For each $t > 0$ there exists some $k(t) \in \mathbb{N}$ and some $s(t) \in [0, 2T)$ such that $t \cdot x = s(t) \cdot x_{k(t)}$. Hence $k(t) \to \infty$ as $t \to \infty$ and

$$\mathrm{dist}(t \cdot x, \gamma) \le |t \cdot x - s(t) \cdot x_0| = |s(t) \cdot x_{k(t)} - s(t) \cdot x_0|.$$

Since $K := \{x_k \mid k \in \mathbb{N}^*\} \cup \{x_0\}$ is compact, and thus also $[0, T] \times K$, and since φ is continuously differentiable, it follows from propositions (6.3) and (6.4) that $\varphi|[0, 2T] \times K$ is uniformly Lipschitz continuous. Hence there exists some $\lambda \in \mathbb{R}_+$ such that

$$\mathrm{dist}(t \cdot x, \gamma) \le \lambda |x_{k(t)} - x_0|, \quad \forall t \ge 0,$$

which implies that $t \cdot x \to \gamma$. Finally, if $y \in U_1$ is chosen arbitrarily, there exist $s \geq 0$ and $x \in \tilde{W}$ with $y = s{\cdot}x$. It therefore follows that $t{\cdot}y = t{\cdot}(s{\cdot}x) = (t+s){\cdot}x \to \gamma$ as $t \to \infty$, which proves that γ is attracting. $\square$

In the next lemma we will establish a relationship between the linearized Poincaré operator and the linearized flow.

(23.6) Lemma.   *If $D\varphi^T(x_0) \in \mathcal{GL}(E)$ leaves H invariant, then $D\pi(x_0) = D\varphi^T(x_0)|H$.*

Proof. Since $\tau(x_0) = T$, as well as $D\pi(x_0) = D\pi_0(0)$ and $\pi_0(y) = \pi(y+x_0)-x_0 = \tau(y + x_0) \cdot (y + x_0) - x_0$, it follows from the chain rule that

$$D\pi(x_0) = D\varphi^T(x_0)|H + D_1\varphi(T, x_0)D\tau(x_0). \tag{13}$$

As in the proof of lemma (23.4), we have

$$g(y, \tau(y)) = 0, \quad \forall y \in W,$$

where $g(y,t) := \langle h, t \cdot y - x_0\rangle$ and $h \in E' \setminus \{0\}$ is such that $\ker h = H$. Hence it follows from the chain rule that

$$D_1g(x_0, T) + D_2g(x_0, T)D\tau(x_0) = 0$$

and thus

$$D\tau(x_0) = -(\langle h, f(x_0)\rangle)^{-1}h \circ D\varphi^T(x_0)|H.$$

Since H is invariant under $D\varphi^T(x_0)$, we have $D\varphi^T(x_0)(H) = H = \ker h$ and so $D\tau(x_0) = 0$. Now the assertion follows from (13). $\square$

From $[\varphi^t(x_0)]^{\cdot} = f(\varphi^t(x_0))$ and the differentiability theorem (9.2) we obtain

$$[D\varphi^t(x_0)]^{\cdot} = Df(\varphi^t(x_0))D\varphi^t(x_0)$$

and

$$D\varphi^0(x_0) = id_E.$$

Therefore $D\varphi^t(x_0)$ is the evolution operator of the linearized equation

$$\dot{y} = Df(t \cdot x_0)y \tag{14}$$

and $D\varphi^T(x_0)$ is the monodromy operator of the T-periodic equation (14).

Let now $x \in \gamma$ be some other point on the T-periodic orbit γ. Then there exists some $s \geq 0$ such that $x = s{\cdot}x_0$. Because $T \cdot x = T \cdot (s \cdot x_0) = s \cdot (T \cdot x_0) = s \cdot x_0 = x$, it follows that

$$D\varphi^T(x_0) = D(\varphi^{-s} \circ \varphi^T \circ \varphi^s)(x_0) = D\varphi^{-s}(x)D\varphi^T(x)D\varphi^s(x_0),$$

and from $\varphi^{-s} \circ \varphi^s(x_0) = x_0$ we obtain

$$D\varphi^{-s}(x) = [D\varphi^s(x_0)]^{-1},$$

and thus

$$D\varphi^T(x_0) = [D\varphi^s(x_0)]^{-1} D\varphi^T(x) D\varphi^s(x_0). \tag{15}$$

In other words, if x_0 and x_1 are arbitrary points on the T-periodic orbit γ, then the monodromy operators $U_{x_0}(T)$ and $U_{x_1}(T)$ of the linearized equations

$$\dot{y} = Df(t \cdot x_0)y \quad \text{and} \quad \dot{y} = Df(t \cdot x_1)y$$

are conjugate operators (cf. section 19). In particular, we have

$$\sigma(U_{x_0}(T)) = \sigma(U_{x_1}(T)),$$

as can be seen immediately from (15). We have therefore shown that:

If γ is a periodic orbit of $\dot{x} = f(x)$ with minimal period $T > 0$, then all Floquet multipliers of the linearized T-periodic equation

$$\dot{y} = Df(t \cdot x)y \tag{16}$$

are independent of $x \in \gamma$. For this reason one can talk about the *Floquet multipliers of the orbit γ* .

By remark (23.2 c), 1 is always a Floquet multiplier of γ. Analogous to the situation with critical points, we call a periodic orbit γ *hyperbolic* if 1 is a simple Floquet multiplier of γ (i.e., an algebraically simple eigenvalue of the monodromy operator corresponding to (16)), and if no other Floquet multiplier of γ lies on the unit circle in the complex plane.

If 1 is a simple Floquet multiplier of γ and if $U := U_x(T)$ is the monodromy operator of (16), then there exists a decomposition

$$E = \mathbb{R}e \oplus H, \tag{17}$$

with

$$\mathbb{R}e = \ker(1 - U)$$

and

$$H := H(x) = \bigoplus_{\beta \in \sigma(U)\backslash\{1\}} \ker\left([\beta - U_\mathbb{C}]^{m(\beta)}\right) \cap E,$$

where $m(\beta)$ denotes the algebraic multiplicity of $\beta \in \sigma(U)$. In particular, the decomposition (17) reduces the monodromy operator and we have

$$\sigma(U|H) = \sigma(U) \backslash \{1\}. \tag{18}$$

By remark (23.2 c) we know that, with $u(t) := t \cdot x$ and $v := \dot{u}$, the function v is a nontrivial solution of the linearized equation (16). Because the monodromy operator U is the time-T-map of equation (16), it follows from remark (20.2 a) that $v(0) = \dot{u}(0) = f(x)$ is a fixed point of U, hence it is an eigenvector of U corresponding to the eigenvalue 1. So we can choose $e = f(x)$, that is,

$$E = \mathbb{R}f(x) \oplus H(x), \tag{19}$$

which, in particular, shows that $f(x)$ is transversal to $H(x)$ (i.e., $f(x) \notin H$). The Poincaré operator π is therefore well-defined with respect to some appropriate local section

$$V \subseteq x + H(x)$$

through x. We simply call it the *Poincaré operator of $H(x)$* or of the decomposition (19).

The following proposition now motivates the concept of a hyperbolic periodic orbit.

(23.7) Proposition. *Let γ be a noncritical periodic orbit of $\dot{x} = f(x)$ and assume that 1 is a simple Floquet multiplier of γ. Also let $x_0 \in \gamma$ be arbitrary. Then γ is hyperbolic if and only if the following holds for the Poincaré operator π corresponding to the decomposition*

$$E = \mathbb{R}f(x_0) \oplus H(x_0) :$$

$D\pi(x_0)$ is a hyperbolic automorphism of $H(x_0)$.

Proof. Since $D\varphi^T(x_0) = U_{x_0}(T)$, the proposition now follows immediately from lemma (23.6) and the considerations above. $\square$

After these preliminaries we at once obtain the following criterion for the asymptotic orbital stability of periodic orbits.

(23.8) Theorem. *Let E be a real finite dimensional Banach space, let X be open in E and let $f \in C^1(X, E)$. Moreover, let γ be a nontrivial hyperbolic periodic orbit of $\dot{x} = f(x)$. Then γ is asymptotically stable if and only if all Floquet multipliers of γ, different from 1, lie in the interior of the unit disc in the complex plane.*

Proof. The sufficiency of the given condition immediately follows from the propositions (23.7) and (23.5). The necessity is a consequence of theorem (23.1). $\square$

Evidently the assumption that $\mathbb{K} = \mathbb{R}$ represents no loss of generality since the complex case can be reduced to the real case by introducing a basis and decomposing into a real and imaginary part.

The following proposition shows that it is possible to determine the stability of periodic orbits in some cases without knowing the Floquet multipliers explicitly.

(23.9) Proposition. *Let* $f \in C^1(X, \mathbb{R}^m)$ *and assume that* γ *is a noncritical periodic orbit with period* T. *For some arbitrary* $x \in \gamma$ *set*

$$\Delta := \int_0^T \operatorname{div} f(t \cdot x) \, dt.$$

If $\Delta > 0$ *then* γ *is unstable. If* $m = 2$, *and if* $\Delta < 0$, *then* γ *is asymptotically stable.*

Proof. Since

$$\operatorname{div} f(t \cdot x) = \operatorname{trace}(Df(t \cdot x)),$$

it follows from Liouville's theorem (11.4) that

$$\det U(T) = e^{\Delta},$$

where $U(T)$ denotes the monodromy operator of the linearized equation

$$\dot{y} = Df(t \cdot x)y.$$

Let μ_j, $j = 1, \ldots, m$, denote the eigenvalues of $U(T)$, i.e., the Floquet multipliers of γ, counted according to their multiplicities. Since $\det U(T) = \mu_1 \cdot \cdots \cdot \mu_m$, it follows from $\Delta > 0$ that at least for one Floquet multiplier μ of γ we have $|\mu| > 1$. If $\Delta < 0$, that is, if $e^{\Delta} < 1$, then for at least one Floquet multiplier μ of γ we have $|\mu| < 1$. Because 1 is always a Floquet multiplier of γ it follows that in case $\dim(E) = 2$ all Floquet multipliers of γ different from 1 lie in the interior of the unit disc. Now the assertion follows from theorem (23.8). □

In particular, it follows from theorem (23.3) and theorem (23.8) that a noncritical, asymptotically stable hyperbolic periodic orbit γ of $\dot{x} = f(x)$ has a neighborhood U such that every $x \in U$ has asymptotically the same period as any other point on γ. The following theorem shows also that every point of U is *asymptotically in phase with* γ. More precisely theorem (23.10) says that every point x of U "behaves asymptotically like a particular point on γ."

(23.10) Theorem. *Let* γ *be a noncritical, asymptotically stable hyperbolic periodic orbit of the real vector field* $f \in C^1(X, E)$. *If* $t \cdot x \to \gamma$ *for some* $x \in X$, *there exists a unique point* $y \in \gamma$ *such that*

$$t \cdot x - t \cdot y \to 0 \quad \text{as } t \to \infty.$$

Proof. Let T denote the minimal period of γ and let $z \in \gamma$ be arbitrary. Moreover, let π denote the Poincaré operator (which is defined in a neighborhood W of z in $z + H(z)$) corresponding to the decomposition

$$E = \mathbb{R}f(z) \oplus H(z).$$

Since $t \cdot x \to \gamma$, it follows that the semiorbit $\gamma^+(x)$ intersects the set W, and so it suffices to consider the case when $x \in W$. We also can choose W so small (an open ball in $u + H(z)$ centered at z) that

$$\|\pi(x) - z\| \le \beta\|x - z\|, \quad \forall y \in W, \tag{20}$$

for some appropriate norm on $H(z)$ and some $\beta < 1$ (cf. (12)). Finally, we may assume, without loss of generality, and because $D\tau(z) = 0$ (cf. the proof of lemma (23.6)), that W is so small that

$$|\tau(y) - \tau(z)| \le \|y - z\|, \quad \forall y \in W, \tag{21}$$

follows from the mean value theorem.

We now define inductively a sequence (t_k) in $\mathbb{R}$ by $t_0 := 0$ and

$$t_{k+1} := t_k + T - \tau(\pi^k(x)), \quad \forall k \in \mathbb{N}. \tag{22}$$

One easily verifies that

$$(kT) \cdot x = t_k \cdot \pi^k(x), \quad \forall k \in \mathbb{N}, \tag{23}$$

holds. Since $T = \tau(z)$, we obtain from (20) – (22) the estimate

$$|t_{k+1} - t_k| = |\tau(\pi^k(x)) - \tau(z)| \le \|\pi^k(x) - z\| \le \beta^k\|x - z\|$$

for all $k \in \mathbb{N}$. Hence we have for arbitrary $k, m \in \mathbb{N}$ that

$$|t_{k+m} - t_k| \le \sum_{j=0}^{m-1} |t_{k+j+1} - t_{k+j}| \le \sum_{j=0}^{m-1} \beta^{k+j}\|x - z\|$$
$$\le \beta^k(1 - \beta)^{-1}\|x - z\|.$$

Consequently (t_k) is a Cauchy sequence in $\mathbb{R}$, and hence there exists some $s \in \mathbb{R}$ such that $t_k \to s$ as $k \to \infty$. Since $\pi^k(x) \to z$ as $k \to \infty$, it follows from (23) that

$$(kT) \cdot x \to s \cdot z =: y \in \gamma \quad \text{as } k \to \infty. \tag{24}$$

Because $(kT) \cdot y = y$ for all $k \in \mathbb{N}$, we obtain the relation

$$t \cdot x - t \cdot y = (t - kT) \cdot ((kT) \cdot x) - (t - kT) \cdot y,$$

for all $t \in [kT, (k+1)T)$. Now φ is uniformly Lipschitz continuous on the compact set $[0, T] \times \{(kT) \cdot x \mid k \in \mathbb{N}\}$ and so there exists a constant λ such that

$$|t \cdot x - t \cdot y| \le \lambda|(kT) \cdot x - y|, \quad \forall k \in \mathbb{N},$$

from which it follows, along with (24), that

$$t \cdot x - t \cdot y \to 0 \quad \text{as } t \to \infty.$$

If $\overline{y} \in \gamma$ is any other point such that $t \cdot x - t \cdot \overline{y} \to 0$ as $t \to \infty$, it follows that

$$t \cdot y - t \cdot \overline{y} = (t \cdot y - t \cdot x) + (t \cdot x - t \cdot \overline{y}) \to 0 \quad \text{as} \quad t \to \infty.$$

Since $y, \overline{y} \in \gamma$, this is only possible if $y = \overline{y}$. $\qquad\qquad\qquad\square$

Theorems (23.3) and (23.10) show graphically that in the neighborhood of a hyperbolic, asymptotically stable periodic orbit γ all points behave, after a long time, almost as if they moved along the orbit γ. Practically speaking, after a long time the movements of these points can no longer be distinguished from the periodic movements along the orbit γ.

The Recurrence Theorem

We have already observed in remark (18.11 b) that a Hamiltonian system has no asymptotically stable trajectories. Here we cannot go deeper into the stability theory of Hamiltonian systems (cf. Abraham-Marsden [1], Arnold [1], Arnold-Avez [1], Moser [1], Siegel-Moser [1]). We will merely prove the Poincaré recurrence theorem which simply says that almost every (in the sense of Lebesgue measure) point of a compact invariant set of a Hamiltonian system returns arbitrarily often arbitrarily close to the initial point.

(23.11) Poincaré Recurrence Theorem. *Let $E = \mathbb{R}^m$ and let $f \in C^1(X, \mathbb{R}^m)$ be divergence free. Moreover, let $M \subseteq X$ be a compact invariant set of the flow induced by f. Finally, let $A \subseteq M$ be an arbitrary measurable set. Then for almost every $x \in A$ there exists a subsequence (n_k) of $\mathbb{N}$ such that*

$$n_k \cdot x \in A, \quad \forall k \in \mathbb{N}.$$

Proof. Since M is compact, M has finite Lebesgue measure: $\lambda_m(M) < \infty$. Because M is compact and invariant, it follows from corollary (10.13) that the flow φ, induced by f, is global on M. And since f is divergence free, we know from corollary (11.9), following Liouville's theorem, that $\lambda_m(t \cdot K) = \lambda_m(K)$ for every $t \in \mathbb{R}$ and every compact subset K of M. The regularity of the Lebesgue measure implies that

$$\lambda_m(B) = \sup\{\lambda_m(K) \mid K \subseteq B, \ K \text{ compact}\}$$

for every measurable subset $B \subseteq \mathbb{R}^m$. If $K \subseteq M$ is compact, then for every $t \in \mathbb{R}$ we have that $t \cdot K$ is also compact, because $\varphi^T | M$ is a homeomorphism from M onto itself. Consequently, K is a compact subset of $t \cdot B$ if and only if $(-t) \cdot K$ is a compact subset of B. Theorem (10.3) implies that $\varphi^t \in C^1(X, \mathbb{R}^m)$ for every $t \in \mathbb{R}$. It is well-known that a C^1-function maps measurable sets into measurable sets (cf. Lang [1], Reiffen-Trapp [1]), and therefore $t \cdot B$ is measurable for every $t \in \mathbb{R}$ and every measurable set $B \subseteq M$. Hence we finally obtain

$$\lambda_m(t \cdot B) = \sup\{\lambda_m(t \cdot K) \mid K \subseteq B, \ K \text{ is compact}\}$$
$$= \sup\{\lambda_m(K) \mid K \subseteq B, \ K \text{ is compact}\} = \lambda_m(B)$$

for every measurable subset B of M and every $t \in \mathbb{R}$, i.e., the flow φ is measure preserving on M.

Evidently, $k \cdot x$ is not in A if and only if $k \cdot x \in M \setminus A$, hence if and only if $x \in (-k) \cdot (M \setminus A)$. Consequently,

$$B_l := A \cap \bigcap_{k \geq l}(-k) \cdot (M \setminus A), \quad l \in \mathbb{N}^*,$$

is the totality of all points in A such that $k \cdot x$ is not in A whenever $k \geq l$ and each B_l is measurable. Therefore

$$B := \bigcup_{l=1}^{\infty} B_l$$

is the set of all points $x \in A$ such that only finitely many points of the form $k \cdot x$, $k \in \mathbb{N}$, also belong to A. We thus have to show that $\lambda_m(B) = 0$. To that end, it suffices to prove that $\lambda_m(B_l) = 0$ for all $l \in \mathbb{N}^*$.

Since $x \in B_l$ and $k \cdot x \in A$ always imply that $k < l$, it follows in particular that $(jl) \cdot x \notin B_l$, and so $((jl) \cdot B_l) \cap B_l = \emptyset$ for $j \geq 1$. Because φ^t restricted to M is a homeomorphism, it follows that

$$((jl) \cdot B_l) \cap (k \cdot B_l) = k \cdot [(jl - k) \cdot B_l \cap B_l] = \emptyset$$

for all $k = nl$ and $0 \leq n < j$. Therefore the sets $(jl) \cdot B_l$, $j \in \mathbb{N}^*$, are mutually disjoint. The fact that φ is measure preserving on M implies that

$$\lambda_m((jl) \cdot B_l) = \lambda_m(B_l), \quad \forall j \in \mathbb{N}^*.$$

Consequently, from the σ-additivity of the measure it follows that $\lambda_m(B_l) = 0$. $\square$

(23.12) Remarks. (a) The Poincaré recurrence theorem evidently remains true if we only assume: $M \subseteq X$ is invariant and measurable with finite measure and for every $x \in M$ we have $t^-(x) = -\infty$ and $t^+(x) = \infty$.

(b) An inspection of the proof of theorem (23.11) shows that we have actually proved a more general theorem. In particular, it suffices to assume that M is a finite measure space and that $\varphi^t : M \to M$, $t \in \mathbb{R}$, is a one-parameter group of measure preserving transformations on M. Such "measure preserving flows" are studied in great depth in ergodic theory (cf. Hopf [1], Jacobs [1]).

(c) If φ is a semiflow on a metric space M, then $x \in M$ is called *Poisson stable* if $t^+(x) = \infty$, and if for every neighborhood U of x and for very $T > 0$, there exists some $t > T$ such that $t \cdot x \in U$. *Evidently, x is Poisson stable if and only if $x \in \omega(x)$, i.e., if and only if x lies in its own ω-limit set* (cf. remark (17.1 b)). If x is Poisson stable, then so is also $t \cdot x$ for all $t \geq 0$. *Therefore x is Poisson stable if and only if $\overline{\gamma^+(x)} = \omega(x)$* (cf. remarks (17.1 d and e)). If $\gamma^+(x)$ is relatively compact, it follows from example (17.4 a) that $\gamma^+(x) = \omega(x)$ if and only if x is periodic. This shows that a Poisson stable point is "nearly" periodic. (For further, related concepts, such as "nonwandering" and "recurring" point, we refer to the literature (e.g., Bhatia-Szegö [1], Nemytskii-Stepanov [1]).) $\square$

(23.13) Corollary. *Under the assumptions of the Poincaré recurrence theorem almost every point in M is Poisson stable.*

Proof. Since M is compact, it follows that for every $k \in \mathbb{N}^*$ there exist finitely many balls $\mathbb{B}(x_j, 1/k)$, with $x_j \in M$ for $0 \leq j \leq n_k$, so that they cover M. We now apply the recurrence theorem to each one of the countably many sets $\mathbb{B}(x_j, 1/k) \cap M$, $0 \leq j \leq n_k$, $k = 0, 1, 2, \ldots$ Then there exists a set of measure zero N (a countable union of sets of measure zero) such that every point $x \in M \backslash N$ returns infinitely many times to every ball $\mathbb{B}(x_j, 1/k)$ in which it lies. For each $x \in M \backslash N$ and every $k \in \mathbb{N}^*$, there exists some $j(x) \in \{0, \ldots, n_k\}$ with $x \in \mathbb{B}(x_{j(x)}, 1/k) =: B_k$. Since x returns infinitely often to B_k, we may choose inductively a subsequence (n_k) of $\mathbb{N}$ such that $n_k \cdot x = B_k$ for all $k \in \mathbb{N}^*$. If now U is an arbitrary neighborhood of x, then there exists some $\delta > 0$ such that $\mathbb{B}(x, \delta) \subseteq U$. Hence from the triangle inequality we have for $2/k < \delta$ that $B_k \subseteq \mathbb{B}(x, \delta) \subseteq U$. Consequently, all but finitely many elements of the sequence $(n_k \cdot x)$ lie in every neighborhood of x, that is, $n_k \cdot x \to x$ as $k \to \infty$. Therefore $x \in \omega(x)$ by remark (17.1 c). $\square$

(23.14) Corollary. *Let $E = \mathbb{R}^{2m}$ and $H \in C^1(X, \mathbb{R})$. Moreover, assume that for some $\alpha \in \mathbb{R}$ the set $H^{-1}(-\infty, \alpha]$ is nonempty and compact. Then almost every point in $H^{-1}(-\infty, \alpha]$ is Poisson stable with respect to the flow induced by the Hamiltonian system*

$$\dot{x} = H_y, \quad \dot{y} = -H_x.$$

Proof. Since H is a Liapunov function for the Hamiltonian flow with $\dot{H} = 0$ (cf. example (18.11 b)), it follows that $H^{-1}(-\infty, \alpha]$ is invariant. Hence the assertion follows from corollary (23.13) and example (11.10 a). $\square$

(23.15) Remarks. (a) The stability theorems (23.1) and (23.8) are statements about the linearizations along periodic orbits ("principle of linearized stability"). One can, of course, obtain stability assertions for periodic orbits also from the general results of section 18 (e.g., from theorem (18.7)) if it is possible to find appropriate Liapunov functions.

(b) As in the case of critical points, one can associate to a hyperbolic periodic orbit a stable and unstable manifold. This result is derived relatively easily from proposition (19.10) by means of the Poincaré operator. However, for simplicity we refer to the literature (e.g. Irwin [1]).

(c) The importance of hyperbolic critical points and hyperbolic periodic orbits lies in the fact that they are "stable" with respect to small perturbations of the vector field f (in some appropriate C^1-topology)(cf. theorem (25.15)).   □

Problems

1. Under what conditions does the linear differential equation $\dot{x} = Ax$, $A \in \mathcal{L}(E)$, have noncritical asymptotically stable periodic orbits?

2. Show that the system

$$\dot{x} = (1 - x^2 - y^2)x - y$$
$$\dot{y} = (1 - x^2 - y^2)x + x$$

possesses exactly one noncritical periodic orbit γ. Calculate the corresponding Poincaré operator at an arbitrary point and show that γ is an asymptotically stable, hyperbolic critical orbit.

3. Show that the linear equation $\dot{x} = A(t)x$, where

$$A(t) = \begin{bmatrix} -1 - 2\cos 4t & 2 + 1\sin 4t \\ -2 + 2\sin 4t & -1 + 2\cos 4t \end{bmatrix}$$

has the solution $x(t) = (e^t \sin 2t, e^t \cos 2t)$. Calculate the eigenvalues $\lambda_{1,2}(t)$ of $A(t)$.

Remark: This problem shows that we cannot deduce the stability of the zero solution from the behavior of the eigenvalues of $A(t)$.

24. Planar Flows

In this section let X be an open subset of $\mathbb{R}^2$, and let φ denote the flow on X induced by the vector field $f \in C^1(X, \mathbb{R}^2)$.

The Poinaré-Bendixson Theorem

We will first show that in this special "planar" situation a point $x \in X$ is Poisson stable if and only if it is periodic. This fact greatly simplifies the study of planar flows and has important consequences for the existence of periodic orbits.

If V is a local section of φ, then V is an open subset of a straight line L. Hence V is an open interval in L whenever V is connected.

In what follows, we mean by a *transversal segment* S a connected local section of φ.

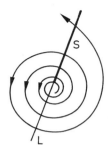

Since every noncritical orbit γ of φ has a natural orientation (by the parametrization $\varphi_x : (t^-(x), t^+(x)) \to \gamma(x)$, it is clear what we mean by an increasing sequence along the orbit γ. If (y_k) is a sequence on a straight line $L \subseteq \mathbb{R}^2$, then we call it increasing if $y_k - y_0 = t_k(y_1 - y_0)$ for $k = 2, 3, \ldots$ and some increasing sequence $t_k \geq 1$.

(24.1) Lemma. *Assume that S is a transversal segment, and let (y_i) be a sequence of points in S which all lie on the same trajectory γ. If the sequence is increasing along γ, then it also increases along S.*

Proof. It suffices to consider three arbitrary points y_k, y_{k+1}, y_{k+2}. If necessary, we may intersperse finitely many points which also lie on $\gamma \cap S$, so that we may assume, without loss of generality, that y_{k+1} is the first return point of y_k on γ in S, i.e., that $y_{k+1} = \tau(y_k) \cdot y_k$ holds.

Now, let Γ denote the simple closed, piecewise C^1-curve, which consists of the part δ of the trajectory γ between y_k and y_{k+1} (i.e., $\delta = [0, \tau(y_k)] \cdot y_k$) and the part T of the segment S between y_k and y_{k+1}. Moreover, let D denote the

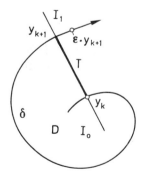

(well-defined) bounded component of Γ. We now assume that γ "leaves" $\overline{D}$ at y_{k+1}, i.e., $(y_{k+1}, f(y_{k+1})) \in T_{y_{k+1}}\mathbb{R}^2$ points away from $\overline{D}$. The set T_- of all those points of T which leave $\overline{D}$ is open in T, by continuity, and it contains y_{k+1}. Also the set T_+ of all points of T which enter $\overline{D}$ is open in T. Since S is a local section, we have $T = T_- \cup T_+$. The connectedness of T implies that $T_+ = \emptyset$. Therefore $X \setminus \overline{D}$ is positively invariant because no trajectory different from γ can intersect δ. It also follows that y_{k+2} lies in $S \setminus T$ since $y_{k+2} = t \cdot y_{k+1}$ for some $t > 0$. The set $S \setminus T$ is the union of two half-open intervals I_0 and I_1 with $y_{k+j} \in \partial I_j$, $j = 0, 1$. If $\epsilon > 0$ is chosen sufficiently small, then $\epsilon \cdot y_{k+1}$ can be connected to I_1 by a continuous curve without meeting Γ. Hence $\text{int}(I_1) \subseteq X \setminus \overline{D}$ follows. Analogously, we have $\text{int}(I_0) \subseteq \text{int}(D)$. Therefore y_{k+1} is between y_k and y_{k+2} on S. If, instead, γ leaves the set $\overline{D}$ in y_k, the argument is similar. $\qquad \square$

(24.2) Lemma. *Let $x \in X$ and $y \in \omega(x)$. Then any transversal segment meets $\omega(x)$ and $\gamma^+(y)$ in at most one point.*

Proof. Since $\omega(x)$ is positively invariant by remark (17.1 c), it follows that $\gamma^+(y) \subseteq \omega(x)$. It suffices therefore to consider $\omega(x)$. Let $y_1, y_2 \in \omega(x)$ be such that $y_1 \neq y_2$ and assume that S is a transversal segment with $y_1, y_2 \in S$. Now let U_j denote disjoint neighborhoods of y_j in X. Then, by remark (17.1 c), there exists a sequence $t_k \to \infty$ such that $t_{2k+1} \cdot x \in U_1$ and $t_{2k} \cdot x \in U_2$ for $k \in \mathbb{N}$. By lemma (23.4) we can find (if we choose U_j sufficiently small) a function $\tau \in C^1(U_1 \cap U_2, \mathbb{R})$ such that

$$\tau(x) \cdot x \in U_j \cap S =: I_j \quad \text{for all} \quad x \in U_j.$$

If we now set

$$a_k := \tau(t_{2k+1} \cdot x) \cdot (t_{2k+1} \cdot x)$$

and

$$b_k := \tau(t_{2k} \cdot x) \cdot (t_{2k} \cdot x),$$

we obtain the sequence

$$a_1, b_1, a_2, b_2, a_3, b_3, \ldots,$$

which is increasing along $\gamma(x)$ and lies on S. By lemma (24.1) the sequence is therefore also increasing along S, which contradicts $a_k \in I_1$ and $b_k \in I_2$. $\qquad \square$

(24.3) Lemma. *If $\gamma^+(x) \cap \omega(x) \neq \emptyset$, then x is periodic.*

Proof. Evidently, we may assume that $\gamma^+(x) \neq \{x\}$. Then $f(y) \neq 0$ for all $y \in \gamma^+(x) \cap \omega(x)$. Hence there exists a transversal segment S through y and, by lemma (24.2), $\gamma^+(y)$ intersects S only in y. Because $y \in \gamma^+(x) \cap \omega(x)$, there exists some $t > 0$ so that $t \cdot x = y$. If U is a sufficiently small neighborhood of y, then $y \in \omega(x)$ implies that there exists some $s > t$ with $s \cdot x \in U$ and $s > t - \tau(s \cdot x)$. Hence we have $\tau(s \cdot x) \cdot (s \cdot x) \in S$. Consequently we have for $\bar{t} := s + \tau(s \cdot x) - t$ that $\bar{t} \cdot y = [s + \tau(s \cdot x) - t] \cdot (t \cdot x) = \tau(s \cdot x) \cdot (s \cdot x) \in S$ and hence $\bar{t} \cdot y = y$. Therefore $\gamma^+(y)$ is periodic and, since φ is a flow, x is also periodic. $\qquad \square$

(24.4) Corollary. *x is Poisson stable if and only if x is periodic.*

Proof. By remark (23.12 c), x is Poisson stable if and only if $\overline{\gamma^+(x)} = \omega(x)$. Therefore the assertion follows from lemma (24.3) and example (17.4 a). $\quad\square$

(24.5) Proposition. *Let $K \subseteq X$ be compact and $\gamma^+(x) \subseteq K$. If $\omega(x)$ contains a noncritical periodic orbit γ, then $\omega(x) = \gamma$.*

Proof. Assume that $\omega(x) \setminus \gamma \neq \emptyset$. Since $\gamma = \gamma(y)$ is the continuous image of the flow through y of the period-interval $[0, T]$, γ is closed (in X and, because $\omega(x) = \overline{\omega(x)}$, also in $\omega(x)$) and therefore $\omega(x) \setminus \gamma$ is open in $\omega(x)$. Since $\omega(x)$ is connected by theorem (17.2), it follows that γ contains an accumulation point z of $\omega(x) \setminus \gamma$. We also have $f(z) \neq 0$ because $z \in \gamma$ and γ is a noncritical orbit. Therefore there exists a transversal segment S through z. Then every neighborhood of z contains some $p \in \omega(x) \setminus \gamma$ so that $\gamma(p)$ intersects the segment S, by lemma (23.4), if p is chosen sufficiently close to z. The invariance of $\omega(x)$ (cf. theorem (17.2)) implies that $\gamma(p) \subseteq \omega(x)$. Therefore the intersection of S and $\omega(x)$ contains two distinct points (namely z and $\tau(p) \cdot p$), which contradicts lemma (24.2). $\quad\square$

After these preliminaries we can now easily prove the following well-known theorem.

(24.6) Theorem *(Poincaré-Bendixson). Let $K \subseteq X$ be compact and $\gamma^+(x) \subseteq K$. If $\omega(x)$ contains no critical point, then $\omega(x)$ is a periodic orbit.*

Proof. By theorem (17.2), $\omega(x)$ is nonempty, compact and invariant. So there exists some $y \in \omega(x)$, and thus $\gamma(y) \subseteq \omega(x)$. Then also $\omega(y)$ is nonempty and $\omega(y) \subseteq \omega(x)$. Let $z \in \omega(y)$ and note that $f(z) \neq 0$, because $\omega(x)$ contains no critical points. Hence there exists a transversal segment S through z. Since $z \in \omega(y)$, we obtain from remark (17.1 c) and lemma (23.4) that $\gamma^+(y)$ intersects the segment S. Lemma (24.2) implies that this intersection contains a unique point, which evidently must be equal to z (because otherwise S would intersect $\omega(x)$ in more than one point). Therefore $z \in \gamma^+(y) \cap \omega(y)$, and so from lemma (24.3) it follows that $\gamma(y)$ is a periodic orbit. The assertion now follows from proposition (24.5). $\square$

(24.7) Corollary. *Let $K \subseteq X$ be nonempty, compact and either positively or negatively invariant. Then K contains either a critical point or a noncritical periodic orbit.*

Proof. If K is positively invariant, the assertion follows immediately from theorem (24.6). If K is negatively invariant, the assertion follows by applying theorem (24.6) to the flow induced by $-f$. $\quad\square$

A noncritical periodic orbit γ is called a *limit cycle* (more precisely: ω-limit cycle or α-limit cycle) if there exists some $x \in X \backslash \gamma$ such that $t \cdot x \to \gamma$ as $t \to \infty$ or $t \to -\infty$.

(24.8) Remarks. (a) *If $K \subseteq X$ is a compact neighborhood of a limit cycle γ, then there exists some $x \in K \backslash \gamma$ such that $t \cdot x$ "spirals" toward the limit cycle as $t \to \infty$ or $t \to -\infty$.* From theorem (17.2) and proposition (24.5) it follows that $t \cdot x \to \gamma$ as $t \to \infty$ or $t \to -\infty$. If S is a transversal segment through an arbitrary point $y \in \gamma$, it follows from lemma (24.1) that $\gamma^+(x)$ meets the segment S infinitely often, and the intersection points increase (resp. decrease) along S toward y as t increases (resp. decreases).   From proposition (19.1) it

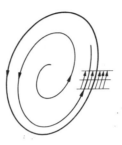

finally follows that $\gamma^+(x)$ crosses the segment S always "in the same direction."

(b) *Limit cycles are "one-sided asymptotically stable."*   In fact, if γ is a limit cycle and if $t \cdot x \to \gamma$, for some $x \in X \backslash \gamma$, as $t \to \infty$ (resp. $t \to -\infty$), it easily follows from the compactness of γ and from proposition (19.1) that there exists a compact neighborhood K of γ so that K contains no critical points and that all trajectories in K have the same orientation.   Now let $y \in \gamma$ be arbitrary and let $S \subseteq K$ be a transversal segment through

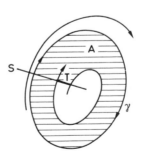

y. Then there exists an interval T in S, which does not intersect γ, and with end points of the form $t_1 \cdot x$ and $t_2 \cdot x$, with $t_1 < t_2$ (resp. $t_1 > t_2$) such that $\gamma(x)$ does not intersect T in any other points. The compact region A, bounded by T, γ and $[t_1, t_2] \cdot x$ (resp. $[t_2, t_1] \cdot x$), is positively (resp. negatively) invariant, and so is $A \backslash \gamma$, and lies completely in K. From this it easily follows (by theorem (24.6)) that $t \cdot y \to \gamma$ as $t \to \infty$ (resp. $t \to -\infty$) for every $y \in A$. Therefore A is a "one-sided neighborhood of γ" such that $\omega(y) = \gamma$ (resp. $\omega^-(y) = \gamma$) for all $y \in A$ (by proposition (24.5)). $\square$

As a simple consequence of the considerations above we obtain the following criterion for the nonexistence of limit cycles.

(24.9) Proposition. *If $H \in C^1(X, \mathbb{R})$ is a first integral which is not constant on nonempty open subsets, then there are no limit cycles.*

Proof. Let γ be a limit cycle and let $a := H(y)$ for some arbitrary (and hence for every) $y \in \gamma$. Moreover, let $x \in X \backslash \gamma$ be such that $t \cdot x \to \gamma$. Since H is constant on trajectories, it follows from continuity that $H(t \cdot x) = a$ for all $t \geq 0$. By remark (24.8 b), H is constant on some appropriate "one-sided neighborhood" of γ and therefore on a nonempty open subset. This contradicts the assumption. $\square$

(24.10) Corollary. *If U is open in $\mathbb{R} \times \mathbb{R}$ and if $H \in C^2(U, \mathbb{R})$ is not constant on nonempty open subsets, then the Hamiltonian system*

$$\dot{x} = H_y, \quad \dot{y} = -H_x$$

has no limit cycles. In particular, the second order equation (with $f \in C^1(X, \mathbb{R})$, $X \subseteq \mathbb{R}$ open)

$$\ddot{x} = f(x),$$

that is, the system

$$\dot{x} = y, \quad \dot{y} = f(x), \tag{1}$$

has no limit cycles.

Proof. The first assertion follows immediately from proposition (24.9) and corollary (3.12). The second assertion follows from the first since the Hamiltonian

$$H(x, y) = \frac{1}{2}y^2 - \int_0^x f(\xi)d\xi,$$

corresponding to (1), is not constant on nonempty open sets because $H_y = y$. $\square$

The next proposition guarantees the existence of limit cycles under certain conditions.

(24.11) Proposition. *Let $K \subseteq X$ be compact and positively invariant and assume that there exists a unique critical point x_0 in K. If $K \neq \{x_0\}$ and if x_0 is a source, i.e., if $\mathrm{Re}\,\sigma(Df(x_0)) > 0$, then there exists at least one limit cycle γ in K. If $\gamma \subseteq \mathrm{int}(K)$ and if γ is the only noncritical periodic orbit in K, then γ is asymptotically stable.*

Proof. The assumptions imply that for every $x \in K \backslash \{x_0\}$ the semiorbit $\gamma^+(x)$ lies in K and contains no critical point. By theorem (24.6), $\gamma := \omega(x)$ is therefore a limit cycle. The last part of the assertion is now clear. $\square$

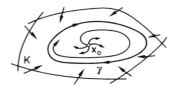

The Winding Number

The goal of the following considerations is to make statements about the existence of critical points and about the nonexistence of periodic orbits by means of topological methods. For this we need a few preliminaries, which are of interest in their own right.

For later purposes we formulate the following definition and the next lemma for the m-dimensional case.

Let $\Omega \subseteq \mathbb{R}^m$ be open and nonempty. We say that Ω *lies locally on one side of* $\Gamma \subseteq \partial\Omega$ if for every point $x \in \Gamma$ and for every sufficiently small open ball U centered at x we have: $U \setminus \Gamma$ consists of two connected components V_1 and V_2 with $V_1 \subseteq \Omega$ and $V_2 \subseteq \overline{\Omega}^c$. We call an open set $\Omega \subseteq \mathbb{R}^m$ a C^k-*domain*, $1 \leq k \leq \infty$, (in symbols: $\Omega \in C^k$) if $\partial\Omega$ is

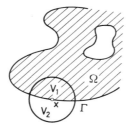

an $(m-1)$-dimensional C^k-manifold and Ω lies locally on one side of $\partial\Omega$. The following figure shows an example of an open set Ω which does not lie locally on one side of $\partial\Omega$.

(24.12) Lemma. *Let $\Omega \subseteq \mathbb{R}^m$ be open and nonempty. Then Ω belongs to C^k if and only if $\overline{\Omega}$ is an m-dimensional C^k-submanifold of $\mathbb{R}^m$ with boundary.*

Proof. The sufficiency follows immediately from the definition of a C^k-manifold with boundary. So let Ω belong to C^k and let $x_0 \in \partial\Omega$ be arbitrary. Then there exists an arbitrarily small open ball U centered at x_0 in $\mathbb{R}^m$, as in the definition above. Moreover, we may assume that $U \cap \partial\Omega$ is the domain of a C^k-chart ψ of $\partial\Omega$. Therefore ψ is a

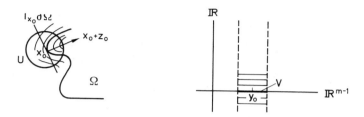

homeomorphism of $U \cap \partial\Omega$ onto an open subset V of $\mathbb{R}^{m-1}$ such that for the mapping $g := \psi^{-1}$ from V into $\mathbb{R}^m$ we have:

$$g \in C^k(V, \mathbb{R}^m) \quad \text{and} \quad Dg(y) \in \mathcal{L}(\mathbb{R}^{m-1}, \mathbb{R}^m) \text{ is injective for every } y \in V.$$

We now define $G \in C^k(V \times \mathbb{R}, \mathbb{R}^m)$ by

$$G(y, \eta) := g(y) + \eta z_0, \quad (y, \eta) \in V \times \mathbb{R},$$

where $z_0 \in \mathbb{R}^m$ is a unit vector orthogonal to $\operatorname{im}(Dg(y_0)) = Dg(y_0)(\mathbb{R}^{m-1})$, with $y_0 := \psi(x_0)$, and such that $x_0 + \eta z_0$ lies in $U \cap \Omega$ for sufficiently small $\eta > 0$. Then

$$DG(y_0, 0) = [Dg(y_0), z_0],$$

and, since $Dg(y_0)$ is injective, the column vectors $D_1 g(y_0), \cdots, D_{m-1} g(y_0)$ are linearly independent. The matrix above has even maximal rank because $z_0 \notin \operatorname{im}(Dg(y_0))$. Hence $DG(y_0, 0) \in \mathcal{GL}(\mathbb{R}^m)$, and so by the inverse function theorem there exists an open neighborhood W of $(y_0, 0)$ in $\mathbb{R}^m$ such that $G|W$ is a C^k-diffeomorphism onto an open neighborhood $\tilde{U} := G(W)$ of x_0. Here we may assume, without loss of generality, that $\tilde{U}$ is contained in U. For $\tilde{\psi} := (G|W)^{-1}$ it follows from $\tilde{\psi}|(\partial\Omega \cap \tilde{U}) = \psi|(\partial\Omega \cap \tilde{U})$ and the construction of $\tilde{\psi}$ that $\tilde{\psi}(\tilde{U} \cap \Omega)$ is contained in $\mathbb{R}^{m-1} \times (0, \infty)$. This shows that $\overline{\Omega}$ is an m-dimensional C^k-submanifold of $\mathbb{R}^m$ with boundary (and that $(\tilde{U}, \tilde{\psi})$ is a chart for $\overline{\Omega}$).$\square$

Finally, we recall the general *Stokes' theorem*: Let M be an oriented m-dimensional C^2-submanifold with boundary of a Euclidean space (more generally: a Riemannian manifold), and let α be a continuously differentiable $(m-1)$-form on M with compact support. Then

$$\int_{\partial M} \omega = \int_M d\omega,$$

where ∂M is oriented by means of the outward normal (with respect to M). As a special case we obtain the *divergence theorem*:

$$\int_M \operatorname{div} g \, d\sigma_M = \int_{\partial M} (g \mid \nu) \, d\sigma_{\partial M},$$

where g is a continuously differentiable vector field on M with compact support, ν is the unit outward normal on ∂M, $(\cdot \mid \cdot)$ is the inner product on the tangent bundle $T(M)$ and σ_M, respectively $\sigma_{\partial M}$, is the volume element on M, respectively ∂M (cf. Fleming [1], Spivak [2], or Bröcker [1]).

After these preliminary considerations "we return to the plane." A planar *Jordan curve* Γ is a continuous curve in $\mathbb{R}^2$ with image homeomorphic to $\mathbb{S}^1$. Since we can think of $\mathbb{S}^1$ as the interval $[0, T]$ with the two end points identified, where $T > 0$ is arbitrary, it follows that Γ is a Jordan curve if and only if there exists a continuous parametrization $c:[0, T] \to \mathbb{R}^2$ of Γ such that $c(0) = c(T)$ and $c(t_1) \neq c(t_2)$ for all $0 < |t_1 - t_2| < T$. (This follows from the fact that a continuous bijection from a compact space onto a Hausdorff space is a homeomorphism (because the image of a closed set is closed).) The well-known *Jordan curve theorem* says: *If Γ is a Jordan curve, then $\mathbb{R}^2 \backslash \Gamma$ consists of exactly two connected components G_1 and G_2 with $\partial G_1 = \partial G_2 = \Gamma$.* For a proof of this we refer, for example, to Dugundji [1]. (Here and in what follows, we do not distinguish between a curve (as equivalence class of paths) and its image.) The compactness of Γ implies that exactly one of the components G_1 and G_2 must be bounded, the so-called *interior of* Γ. Evidently the interior is simply connected. The unbounded component is called the *exterior*.

A curve Γ in $\mathbb{R}^2$ is called a C^k-*Jordan curve*, $1 \leq k \leq \infty$, if it is C^k-diffeomorphic to the unit circle $\mathbb{S}^1$.

(24.13) Remarks. (a) $\Gamma \subseteq \mathbb{R}^2$ is a C^k-Jordan curve, $1 \leq k \leq \infty$, if and only if there exists a C^k-parametrization $c : [0, T] \to \mathbb{R}^2$, $T > 0$, of Γ such that $D^j c(0) = D^j c(T)$ for all $0 \leq j \leq k$, $Dc(t) \neq 0$ for all $t \in [0, T]$, and $c(t_1) \neq c(t_2)$ for all $0 < |t_1 - t_2| < T$.

(b) *If γ is a noncritical periodic orbit of the differential equation $\dot{x} = f(x)$ and if $f \in C^k(X, \mathbb{R}^2)$, $1 \leq k \leq \infty$, then γ is a C^{k+1}-Jordan curve and the interior Ω lies locally on one side of γ.*

Indeed, if T is the minimal period of γ, then for an arbitrary $x \in \gamma$, $\varphi_x : [0, T] \to \mathbb{R}^2$ is a C^{k+1}-parametrization of γ. By theorem (10.3) it follows that $\varphi \in C^k$. Hence $\dot{\varphi}_x = f(\varphi_x) \in C^k(\mathbb{R}, \mathbb{R}^2)$ and therefore $\varphi_x \in C^{k+1}(\mathbb{R}, \mathbb{R}^2)$. Moreover, φ_x clearly satisfies the conditions of (a). Finally, from the linearization theorem (19.1) it follows that Ω lies locally on one side of γ.   □

After these preliminaries we can easily prove the following nonexistence theorem.

(24.14) Proposition. *Let $X \subseteq \mathbb{R}^2$ be simply connected and assume that $f \in C^1(X, \mathbb{R}^2)$ and $\varrho \in C^1(X, \mathbb{R})$. If either*

$$\text{div}(\varrho f) > 0, \quad \text{or} \quad \text{div}(\varrho f) < 0, \quad \lambda_2\text{-a.e.},$$

then the differential equation $\dot{x} = f(x)$ has no noncritical periodic orbits.

Proof. Let γ be a noncritical periodic orbit. Then, by (24.13 c), γ is a C^2-Jordan curve. If Ω denotes the interior of γ, then $\Omega \subseteq X$ because X is simply connected and $\overline{\Omega}$ is a compact two-dimensional C^2-manifold with boundary by (24.13 b) and lemma (24.12). Since the vector field $\varrho f \in C^1(\overline{\Omega}, \mathbb{R}^2)$ is tangential along $\gamma = \partial\Omega$, it follows from the divergence theorem that

$$\int_{\overline{\Omega}} \operatorname{div}(\varrho f)dx = 0,$$

which contradicts our assumptions. $\square$

We say that Γ is an oriented *Jordan arc* if there exists a compact, injective, oriented, continuous planar curve. That is, there exists an injective continuous parametrization $c : [\alpha, \beta] \to \mathbb{R}^2$ with $-\infty < \alpha < \beta < \infty$, in short, a permissible parametrization, and Γ is the equivalence class of all permissable parametrizations, where, as usual, the equivalence relation is defined by strictly increasing reparametrizations. Without risk of confusion, we will denote by Γ the curve as well as the image. We then interpret a function $a \in C(\Gamma, \mathbb{R}^2)$ as a *continuous planar vector field along* Γ, i.e., for each $x \in \Gamma$ we identify $a(x)$ with the tangent vector $(x, a(x)) \in T_x\mathbb{R}^2$.

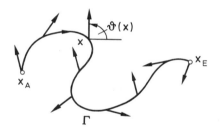

Assume that $a \in C(\Gamma, \mathbb{R}^2 \setminus \{0\})$, that is, a is a nowhere vanishing continuous planar vector field along Γ. One can define a unique continuous function

$$\vartheta_a \in C(\Gamma, \mathbb{R}),$$

the *angle function of the vector field along* Γ, by the following equations

$$\cos \vartheta_a(x) = \frac{a^1(x)}{|a(x)|}, \quad \sin \vartheta_a(x) = \frac{a^2(x)}{|a(x)|}, \tag{2}$$

that is,

$$\vartheta_a(x) := \tan^{-1}\left(\frac{a^2(x)}{a^1(x)}\right), \quad \text{if } a^1(x) \neq 0, \tag{3}$$

and

$$\vartheta_a(x) := \cot^{-1}\left(\frac{a^1(x)}{a^2(x)}\right), \quad \text{if } a^2(x) \neq 0, \tag{4}$$

as well as the additional condition

$$\vartheta_a(x_S) \in [0, 2\pi),$$

where x_S denotes the starting point of Γ. Evidently $\vartheta_a(x)$ measures the angle between the positive x^1-axis and the vector $a(x)$ in the mathematically positive direction. Here we have normalized $\vartheta_a(x)$ so that the angle at the starting point x_S is in the interval $[0, 2\pi)$ and depends continuously on $x \in \Gamma$. If x_E is the end point of Γ, we call the growth of the angle function along Γ, measured in units of full rotations of the vector field a, the *rotation* or *winding number* $w(a, \Gamma)$ *of the vector field a along Γ*. That is

$$w(a, \Gamma) = \frac{\vartheta_a(x_E) - \vartheta_a(x_S)}{2\pi}. \tag{5}$$

Evidently, this definition makes also sense when Γ is a (closed) Jordan curve, because the right-hand side of (5) is independent of the choice of the starting point. Hence we can also define the *rotation of a along a (oriented) Jordan curve Γ* by (5), where the starting point $x_S(= x_E)$ is arbitrary. Clearly, *the rotation of a continuous vector field a along a Jordan curve Γ is always an integer. It tells us how often the closed continuous path $a : \Gamma \to \mathbb{R}^2\backslash\{0\}$ "winds around" the origin in the mathematically positive sense.*

A Jordan curve Γ is called *positively oriented* if the interior Ω of Γ always lies on the left-hand side as Γ is traversed in the positive direction. If Γ is a C^1-Jordan curve, then Γ is clearly positively oriented if and only if it is positively oriented as the boundary of the canonically oriented C^1-manifold $\overline{\Omega}$, i.e., if it is oriented by the outward normal.

(24.15) Theorem (*"Umlaufsatz"*). *Let Γ be a positively oriented C^1-Jordan curve and let $a \in C(\Gamma, \mathbb{R}^2\backslash\{0\})$ be a positively oriented tangent vector field along Γ (i.e., at every point $x \in \Gamma$, $a(x)$ has the direction of the positive unit vector tangent to Γ). Then*

$$w(a, \Gamma) = 1.$$

Proof. Since the definition of the winding number is independent of the length $|a|$ of the vector field, we may assume that $|a| = 1$. Since clearly $w(a, \Gamma)$ can be calculated for an arbitrary parametrization, we may use the parametrization by arc length $c:[0, L] \to \mathbb{R}^2$. Then, as is well-known, $\dot{c}(t)$ is simply the positively oriented tangent unit vector at the point $c(t) \in \Gamma$, i.e., $a = \dot{c} \circ c^{-1}$. We now set $\vartheta(t) := \vartheta_a(c(t))$, where ϑ_a is the angle function of the vector field a corresponding to the starting point $x_S = c(0)$. We then choose a subdivision

$$0 = t_0 < t_1 < \cdots < t_m = L$$

with the following properties:

(i) $\Delta\vartheta(t_i) := \vartheta(t_{i+1})-\vartheta(t_i)$ is equal to the angle measured (in the mathematically positive sense) between the tangent vectors $\dot{c}(t_i)$ and $\dot{c}(t_{i+1})$, and $|\Delta\vartheta(t_i)| < \pi$ for all $i = 0, 1, \ldots, m-1$.

(ii) A closed, nonintersecting polygon is formed by the segments T_i of the tangents through the points $c(t_i)$, which are determined by the intersection

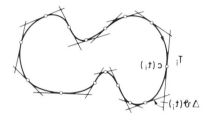

points of the neighboring tangents through the points $c(t_{i-1})$ and $c(t_{i+1})$, where $t_{m+1} := t_1$ and $t_{-1} := t_{m-1}$. Such a subdivision is possible because of the uniform continuity of c and $\dot{c}$.

Now it is easily seen (by induction and based on the fact that the angle sum in a triangle is π) that

$$\sum_{i=0}^{m-1} \Delta\vartheta(t_i) = 2\pi.$$

And since

$$2\pi w(a,\Gamma) = \vartheta(t_m) - \vartheta(t_0) = \sum_{i=0}^{m-1} \Delta\vartheta(t_i),$$

the assertion follows.   □

The following proposition gives a different representation of the winding number in the C^1-case, which can be generalized to higher dimensions. We will then use this representation to make an important connection between the winding number and the Brouwer degree.

(24.16) Proposition. *Let Γ be an oriented C^1-Jordan curve and $a \in C^1(\Gamma, \mathbb{R}^2\backslash\{0\})$. Also set*

$$\alpha := \frac{x\,dy - y\,dx}{x^2 + y^2}, \quad \text{on } \mathbb{R}^2\backslash\{0\}.$$

Then

$$w(a,\Gamma) = \frac{1}{2\pi}\int_\Gamma a^*\alpha = \frac{1}{2\pi}\int_\Gamma \frac{a^1\,da^2 - a^2\,da^1}{|a|^2}. \tag{6}$$

Proof. From either (2) or (3) and from (4) one can see that ϑ_a is continuously differentiable. If now $c : [0, T] \to \mathbb{R}^2$ is a regular C^1-parametrization of Γ, it follows from (3), (4) and a short calculation that

$$\dot{\vartheta} dt = c^* \frac{a^1 da^2 - a^2 da^1}{|a|^2} = c^* a^* \alpha,$$

where $\vartheta(t) := \vartheta_a(c(t))$. From the fundamental theorem of calculus we therefore obtain

$$2\pi w(a, \Gamma) = \vartheta(T) - \vartheta(0) = \int_0^T \dot{\vartheta}(t) dt = \int_0^T c^* a^* \alpha$$

$$= \int_\Gamma a^* \alpha = \int_\Gamma \frac{a^1 da^2 - a^2 da^1}{|a|^2}. \qquad \square$$

(24.17) Remark. Let Γ be a positively oriented C^2-Jordan curve and let $a \in C^1(\Gamma, \mathbb{S}^1)$ denote the positively oriented unit vector field tangent to Γ. Using a parametrization $c : [0, L] \to \mathbb{R}^2$ of Γ by arc length, we have, because $a = \dot{c} \circ c^{-1}$, that

$$a^* \alpha = a^1 da^2 - a^2 da^1 = [\dot{c}^1 \ddot{c}^2 - \dot{c}^2 \ddot{c}^1] dt = \kappa(t) dt,$$

where κ denotes the curvature of Γ. Therefore we obtain from theorem (24.15) the formula

$$\int_\Gamma \kappa(t) dt = 2\pi.$$

That is, *the "total curvature" of a positively oriented smooth Jordan curve is 2π*. This is the simplest case of the *Gauss-Bonnet theorem* from global differential geometry (cf. Klingenberg [1], Walter [1]). $\qquad \square$

As is well-known, the volume element $\omega_{\mathbb{S}^1}$ of the 1-sphere $\mathbb{S}^1 \subseteq \mathbb{R}^2$ is given by

$$\omega_{\mathbb{S}^1} = x^1 dx^2 - x^2 dx^1 \qquad (7)$$

(where the right-hand side is the restriction of the differential form $x^1 dx^2 - x^2 dx^1$ on $\mathbb{R}^2$ to $\mathbb{S}^1$. With the natural inclusion $i : \mathbb{S}^1 \hookrightarrow \mathbb{R}^2$ the precise formulation of (7) becomes

$$\omega_{\mathbb{S}^1} = i^*(x^1 dx^2 - x^2 dx^1).$$

We will use the less precise form (7) without fear of confusion, as is usual, for instance, with Stokes' theorem). With the radial retraction

$$r : \mathbb{R}^2 \backslash \{0\} \to \mathbb{S}^1, \qquad x \mapsto \frac{x}{|x|}$$

it follows from

$$dr^i = \frac{dx^i}{|x|} - \frac{x^i}{|x|^3} \sum_{k=1}^2 x^k dx^k, \qquad i = 1, 2,$$

that

$$r^* \omega_{\mathbb{S}^1} = r^1 dr^2 - r^2 dr^1 = \frac{x^1 dx^2 - x^2 dx^1}{|x|^2} = \alpha.$$

Under the assumptions of proposition (24.16) we thus obtain from (6) that

$$w(a, \Gamma) = \frac{1}{2\pi} \int_\Gamma a^* r^* \omega_{\mathbb{S}^1} = \frac{1}{\text{vol}(\mathbb{S}^1)} \int_\Gamma (r \circ a)^* \omega_{\mathbb{S}^1}.$$

This formula suggests the following generalization of the winding number:

Let $\Omega \subseteq \mathbb{R}^m$ be a bounded region of class C^1 and let

$$r : \mathbb{R}^m \backslash \{0\} \to \mathbb{S}^{m-1}, \quad x \mapsto \frac{x}{|x|},$$

be the radial retraction. Moreover, let $a \in C^1(\partial\Omega, \mathbb{R}^m \backslash \{0\})$ be a nowhere vanishing m-dimensional vector field on $\partial\Omega$. Then we define the *winding number* (the *rotation*) $w(a, \partial\Omega)$ *of the vector field* a *on* $\partial\Omega$ by the so-called *Kronecker integral*

$$w(a, \partial\Omega) := \frac{1}{\text{vol}(\mathbb{S}^{m-1})} \int_{\partial\Omega} (r \circ a)^* \omega_{\mathbb{S}^{m-1}}. \tag{8}$$

(24.18) Remarks. (a) As in the planar case (but substantially more difficult), one can show here also that

$$r^* \omega_{\mathbb{S}^{m-1}} = r^* \sum_{i=1}^m (-1)^{i+1} x^i dx^1 \wedge \cdots \wedge \widehat{dx^i} \wedge \cdots \wedge dx^m$$

$$= \sum_{i=1}^m (-1)^{i+1} \frac{x^i}{|x|^m} dx^1 \wedge \cdots \wedge \widehat{dx^i} \wedge \cdots \wedge dx^m,$$

holds, where the symbol $dx^1 \wedge \cdots \wedge \widehat{dx^i} \wedge \cdots \wedge dx^m$ means that the term dx^i is omitted (cf. Spivak [1]). Hence from

$$(r \circ a)^* \omega_{\mathbb{S}^{m-1}} = a^*(r^* \omega_{\mathbb{S}^{m-1}})$$

follows that

$$w(a, \partial\Omega) = \frac{1}{\text{vol}(\mathbb{S}^{m-1})} \int_{\partial\Omega} \sum_{i=1}^m (-1)^{i+1} \frac{a^i}{|a|^m} da^1 \wedge \cdots \wedge \widehat{da^i} \wedge \cdots \wedge da^m,$$

which corresponds to the second part of formula (6).

(b) Intuitively $w(a, \partial\Omega)$ measures how often the $(m-1)$-sphere $\mathbb{S}^{m-1}$ is covered by the image of $\partial\Omega$ under the mapping $r \circ a : \partial\Omega \to \mathbb{S}^{m-1}$ (with the orientation taken into consideration). $\quad\square$

The following theorem shows that, in principle, we already know the winding number.

(24.19) Theorem. *Let* $\Omega \subseteq \mathbb{R}^m$ *be a bounded* C^2-*domain and let* $a \in C^1(\overline{\Omega}, \mathbb{R}^m)$ *be such that* $0 \notin a(\partial\Omega)$. *Then*

$$w(a, \partial\Omega) = \deg(a, \Omega, 0).$$

Proof. Since there exists no nontrivial m-form on $\mathbb{S}^{m-1}$, we have (with $\omega := \omega_{\mathbb{S}^{m-1}}$) that

$$d(r^*\omega) = r^*d\omega = 0,$$

i.e., the $(m-1)$-form $\alpha := r^*\omega$ is defined on $\mathbb{R}^m \backslash \{0\}$ and is closed.

Let $\epsilon := \mathrm{dist}(a(\partial\Omega), 0)$. Then, by lemma (21.1), there exists some $b \in \overline{C}_r^\infty(\Omega, \mathbb{R}^m)$ such that $\|a - b\|_\infty < \epsilon$. Consequently we have that

$$h : [0,1] \times \overline{\Omega} \to \mathbb{R}^m, \quad (t, x) \mapsto ta(x) + (1-t)b(x)$$

is continuously differentiable and $0 \notin h([0,1] \times \partial\Omega)$. Therefore

$$\deg(a, \Omega, 0) = \deg(b, \Omega, 0),$$

using the homotopy invariance of the Brouwer degree. Since α is closed, it follows from the usual proof of the Poincaré lemma (e.g., from the "chain homotopy theorem" of Bröcker [1]) that $h_1^*\alpha - h_0^*\alpha$ is exact on $\partial\Omega$, where $h_t := h(t, \cdot)|\partial\Omega$, $0 \le t \le 1$. Therefore, by Stokes' theorem, the integral of $h_1^*\alpha - h_0^*\alpha = (r \circ a)^*\omega - (r \circ b)^*\omega$ over the closed C^2-manifold $\partial\Omega$ vanishes. From this it follows that

$$w(a, \partial\Omega) = w(b, \partial\Omega).$$

We therefore may assume, without loss of generality, that 0 is a regular value of a and that a is smooth. If $a^{-1}(0) = \emptyset$, then $\beta := (r \circ a)^*\omega = a^*\alpha$ is defined on $\overline{\Omega}$ and, because $d\beta = a^*d\alpha - 0$, it is closed. Hence Stokes' theorem implies that

$$\mathrm{vol}(\mathbb{S}^{m-1})w(a, \partial\Omega) = \int_{\partial\Omega} \beta = \int_\Omega d\beta = 0 = \deg(a, \Omega, 0),$$

where the last equality follows from the solution property of the degree.

Now let $x_1, \ldots, x_n$ be the zeros of a in Ω and choose $\epsilon > 0$ so small that the balls $\overline{\mathbb{B}}(x_j, \epsilon)$, $j = 1, \ldots, n$, are mutually disjoint and all lie in Ω. Then $M := \overline{\Omega} \backslash \cup_{j=1}^n \mathbb{B}(x_j, \epsilon)$

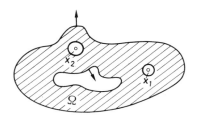

is an oriented C^2-manifold with boundary $\partial M = \partial\Omega \cup \bigcup_{j=1}^n (x_j + \epsilon\mathbb{S}^{m-1})$, where $\mathbb{S}^{m-1}_-$ means that $\mathbb{S}^{m-1}$ has the orientation which is the opposite of the usual one, i.e., $\mathbb{S}^{m-1}_-$ is oriented by the *inward* normal. Since β is closed on M, it follows from Stokes' theorem that

$$0 = \int_M d\beta = \int_{\partial M} \beta - \sum_{j=1}^n \int_{x_j + \epsilon\mathbb{S}^{m-1}} \beta. \tag{9}$$

If $\epsilon > 0$ is chosen small enough, then, by the inverse function theorem, $a|\mathbb{B}(x_j, 2\epsilon)$ is a C^∞-diffeomorphism from $\mathbb{B}(x_j, 2\epsilon)$ onto a neighborhood of $0 \in \mathbb{R}^m$. This diffeomorphism preserves or reverses the orientation depending on whether sign $\det Da(x_j) = 1$ or $= -1$.

Hence $N_j := a(\overline{\mathbb{B}}(x_j, \epsilon))$ is an oriented C^∞-manifold with boundary $\partial N_j = a(x_j + \epsilon \mathbb{S}^{m-1})$ and $0 \in N_j$. The orientation corresponds to the one induced by the natural orientation on $\mathbb{R}^m$ if and only if sign det $Da(x_j) = 1$. Hence we get (with $\mathbb{S} := \mathbb{S}^{m-1}$) that

$$\int_{x_j + \epsilon \mathbb{S}} \beta = \int_{x_j + \epsilon \mathbb{S}} a^* \alpha = \text{sign det} Da(x_j) \int_{\partial N_j} \alpha, \tag{10}$$

where ∂N_j is oriented by the outward normal. If $\varrho > 0$ is chosen small enough, then

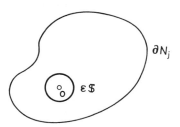

$$\partial N_j$$

$$\epsilon \mathbb{S}$$

$\varrho\overline{\mathbb{B}} \subseteq N_j$. Consequently Stokes' theorem can be applied to $N_j \setminus \varrho\mathbb{B}$ and, since α is defined and closed on $N_j \setminus \varrho\mathbb{B}$, it follows that

$$0 = \int_{N_j \setminus \varrho\mathbb{B}} d\alpha = \int_{\partial N_j} \alpha - \int_{\varrho \mathbb{S}} \alpha. \tag{11}$$

Since $r : \varrho\mathbb{S} \to \mathbb{S}$ is evidently a diffeomorphism, we finally obtain

$$\int_{\varrho \mathbb{S}} \alpha = \int_{\varrho \mathbb{S}} r^* \omega = \int_{\mathbb{S}} \omega = \text{vol}(\mathbb{S}),$$

and so from (9), (10) and (11) it follows that

$$\text{vol}(\mathbb{S}) w(a, \partial\Omega) = \int_{\partial\Omega} a^* \alpha = \sum_{j=1}^{n} \text{sign det} Da(x_j) \text{vol}(\mathbb{S}).$$

The assertion now follows from corollary (21.9). $\qquad\qquad\square$

(24.20) Remark. Since the degree only depends on the boundary values, one can define a *global degree* for any continuous function $g : \partial\Omega \to \mathbb{S}^{m-1}$ by

$$\deg(g) := \deg(g, \partial\Omega, \mathbb{S}^{m-1}) := \deg(\tilde{g}, \Omega, 0),$$

where $\tilde{g} \in C(\overline{\Omega}, \mathbb{R}^m)$ is any continuous extension of g. Then theorem (24.19) says that for $g \in C^1(\partial\Omega, \mathbb{S}^{m-1})$ we have

$$\int_{\partial\Omega} g^* \omega_{\mathbb{S}^{m-1}} = \deg(g) \int_{\mathbb{S}^{m-1}} \omega_{\mathbb{S}^{m-1}}.$$

One can show that this formula remains true if $\omega_{\mathbb{S}^{m-1}}$ is replaced by an arbitrary $(m-1)$-form.

More generally one can show: If M and N are smooth, compact m-dimensional manifolds with N connected, then for every smooth mapping $g : M \to N$ there exists a unique integer $\deg(g)$, the *(global) degree of g*, such that

$$\int_M g^* \omega = \deg(g) \int_N \omega$$

holds for every m-form ω on N (cf. Greub, Halperin and Vanstone[1]). $\square$

The following statements again refer to the flow induced by a planar vector field $f \in C^1(X, \mathbb{R}^2)$.

(24.21) Theorem. *Let γ be a noncritical periodic orbit of the differential equation $\dot{x} = f(x)$ and assume that its interior Ω lies completely in X. Then*

$$\deg(f, \Omega, 0) = 1.$$

Proof. Without loss of generality we may assume that the orientation of the orbit γ agrees with the positive orientation of γ as boundary of the C^2-manifold $\overline{\Omega}$ (cf. lemma (24.12) and remark (24.13 b)). For otherwise, we could consider the flow induced by $-f$ and use the fact that $\deg(f, \Omega, 0) = \deg(-f, \Omega, 0)$ (cf. problem 6 of section 21). Now the assertion follows from theorems (24.15) and (24.19). $\square$

(24.22) Corollary. *Under the assumptions of theorem (24.21), there exists at least one critical point.*

Proof. This follows from the solution property of the degree. $\square$

The following proposition shows how these results can be used to prove nonexistence statements for periodic orbits.

(24.23) Proposition. *Let X be simply connected and assume that f has exactly one zero x_0. If x_0 is a saddle point, then no noncritical periodic orbit exists.*

Proof. If γ is a noncritical periodic orbit and if Ω is its interior, then $\Omega \subseteq X$ and, by corollary (24.22), $x_0 \in \Omega$. Theorem (24.21) implies that

$$1 = \deg(f, \Omega, 0) = i_0(f, x_0).$$

Since x_0 is a saddle point, it follows from proposition (21.10) that $i_0(f, x_0) = -1$. This, however, is a contradiction. $\square$

In connection with the Poincaré-Bendixson theorem, one can draw conclusions from corollary (24.22) about the existence of critical points in more general situations.

(24.24) Proposition. *Let $K \subseteq X$ be compact, nonempty, simply connected and either positively or negatively invariant. Then there exists at least one critical point in K.*

Proof. Let K be positively invariant and choose any $x \in K$. If $\omega(x)$ contains no critical point, then $\omega(x)$ is a periodic orbit in K by theorem (24.6). Since K is simply connected, it follows that the interior Ω of $\omega(x)$ also lies in K. Hence corollary (24.22) implies that Ω contains a critical point. If K is negatively invariant, the assertion follows if we apply the above proof to the flow induced by $-f$. $\qquad\square$

(24.25) Examples. (a) We consider the Liénard equation

$$\ddot{x} + g(x, \dot{x})\dot{x} + h(x) = 0 \qquad (12)$$

under the following assumptions:

(i) $g \in C^1(\mathbb{R} \times \mathbb{R}, \mathbb{R}), \quad h \in C^1(\mathbb{R}, \mathbb{R})$;
(ii) $g(0,0) < 0, \qquad g(x,y) > 0, \quad$ if $|(x,y)| \geq r > 0$;
(iii) $xh(x) > 0, \quad$ if $x \neq 0$;
(iv) $H(x) := \int_0^x h(\xi)\, d\xi \to \infty \quad$ as $|x| \to \infty$.

The equation (12) is equivalent to the system

$$\dot{x} = y, \quad \dot{y} = -g(x,y)y - h(x) \qquad (13)$$

in the plane $\mathbb{R}^2$.

If we had $g = 0$, then (13) would be a Hamiltonian system with Hamiltonian (= total energy)

$$\Phi(x,y) := \frac{y^2}{2} + H(x).$$

From (iii) and (iv) it follows that Φ is coercive, positive for all $(x,y) \neq (0,0)$ and that $(0,0)$ is the only critical point. Since Φ is strictly increasing along all rays emanating from $(0,0)$, it follows that the level curves $\Phi^{-1}(c), c > 0$, are C^2-Jordan curves which include the origin in their interior. With

$$f(x,y) := (y, -g(x,y)y - h(x))$$

we get

$$(\nabla\Phi(x,y) \mid f(x,y)) = -y^2 g(x,y).$$

Now, let $c_1 > 0$ be chosen so small that $\Phi^{-1}(c)$ lies in the region where g is negative. We then have

$$(\nabla\Phi_1(x,y) \mid f(x,y)) \leq 0, \quad \forall(x,y) \in \Phi_1^{-1}(0),$$

where $\Phi_1 := c_1 - \Phi$. On the other hand, if we set $\Phi_2 := \Phi - c_2$ with c_2 sufficiently large, it follows from the coercivity of Φ and from (ii) that

$$(\nabla\Phi_2(x,y) \mid f(x,y)) \leq 0, \quad \forall(x,y) \in \Phi_2^{-1}(0).$$

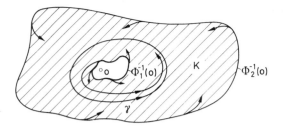

Hence $K := \Phi_1^{-1}(-\infty, 0] \cap \Phi_2^{-1}(-\infty, 0]$ is positively invariant by corollary (16.10). Since Φ_2 is coercive, K is compact. Evidently $K \neq \emptyset$ and it contains no critical point (because K is an "annular region" around the origin and $(0,0)$ is the only critical point of (13)). By the Poincaré-Bendixson theorem there exists a periodic orbit in K. There even exists a limit cycle γ in K because ∂K is not invariant. Since $(0,0)$ is the only critical point, it follows from corollary (24.22) that $(0,0)$ must lie in the interior of γ.

(b) The special Liénard equation

$$\ddot{x} + g(x)\dot{x} + h(x) = 0,$$

where $g, h \in C^1(\mathbb{R}, \mathbb{R})$, can have no periodic orbit such that its interior region lies completely in either the set $g^{-1}(-\infty, 0)$ or the set $g^{-1}(0, \infty)$.

In fact, the vector field

$$f(x, y) := (y, -g(x)y - h(x))$$

satisfies div $f = -g$. Hence the assertion follows from proposition (24.14).

(c) For further examples and a more detailed analysis of planar systems with special conditions we refer to Sansone-Conti [1].

(d) The Poincaré-Bendixson theorem is false in $\mathbb{R}^m$, $m \geq 3$. For a counter example we refer to D'Heedene [1]. In higher dimensions one knows relatively little about the existence of periodic solutions of general systems. Some general principles, which can also be applied to higher order equations, as well as detailed discussions of special higher order equations, are given in Reissig-Sansone-Conti [1].

An obvious method to prove the existence of noncritical periodic orbits consists in searching for a positively invariant torus T^m which contains no critical points. If it is also possible to find a "cross section" S^{m-1} of T^m so that S^{m-1} is a local transversal section, then it is possible to define an operator which is analogous to the Poincaré operator. The fixed points of this operator then correspond to periodic orbits in T^m. For applications

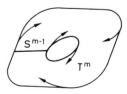

of this procedure we refer to Reissig-Sansone-Conti [1], as well as Hastings-Murray [1] (for the proof of periodic solutions of the Field-Noyes equations) and Hastings-Tyson-Webster [1] (for the proof of periodic solutions of the (generalized) Goodwin equations (cf. problems 2 and 3 in section 15)). Also see Tyson [1] and Li [1].   □

Problems

1. Let $g \in C^1(\mathbb{R}_+, \mathbb{R})$ and consider the system

$$\dot{x} = -y + xg(x^2 + y^2), \quad \dot{y} = x + yg(x^2 + y^2).$$

Determine the periodic solutions and limit cycles, as well as the one-sided, respectively, two-sided stability of the limit cycles of this system in the following cases:

(a) $g := 0$,
(b) $g(\xi) := 1 - \xi$,
(c) $g(\xi) := (\xi - 1)^2 \sqrt{\xi}$,
(d) $g(\xi) := \xi^{-1}(\xi - 1)^2 \ln(\xi - 1)$, if $\xi > 1$, and $g(\xi) := 0$, if $\xi \le 1$,
(e) $g(\xi) := (\xi - 1)^2 \sin(\xi - 1)^{-1}$, if $\xi \ne 1$, and $g(1) := 1$,
(f) $g(\xi) := \xi^2 \sin(1/\xi)$, if $\xi > 0$, and $g(0) := 0$.

(*Hint:* Polar coordinates.)

2. Let $X \subseteq \mathbb{R}^2$ be a doubly connected domain, let $f \in C^1(X, \mathbb{R}^2)$ and let $\varrho \in C^1(X, \mathbb{R})$. Show that: If $\operatorname{div}(\varrho f)(x) \ne 0$ for all $x \in X$, then the equation $\dot{x} = f(x)$ has at most one periodic solution.

3. Show that the system

$$\dot{x} = -y - x + (x^2 + 2y^2)x$$
$$\dot{y} = x - y + (x^2 + 2y^2)y$$

has exactly one nontrivial periodic solution.

(*Hint:* Poincaré-Bendixson theorem and problem 2.)

4. (a) Let $c : [0, T] \to \mathbb{R}^2$ be a C^1-parametrization of an oriented C^1-Jordan curve Γ and let $a \in C^1(\Gamma, \mathbb{R}^2 \setminus \{0\})$. Assume that a^1 has only finitely many zeros and set $b := a \circ c$. Show that: If p (resp. q) is the number of zeros t of b^1 such that $b^2(t) > 0$ and $\dot{b}^1(t) > 0$ (resp. $\dot{b}^1(t) < 0$), we have

$$w(a, \Gamma) = q - p.$$

(b) Calculate the local index $i_0(f, 0)$ of the vector field

$$f(x, y) := (y - \frac{1}{2}xy - 3x^2, -xy - \frac{3}{2}y^2), \quad (x, y) \in \mathbb{R}^2.$$

5. The following system

$$\dot{x} = a - bx + x^2 y - x$$
$$\dot{y} = bx - x^2 y,$$

where a and b are positive constants, models a chemical reaction studied by Prigogine and Lefever. Show that for $b > 1 + a^2$ these equations have at least one limit cycle.

(*Hint:* Consider a sub-region of $\mathbb{R}_+^2$ which is bounded by two appropriate lines of the form $x + y = \alpha$ and $y - x = \beta$, $\alpha, \beta > 0$.)

Chapter VI
Continuation and Bifurcation Problems

In this last chapter we will study the existence of periodic solutions – and of critical points – of parameter-dependent differential equations. This problem area naturally breaks down into two subparts, into continuation and bifurcation problems.

Continuation methods deal with the "continuation" of a periodic solution u_{λ_0} of a parameter-dependent equation $\dot{x} = f(\lambda_0, t, x)$ "along the parameter λ." In other words, one assumes that the equation $\dot{x} = f(\lambda_0, t, x)$ is so simple that at least the existence of one periodic solution is ascertained – e.g., by the methods of the last chapter. Then one tries to obtain information about the existence of periodic solutions of $\dot{x} = f(\lambda, t, x)$ for small perturbations of the parameter.

The characteristic feature of continuation methods lies in the fact that the local continuation of the periodic solution $u(\lambda_0)$ can be done uniquely and that this does not change the stability properties. These facts will be proved in the first section of this chapter.

In sharp contrast to the continuation methods are the bifurcation problems, which will be analyzed in the second section of this chapter. Here we deal exactly with those phenomena which occur when the unique continuation is impossible. One finds bifurcation problems in many other areas of analysis, e.g., in nonlinear boundary value problems, in nonlinear integral equations or in nonlinear partial differential equations, to name a few. The common background of these questions comes to light best in the abstract framework of nonlinear functional analysis. For simplicity's sake we restrict ourselves to the finite dimensional case, which suffices for our purposes. However, all proofs remain true – with the obvious modifications, which will be pointed out occasionally – in the infinite dimensional case.

The central, abstract result of the second section is a "preparation theorem" which reduces the problem to the study of a so-called bifurcation equation by means of a Liapunov-Schmidt reduction. The significance of this preparation theorem is the fact that the bifurcation equation is brought into a normal form from which the structure of the solution set can nearly be "read off" directly.

As the most important application of the abstract preparation theorem we will give a simple, geometric proof of the fundamental theorem about Hopf bifurcation, that is, the problem of the branching of a periodic orbit from a critical point of a one-parameter vector field. As a simple consequence of the Hopf bifurcation theorem we will prove the Liapunov center theorem for the existence of periodic orbits of autonomous Hamiltonian systems.

Finally, the last section is devoted to stability questions of bifurcation problems. Next to the known geometric criteria, which serve to deduce the stability of the bifurcating solution from the direction of the bifurcation branches, we will also give numerically verifiable stability criteria. To keep the calculations within reasonable limits, we will not consider the most general case but, instead, make a simplifying assumption. However, the proof for the general case will be sketched, so that it can be carried out by the interested reader himself.

25. Continuation Methods

In this section let E and F denote finite dimensional real Banach spaces, and assume that $X \subseteq E$ and $\Lambda \subseteq F$ are open and

$$f \in C^k(\Lambda \times \mathbb{R} \times X, E)$$

for some $k \geq 1$. Moreover, let

$$T \in C^k(\Lambda, \mathbb{R})$$

be a function such that for every $(\lambda, x) \in \Lambda \times X$, $f(\lambda, \cdot, x) : \mathbb{R} \to E$ is periodic with period $T(\lambda)$.

We consider the parameter-dependent differential equation

$$\dot{x} = f(\lambda, t, x) \tag{1}_\lambda$$

and we are interested in the existence of $T(\lambda)$-periodic solutions as the parameter $\lambda \in \Lambda$ is varied.

First we assume that equation $(1)_\lambda$ has a $T(\lambda_0)$-periodic solution $u(\lambda_0)$ for some $\lambda_0 \in \Lambda$ and we pose the question:

For λ near λ_0, does equation $(1)_\lambda$ also have $T(\lambda)$-periodic solutions near $u(\lambda_0)$?

(25.1) Examples. We consider the second order equation

$$\ddot{y} + \alpha \dot{y} + \beta y + \gamma y^3 + \delta \sin y = \epsilon \sin^3(\omega t), \tag{2}$$

where $\alpha, \beta, \gamma, \delta, \epsilon, \omega \in \mathbb{R}$. As is well-known, (2) is equivalent to a system of type $(1)_\lambda$ in $\mathbb{R}^2$. There are many possible choices for the parameter space Λ, depending on which of the parameter(s) $\alpha, \ldots, \omega$ one wants to vary. In what follows we will consider several possibilities:

(a) $\lambda := (\alpha, \beta, \gamma, \delta, \epsilon, \omega) \in \Lambda := \mathbb{R}^6$.   In this case $T(\lambda) := pr_6(\lambda) = \omega$, and for every $\lambda_0 := (0, \ldots, 0, \omega_0) \in \{0\} \times \mathbb{R}$, equation (2) has a one-parameter family of ω_0-periodic solutions, $u(\lambda_0) =$ constant. If $\lambda_0 := (\alpha_0, \beta_0, 0, 0, 0, \omega_0)$, then $u(\lambda_0) = 0$ is the only ω_0-periodic solution of (2), assuming that either $\alpha_0 \neq 0$ and $\beta_0 \neq 0$, or $\beta_0 < 0$. However, if $\lambda_0 := (0, \beta_0, 0, 0, 0, \omega_0)$ and $\beta_0 > 0$, then (2) has the two-parameter family of ω_0-periodic solutions

$$a\cos(\omega_0 t) + b\sin(\omega_0 t),$$

provided that $\omega_0 = \sqrt{\beta_0}$ (cf. example (14.7)).

(b) $\alpha = \delta = 0$, $\beta, \gamma, \epsilon > 0$, $\lambda = \omega$. In this case the equation becomes

$$\ddot{y} + \beta y + \gamma y^3 = \epsilon \sin^3(\omega t). \tag{3}$$

For $\omega_0 = 1/\sqrt{\beta}$ it has the ω_0-periodic solution $y(t) = \sqrt[3]{\epsilon/\gamma}\, \sin(\omega_0 t)$.

(c) $\beta = \gamma = \epsilon = 0$, $\delta > 0$, $\lambda = \alpha$. For this choice of constants we obtain the equation of the damped (if $\alpha > 0$) or excited (if $\alpha < 0$) simple pendulum

$$\ddot{y} + \alpha \dot{y} + \delta \sin y = 0. \tag{4}$$

By example (3.4 c) (and subsequent mathematical elucidations), the problem possesses a two-parameter family of periodic solutions if $\alpha = 0$. In fact, there exists a whole neighborhood U of 0 in the phase plane such that through every point of U there passes exactly one periodic orbit.

Since (4) is an autonomous problem, no natural choice for the period T can be made, that is, in the search for periodic solutions of autonomous parameter dependent problems, the period is an additional unknown. For this reason, the study of autonomous differential equations is, in general, more difficult than the investigation of differential equations with periodic coefficients.

The total energy of the frictionless pendulum,

$$V(y, \dot{y}) = \dot{y}^2/2 + \delta(1 - \cos y),$$

is a natural candidate for a Liapunov function for (4), i.e., for the system

$$\begin{aligned} \dot{x}^1 &= x^2 \\ \dot{x}^2 &= -\alpha x^2 - \delta \sin x^1. \end{aligned} \tag{5}$$

For this we have

$$\dot{V}(x) = (\nabla V(x) \mid \dot{x}) = -\alpha(x^2)^2.$$

Hence for $\alpha \neq 0$, $\mathrm{sgn}(\alpha)V$ is a Liapunov function for (5). Now let γ be a periodic orbit of (5). Then it follows from example (17.4 b) that $\gamma = \omega(x)$ for every $x \in \gamma$. From the invariance principle (18.3) we deduce that

$$\gamma = \omega(x) \subseteq \{x \in \mathbb{R}^2 \mid \dot{V}(x) = 0\} = \mathbb{R} \times \{0\},$$

which, together with (5), finally implies that $\gamma(x) = \{x\}$ and $x \in 2\pi\mathbb{Z} \times \{0\}$. For $\alpha \neq 0$, problem (4) has therefore no nontrivial periodic solution. This shows that our question has, in general, no affirmative answer. $\qquad\square$

A simple condition, producing a positive result for our problem, follows from the implicit function theorem. In principle, the following result was already known to Poincaré and it is sometimes called the "Poincaré continuation method."

(25.2) Theorem. *Let $u(\lambda_0)$ denote a $T(\lambda_0)$-periodic solution of the equation $\dot{x} = f(\lambda_0, t, x)$ and assume that the linearized problem*

$$\dot{y} = D_3 f(\lambda_0, t, u(\lambda_0)(t))y \tag{6}$$

possesses no nontrivial $T(\lambda_0)$-periodic solution. Then there exists some neighborhood V of λ_0 in Λ and some $\epsilon > 0$ such that for every $\lambda \in V$, there exists a unique $T(\lambda)$-periodic solution $u(\lambda)$ of

$$\dot{x} = f(\lambda, t, x) \tag{7}$$

satisfying

$$|u(tT(\lambda), \lambda) - u(tT(\lambda_0), \lambda_0)| < \epsilon, \quad \forall t \in \mathbb{R}, \tag{8}$$

where we have set $u(t, \lambda) := u(\lambda)(t)$. Moreover,

$$[(t, \lambda) \mapsto u(t, \lambda)] \in C^k(\mathbb{R} \times V, E).$$

Proof. We let $(t, \xi, \lambda) \mapsto u(t, 0, \xi, \lambda)$ denote the global solution of the IVP

$$\dot{x} = f(\lambda, t, x), \quad x(0) = \xi.$$

It follows from theorem (8.3) that the domain of this function is open in $\mathbb{R} \times X \times \Lambda$ and, by assumption, we have $\mathbb{R} \times \{\xi_0\} \times \{\lambda_0\} \subseteq \mathbb{R} \times X \times \Lambda$ for $\xi_0 := u(\lambda_0)(0)$. For reasons of compactness and continuity, there exists a neighborhood $W_1 \times V_1$ of $(\xi_0, \lambda_0) \in X \times \Lambda$ such that $u(\cdot, 0, \xi, \lambda)$ exists on $[0, T(\lambda)]$ for each $(\xi, \lambda) \in W_1 \times V_1$.

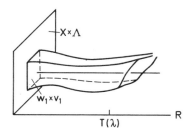

Hence the time-$T(\lambda)$-map

$$(\xi, \lambda) \mapsto u(T(\lambda), 0, \xi, \lambda)$$

is well-defined on $W_1 \times V_1$ and it belongs to $C^k(W_1 \times V_1, E)$ by theorem (9.5). It follows from theorem (20.1) that for every $\lambda \in \Lambda$, (7) has a $T(\lambda)$-periodic solution if and only if the mapping $\xi \mapsto u(T(\lambda), 0, \xi, \lambda)$ has a fixed point. With

$$g(\xi, \lambda) := \xi - u(T(\lambda), 0, \xi, \lambda), \quad \forall (\xi, \lambda) \in W,$$

and using remark (20.2), we can state more precisely that: $g(\xi, \lambda) = 0$ if and only if $u(\cdot, 0, \xi, \lambda)$ is a $T(\lambda)$-periodic solution which passes through ξ at time $t = 0$. In particular, our assumption implies that $g(\xi_0, \lambda_0) = 0$. In addition,

$$D_1 g(\xi_0, \lambda_0) = id_E - D_3 u(T(\lambda_0), 0, \xi_0, \lambda_0),$$

and from the differentiability theorem (9.2) it follows that $D_3 u(T(\lambda_0), 0, \xi_0, \lambda_0)$ is the monodromy operator of equation (6). Since 1 is not a Floquet multiplier of (6), by proposition (20.12), it follows that $0 \notin \sigma(D_1 g(\xi_0, \lambda_0))$, i.e., $D_1 g(\xi_0, \lambda_0) \in \mathcal{GL}(E)$. Hence by the implicit function theorem, there exists a neighborhood $W_0 \times V_0$ of (ξ_0, λ_0) in $X \times \Lambda$ and a function $h \in C^k(V_0, W_0)$ such that

$$(\xi, \lambda) \in W_0 \times V_0 \quad \text{and} \quad g(\xi, \lambda) = 0 \quad \Leftrightarrow \quad \lambda \in V_0 \quad \text{and} \quad \xi = h(\lambda).$$

In other words: For every $\lambda \in V_0$, there exists a unique $T(\lambda)$-periodic solution $u(\lambda) := u(\cdot, \lambda) := u(\cdot, 0, h(\lambda), \lambda)$ of $\dot{x} = f(\lambda, t, x)$ such that $u(0, \lambda) \in W_0$. It follows immediately from the differentiability theorem that $u \in C^k(\mathbb{R} \times V_0, E)$. Setting

$$v(t, \lambda) := u(tT(\lambda), \lambda),$$

we have $v \in C^k(\mathbb{R} \times V_0, E)$, and $v(\cdot, \lambda)$ is periodic with period 1 for every $\lambda \in V_0$. Now we choose $\epsilon > 0$ so that $\mathbb{B}(\xi_0, \epsilon) \subseteq W_0$. By the compactness of $[0, 1]$, we can then find a neighborhood V of λ_0 in V_0 such that $|v(t, \lambda) - v(t, \lambda_0)| < \epsilon$ for all $t \in [0, 1]$ and all $\lambda \in V$. Because $v(\cdot, \lambda)$ is 1-periodic, this estimate holds for all $t \in \mathbb{R}$ and so (8) is proved.   □

(25.3) Remark. The proof above shows that the mapping

$$V \to BC(\mathbb{R}, E), \quad \lambda \mapsto v(\cdot, \lambda) := u(\cdot T(\lambda), \lambda),$$

is continuous. Because $v \in C^k(\mathbb{R} \times V, E)$, a similar assertion holds for $D_1^i D_2^j v$ for all $i, j \in \mathbb{N}$ satisfying $i + j \leq k$. With

$$BC^k(\mathbb{R}, E) := \{u \in C^k(\mathbb{R}, E) \mid \|u\|_{C^k} := \sum_{j=0}^{k} \|D^j u\|_{\infty} < \infty\},$$

we even have

$$[\lambda \mapsto v(\cdot, \lambda)] \in C^k(V, BC^k(\mathbb{R}, E)).$$

Indeed, from the mean value theorem we obtain

$$v(t, \lambda_1 + \lambda) - v(t, \lambda_1) - D_2 v(t, \lambda_1)\lambda = \int_0^1 [D_2 v(t, \lambda_1 + \tau\lambda) - D_2 v(t, \lambda)]\lambda \, d\tau \quad (9)$$

for all $t \in \mathbb{R}$, $\lambda_1 \in V$ and $\lambda \in F$ with $\lambda_1 + \lambda \in V$. Since $D_2 v$ is 1-periodic with respect to t and is continuous in both variables, there exists for every $\epsilon > 0$ some $\delta > 0$ such that

$|D_2 v(t, \lambda_1 + \tau\lambda) - D_2 v(t, \lambda)| < \epsilon$ for all $t \in \mathbb{R}$ and all $\lambda \in F$ satisfying $|\lambda| < \delta$. Hence from (9) we deduce

$$\|v(\cdot, \lambda_1 + \lambda) - v(\cdot, \lambda_1) - D_2 v(\cdot, \lambda_1)\lambda\|_\infty \le \epsilon|\lambda|$$

for all sufficiently small $\lambda \in F$. This proves the assertion for the case $k = 1$. The general case is proved similarly by induction.

The reparametrization of the $T(\lambda)$-periodic solution $u(\cdot, \lambda)$, i.e., the passage to the function $t \mapsto u(tT(\lambda), \lambda)$, is essential. If $T(\lambda) \ne T(\lambda_1)$, one easily sees that the integral curves $t \mapsto u(t, \lambda)$ and $t \mapsto u(t, \lambda_1)$ cannot be uniformly (that is, for all time) adjacent, while the preceding assertion implies that the orbits $\gamma(u(\lambda), 0) := \{u(t, \lambda) \mid t \in \mathbb{R}\}$ and $\gamma(u(\lambda_1), 0)$ are close together for λ near λ_1. $\square$

(25.4) Remarks and Examples. (a) If 1 is not a Floquet multiplier of (6), then theorem (25.2) implies, in particular, that $u(\lambda_0)$ is an isolated $T(\lambda_0)$-periodic solution of $\dot{x} = f(\lambda_0, t, x)$. Hence theorem (25.2) is not applicable to example (25.1 a) if $\lambda_0 = (0, \dots, 0, \omega_0)$.

(b) If $u(\lambda_0)$ is a nonconstant $T(\lambda_0)$-periodic solution of $\dot{x} = f(\lambda_0, t, x)$ *and if the equation*

$$\dot{x} = f(\lambda_0, t, x)$$

is autonomous, then, by remark (23.2 c), 1 is a Floquet multiplier of (6). *Hence theorem (25.2) can be applied neither in this case, nor to autonomous equations $\dot{x} = f(\lambda, x)$ with nonconstant periodic solutions.*

(c) We consider the equation

$$\ddot{y} + \alpha\dot{y} + \beta y + \gamma y^3 + \delta \sin y = \epsilon \sin^3(\omega t) \tag{10}$$

of example (25.1) with $\lambda = (\alpha, \beta, \gamma, \delta, \epsilon, \omega) \in \Lambda := \mathbb{R}^6$. Moreover, let $\lambda_0 := (\alpha_0, \beta_0, 0, 0, 0, \omega_0)$ be such that $\omega_0 \ne 0$ and either $\alpha_0 \ne 0$ and $\beta_0 \ne 0$, or $\beta_0 < 0$. Then, if $\lambda = \lambda_0$, 0 is the only periodic solution of (10) and the corresponding linearized equation is (written in scalar form) given by

$$\ddot{y} + \alpha_0\dot{y} + \beta_0 y = 0. \tag{11}$$

By assumption, (11) has no nontrivial ω_0-periodic solution (cf. example (14.7)). Hence there exists a neighborhood V of λ_0 in $\mathbb{R}^6$ such that for every $\lambda \in V$, equation (10) has a unique ω-periodic solution $u(\lambda) = u(\cdot, \lambda)$ (where $\omega = pr_6 \lambda$) satisfying

$$u \in C^\infty(\mathbb{R} \times V, \mathbb{R})$$

and $u(\lambda_0) = 0$.

(d) Suppose $\omega > 0$ and $f \in C^k(\mathbb{R}^4, \mathbb{R})$, and let f be T-periodic in the last variable. We then consider the T-periodic equation

$$\ddot{y} + \omega^2 y = \lambda f(\lambda, y, \dot{y}, t). \tag{12}$$

In this case we have $\Lambda = \mathbb{R}$ and $T(\lambda) = T$. If $\lambda = 0$ and $T \notin (2\pi/\omega)\mathbb{Z}$, equation (12) has only the trivial T-periodic solution. Hence the corresponding linearized equation $\ddot{\eta} + \omega^2\eta = 0$ has no nontrivial T-periodic solution either (if $T \notin (2\pi/\omega)\mathbb{Z}$). Thus from theorem (25.2) and remark (25.3) we deduce the existence of a neighborhood V of 0 in $\mathbb{R}$ and some unique function $u(\cdot) \in C^k(V, BC^k(\mathbb{R}, \mathbb{R}))$ satisfying: $u(0) = 0$ and $u(\lambda)$ is a T-periodic solution of (12) for all $\lambda \in V$.

If $T \in (2\pi/\omega)\mathbb{Z}^*$, theorem (25.2) cannot be applied because then 1 is a Floquet multiplier of the linearized equation at $\lambda = 0$, $u = 0$; in fact its multiplicity is 2. $\square$

Perturbation Theory of Eigenvalues

If $u(\lambda_0)$ is a $T(\lambda_0)$-periodic solution of the equation $\dot{x} = f(\lambda_0, t, x)$, then by theorem (23.1), $u(\lambda_0)$ is asymptotically Liapunov stable [resp. unstable] if for all Floquet multipliers μ of the linearized equation

$$\dot{y} = D_3 f(\lambda_0, t, u(\lambda_0)(t))y$$

we have $|\mu| < 1$ [resp. if for some μ we have $|\mu| > 1$]. We will show now that, in this case, the solution $u(\lambda)$ – at least for λ close to λ_0 – "inherits" the stability features of $u(\lambda_0)$. For this we need some preliminaries which are of importance in their own right.

(25.5) Proposition. *The spectrum*

$$\sigma(\cdot) : \mathcal{L}(E) \longrightarrow 2^{\mathbb{C}}$$

is upper semicontinuous, that is, for every $T_0 \in \mathcal{L}(E)$ and every neighborhood U of $\sigma(T_0)$ in $\mathbb{C}$ there exists a neighborhood V of T_0 in $\mathcal{L}(E)$ such that $\sigma(T) \subseteq U$ for all $T \in V$.

Proof. Since $\sigma(T) = \sigma(T_{\mathbb{C}})$, we may assume, for this proof, that E is a complex Banach space. Hence $\lambda \notin \sigma(T)$ if and only if $\lambda - T \in \mathcal{GL}(E)$.

Let $S \in \mathcal{L}(E)$ and $S_0 \in \mathcal{GL}(E)$ and assume that $\|S - S_0\| < 1/\|S_0^{-1}\|$. From $S = S_0 - (S_0 - S) = S_0[I - S_0^{-1}(S_0 - S)]$ and $\|S_0^{-1}(S_0 - S)\| \leq \|S_0^{-1}\|\|S - S_0\| < 1$ it follows that $S \in \mathcal{GL}(E)$ and $S^{-1} = [I - S_0^{-1}(S_0 - S)]^{-1}S_0^{-1}$, because for every $B \in \mathcal{L}(E)$ with $\|B\| < 1$, the geometric series ("Neumann series") $\sum_{j=0}^{\infty} B^j$ converges in $\mathcal{L}(E)$ and, since

$$(I - B)\sum_{j=0}^{\infty} B^j = \sum_{j=0}^{\infty} B^j(I - B) = I,$$

it represents the operator $(I - B)^{-1}$. Based on

$$
\begin{aligned}
S^{-1} - S_0^{-1} &= \{[I - S_0^{-1}(S_0 - S)]^{-1} - I\}S_0^{-1} \\
&= [I - S_0^{-1}(S_0 - S)]^{-1}S_0^{-1}(S_0 - S)S_0^{-1},
\end{aligned}
$$

we finally obtain the estimate

$$\|S^{-1} - S_0^{-1}\| \le \frac{\|S_0^{-1}\|^2 \, \|S - S_0\|}{1 - \|S_0^{-1}\| \, \|S - S_0\|},$$

because if $\|B\| < 1$, we deduce from the geometric series the estimate $\|(I - B)^{-1}\| \le 1/(1 - \|B\|)$. This estimate shows that $\mathcal{GL}(E)$ is open and that the "inverse mapping" $T \mapsto T^{-1}$ is a continuous function from $\mathcal{GL}(E)$ onto itself. More precisely, for every $S_0 \in \mathcal{GL}(E)$, the ball $\mathbb{B}_{\mathcal{L}(E)}(S_0, \|S_0^{-1}\|^{-1})$ also belongs to $\mathcal{GL}(E)$.

If $T \in \mathcal{L}(E)$ and $|\lambda| > \|T\|$, then $\|T/\lambda\| < 1$, hence $\lambda - T = \lambda(I - T/\lambda) \in \mathcal{GL}(E)$. Consequently

$$\sigma(T) \subseteq \overline{\mathbb{B}}_{\mathbb{C}}(0, \|T\|), \quad \forall T \in \mathcal{L}(E). \tag{13}$$

Let now $T_0 \in \mathcal{L}(E)$ be arbitrary and let U be some bounded open neighborhood of $\sigma(T_0)$ in $\mathbb{C}$. In addition, let $r > \|T_0\|$ be a real number such that

$$U \subseteq \mathbb{B}_{\mathbb{C}}(0, r).$$

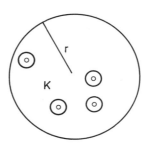

Then $K := \overline{\mathbb{B}}_{\mathbb{C}}(0, r) \setminus U$ is compact, nonempty and $K \subseteq \rho(T_0) := \mathbb{C} \setminus \sigma(T_0)$. Since by the preceding considerations the mapping

$$\rho(T_0) \to \mathcal{GL}(E), \quad \lambda \mapsto (\lambda - T_0)^{-1},$$

is continuous, there exists some $\alpha \in \mathbb{R}$ such that $\|(\lambda - T_0)^{-1}\| \le \alpha$ for all $\lambda \in K$. Then for each $T \in \mathcal{L}(E)$ satisfying $\|T - T_0\| < \min\{r - \|T_0\|, 1/\alpha\} =: \beta$, we have

$$\|(\lambda - T) - (\lambda - T_0)\| = \|T - T_0\| < \alpha^{-1} \le 1/\|(\lambda - T_0)^{-1}\|$$

for all $\lambda \in K$. By the preceding considerations it follows that $\lambda - T$ belongs to $\mathcal{GL}(E)$ for every $\lambda \in K$, i.e., $K \subseteq \rho(T)$. From $\|T\| \le \|T_0\| + \|T - T_0\| < r$ and (13) it even follows that $U^c \subseteq \rho(T)$, i.e., $\sigma(T) \subseteq U$ for every $T \in V := \{T \in \mathcal{L}(E) \mid \|T - T_0\| < \beta\}$, hence the assertion.   □

(25.6) Remarks.   (a)   An inspection of the foregoing proof shows that we have proved the following, general facts:

Let Y be a Banach space of any dimension over $\mathbb{K}$. Then, if $S_0 \in \mathcal{GL}(Y)$, the entire ball $\mathbb{B}_{\mathcal{L}(Y)}(S_0, \|S_0^{-1}\|^{-1})$ belongs to $\mathcal{GL}(Y)$. Therefore $\mathcal{GL}(Y)$ is open and the inverse mapping $S \mapsto S^{-1}$ is continuous. These assertions, along with their proofs, remain correct if $\mathcal{L}(E)$ is replaced by an arbitrary Banach algebra.

(b) *Let Y be an arbitrary complex Banach space. For $T \in \mathcal{L}(E)$, one defines the resolvent set of T, $\rho(T)$, by*

$$\rho(T) := \{\lambda \in \mathbb{C} \mid \lambda - T \in \mathcal{GL}(Y)\}$$

and the *spectrum of T, $\sigma(T)$, by*

$$\sigma(T) := \mathbb{C} \setminus \rho(T).$$

Then the proof of the previous proposition shows that the following holds: $\rho(T)$ *is open and*

$$\sigma(T) \subseteq \overline{\mathbb{B}}_{\mathbb{C}}(0, \|T\|).$$

Moreover, the spectrum

$$\sigma(\cdot) : \mathcal{L}(Y) \to 2^{\mathbb{C}}$$

is upper semicontinuous.

Thus, in particular, the spectrum of $T \in \mathcal{L}(Y)$ is compact. In the infinite dimensional case, $\sigma(T)$ is, in general, no longer a finite set.

The upper semicontinuity of the spectrum means graphically that a continuous change of T does not abruptly "expand" the spectrum. It is quite possible, however, that it suddenly "shrinks." For an example of such a situation we refer to Kato [1, Example IV.3.8]. □

In the next proposition we give an important representation of the winding number of certain planar vector fields along a closed Jordan curve. To do this, we use the canonical identification of $\mathbb{R}^2$ with $\mathbb{C}$ ($= \mathbb{R} + i\mathbb{R}$) and we sometimes interpret a function $f : U \subseteq \mathbb{R}^2 \to \mathbb{R}^2$ as a complex function $f : U \subseteq \mathbb{C} \to \mathbb{C}$, without expressing this by means of some other notation. The context will always make clear whether we consider f as a mapping in $\mathbb{R}^2$ or in $\mathbb{C}$.

(25.7) Proposition. *Let Γ be a positively oriented planar C^1-Jordan curve and let U be an open neighborhood of Γ. Moreover, let $f : U \to \mathbb{C}$ be analytic and $f(z) \neq 0$ for all $z \in \Gamma$. Then for the winding number we have*

$$w(f, \Gamma) = \frac{1}{2\pi i} \int_{\Gamma} \frac{f'(z)}{f(z)} \, dz.$$

Proof. Let $f = u + iv$, where $u := Re(f)$ and $v := Im(f)$. Since f is differentiable at z_0, we have, setting $f'(z_0) =: \alpha + i\beta$,

$$f(z_0 + h) - f(z_0) - (\alpha + i\beta)(\xi + i\eta) = o(h)$$

as $h = \xi + i\eta \to 0$. By partitioning into real and imaginary parts, we obtain from this

$$F(x_0 + \xi, y_0 + \eta) - F(x_0, y_0) - DF(x_0, y_0)(\xi, \eta) = o(|(\xi, \eta)|)$$

as $(\xi, \eta) \to 0$, where $F := (u, v)$ and

$$[DF] = \begin{bmatrix} u_x & u_y \\ v_x & v_y \end{bmatrix} = \begin{bmatrix} \alpha & -\beta \\ \beta & \alpha \end{bmatrix}.$$

Hence, in particular, we obtain the Cauchy-Riemann equations $\alpha = u_x = v_y$ and $\beta = v_x = -u_y$. Consequently

$$f' = u_x + iv_x = u_x - iu_y$$

and therefore

$$\frac{f' dz}{f} = \frac{\overline{f} f' dz}{|f|^2} = \frac{(u - iv)(u_x + iv_x)(dx + idy)}{u^2 + v^2}$$

$$= \frac{u\,du + v\,dv}{u^2 + v^2} + i\frac{u\,dv - v\,du}{u^2 + v^2},$$

where we have made use of the Cauchy-Riemann equations. Using

$$\alpha := x\,dx + y\,dy = \frac{1}{2}d(x^2 + y^2) \quad \text{and} \quad \omega := \omega_{\mathbb{S}^1} = x\,dy - y\,dx,$$

as well as the Euclidean retraction $r : \mathbb{R}^2 \setminus \{0\} \to \mathbb{S}^1$, we have along Γ that

$$\frac{f' dz}{f} = F^* r^* \alpha + i F^* r^* \omega = (r \circ F)^* \alpha + i(r \circ F)^* \omega.$$

Since $(r \circ F)^* \alpha = (r \circ F)^* dg = d(r \circ F)^* g$ is exact (where $g(x, y) := (x^2 + y^2)/2$), we obtain, by integrating over the closed curve Γ and utilizing proposition (24.16), that

$$\int_\Gamma \frac{f' dz}{f} = i \int_\Gamma (r \circ F)^* \omega = 2\pi i w(F, \Gamma) = 2\pi i w(f, \Gamma),$$

which was to be proved.   □

(25.8) Corollary. *Let* $U \subseteq \mathbb{C}$ *be open and* $f : U \to \mathbb{C}$ *be analytic. In addition, assume that* Ω *is a bounded* C^2*-domain such that* $\overline{\Omega} \subseteq U$ *and* $0 \notin f(\partial\Omega)$. *Then*

$$\deg(f, \Omega, 0) = \frac{1}{2\pi i} \int_{\partial\Omega} \frac{f'(z)}{f(z)}\,dz.$$

Proof. This follows from proposition (25.7) and theorem (24.19).   □

(25.9) Remark. If $f(z) \neq 0$, then

$$\frac{f'(z)dz}{f(z)} = d(\ln f(z)) = d(\ln|f(z)| + i \arg f(z)),$$

where the *argument* of the complex number $w \neq 0$, $\arg w \in \mathbb{R}$, is defined by $w = |w|e^{i \arg w}$ modulo 2π. Hence the proof of proposition (25.7) shows that

$$w(f, \Gamma) = \frac{1}{2\pi} \int_\Gamma d(\arg f(z)).$$

(Caution: $d(\arg f(z))$ is, despite the notation, not an exact 1-form on $\mathbb{R}^2 \setminus \{0\}$.) From this representation of the winding number one can again read off the intuitive meaning of $w(f, \Gamma)$, because for every "rotation" of w around $0 \in \mathbb{C}$ in the positive sense, the argument of w experiences an increase of 2π.   □

As a first application of this result we will prove the "argument principle" of complex analysis.

(25.10) Proposition. *Let $U \subseteq \mathbb{C}$ be open and $f : U \to \mathbb{C}$ be analytic. Moreover, let Ω be a bounded C^2-domain such that $\overline{\Omega} \subseteq U$ and $0 \notin f(\partial\Omega)$. Then*

$$\deg(f, \Omega, 0) = w(f, \partial\Omega) = \frac{1}{2\pi i} \int_{\partial\Omega} \frac{f'(z)}{f(z)} \, dz = N,$$

where N denotes the zeros of f in Ω, counted according to their multiplicities.

Proof. Since f is analytic, its zeros are isolated. Hence there are at most finitely many in $\overline{\Omega}$ – and thus in Ω. If $z_1, \ldots, z_m$ are the distinct zeros of f in Ω, it follows from remark (21.7 c) that

$$\deg(f, \Omega, 0) = \sum_{j=1}^{m} i_0(f, z_j). \tag{14}$$

Let $\epsilon > 0$ be so small that the discs $\overline{\mathbb{B}}_{\mathbb{C}}(z_j, \epsilon)$, $j = 1, \ldots, m$, are mutually disjoint. Then we deduce from the Taylor series of f around z_j that

$$f(z) = a_j(z)(z - z_j)^{m_j}, \quad \forall z \in \overline{\mathbb{B}}(z_j, \epsilon),$$

for some appropriate $m_j \in \mathbb{N}^*$ and some analytic function a_j for which we have $a_j(z) \neq 0$ for all $z \in \overline{\mathbb{B}}(z_j, \epsilon)$. Therefore

$$\frac{f'(z)}{f(z)} = \frac{m_j}{z - z_j} + \frac{a_j'(z)}{a_j(z)}, \quad \forall z \in \overline{\mathbb{B}}(z_j, \epsilon),$$

where the second summand is analytic in some neighborhood of z_j. Hence from the Cauchy integral formula (for $\Gamma_j := \partial\mathbb{B}(z_j, \epsilon)$ with ϵ appropriately reduced) we obtain

$$\frac{1}{2\pi i} \int_{\Gamma_j} \frac{f'(z) \, dz}{f(z)} = \frac{m_j}{2\pi i} \int_{\Gamma_j} \frac{dz}{z - z_j} = m_j.$$

By corollary (25.8),

$$i_0(f, z_j) = \deg(f, \mathbb{B}(z_j, \epsilon), 0) = \frac{1}{2\pi i} \int_{\Gamma_j} \frac{f'(z)}{f(z)} \, dz = m_j,$$

and hence $\deg(f, \Omega, 0) = \sum_{j=1}^{m} m_j = N$ by (14). Now the assertion follows from proposition (25.7) and corollary (25.8). If f has no zeros in Ω, then f'/f is analytic in Ω and then the assertion follows from Cauchy's integral formula, proposition (25.7) and corollary (25.8), as well as the solution property of the degree. $\qquad\square$

(25.11) Corollary. *(Rouché's theorem): Let $\Omega \subseteq \mathbb{R}^2$ be a bounded C^2-domain and let U be an open neighborhood of $\overline{\Omega}$. Moreover, assume that f and g are analytic functions on U satisfying*

$$|f(z) - g(z)| < |f(z)|, \quad \forall z \in \partial\Omega.$$

Then f and g have the same number of zeros (counted according to their multiplicities) in Ω.

Proof. For $0 \leq t \leq 1$ and $z \in \partial\Omega$ we have

$$|(1-t)f(z) + tg(z)| \geq |f(z)| - t|f(z) - g(z)| > 0.$$

Hence the assertion follows from the homotopy invariance of the degree and from proposition (25.10). $\qquad\qquad\square$

Qualitatively Rouché's theorem says that the total number of zeros of an analytic function on a bounded domain is stable with respect to analytic perturbations if in the process no zeros escape across the boundary.

It should be remarked that the assumptions "$\Omega \in C^2$" can be weakened substantially, e.g., by an approximation argument. For more general formulations and different proofs of Rouché's theorem as well as the "argument principle," we refer to books on complex analysis (e.g., Conway [1], Behnke-Sommer [1]). $\qquad\qquad\square$

As an application of the preceding results, we will now sharpen considerably the perturbation theorem (25.5) for the spectrum (in the finite dimensional case).

(25.12) Proposition. *Let $T_0 \in \mathcal{L}(E)$ and $\sigma(T_0) = \{\lambda_j \mid j = 1, \ldots, n\}$. Moreover, let m_j denote the multiplicity of the eigenvalue λ_j. For each $j = 1, \ldots, n$, let U_j be a bounded neighborhood of λ_j such that $\overline{U}_j \cap \overline{U}_k = \emptyset$ if $j \neq k$ and set $U := \bigcup_{j=1}^n U_j$. Finally, let V be a connected neighborhood of T_0 in $\mathcal{L}(E)$ such that $\sigma(T) \subseteq U$ for all $T \in V$ (cf. proposition (25.5)). Then*

$$\sigma(T) \cap U_j \neq \emptyset, \quad \forall T \in V, \ j = 1, \ldots, n,$$

and the sum of the multiplicities of the eigenvalues in $\sigma(T) \cap U_j$ equals m_j.

Proof. The eigenvalues of $T \in V$ are the zeros of the characteristic polynomial

$$f_T(z) = \det(z - T), \quad z \in \mathbb{C}.$$

If $z_1, \ldots, z_{n_j}$ are the eigenvalues of T in U_j, it follows from proposition (25.10) and the excision property of the degree that

$$\deg(f_T, U_j, 0) = N_j,$$

where N_j denotes the sum of the multiplicities of $z_1, \ldots, z_{n_j}$. One easily verifies that f_T is a continuous function of $T \in \mathcal{L}(E)$. Hence, based on the continuity of the degree, the function $T \mapsto \deg(f_T, U_j, 0)$ is locally constant. Since V is connected, $T \mapsto \deg(f_T, U_j, 0)$ is constant on V. Thus we have $N_j = m_j$ for $j = 1, \ldots, n$. $\qquad\qquad\square$

Questions of Stability

After these preliminaries we can easily prove the announced stability assertion. To simplify the presentation, we say that the $T(\lambda)$-periodic solution $u(\lambda)$ of $\dot{x} =$

$f(\lambda, t, x)$ is *strongly asymptotically Liapunov stable* [resp., *strongly unstable*] if all Floquet multipliers of the linearized equation

$$\dot{y} = D_3 f(\lambda, t, u(\lambda)(t))y \tag{15}$$

lie in the interior of the unit disc [resp., if (15) has a Floquet multiplier μ satisfying $|\mu| > 1$].

(25.13) Theorem. *Let $u(\lambda_0)$ be a strongly asymptotically Liapunov stable [resp., strongly unstable] $T(\lambda_0)$-periodic solution of the equation $\dot{x} = f(\lambda_0, t, x)$ [and suppose the linearized problem*

$$\dot{y} = D_3 f(\lambda_0, t, u(\lambda_0)(t))y$$

has no nontrivial $T(\lambda_0)$-periodic solution]. Then there exists a neighborhood V_0 of λ_0 in V such that the $T(\lambda)$-periodic solution $u(\lambda)$, obtained from theorem (25.2), is strongly asymptotically Liapunov stable [resp., strongly unstable] for every $\lambda \in V_0$.

Proof. For each $\lambda \in V$, let $M(\lambda)$ denote the monodromy operator of equation (15). We then deduce from the continuity theorem (8.3) and from $[(t, \lambda) \mapsto u(\lambda, t)] \in C(\mathbb{R} \times V, E)$ that

$$M \in C(V, \mathcal{L}(E)).$$

If $u(\lambda_0)$ is strongly asymptotically Liapunov stable, then $\sigma(M(\lambda_0)) \subseteq \mathbb{B}_{\mathbb{C}}$. Hence by proposition (25.5), there exists a complete neighborhood V_0 of λ_0 in V for which $\sigma(M(\lambda)) \subseteq \mathbb{B}_{\mathbb{C}}$.

If $u(\lambda_0)$ is strongly unstable, there exists some $\mu \in \sigma(M(\lambda_0))$ such that $|\mu| > 1$. Moreover, we may choose a neighborhood U of μ in $\mathbb{C}$ so that $\sigma(M(\lambda_0)) \cap U = \{\mu\}$ and $\overline{U} \cap \overline{\mathbb{B}}_{\mathbb{C}} = \emptyset$ hold. Then it follows from (25.12) that there exists a neighborhood V_0 of λ_0 in V satisfying $\sigma(M(\lambda)) \cap U \neq \emptyset$ for all $\lambda \in V$. Hence $u(\lambda)$ is strongly unstable for all $\lambda \in V_0$. $\qquad\qquad\square$

(25.14) Examples. (a) We consider the equation

$$\ddot{y} + \alpha \dot{y} + \beta y + \gamma y^3 + \delta \sin y = \epsilon \sin^3(\omega t) \tag{16}$$

from example (25.4 c), where $\lambda := (\alpha, \beta, \gamma, \delta, \epsilon, \omega) \in \mathbb{R}^6$ and $\lambda_0 := (\alpha_0, \beta_0, 0, \ldots, 0, \omega_0)$ are such that $\omega_0 \neq 0$ and either $\alpha_0 \neq 0$ and $\beta_0 \neq 0$, or $\beta_0 < 0$.

If we have $\alpha_0 > 0$ and $\beta_0 > 0$, then the zeros of the characteristic polynomial of the linearized equation

$$\ddot{y} + \alpha_0 \dot{y} + \beta_0 y = 0 \tag{17}$$

lie in the open left half plane (cf. example (14.7)). It then follows from the spectral theorem (lemma (19.3)) that all Floquet multipliers of the system, corresponding to (17), lie in the interior of the unit disc. (For an autonomous system $\dot{x} = Ax$, the Floquet multipliers of a T-periodic solution are precisely the eigenvalues of e^{TA}.) In this case, the ω_0-periodic solution $u(\lambda_0) = 0$ of (16) is therefore strongly asymptotically Liapunov stable. If $\beta_0 < 0$, the system corresponding to (17) has a proper saddle (example (14.7), case 5). Therefore the λ_0-periodic solution $u(\lambda_0) = 0$ of (16) is strongly unstable. Hence from theorem (25.13) we deduce the following:

There exist a neighborhood V of λ_0 in $\mathbb{R}^6$ and a unique ω-periodic solution $u(\lambda)$ of (16) with the following properties:

(i) $[(t, \lambda) \mapsto u(\lambda)(t)] \in C^\infty(\mathbb{R} \times V, \mathbb{R})$ and $u(\lambda_0) = 0$;
(ii)   for every $\lambda \in V$, $u(\lambda)$ is (strongly) asymptotically Liapunov stable if $\alpha_0 > 0$ and $\beta_0 > 0$, and (strongly) unstable if we have $\beta_0 < 0$.

(b) The T-periodic equation

$$\ddot{y} + \omega^2 y = \lambda f(\lambda, y, \dot{y}, t)$$

of example (25.4 d) has the linearization $\ddot{y} + \omega^2 y = 0$ at $\lambda_0 = 0$, which has the Floquet multipliers $e^{\pm i\omega T}$. Thus theorem (25.13) cannot be applied to this case. $\square$

We now consider the autonomous case

$$f \in C^k(\Lambda \times X, E), \quad k \geq 1,$$

and assume that $\gamma(\lambda_0)$ is a hyperbolic periodic orbit of $\dot{x} = f(\lambda_0, x)$ (cf. section 23). With the aid of the Poincaré operator, it is not difficult to prove analogues of theorems (25.2) and (25.13). To formulate and prove the continuous dependence of the orbits $\gamma(\lambda)$ on the parameter $\lambda \in \Lambda$ we need the concept of Hausdorff metric.

Let $M = (M, d)$ be a metric space. We define the *Hausdorff distance* of two subsets A and B of M by

$$\delta(A, B) := \max\{ \sup_{x \in A} d(x, B), \ \sup_{y \in B} d(y, A)\}.$$

It is not difficult to see that δ is a metric on the set of all bounded, closed subsets of M. For this reason, we simply speak of the *Hausdorff metric* on 2^M (although for

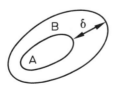

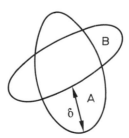

arbitrary, nonclosed sets $A, B \in 2^M$, we cannot deduce $A = B$ from $\delta(A, B) = 0$ and $\delta(A, B) = \infty$ may hold).

(25.15) Theorem. *Let $\gamma(\lambda_0)$ be a noncritical, hyperbolic periodic orbit of the equation $\dot{x} = f(\lambda_0, x)$. Then there exist neighborhoods, V of λ_0 in Λ and W of $\gamma(\lambda_0)$ in X, satisfying:*

(i)   *for every $\lambda \in V$ there exists a unique periodic orbit $\gamma(\lambda)$ of $\dot{x} = f(\lambda, x)$ in W;*

(ii)   *$\gamma(\lambda)$ is a noncritical hyperbolic orbit and possesses the same stability features as $\gamma(\lambda_0)$;*

(iii)   *the mapping*

$$V \to 2^X, \quad \lambda \mapsto \gamma(\lambda),$$

is continuous with respect to the Hausdorff metric.

Proof. Let $x_0 \in \gamma(\lambda_0)$ be arbitrary and assume that

$$E = \mathbb{R}f(\lambda_0, x_0) \oplus H$$

is the decomposition of E which reduces the monodromy operator $U(\lambda_0)$ of

$$\dot{y} = D_2 f(\lambda_0, t \cdot x_0)y$$

(cf. (19) of section (23)). For the corresponding Poincaré operator $\pi(\lambda_0, \cdot) :$ $H_{x_0} \to H_{x_0}$ (where $\pi(\lambda_0, \cdot)$ is only defined near x_0 and we have set $H_{x_0} := x_0 + H$) we then have:

x_0 is a fixed point of $\pi(\lambda_0, \cdot)$ and the linearization $D_2 \pi(\lambda_0, x_0)$ is a hyperbolic automorphism on H (cf. proposition (23.7)).

Since $f \in C^k(\Lambda \times X, E)$, it follows from the proof of lemma (23.4) that there exist neighborhoods, V of λ_0 in Λ and S of $x_0 \in H_{x_0}$, such that S is a local section of the flow induced by $f(\lambda, \cdot), \lambda \in V$, and the Poincaré operator satisfies

$$\pi \in C^k(V \times S, H_{x_0}).$$

For this one simply must take into account the dependence on λ when applying the implicit function theorem in the proof of lemma (23.4).

Now, $x \in S$ lies on a periodic orbit $\gamma(\lambda)$ of $f(\lambda, \cdot)$ if and only if x is a fixed point of $\pi(\lambda, \cdot)$, that is, if and only if

$$g(\lambda, x) := x - \pi(\lambda, x) = 0.$$

Since $\gamma(\lambda_0)$ is a noncritical hyperbolic orbit, it follows that $D_2 g(\lambda_0, x_0) = I - D_2 \pi(\lambda_0, x_0) \in \mathcal{GL}(H)$. Hence we deduce from $g(\lambda_0, x_0) = 0$ and the implicit function theorem that there exist neighborhoods, V_0 of λ_0 in V and W_0 of x_0 in S, as well as a function $h \in C^k(V_0, W_0)$ such that

$$(\lambda, x) \in V_0 \times W_0 \text{ and } g(\lambda, x_0) = 0 \quad \Leftrightarrow \quad \lambda \in V_0 \text{ and } x = h(\lambda).$$

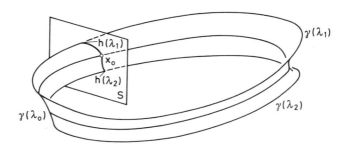

By shrinking V, we may assume that $V = V_0$. Hence for every $\lambda \in V$, there exists a unique periodic orbit $\gamma(\lambda)$ corresponding to $f(\lambda, \cdot)$ which meets S in W_0, namely, the orbit which passes through the point $h(\lambda)$. We now choose an arbitrary neighborhood W of $\gamma(\lambda_0)$ in X such that it contains all periodic orbits $\gamma(\lambda)$, $\lambda \in V$, and so that $W \cap H = W_0$. Then clearly (i) is satisfied.

By (the modified proof of) lemma(23.4), the Poincaré operator $\pi(\lambda, x)$ has the form $\tau(\lambda, x) \cdot x$ and "return time" $\tau \in C^k(V \times S, \mathbb{R})$. Hence for the minimal period $T(\lambda)$ of the orbit $\gamma(\lambda)$ we have

$$T(\cdot) = \tau(\lambda, h(\cdot)) \in C^k(V, \mathbb{R}), \tag{18}$$

and, since $T(\lambda_0) > 0$, we may therefore assume that $T(\lambda) > 0$ for all $\lambda \in V$ (by possibly shrinking V). Consequently, $\gamma(\lambda)$ is not critical for any $\lambda \in V$.

For every $\lambda \in V$, we let $u(\lambda)$ denote the $T(\lambda)$-periodic solution of $\dot{x} = f(\lambda, x)$ satisfying $u(\lambda)(0) = h(\lambda) \in W_0$. Then the Floquet multipliers of the periodic orbit $\gamma(\lambda)$ are the eigenvalues of the monodromy operator corresponding to the equation

$$\dot{y} = D_2 f(\lambda, u(\lambda)(t))y.$$

By (18), we now obtain the remaining assertions of (ii) by means of the perturbation theorems (25.5) and (25.12), similar to the proof of theorem (25.13).

Since $[(t, \lambda) \mapsto u(\lambda)(t)] \in C^k(\mathbb{R} \times V, E)$, we deduce from remark (25.3) that $[\lambda \mapsto v(\cdot, \lambda)] \in C^k(V, BC^k(\mathbb{R}, E))$, where $v(t, \lambda) := u(tT(\lambda), \lambda)$. This implies (iii).   □

(25.16) Remarks. (a) It is worth noting that we have proved the following, sharper continuity result:

If for every $\lambda \in V$, $u(\lambda)$ denotes the unique $T(\lambda)$-periodic solution of $\dot{x} = f(\lambda, \cdot)$ satisfying $u(\lambda)(0) = h(\lambda) \in W_0$, which exists by theorem (25.15), we have: $T(\cdot) \in C^k(V, \mathbb{R})$ and $[(t, \lambda) \mapsto u(\lambda)(t)] \in C^k(\mathbb{R} \times V, E)$. For $v(t, \lambda) := u(\lambda)(tT(\lambda))$ we even have $[\lambda \mapsto v(\cdot, \lambda)] \in C^k(V, BC^k(\mathbb{R}, E))$.

(b) *Let $\gamma(\lambda_0)$ be a noncritical periodic orbit of $\dot{x} = f(\lambda_0, x)$ and assume that 1 is a simple Floquet multiplier of γ. Then there exist neighborhoods, V of λ_0 in Λ and W of $\gamma(\lambda_0)$ in X, satisfying conditions* (i) *and* (iii) *of theorem* (25.15) *and such that $\gamma(\lambda)$ is noncritical for all $\lambda \in V$. Moreover, the continuity assertion* (a) *holds.*

This follows from the proof of theorem (25.15) since the assumption already implies that $D_2 g(\lambda_0, x_0) \in \mathcal{GL}(H)$. $\square$

(25.17) Example. Let $\Lambda \subseteq F$ be a neighborhood of 0 and suppose $g \in C^k(\Lambda \times \mathbb{R}^2, \mathbb{R}^2)$. Moreover, assume that

$$g(0, x, y) := (xh(x^2 + y^2), yh(x^2 + y^2)),$$

where $h(\xi) = 1 - \xi$, and consider the system

$$\dot{x} = f^1(\lambda, x, y) := -y + g^1(\lambda, x, y)$$
$$\dot{y} = f^2(\lambda, x, y) := x + g^2(\lambda, x, y). \tag{19$_\lambda$}$$

If $\lambda = 0$, then $(19)_0$ becomes in polar coordinates:

$$\dot{r} = rh(r^2), \quad \dot{\varphi} = 1.$$

Hence $(19)_0$ has exactly one noncritical periodic orbit, namely $\gamma(0) := \mathbb{S}^1$.

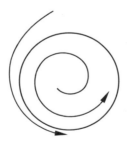

This orbit has minimal period 2π, it is hyperbolic and asymptotically stable. The last assertion is easily seen by calculating div $f(t \cdot x)$ for $x \in \gamma(0)$. For this we find that

$$\text{div } f = \frac{\partial f}{\partial r} + \frac{1}{r} f = 2h(r^2) + 2r^2 h'(r^2) = 2(h(r^2) - r^2) = 2(1 - 2r^2).$$

Hence div $f = -2$ along the orbit $\gamma(0)$. From this we obtain, as in the proof of proposition (23.9), that the Floquet multiplier of $\gamma(0)$ which is distinct from 1, has the value $e^{-4\pi} < 1$. It now follows from theorem (25.15) that there exists a neighborhood V of 0 in Λ such that for every $\lambda \in V$, the system $(19)_\lambda$ has a noncritical, asymptotically stable periodic orbit $\gamma(\lambda)$ and that $\gamma(\lambda) \to \gamma(0) = \mathbb{S}^1$ (in the Hausdorff metric). In addition, there exists a neighborhood U of $\gamma(0)$ such that $\gamma(\lambda)$ is the only periodic orbit of $(19)_\lambda$ in U. Finally, for the minimal period $T(\lambda)$ of $\gamma(\lambda)$ we have: $T(\cdot) \in C^k(V, \mathbb{R})$ and $T(0) = 2\pi$. $\square$

Problems

1. Suppose the assumptions of theorem (25.2) are satisfied and let $\dim(\Lambda) = 1$. Show that there exists a maximal interval $\Lambda_0 \subseteq \Lambda$ for which $\gamma_0 \in \Lambda_0$ and such that Λ_0 is open and $u(\lambda_0)$ "can be extended to all of Λ_0."

2. Let Y be a Banach space over $\mathbb{K}$. For $T \in \mathcal{L}(Y)$, we define the *spectral radius of T*, $r(T)$, by

$$r(T) := \lim_{k \to \infty} \|T^k\|^{1/k}.$$

Show that:
(i)   This definition makes sense (i.e., the limit exists) and we have: $r(T) \leq \|T\|$.
(ii) $r(T)$ is independent of the particular norm.
(iii)   If $\dim(Y) < \infty$, then

$$r(T) = \max\{|\lambda| \mid \lambda \in \sigma(T)\}.$$

(*Hint* for (iii): Consider (ii) and the proof of lemma (13.1).)

3. Let Y be a complex Banach space. For each $T \in \mathcal{L}(Y)$, we define the *resolvent of T*, $R(\cdot, T)$, by

$$R(\cdot, T) : \sigma(T) \to \mathcal{L}(Y), \quad \lambda \mapsto (\lambda - T)^{-1}.$$

Show that:
(i)   The resolvent is an analytic function of $\lambda \in \rho(T)$, that is, locally it can be represented by a power series in $\mathcal{L}(E)$.
(ii)   The *resolvent equation*

$$R(\lambda, T) - R(\mu, T) = (\mu - \lambda)R(\lambda, T)R(\mu, T), \quad \forall \lambda, \mu \in \sigma(T),$$

is satisfied.

(*Hint* for (i): Consider the proof of proposition (25.5).)

26. Bifurcation Problems

We again let E, F and G denote real finite dimensional Banach spaces and assume that $X \subseteq E$ and $\Lambda \subseteq F$ are open and that

$$f \in C^k(\Lambda \times \mathbb{R} \times X, E)$$

for some $k \geq 1$.

In this section we will study the parameter-dependent differential equation

$$\dot{x} = f(\lambda, t, x)$$

in situations in which the continuation method does not apply. To this end, we assume that a family $u(\lambda)$ of $T(\lambda)$-periodic solutions is already known explicitly. More precisely, let

$$T(\cdot) \in C^k(\Lambda, \mathbb{R}_+^*)$$

and

$$u \in C^k(\mathbb{R} \times \Lambda, E)$$

be given such that for every $\lambda \in \Lambda$, $u(\lambda) := u(\cdot, \lambda)$ is a $T(\lambda)$-periodic solution of the $T(\lambda)$-periodic differential equation $\dot{x} = f(\lambda, t, x)$. Then the function $\lambda \mapsto u(\lambda)$ in $BC(\mathbb{R}, E)$ describes a continuum of periodic solutions (namely, the graph of u) – in the simplest case, a curve. A point $(\lambda_0, u(\lambda_0))$ is then called a *bifurcation point of periodic solutions* if every neighborhood of $(\lambda_0, u(\lambda_0))$ contains periodic solutions not belonging to this continuum.

Before we study this problem in more detail, we will first introduce a normalization. For this we set

$$\hat{f}(\lambda, t, x) := f(\lambda, t, x + u(\lambda)(t)) - f(\lambda, t, u(\lambda)(t))$$

and

$$\tilde{f}(\lambda, t, x) := (T(\lambda)/2\pi)\hat{f}(\lambda, tT(\lambda)/2\pi, x)$$

for all $(\lambda, t) \in \Lambda \times \mathbb{R}$ and all $x \in E$ for which the right-hand side is defined. Then the function $\tilde{f}$ is 2π-periodic in t and we have $\tilde{f}(\lambda, t, 0) = 0$ for all $(\lambda, t) \in \Lambda \times \mathbb{R}$. Moreover, $t \mapsto \tilde{u}(t)$ is a solution of the equation $\dot{y} = \tilde{f}(\lambda, t, y)$ if and only if $t \mapsto \tilde{u}(2\pi t/T(\lambda)) - u(\lambda)(t)$ is a solution of $\dot{x} = f(\lambda, t, x)$. Finally, let $\lambda_0 \in \Lambda$ be a possible bifurcation point and assume that V is a compact neighborhood of λ_0 in Λ. Then there exists some $T_0 > 0$ such that $T(V) \subseteq [0, T_0]$, and, since the function $(\lambda, t) \mapsto u(\lambda, t) := u(\lambda)(t)$ is continuous, we have that $u(V \times [0, T_0])$ is compact and is contained in X. Therefore $\text{dist}(u(V \times [0, T_0]), X^c) > 0$ and for some $r > 0$ we have $\mathbb{B}_E(u(\lambda, t), r) \subseteq X$ for all $\lambda \in V$ and $t \in \mathbb{R}$. This shows that the function $\tilde{f}$ is defined on $V \times \mathbb{R} \times \mathbb{B}_E(0, r)$. Since we are interested in local properties (i.e., in some neighborhood of $(\lambda_0, u(\lambda_0, 0))$) in $\Lambda \times E$, and since f is a C^k-function on $V \times \mathbb{R} \times \mathbb{B}_E(0, r)$, *we may assume, without loss of generality, that*

$$\begin{array}{lll} \text{(i)} & f \in C^k(\Lambda \times \mathbb{R} \times X, E), & k \geq 1; \\ \text{(ii)} & f \text{ is } 2\pi\text{-periodic in } t; & \quad (1) \\ \text{(iii)} & f(\cdot, \cdot, 0) = 0. & \end{array}$$

From theorem (25.2) we immediately obtain a *necessary condition* for $\lambda_0 \in \Lambda$ to be a *"point of bifurcation from the trivial solution."*

(26.1) Proposition. *Assume that $(\lambda_0, 0)$ is a bifurcation point of 2π-periodic solutions of the equation $\dot{x} = f(\lambda, t, x)$. Then the linearized equation*

$$\dot{y} = D_3 f(\lambda_0, t, 0)y \qquad (2)$$

possesses nontrivial 2π-periodic solutions.

Proof. If this condition is not satisfied, then, by theorem (25.2) and remark (25.3), there exists a neighborhood V of λ_0 in Λ and a neighborhood U of 0 in $BC(\mathbb{R}, E)$ such that the equation $\dot{x} = f(\lambda, t, x)$ has a unique 2π-periodic solution $u(\lambda)$ in U for all $\lambda \in V$. Hence we must have $u(\lambda) = 0$ for all $\lambda \in V$ and so $(\lambda_0, 0)$ is not a bifurcation point. $\qquad\square$

The condition in proposition (26.1) is equivalent to the stipulation that 1 is a Floquet multiplier of the linearized equation (2). That this condition is not sufficient for $(\lambda_0, 0)$ to be a bifurcation point follows, for instance, from theorem (25.15).

Again, we let $u(\cdot, 0, \xi, \lambda)$ denote the global solution of the IVP

$$\dot{x} = f(\lambda, t, x), \quad x(0) = \xi,$$

and we set $g(\lambda, \xi) := \xi - u(2\pi, 0, \xi, \lambda)$. By shrinking X, if necessary, we may assume, as in the proof of theorem (25.2), that

$$g \in C^k(\Lambda \times X, E),$$

and, by remark (20.2), ξ_0 is a zero of $g(\lambda, \cdot)$ if and only if $u(\cdot, 0, \xi_0, \lambda)$ is a 2π-periodic solution of $\dot{x} = f(\lambda, t, x)$. Using assumption (1. iii), we also have

$$g(\cdot, 0) = 0.$$

The bifurcation problem for periodic solutions of the equation $\dot{x} = f(\lambda, t, x)$ is therefore equivalent to the bifurcation problem for "small solutions" of the parameter-dependent equation $g(\lambda, \xi) = 0$. Both problems are special cases of a more general situation, which we will examine in more detail.

Let $\mathbb{E}, \mathbb{F}$ and $\mathbb{G}$ be arbitrary Banach spaces and assume that $X \subseteq \mathbb{E}$ and $\Lambda \subseteq \mathbb{F}$

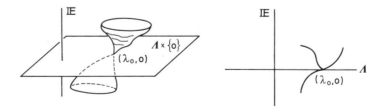

are open and $0 \in X$. Moreover, let

$$\Phi : \Lambda \times X \to \mathbb{G} \quad \text{be such that} \quad \Phi(\lambda, 0) = 0, \quad \forall \lambda \in \Lambda.$$

Then $(\lambda_0, 0) \in \Lambda \times X$ is said to be a *bifurcation point* (or λ_0 is called a *bifurcation point with respect to the trivial solution*) of the equation $\Phi(\lambda, x) = 0$ if every neighborhood of $(\lambda_0, 0)$ in $\Lambda \times \mathbb{X}$ contains a solution (λ, x) of $\Phi(\lambda, x) = 0$ such that $x \neq 0$.

(26.2) Remark. One easily verifies that

$$C_{2\pi}^k(E) := \{u \in BC^k(\mathbb{R}, E) \mid u(t + 2\pi) = u(t), \quad \forall t \in \mathbb{R}\}$$

is a closed vector subspace of $BC^k(\mathbb{R}, E)$ for each $k \in \mathbb{N}$. Since $BC^k(\mathbb{R}, E)$ is a Banach space by the theorem about the differentiability of uniformly convergent sequences of differentiable functions, it follows that $C_{2\pi}^k(E)$ is also a Banach space. We now set

$$\mathbb{E} := C_{2\pi}^1(E), \quad \mathbb{F} := F, \quad \mathbb{G} := C_{2\pi}(E) := C_{2\pi}^0(E),$$

as well as $\mathbf{\Lambda} := \Lambda$ and

$$\mathbb{X} := \{u \in \mathbb{E} \mid u(\mathbb{R}) \subseteq X\}.$$

Then it is easily seen that $\mathbb{X}$ is open in $\mathbb{E}$ (cf. lemma (2.4)). Finally, we set

$$\Phi(\lambda, u)(t) := Du(t) - f(\lambda, t, u(t)), \quad \forall t \in \mathbb{R}.$$

Then

$$\Phi \in C^k(\mathbf{\Lambda} \times \mathbb{X}, \mathbb{G}) \quad \text{and} \quad \Phi(\cdot, 0) = 0.$$

Moreover, (λ_0, u_0) is a solution of the equation $\Phi(\lambda, u) = 0$ if and only if u_0 is a 2π-periodic solution of the differential equation $\dot{x} = f(\lambda_0, t, x)$. Consequently, the bifurcation problem for periodic solutions is – as claimed – a special case of the abstract problem. $\square$

The following, necessary criterion for the existence of bifurcation points is analogous to (more precisely: a generalization of) proposition (26.1).

(26.3) Proposition. *Let* $\Phi \in C^1(\mathbf{\Lambda} \times \mathbb{X}, \mathbb{G})$. *If* $(\lambda_0, 0)$ *is a bifurcation point of the equation* $\Phi(\lambda, x) = 0$, *then* $D_2\Phi(\lambda_0, 0) \in \mathcal{L}(\mathbb{E}, \mathbb{G})$ *is not bijective.*

Proof. If $D_2\Phi(\lambda_0, 0)$ is bijective, then Banach's open mapping theorem (e.g., Yosida [1]) implies that $D_2\Phi(\lambda_0, 0)$ is a topological isomorphism. Hence, by the implicit function theorem (e.g., Dieudonné [1]), there exists a neighborhood $V \times U$ of $(\lambda_0, 0)$ in $\mathbf{\Lambda} \times \mathbb{X}$ such that the equation $\Phi(\lambda, x) = 0$ has a unique solution in U for every $\lambda \in V$. Therefore $(\lambda_0, 0)$ cannot be a bifurcation point. $\square$

(26.4) Corollary. *Assume that* $\mathbb{E} = \mathbb{G}$ *and* $\dim \mathbb{E} < \infty$. *If* $(\lambda_0, 0)$ *is a bifurcation point of the equation* $\Phi(\lambda, x) = 0$, *then* 0 *is an eigenvalue of* $D_2\Phi(\lambda_0, 0) \in \mathcal{L}(\mathbb{E})$, *that is,* $\ker[D_2\Phi(\lambda_0, 0)] \neq \{0\}$.

For the special case $\mathbb{E} = \mathbb{G}$ and $\dim \mathbb{E} < \infty$, it is possible to give an important sufficient condition for the existence of bifurcation points with the aid of the Brouwer degree. This was first discovered by Krasnosel'skii.

(26.5) Theorem. *Let $\Lambda \subseteq \mathbb{R}$ and $\Phi \in C(\Lambda \times X, E)$, and assume that $\lambda_0 \in \Lambda$ and $\epsilon > 0$ are such that the following holds:*

(i) $\Phi(\cdot, 0) = 0$;

(ii)   *for every $\lambda \in [\lambda_0 - \epsilon, \lambda_0 + \epsilon] \setminus \{\lambda_0\} =: \dot{J}$ there exists*
       a neighborhood $J_\lambda \times U_\lambda$ of $(\lambda, 0)$ in $\dot{J} \times X$ such that $\qquad$ (3)
       0 is the only zero of $\Phi(\mu, \cdot)$ in U_λ for all $\mu \in J_\lambda$.

Then $(\lambda_0, 0)$ is a bifurcation point of the equation $\Phi(\lambda, x) = 0$ if the local index

$$i_\Phi(\lambda) := i_0(\Phi(\lambda, \cdot), 0)$$

satisfies

$$i_\Phi(\lambda_0 - \epsilon) \neq i_\Phi(\lambda_0 + \epsilon). \qquad (4)$$

Proof. Assume that $(\lambda_0, 0)$ is not a bifurcation point. Then for every $\lambda \in J := \overline{\dot{J}}$, there exists a neighborhood $J_\lambda \times U_\lambda$ of $(\lambda, 0)$ in $\Lambda \times X$ satisfying $\Phi(\lambda, x) \neq 0$ for all $(\lambda, x) \in J_\lambda \times (\overline{U}_\lambda \setminus \{0\})$. By the compactness of J, we can find finitely many points $\lambda_1, \ldots, \lambda_m \in J$ so that $J_{\lambda_1}, \ldots, J_{\lambda_m}$ cover the interval J.

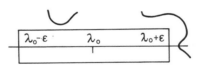

Now set $U := \cap_{j=1}^m U_{\lambda_j}$. Then $\Phi(\lambda, x) = 0$ has no solutions in $\overline{J} \times \overline{U}$ satisfying $x \neq 0$. Hence we deduce from the definition of the local index and the homotopy invariance of the degree that

$$i_\Phi(\lambda_0 - \epsilon) = \deg(\Phi(\lambda_0 - \epsilon, \cdot), U, 0)$$
$$= \deg(\Phi(\lambda_0 + \epsilon, \cdot), U, 0) = i_\Phi(\lambda_0 + \epsilon),$$

which contradicts our assumption. $\qquad\qquad\square$

(26.6) Corollary. *Let $\Lambda \subseteq \mathbb{R}$ and $\Phi \in C^{0,1}(\Lambda \times X, E)$, and assume that*

(i) $\Phi(\cdot, 0) = 0$;

(ii)   *there exists some $\lambda_0 \in \Lambda$ and some $\epsilon > 0$ such that $\ker[D_2\Phi(\lambda, 0)] = \{0\}$ for all $0 < |\lambda - \lambda_0| \leq \epsilon$.*

If the function

$$\lambda \mapsto \det[D_2\Phi(\lambda, 0)] \qquad (5)$$

changes sign at λ_0, then $(\lambda_0, 0)$ is a bifurcation point of the equation $\Phi(\lambda, x) = 0$.

Proof. Since $D_2\Phi(\lambda, 0) \in \mathcal{GL}(E)$ for all $0 < |\lambda - \lambda_0| \leq \epsilon$, it follows from proposition (21.8) that 0 is an isolated zero of $\Phi(\lambda, \cdot)$ and

$$i_\Phi(\lambda) = i_0(\Phi(\lambda, 0), 0) = i_0(D_2\Phi(\lambda, 0), 0)$$
$$= \text{sign det}[D_2\Phi(\lambda, 0)]$$

for all $0 < |\lambda - \lambda_0| \leq \epsilon$. Hence (5) implies condition (4). $\square$

(26.7) Remark. If $n(\lambda)$ denotes the number of negative real eigenvalues of $D_2\Phi(\lambda, 0)$ (counted according to their multiplicities), then

$$\text{sign det}[D_2\Phi(\lambda, 0)] = (-1)^{n(\lambda)}$$

(cf. the proof of proposition (21.10)). Hence (4) is equivalent to

$$(\lambda_0 + \epsilon) - n(\lambda_0 - \epsilon) \equiv 1 (\text{mod } 2).$$ $\square$

(26.8) Examples. (a) Let $f \in C^1(\Lambda \times X, E)$ be such that $f(\cdot, 0) = 0$. Then for every $\lambda \in \Lambda$, $x = 0$ is a critical point of the equation $\dot{x} = f(\lambda, x)$. If $0 \notin \sigma(D_2 f(\lambda_0, 0))$, the implicit function theorem implies that for all λ in some neighborhood of λ_0, the zero solution is an isolated critical point of $\dot{x} = f(\lambda, x)$. If $(\lambda_0, 0)$ is a bifurcation point of the equation $f(\lambda, x) = 0$, then $(\lambda_0, 0)$ is a *point of bifurcation for critical points* of the equation $\dot{x} = f(\lambda, x)$ (or of the parameter-dependent flow induced by $f(\lambda, \cdot)$). It is useful to have a geometric picture of the qualitative behavior of the flow as the parameter changes, as will be done in the following simple cases.

(b) Let $\Lambda = \mathbb{R}$ and $E = \mathbb{R}^2$, and set

$$f(\lambda, x, y) := (\lambda a x - a x^3, by),$$

where $a, b \in \mathbb{R}^*$. Then we clearly have $f(\lambda, 0, 0) = 0$ for all $\lambda \in \mathbb{R}$ and

$$D_2 f(\lambda, 0, 0) = \begin{bmatrix} \lambda a & 0 \\ 0 & b \end{bmatrix}.$$

Hence $0 \notin \sigma(D_2 f(\lambda, 0, 0))$ if $\lambda \neq 0$, and the function $\lambda \mapsto \det D_2 f(\lambda, 0, 0)$ changes sign at $\lambda = 0$. Consequently $\lambda_0 = 0$ is a point of bifurcation from the trivial solution of the equation $f(\lambda, x, y) = 0$, hence $\lambda_0 = 0$ is a bifurcation point of critical points of the planar flow induced by $f(\lambda, \cdot, \cdot)$. Here the critical point $(0, 0)$ of $f(\lambda, \cdot, \cdot)$ changes its stability properties as λ passes through zero (for instance, a saddle changes into a source).

More precise information about the bifurcation is obtained if one observes that we deal with a gradient flow. Setting

$$V_\lambda(x, y) := a\left(\frac{x^4}{4} - \lambda\frac{x^2}{2}\right) - b\frac{y^2}{2},$$

we, of course, have $f(\lambda, \cdot, \cdot) = -\text{grad } V_\lambda$. Based on this fact, we obtain the following assertions (cf. example (18.11 a)):

(α) $a > 0, b > 0$.   In this case we have a saddle if $\lambda < 0$, which, as λ passes through zero, becomes a source, from which two saddles have separated.   The corresponding

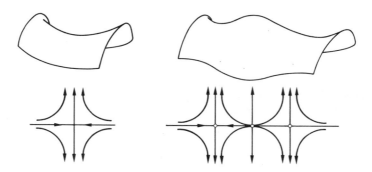

bifurcation diagram (i.e., the set of zeros of f in a neighborhood of the bifurcation point) has the following qualitative characteristics.

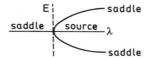

(β) $a > 0$, $b < 0$.   In this case a sink splits off into a saddle and two sinks. The corresponding bifurcation diagram again has – as it also has in cases (γ) and (δ) – the form of a "fork."

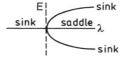

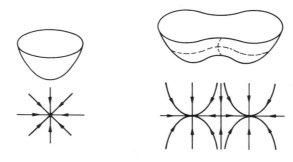

(γ) $a < 0$, $b > 0$.   In this case V_λ is obtained by reflecting the Liapunov function corresponding to the case (β) in the (x, y) plane. Hence a source is split off into a saddle and two sources.

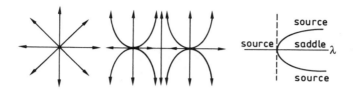

(δ) $a < 0$, $b < 0$.   Now the vector field $f(\lambda, \cdot, \cdot)$ is the reverse of the vector field in case (α). Hence the phase portrait is obtained from that in case (α) by reversing the arrows. Consequently, a saddle is split off into a sink and two saddles.

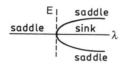

(c) We again let $\Lambda = \mathbb{R}$ and $E = \mathbb{R}^2$, while

$$f(\lambda, x, y) := (\lambda a x - a x^3, \lambda b y - b y^3),$$

where $a, b \in \mathbb{R}^*$. Then

$$D_2 f(\lambda, 0, 0) = \begin{bmatrix} \lambda a & 0 \\ 0 & \lambda b \end{bmatrix}.$$

So if $\lambda \neq 0$, then $0 \notin \sigma(D_2 f(\lambda, 0, 0))$, i.e., if $\lambda \neq 0$, then $(0, 0)$ is an isolated critical point of the flow induced by $f(\lambda, \cdot, \cdot)$. However, the function $\lambda \mapsto \det D_2 f(\lambda, 0, 0) = \lambda^2 ab$ does not change its sign at $\lambda = 0$. Therefore corollary (26.6) is not applicable.

If

$$V_\lambda(x, y) := a \left(\frac{x^4}{4} - \lambda \frac{x^2}{2} \right) + b \left(\frac{y^4}{4} - \lambda \frac{y^2}{2} \right),$$

we again have $f(\lambda, \cdot, \cdot) = -\mathrm{grad}\, V_\lambda$. It follows that we can, also in this case, study the qualitative features of the phase portrait as the parameter λ is changed.

(α) $a > 0$, $b > 0$.   Now a sink splits off into a source, four saddles and four sinks.

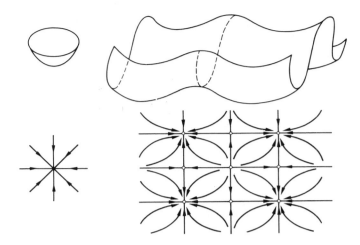

The bifurcation diagram has the (schematic) form:

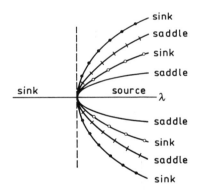

There are four smooth curves which go through the origin in $\mathbb{R} \times E$ and cross the λ-axis transversely. In fact, the set of all zeros of the function f consists exactly of the parabolas given by the parametrization $\mathbb{R} \to \Lambda \times \mathbb{R}^2$

$$t \mapsto (t^2, 0, t), \quad t \mapsto (t^2, t, 0)$$

and

$$t \mapsto (t^2, t, t), \quad t \mapsto (t^2, t, -t)$$

which lie in different planes, as well as the λ-axis $\mathbb{R} \times \{(0, 0)\}$. Projecting the set of zeros of f parallel to the λ-axis and into the E-plane, we obtain a family of straight lines which gives us a better understanding of the preceding bifurcation diagram.

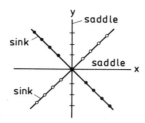

(β) $a > 0$, $b < 0$.   The saddle, which exists if $\lambda < 0$, splits off into five saddles, two sinks and two sources, as is easily deduced from the form of the Liapunov function V.

The bifurcation diagram, again, has the (schematic) form of a nine-pronged fork.

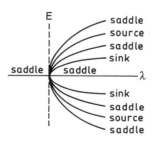

(γ) The case $a < 0$, $b < 0$ [resp. $a < 0$, $b > 0$] is obtained from the case (α) [resp. (β)] by reversing the arrows. Hence in the corresponding bifurcation diagrams the sinks will be replaced by sources and conversely.

(d) Let $\Lambda = \mathbb{R}$ and $E = \mathbb{R}^2$, and assume that

$$f(\lambda, x) := \lambda x - |x|^2 x.$$

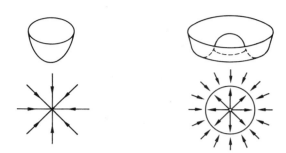

Then $D_2 f(\lambda, 0) = \lambda \, id_E$. So if $\lambda \neq 0$, then 0 is an isolated critical point of the vector field $f(\lambda, \cdot)$ and $\lambda \mapsto \det D_2 f(\lambda, 0) = \lambda^2$ does not change its sign at $\lambda = 0$.   Since $f(\lambda, \cdot) = -\mathrm{grad}\, V_\lambda$, where $V_\lambda(x) = \frac{|x|^4}{4} - \lambda \frac{|x|^2}{2}$, one easily verifies that $(0,0) \in \mathbb{R} \times E$ is a bifurcation point of critical points for $\dot{x} = f(\lambda, x)$. If $\lambda < 0$, then 0 is a sink which, for $\lambda > 0$, splits off into a source and a continuum of critical points which attracts all points in $E \setminus \{0\}$. The corresponding bifurcation diagram shows a highly degenerate behavior which, of course, is based on the symmetry inherent to the problem.

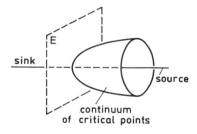

(e) If in the examples above we replace the parameter λ by $-\lambda$, we obtain "*bifurcation to the left*" instead of "*bifurcation to the right*," i.e., the bifurcation diagrams arise from the ones above by a reflection in the hyperplane $\{0\} \times E$.

(f) Let $\Lambda = \mathbb{R}$ and $E = \mathbb{R}^2$, and assume that

$$f(\lambda, x, y) := (\lambda x - x^4, -y).$$

Then $f(\lambda, \cdot, \cdot) = -\mathrm{grad}\, V_\lambda$, where

$$V_\lambda(x, y) = \frac{x^5}{5} - \lambda \frac{x^2}{2} + \frac{y^2}{2}.$$

In addition, $\det D_2 f(\lambda, 0) = \lambda$. Therefore $\lambda_0 = 0$ is a point of bifurcation from the trivial solution of the equation $f(\lambda, x, y) = 0$. In this case the bifurcation diagram consists of the two curves $\Lambda \times \{0\}$ and $t \mapsto (t^3, t, t^3)$. Hence we have *two-sided bifurcation*. By considering the "*potential surfaces*" graph(V_λ), one easily finds the indicated distribution of sinks and saddles.

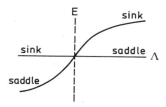

(g) For $\lambda \in \Lambda := \mathbb{R}$ we consider the scalar equation

$$\ddot{z} = \lambda z - z^3,$$

which is equivalent to the system $(x, y)^{\cdot} = f(\lambda, x, y)$, where

$$f(\lambda, x, y) := (y, \lambda x - x^3).$$

Evidently, for every $\lambda \in \mathbb{R}$, this equation possesses the trivial solution and

$$D_2 f(\lambda, 0, 0) = \begin{bmatrix} 0 & 1 \\ \lambda & 0 \end{bmatrix}.$$

Hence the assumptions of corollary (26.6) are satisfied and, since the function $\lambda \mapsto \det D_2 f(\lambda, 0, 0)$ changes its sign at $\lambda_0 = 0$, $\lambda_0 = 0$ is a point of bifurcation from the trivial solution of the equation $f(\lambda, x, y) = 0$. Typical phase portraits and the bifurcation diagram for this example were given in Figures 4 and 5 of section 3.

(h) Finally, we consider the planar system

$$\left. \begin{array}{ccc} \dot{x} & = & -\lambda y + x^3 \\ \dot{y} & = & \lambda x + y^3 \end{array} \right\} =: f(\lambda, x, y),$$

where $\lambda \in \mathbb{R}$. Since $\det D_2 f(\lambda, 0, 0) = \lambda^2$, it follows that for all $\lambda \neq 0$, $(0, 0)$ is an isolated critical point for this system. Corollary (26.6) does not apply, because the determinant of $D_2 f(\lambda, 0, 0)$ does not change its sign at $\lambda_0 = 0$. If we have $f(\lambda, x, y) = 0$, then multiplying the first equation by x, the second by y and subsequent addition, we obtain $x^4 + y^4 = 0$. Hence for every $\lambda \in \mathbb{R}$, $(0, 0)$ is the only critical point and so no bifurcation takes place.□

(26.9) Remark. That condition (4) of theorem (26.5) is not necessary is shown by examples (26.8 c and d). It follows from example (26.8 h) that without restrictions, in general, bifurcation cannot be expected to occur. Examples (26.8 c and d) also show that in the case of *bifurcation from multiple eigenvalues* the situation may be most complicated.   □

The Liapunov-Schmidt Method

For the subsequent investigations of bifurcation problems we need some elementary facts about higher order derivatives.

For this we let Y and Z denote arbitrary Banach spaces, we assume that U is open in Y and that $g \in C^k(U, Z)$ for some $k \in \mathbb{N}^*$. Then we have

$$Dg \in C^{k-1}(U, \mathcal{L}(Y, Z)), \quad D^2 g \in C^{k-2}(U, \mathcal{L}(Y, \mathcal{L}(Y, Z)))$$

and

$$D^j g \in C^{k-j}(U, \mathcal{L}(Y, \mathcal{L}(Y, \dots, \mathcal{L}(Y, Z) \dots)))$$

for all $j \leq k$, where the symbol $\mathcal{L}$ on the right-hand side occurs j-times.

If $Y_1 \times \dots \times Y_m$ is a product Banach space, then

$$\mathcal{L}^m(Y_1 \times \dots \times Y_m, Z)$$

denotes the *vector space* (with respect to the pointwise operations) *of all m-linear continuous mappings* $T : Y_1 \times \dots \times Y_m \to Z$. We set

$$\mathcal{L}^m(Y, Z) := \mathcal{L}^m(Y^m, Z),$$

so, in particular, $\mathcal{L}^1(Y, Z) := \mathcal{L}(Y, Z)$. As in the case $m = 1$, one easily sees that an m-linear map $T : Y_1 \times \dots \times Y_m \to Z$ is continuous if and only if there exists a constant $\alpha \in \mathbb{R}_+$ such that

$$\|T[y_1, \dots, y_m]\| \leq \alpha \|y_1\| \dots \|y_m\|, \quad \forall (y_1, \dots, y_m) \in Y_1 \times \dots \times Y_m.$$

The infimum of all such constants is the *norm* of T, $\|T\|$, that is,

$$\|T\| := \sup\{\|T[y_1, \dots, y_m]\|_Z \mid \|y_i\|_{Y_i} \leq 1, \, i = 1, \dots, m\},$$

and it is not difficult to see that $\mathcal{L}^m(Y_1 \times \dots \times Y_m, Z)$ with this norm – which we will always use – is a Banach space.

Let $B \in \mathcal{L}^2(Y, Z)$ and

$$A_B(y_1) y_2 := B[y_1, y_2], \quad \forall y_1, y_2 \in Y.$$

Then $A_B \in \mathcal{L}(Y, \mathcal{L}(Y, Z))$ and

$$\|A_B(y_1) y_2\| \leq \|B\| \, \|y_1\| \, \|y_2\|,$$

hence $\|A_B(y_1)\|_{\mathcal{L}(Y,Z)} \leq \|B\| \, \|y_1\|$ and therefore

$$\|A_B\|_{\mathcal{L}(Y, \mathcal{L}(Y,Z))} \leq \|B\|.$$

Conversely, if $A \in \mathcal{L}(Y, \mathcal{L}(Y, Z))$, we set

$$B_A[y_1, y_2] := A(y_1) y_2, \quad \forall y_1, y_2 \in Y.$$

Then $B_A \in \mathcal{L}^2(Y, Z)$ and we have

$$\|B_A\| \leq \|A\|_{\mathcal{L}(Y, \mathcal{L}(Y,Z))}.$$

This shows that the mapping

$$\mathcal{L}(Y, \mathcal{L}(Y, Z)) \to \mathcal{L}^2(Y, Z), \quad A \mapsto B_A,$$

is a norm-isomorphism, the *canonical norm-isomorphism*. Analogously, one shows that there exists a canonical norm-isomorphism

$$\mathcal{L}(Y, \mathcal{L}(Y, \ldots, \mathcal{L}(Y, Z) \ldots)) \to \mathcal{L}^j(Y, Z)$$

(where the symbol $\mathcal{L}$ occurs j-times on the left-hand side). *In what follows, we will therefore always identify* $\mathcal{L}(Y, \mathcal{L}(Y, \ldots, \mathcal{L}(Y, Z) \ldots))$ *with* $\mathcal{L}^j(Y, Z)$ *by means of the canonical isomorphism*. Thus we have

$$D^j g \in C^{k-j}(U, \mathcal{L}^j(Y, Z)), \quad 0 \le j \le k.$$

Moreover, it is well-known that $D^j g(x) \in \mathcal{L}^j(Y, Z)$ *is symmetric for every* $x \in U$ (e.g., Dieudonné [1], Lang [1]).

For a deeper study of bifurcation problems, we will apply the *method of Liapunov-Schmidt* (the Liapunov-Schmidt reduction). To this end, we assume that

$$g \in C^k(\Lambda \times X, G) \quad \text{satisfies} \quad g(0,0) = 0.$$

We set

$$N := \ker[D_2 g(0,0)] \quad \text{and} \quad R := \operatorname{im}[D_2 g(0,0)],$$

and choose complimentary subspaces N_c and R_c such that

$$E = N \oplus N_c \quad \text{and} \quad G = R \oplus R_c.$$

Then every $x \in E$ has the unique decomposition

$$x = \xi + \xi_c, \quad \text{where} \quad \xi \in N \quad \text{and} \quad \xi_c \in N_c,$$

and the projections $E \to N$, $x \mapsto \xi$, and $E \to N_c$, $x \mapsto \xi_c$, are continuous linear operators. We let P and $P_c = id_G - P$ denote the canonical projections

$$P : G \to R \quad \text{and} \quad P_c : G \to R_c,$$

and assume that X is a convex product neighborhood of 0, that is,

$$X = U \oplus U_c, \quad \text{where} \quad U \subseteq N \quad \text{and} \quad U_c \subseteq N_c \quad \text{are convex}.$$

We then define functions

$$h^1 \in C^k(\Lambda \times U \times U_c, R_c) \quad \text{and} \quad h^2 \in C^k(\Lambda \times U \times U_c, R)$$

by

$$h^1(\lambda, \xi, \xi_c) := P_c g(\lambda, \xi + \xi_c)$$

and

$$h^2(\lambda, \xi, \xi_c) := P g(\lambda, \xi + \xi_c).$$

Evidently, the equation

$$g(\lambda, x) = 0$$

is equivalent to the system

$$h^1(\lambda, \xi, \xi_c) = 0, \quad h^2(\lambda, \xi, \xi_c) = 0, \tag{6}$$

and we have

$$h^1(0,0,0) = 0, \quad h^2(0,0,0) = 0.$$

It follows from the definition of h^2 that

$$D_3 h^2(0,0,0) = P D_2 g(0,0)|N_c.$$

Therefore $D_3 h^2(0,0,0) \in \mathcal{L}(N_c, R)$ is an isomorphism and, by the implicit function theorem, there exists a neighborhood of $(0,0,0) \in \Lambda \times U \times U_c$ – by possibly shrinking, we can take it to be $\Lambda \times U \times U_c$ – and a unique function $\eta \in C^k(\Lambda \times U, U_c)$ such that

$$h^2(\lambda, \xi, \xi_c) = 0 \quad \Leftrightarrow \quad \xi_c = \eta(\lambda, \xi).$$

In other words: In a sufficiently small neighborhood of $(0,0,0)$, the equation $h^2(\lambda, \xi, \xi_c) = 0$ can be solved uniquely for ξ_c. If we now substitute this solution into the first equation, we see that *in a sufficiently small neighborhood of $(0,0) \in \Lambda \times X$ – which we may again assume to be, without loss of generality, $\Lambda \times X$ – the equation $g(\lambda, x) = 0$ is equivalent to the <u>bifurcation equation</u>*

$$h(\lambda, \xi) := h^1(\lambda, \xi, \eta(\lambda, \xi)) = 0,$$

where we have

$$h \in C^k(\Lambda \times U, R_c) \quad and \quad h(0,0) = 0.$$

Since, in general, we have $\dim(N) < \dim(E)$, we have reduced our original problem (i.e., the investigation of the equation $g(\lambda, x) = 0$ in a neighborhood of $(0,0) \in \Lambda \times X$) to an equivalent problem (i.e., the study of the equation $h(\lambda, \xi) = 0$ in a neighborhood of $(0,0) \in \Lambda \times U$), where the new problem has, in general, a smaller dimension (that is, fewer dependent and independent variables).

(26.10) Remark. *If E, F and G are Banach spaces of arbitrary – finite or infinite – dimension, then the reduction method of Liapunov-Schmidt, just described, remains correct with exactly the same proof, if we <u>assume that</u>:*

$$E = N \oplus N_c \quad and \quad G = R \oplus R_c, \tag{7}$$

where $N := \ker[D_2 g(0,0)]$ and $R := \mathrm{im}[D_2 g(0,0)]$, and $\oplus$ denotes the topological direct sum. Here we call a direct sum decomposition $E = E_1 \oplus E_2$ of a Banach space a topological direct sum if the corresponding projections $P_i : E \to E_i$, $i = 1, 2$, are continuous. It is a simple consequence of the closed graph theorem (e.g., Yosida [1]) that a direct sum decomposition $E = E_1 \oplus E_2$ of a Banach space E is a topological direct sum if and only if E_1 and E_2 are closed subspaces of E.

While condition (7) can always be satisfied in the finite dimensional case, it represents an additional condition if E and G are infinite dimensional Banach spaces. In particular, the image of the linear operator $D_2 g(0,0) \in \mathcal{L}(E, G)$ must be closed.

The method of Liapunov-Schmidt represents one of the most important tools for the local analysis of nonlinear equations, and it has numerous applications in nonlinear functional analysis.   □

We now assume that

$$g(\cdot, 0) = 0,$$

which implies $h^i(\cdot, 0, 0) = 0$. From this and from the fact that the equation $h^2(\lambda, \xi, \xi_c) = 0$ can be solved uniquely for ξ_c, we obtain

$$\eta(\cdot, 0) = 0 \qquad (8)$$

and therefore

$$h(\cdot, 0) = 0. \qquad (9)$$

It follows from the definition of h^2 and η that

$$Pg(\lambda, \xi + \eta(\lambda, \xi)) = 0, \qquad \forall (\lambda, \xi) \in \Lambda \times U. \qquad (10)$$

Differentiating this identity, we get

$$PD_2g(\lambda, \xi + \eta(\lambda, \xi))[\hat{\xi} + D_2\eta(\lambda, \xi)\hat{\xi}] = 0, \qquad \forall \hat{\xi} \in N. \qquad (11)$$

Since $D_2g(0,0)|N = 0$, we obtain from this that $PD_2g(0,0)D_2\eta(0,0)\hat{\xi} = 0$, hence

$$D_2\eta(0,0) = 0. \qquad (12)$$

From the definition of h we obtain

$$D_2h(\lambda, \xi)\hat{\xi} = P_cD_2g(\lambda, \xi + \eta(\lambda, \xi))[\hat{\xi} + D_2\eta(\lambda, \xi)\hat{\xi}], \qquad \forall \hat{\xi} \in N, \qquad (13)$$

which, together with (12), implies

$$D_2h(0,0) = 0. \qquad (14)$$

After these preliminaries we will prove the following fundamental theorem.

(26.11) Theorem. *Assume $k \geq 2$ and let $g \in C^k(\Lambda \times X, G)$ be such that $g(\cdot, 0) = 0$. Furthermore, set*

$$N := \ker D_2g(0,0) \quad and \quad R := \operatorname{im} D_2g(0,0)$$

and let N_c and R_c denote the complementary spaces of N and R, respectively, that is, we have

$$E = N \oplus N_c \quad and \quad G = R \oplus R_c.$$

Finally, let $P : G \to R$ denote the projection on R along R_c and set $P_c := id_G - P$.

Then there exists a neighborhood Λ_0 of 0 in Λ and convex neighborhoods U and U_c of 0 in N and N_c, respectively, as well as functions

$$h \in C^k(\Lambda_0 \times U, R_c) \quad and \quad \eta \in C^k(\Lambda_0 \times U, U_c)$$

with the following properties:

(i) $U \oplus U_c \subseteq X$.

(ii)   *Statements (α) and (β) below are equivalent*:
    (α) $(\lambda, x) \in \Lambda_0 \times (U \oplus U_c)$ *and* $g(\lambda, x) = 0$;
    (β) $(\lambda, \xi) \in \Lambda_0 \times U$, $h(\lambda, \xi) = 0$ *and* $x = \xi + \eta(\lambda, \xi)$.
(iii) *There exists some* $\eta_0 \in C^{k-1}(\Lambda_0 \times U, \mathcal{L}(N, N_c))$ *satisfying*

$$\eta(\lambda, \xi) = \eta_0(\lambda, \xi)\xi \quad and \quad \eta_0(0,0) = 0$$

and we have

$$Pg(\lambda, \xi + \eta(\lambda, \xi)) = 0, \quad \forall(\lambda, \xi) \in \Lambda_0 \times U. \tag{15}$$

(iv) *There exists a function*

$$h_0 \in C^{k-1}(\Lambda_0 \times U, \mathcal{L}(N, R_c))$$

such that

$$h(\lambda, \xi) = h_0(\lambda, \xi)\xi. \tag{16}$$

Moreover, we have

$$h_0(\lambda, \xi) = h_{01}(\lambda)\lambda + h_{02}(\lambda, \xi)\xi, \tag{17}$$

where

$$D_1 h_0(0, 0) = h_{01}(0) = P_c D_1 D_2 g(0, 0)|F \times N \tag{18}$$

and

$$D_2 h_0(0, 0) = h_{02}(0, 0) = \frac{1}{2} P_c D_2^2 g(0, 0)|N \times N, \tag{19}$$

as well as

$$h_{01} \in C^{k-2}(\Lambda_0, \mathcal{L}^2(F \times N, R_c))$$

and

$$h_{02} \in C^{k-2}(\Lambda_0 \times U, \mathcal{L}^2(N, R_c)).$$

Proof. (i) and (ii) follow from the Liapunov-Schmidt reduction.

(iii): We obtain from (8) and the mean value theorem that

$$\eta(\lambda, \xi) = \int_0^1 D_2\eta(\lambda, t\xi)\, dt\xi.$$

Hence

$$\eta_0(\lambda, \xi) := \int_0^1 D_2\eta(\lambda, t\xi)\, dt$$

has the asserted properties, because

$$\eta_0(0, 0) = D_2\eta(0, 0) = 0$$

follows from (12). Formula (15) is a duplication of (10).

(iv): From (9) and the mean value theorem we obtain (16), where

$$h_0(\lambda, \xi) := \int_0^1 D_2 h(\lambda, t\xi) \, dt. \tag{20}$$

Therefore h_0 has the asserted continuity properties. Since

$$h_0(\lambda, \xi) = D_2 h(\lambda, 0) + \int_0^1 [D_2 h(\lambda, t\xi) - D_2 h(\lambda, 0)] \, dt,$$

we deduce from the mean value theorem that

$$h_0(\lambda, \xi) = D_2 h(\lambda, 0) + \int_0^1 \int_0^1 D_2^2 h(\lambda, st\xi) t \, ds \, dt\xi$$

$$= D_2 h(\lambda, 0) + \int_0^1 \int_0^t D_2^2 h(\lambda, \tau\xi) \, d\tau \, dt\xi$$

$$= D_2 h(\lambda, 0) + \int_0^1 (1 - t) D_2^2 h(\lambda, t\xi) \, dt\xi,$$

where the last equality was obtained by integration by parts. From (14) we obtain, again, using the mean value theorem,

$$D_2 h(\lambda, 0) = h_{01}(\lambda)\lambda,$$

where

$$h_{01}(\lambda) := \int_0^1 D_1 D_2 h(t\lambda, 0) \, dt. \tag{21}$$

If we set

$$h_{02}(\lambda, \xi) := \int_0^1 (1 - t) D_2^2 h(\lambda, t\xi) \, dt, \tag{22}$$

then (17) holds, and the functions h_{01} and h_{02} have the asserted continuity properties.

Differentiating (20) and using (21) and (22), we obtain

$$D_1 h_0(0, 0) = D_1 D_2 h(0, 0) = h_{01}(0) \tag{23}$$

and

$$2 D_2 h_0(0, 0) = D_2^2 h(0, 0) = 2 h_{02}(0, 0). \tag{24}$$

Since (8) implies

$$D_1^i \eta(\cdot, 0) = 0 \quad \text{for all} \quad 0 \le i \le k,$$

we obtain, by differentiating (13), that

$$D_1 D_2 h(\lambda, 0)[\hat{\lambda}, \hat{\xi}] = P_c D_1 D_2 g(\lambda, 0)[\hat{\lambda}, \hat{\xi} + D_2 \eta(\lambda, 0)\hat{\xi}]$$
$$+ P_c D_2 g(\lambda, \eta) D_1 D_2 \eta(\lambda, 0)[\hat{\lambda}, \hat{\xi}] \tag{25}$$

for all $\hat{\lambda} \in F$ and $\hat{\xi} \in N$. Using $P_c D_2 g(0,0) = 0$ and (12), we deduce from this that

$$D_1 D_2 h(0,0) = P_c D_1 D_2 g(0,0)|F \times N,$$

which, together with (23), proves assertion (18).

Again, by differentiating (13), we get

$$D_2^2 h(\lambda, \xi)[\hat{\xi}_1, \hat{\xi}_2] = P_c D_2^2 g(\lambda, \xi + \eta)[\hat{\xi}_1 + D_2 \eta(\lambda, \xi)\hat{\xi}_1, \hat{\xi}_2 + D_2 \eta(\lambda, \xi)\hat{\xi}_2] \tag{26}$$
$$+ P_c D_2 g(\lambda, \xi + \eta) D_2^2 \eta(\lambda, \xi)[\hat{\xi}_1, \hat{\xi}_2]$$

for all $\hat{\xi}_1, \hat{\xi}_2 \in N$. For $(\lambda, \xi) = (0,0)$, and using $\operatorname{im}(D_2 g(0,0)) = \ker(P_c)$, this reduces to

$$D_2^2 h(0,0) = P_c D_2^2 g(0,0)|N \times N.$$

This, together with (24), implies (19).   □

We now make the additional *assumption* that

$$E = G.$$

Since we have

$$\dim \ker(A) = \dim \operatorname{coker}(A)$$

for every $A \in \mathcal{L}(E)$, where $\operatorname{coker}(A) = E/\operatorname{im}(A)$, we deduce from $\dim \operatorname{coker}(A) = \dim(R_c)$ that

$$\dim(N) = \dim(R_c). \tag{27}$$

In other words:

The dimension of the bifurcation equation (that is to say, the number of equations one obtains if $h(\lambda, x) = 0$ is written as a system of equations in terms of a basis for R_c) *is equal to the dimension of the kernel* N.

To abbreviate, we set, in what follows,

$$A := D_2 g(0,0) \in \mathcal{L}(E)$$

and

$$B := D_1 D_2 g(0,0) \in \mathcal{L}^2(F \times E, E).$$

Since $g(\cdot, 0) = 0$, we have

$$g(\lambda, x) = Ax + B[\lambda, x] + r(\lambda, x), \tag{28}$$

where

$$r(\cdot, 0) = 0, \quad D_2 r(0,0) = 0 \quad \text{and} \quad D_1 D_2 r(0,0) = 0. \tag{29}$$

In the special case of *one-parameter problems* $F = \mathbb{R}$, we have

$$T[\lambda, x] = \lambda T[1, x], \quad \forall \lambda \in \mathbb{R}, \quad x \in E,$$

and $T[1, \cdot] \in \mathcal{L}(E)$ for every $T \in \mathcal{L}^2(\mathbb{R} \times E, E)$. Conversely, if $S \in \mathcal{L}(E)$, then $\hat{S}[\lambda, x] := \lambda S x$ defines some $\hat{S} \in \mathcal{L}^2(\mathbb{R} \times E, E)$. Hence the correspondence $T \mapsto T[1, \cdot]$ is a bijection from $\mathcal{L}^2(\mathbb{R} \times E, E)$ onto $\mathcal{L}(E)$, and it is clear that this bijection is a norm-isomorphism. Consequently, we may *identify* $\mathcal{L}^2(\mathbb{R} \times E, E)$ *with* $\mathcal{L}(E)$ by means of this "canonical norm-isomorphism." This will always be done. Hence, if $F = \mathbb{R}$, equation (28) has the form

$$g(\lambda, x) = Ax + \lambda Bx + r(\lambda, x), \tag{30}$$

where

$$B := D_1 D_2 g(0, 0) \in \mathcal{L}(E).$$

For each subset M of E, we define the *annihilator of* M, $M^\perp$, by

$$M^\perp := \{x' \in E' \mid \langle x', m \rangle = 0, \quad \forall m \in M\}.$$

Clearly $M^\perp$ is a vector subspace of the dual space E'. For each $T \in \mathcal{L}(E)$, we deduce the well-known relation

$$\text{im}(T)^\perp = \ker(T') \tag{31}$$

from $\langle x', Tx \rangle = \langle T'x', x \rangle$. Since $\dim \ker(T) = \dim \ker(T')$, we obtain, in particular, that

$$\dim[\text{im}(T)^\perp] = \dim \ker(T), \quad \forall T \in \mathcal{L}(E). \tag{32}$$

Now we choose an arbitrary basis $\{e_1', \ldots, e_n'\}$ for $\ker(A')$. As is well-known, there exist n elements $e_1, \ldots, e_n$ in E such that

$$\langle e_i', e_j \rangle = \delta_{ij}, \quad i, j = 1, \ldots, n. \tag{33}$$

We set

$$R_c := \text{span}\{e_1, \ldots, e_n\} \tag{34}$$

and obtain, using (31) and (33),

$$E = \text{im}(A) \oplus R_c. \tag{35}$$

Finally, we define $P_c \in \mathcal{L}(E)$ by

$$P_c(x) := \sum_{i=1}^n \langle e_i', x \rangle e_i, \quad \forall x \in E. \tag{36}$$

Then one immediately verifies that P_c is the projection from E onto R_c, corresponding to the decomposition (35).

(26.12) Lemma. *Let* $E = G$ *and* $A := D_2 g(0, 0) \in \mathcal{L}(E)$. *If* $\{e_1', \ldots, e_n'\}$ *is an arbitrary basis for* $\ker(A')$, *then the bifurcation equation* $h(\lambda, \xi) = 0$ *is equivalent to the system of equations*

$$\langle e_i', h(\lambda, \xi) \rangle = 0, \quad i = 1, \ldots, n.$$

Proof. This immediately follows from (36). $\qquad\square$

Bifurcation from Simple Eigenvalues

Following these general considerations, we will analyze some simple special cases. We begin with the simplest situation.

Case 1: $\Lambda \subseteq \mathbb{R}$, $\dim(N) = 1$.   In this case we have

$$N = \ker(A) = \mathbb{R}e \quad \text{and} \quad \ker(A') = \mathbb{R}e'$$

for some $e \in E \setminus \{0\}$ and $e' = E' \setminus \{0\}$. Hence by lemma (26.12) and theorem (26.11 iv), the bifurcation equation is equivalent to the equation

$$\varphi(\lambda, s)s = 0 \tag{37}$$

in some neighborhood $\Lambda_0 \times U$ of $(0,0)$ in $\mathbb{R} \times \mathbb{R}$, where we have set

$$\varphi(\lambda, s) := \langle e', h_0(\lambda, se)e \rangle.$$

From formula (18) of theorem (26.11) it follows, furthermore, that

$$D_1\varphi(0,0) = \langle e', Be \rangle, \tag{38}$$

where $B := D_1 D_2 g(0,0) \in \mathcal{L}(E)$. Evidently, we have

$$\langle e', Be \rangle \neq 0 \quad \Leftrightarrow \quad B[\ker(A)] \not\subseteq \operatorname{im}(A).$$

If this condition is satisfied, it follows from (38) and from the implicit function theorem that the equation $\varphi(\lambda, s) = 0$ can be solved uniquely for λ in some neighborhood of $(0,0)$. From this we obtain the theorem about *bifurcation from simple eigenvalues*.

(26.13) Theorem. *Let $\Lambda \subseteq \mathbb{R}$ and $g \in C^k(\Lambda \times X, E)$, $k \geq 2$, be such that $g(\cdot, 0) = 0$. Assume that*

$$A := D_2 g(0,0) \in \mathcal{L}(E) \quad \text{and} \quad B := D_1 D_2 g(0,0) \in \mathcal{L}(E)$$

satisfy

$$B[\ker(A)] \not\subseteq \operatorname{im}(A).$$

Then $(0,0)$ is a bifurcation point of the equation $g(\lambda, x) = 0$. In some neighborhood of $(0,0)$, the solution set of $g(\lambda, x) = 0$ consists exactly of the λ-axis $\Lambda \times \{0\}$ and a C^{k-1}-curve Γ which crosses the λ-axis at the origin. More precisely, there exists a C^{k-1}-parametrization

$$(\lambda(\cdot), x(\cdot)) : (-\epsilon, \epsilon) \to \Lambda \times X$$

with the following properties:

(i) $(\lambda(0), x(0)) = (0,0)$.
(ii)   *If* $\ker(A) = \mathbb{R}e$, *we have*

$$x(s) = s(e + y(s)), \quad \forall s \in (-\epsilon, \epsilon),$$

where $y(0) = 0$ and $y(\cdot) \in C^{k-1}((-\epsilon, \epsilon), N_c)$.

Proof. Since, by theorem (26.11), the function φ is of class C^{k-1}, it follows from the preceding considerations that there exists some function $\lambda(\cdot) \in C^{k-1}((-\epsilon, \epsilon), \mathbb{R})$ such that all solutions of $\varphi(\lambda, s) = 0$ near $(0, 0)$ are given by $\{(\lambda(s), s) \mid -\epsilon < s < \epsilon\}$. Using the notation of theorem (26.11), we set

$$y(s) := \eta_0(\lambda(s), se)e.$$

Now the assertion follows from (37) and theorem (26.11). $\square$

(26.14) Remarks. (a) In this case, the bifurcation diagram has the following qualitative features. From $x(s) = s(e + y(s))$ and $y(0) = 0$ we obtain $\dot{x}(0) = e$. This means that the

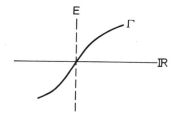
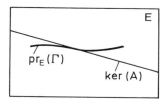

projection of the curve Γ into the space E at the origin is tangent to $\ker(A)$.

(b) We have merely assumed that 0 is an eigenvalue of A of *geometric multiplicity* 1. No assumptions about the algebraic multiplicity of $0 \in \sigma(A)$ have been made.

(c) *The nondegeneracy assumption $B[\ker(A)] \not\subseteq \operatorname{im}(A)$ is, of course, equivalent to*

$$B[\ker(A)] \oplus \operatorname{im}(A) = E. \tag{39}$$

It is also equivalent to $d'(0) \neq 0$, where

$$d(\lambda) := \det[D_2 g(\lambda, 0)],$$

that is to say, condition (39) is satisfied if and only if the function $\det[D_2 g(\cdot, 0)]$ has a simple zero at $\lambda = 0$.

To see this, let $\{e_1, \ldots, e_m\}$ be an arbitrary basis for E and set

$$e_j(\lambda) := D_2 g(\lambda, 0)e_j, \quad j = 1, \ldots, m.$$

Then

$$e_j(0) = Ae_j =: a_j \quad \text{and} \quad \dot{e}_j(0) = Be_j =: b_j,$$

and, by differentiating the determinant at $\lambda = 0$, we obtain

$$d'(0) = \sum_{j=1}^{m} \det[a_1, \ldots, a_{j-1}, b_j, a_{j+1}, \ldots, a_m],$$

which is written in terms of the columns (cf. the proof of proposition (11.4)). We now choose the basis so that e_1 spans $\ker(A)$. Then $Ae_1 = 0$, from which we obtain

$$d'(0) = \det[b_1, a_2, \ldots, a_m].$$

We set $E_0 := \ker(A^m)$, i.e., E_0 denotes the algebraic eigenspace corresponding to $0 \in \sigma(A)$. Then there exists a direct sum decomposition $E = E_0 \oplus E_1$ which reduces A, and we may choose the basis for E in such a way that $\{e_1, \ldots, e_n\}$ is a basis for E_0, $\{e_{n+1}, \ldots, e_m\}$ is a basis for E_1 and such that $Ae_j = e_{j-1}$ for all $j = 2, \ldots, n$ (cf. section 12). Since $A|E_1 \in \mathcal{GL}(E_1)$, $\{a_{n+1}, \ldots, a_m\}$ is a basis for E_1 and we have

$$d'(0) = \det[b_1, e_1, e_2, \ldots, e_{n-1}, a_{n+1}, \ldots, a_m].$$

From this one reads off the assertion. $\qquad\qquad\qquad\qquad\qquad\qquad\qquad\qquad\quad\square$

(26.15) Examples. (a) We consider the mth order linear equation

$$\sum_{j=0}^{m} a_j(\lambda)D^j u = f(\lambda, u, \ldots, D^{m-1}u), \quad a_m \equiv 1,$$

where $a_j \in C^2(\mathbb{R}, \mathbb{R})$, $f \in C^2(\mathbb{R} \times \mathbb{R}^m, \mathbb{R})$ and $f(\cdot, 0, \ldots, 0) = 0$. If this equation is written as a system

$$\dot{x} = g(\lambda, x),$$

where

$$g(\lambda, x) = \left(x^2, \ldots, x^m, -\sum_{j=0}^{m-1} a_j(\lambda)x^{j+1} + f(\lambda, x^1, \ldots, x^m) \right),$$

then 0 is a critical point and

$$d(\lambda) := \det[D_2 g(\lambda, 0)] = \pm[-a_0(\lambda) + D_2 f(\lambda, 0, \ldots, 0)]$$

(cf. section 14). If we assume that $a_0(0) = D_2 f(0, 0, \ldots, 0)$, then

$$d(0) = 0 \quad \text{and} \quad d'(0) = \pm[-a_0'(0) + D_1 D_2 f(0, 0, \ldots, 0)].$$

If $d'(0) \neq 0$, it follows from remark (26.14 c) and theorem (26.13) that, in some neighborhood of $(\lambda, x) = (0, 0)$, there exists a unique C^1-curve $s \mapsto (\lambda(s), x(s))$ of nontrivial critical points of $\dot{x} = g(\lambda, x)$ satisfying $(\lambda(0), x(0)) = (0, 0)$. With regard to the mth order equation, this means that there exists exactly one pair of functions $\lambda(\cdot), \xi(\cdot) \in C^1((-\epsilon, \epsilon), \mathbb{R})$ such that $\xi(0) = 0$ and $\xi(s) \neq 0$ for all $s \neq 0$ and so that the following holds:

$$a_0(\lambda(s))\xi(s) = f(\lambda(s), \xi(s), 0, \ldots, 0),$$

i.e., $\xi(s)$ is a constant solution.

If, in addition, we have $-a_j(0) - D_{j+2}f(0, 0, \ldots, 0) = 0$ for $j = 1, \ldots, m-1$, then the matrix of $A := D_2 g(0, 0)$ has the form

$$\begin{bmatrix} 0 & 1 & 0 & 0 & \cdots & 0 \\ 0 & 0 & 1 & 0 & \cdots & 0 \\ 0 & 0 & 0 & 1 & \cdots & 0 \\ \vdots & \vdots & \vdots & \vdots & \ddots & \vdots \\ 0 & 0 & 0 & 0 & \cdots & 1 \\ 0 & 0 & 0 & 0 & \cdots & 0 \end{bmatrix}$$

Hence in this case, 0 is an eigenvalue of A with geometric multiplicity 1 and algebraic multiplicity m.

If $d'(0) = 0$, theorem (26.13) is not applicable. However, if, for instance, we have $d'(0) = d''(0) = 0$ and $d'''(0) \neq 0$, then the function $d(\cdot)$ changes sign at $\lambda = 0$ and, by corollary (26.6), $(0,0)$ is a point of bifurcation for critical points of the system $\dot{x} = g(\lambda, x)$.

(b) Let $f \in C^k(\Lambda \times \mathbb{R} \times X, E)$, $k \geq 2$, be such that $f(\cdot, \cdot, 0) = 0$ and suppose f is 2π-periodic in t. Then we set

$$g(\lambda, \xi) := \xi - u(2\pi, 0, \xi, \lambda + \lambda_0),$$

where $u(\cdot, 0, \xi, \lambda)$ again denotes the global solution of the IVP $\dot{x} = f(\lambda, t, x)$, $x(0) = \xi$. By possibly shrinking Λ and X, we may assume, without loss of generality, that g is defined on all of $\Lambda_0 \times X$, where $\Lambda_0 := \Lambda - \lambda_0$. We have

$$A := D_2 g(0,0) = id_E - D_3 u(2\pi, 0, 0, \lambda_0)$$

and

$$B := D_1 D_2 g(0,0) = -D_4 D_3 u(2\pi, 0, 0, \lambda_0).$$

Since

$$\dot{u}(t, 0, \xi, \lambda) = f(\lambda, t, u(t, 0, \xi, \lambda)),$$

we deduce from the differentiability theorem (9.2) that

$$D_3 \dot{u}(t, 0, \xi, \lambda) = D_3 f(\lambda, t, u(t, 0, \xi, \lambda)) D_3 u(t, 0, \xi, \lambda)$$
$$D_3 u(0, 0, \xi, \lambda) = id_E \tag{40}$$

and – since $u(t, 0, 0, \cdot) = 0$ implies $D_4 u(t, 0, 0, \cdot) = 0$ – we have

$$D_4 D_3 \dot{u}(t, 0, 0, \lambda_0) = D_1 D_3 f(\lambda_0, t, 0) D_3 u(t, 0, 0, \lambda_0)$$
$$+ D_3 f(\lambda_0, t, 0) D_4 D_3 u(t, 0, 0, \lambda_0) \tag{41}$$
$$D_4 D_3 u(0, 0, 0, \lambda_0) = 0.$$

We let $U(\lambda, t, \tau)$ denote the evolution operator of the linear equation

$$\dot{y} = D_3 f(\lambda, t, 0) y.$$

Then

$$A = id_E - U(\lambda_0, 2\pi, 0)$$

and, since $D_3 u(t, 0, 0, \lambda_0) = U(\lambda_0, t, 0)$, it follows from (41) (cf. theorem (11.13)) that

$$B = -D_4 D_3 u(2\pi, 0, 0, \lambda_0)$$

$$= -\int_0^{2\pi} U(\lambda_0, 2\pi, \tau) D_1 D_3 f(\lambda_0, \tau, 0) U(\lambda_0, \tau, 0) \, d\tau.$$

Hence for $e' \in E'$ and $e \in E$ we have

$$\langle e', Be \rangle = - \int_0^{2\pi} \langle U'(\lambda_0, 2\pi, \tau)e', D_1 D_3 f(\lambda_0, \tau, 0)U(\lambda_0, \tau, 0)e \rangle \, d\tau.$$

It follows from proposition (11.15) that $w(t) := U'(\lambda_0, 2\pi, t)e'$ is the solution of the dual IVP

$$\dot{z} = -[D_3 f(\lambda_0, t, 0)]'z, \quad z(2\pi) = e'.$$

From

$$A = id_E - U(\lambda_0, 2\pi, 0)$$

we obtain

$$A' = id_{E'} - U'(\lambda_0, 2\pi, 0).$$

Therefore $e' \in E'$ is an element of $\ker(A')$ and distinct from zero if and only if the dual linear equation

$$\dot{z} = -[D_3 f(\lambda_0, t, 0)]'z \quad \text{in} \quad E'$$

possesses a nontrivial 2π-periodic solution w with $w(0) = e'$. In that case

$$w(t) = U'(\lambda_0, 2\pi, t)e',$$

and we obtain the relation

$$\langle e', Be \rangle = - \int_0^{2\pi} \langle w(t), D_1 D_3 f(\lambda_0, t, 0)v(t) \rangle \, dt, \tag{42}$$

where v denotes the unique 2π-periodic solution of

$$\dot{y} = D_3 f(\lambda_0, t, 0)y$$

satisfying $v(0) = e$ and w denotes the unique 2π-periodic solution of the dual equation

$$\dot{z} = -[D_3 f(\lambda_0, t, 0)]'z$$

satisfying $w(0) = e'$. It should be remarked here that these considerations are independent of the dimension of the parameter space. □

From theorem (26.13) we therefore obtain the following proposition.

(26.16) Proposition. *Let $f \in C^k(\Lambda \times \mathbb{R} \times X, E)$, $k \geq 2$, and $\Lambda \subseteq \mathbb{R}$ and assume that f is 2π-periodic in t and $f(\cdot, \cdot, 0) = 0$. Suppose that for some $\lambda_0 \in \Lambda$, the linearized equation*

$$\dot{y} = D_3 f(\lambda_0, t, 0)y$$

has a unique linearly independent 2π-periodic solution $v \in C^1(\mathbb{R}, E)$. If w is a nontrivial 2π-periodic solution of the dual equation

$$\dot{z} = -[D_3 f(\lambda_0, t, 0)]'z$$

and if we have

$$\int_0^{2\pi} \langle w(t), D_1 D_3 f(\lambda_0, t, 0)v(t)\rangle \, dt \neq 0, \tag{43}$$

then $(\lambda_0, 0)$ is a bifurcation point of 2π-periodic solutions of the equation

$$\dot{x} = f(\lambda, t, x). \tag{44}$$

If (43) is satisfied, there exists a C^{k-1}-curve Γ through $(\lambda_0, 0)$ in $\Lambda \times BC^k(\mathbb{R}, E)$,

$$(-\epsilon, \epsilon) \to \Lambda \times BC^k(\mathbb{R}, E), \quad s \mapsto (\lambda(s), u(s)),$$

so that $u(s) \neq 0$ if $s \neq 0$ and such that in a sufficiently small neighborhood of $(\lambda_0, 0)$ in $\Lambda \times BC(\mathbb{R}, E)$, the set of all 2π-periodic solutions consists exactly of Γ and the trivial solutions $\Lambda \times \{0\}$.

Proof. By theorem (26.13) and example (26.15 b), there exists a C^{k-1}-curve Γ_0 in $\Lambda_0 \times X$,

$$(\lambda(\cdot), \xi(\cdot)) \in C^{k-1}((-\epsilon, \epsilon), \Lambda_0 \times X),$$

satisfying $(\lambda(0), \xi(0)) = (0, 0)$ and $\xi(s) \neq 0$ for all $s \neq 0$, and such that in some sufficiently small neighborhood of $(0, 0) \in \Lambda_0 \times X$, the set of all solutions of the equation $g(\lambda, \xi) = 0$ consists of Γ_0 and the Λ_0-axis $\Lambda_0 \times \{0\}$. Setting

$$u(s) := u(\cdot, 0, \xi(s), \lambda(s) + \lambda_0), \quad -\epsilon < s < \epsilon,$$

we deduce from this, as well as from remark (20.2) and the differentiability theorem (9.5), that the assertion holds. $\qquad \square$

(26.17) Remarks. (a) If $(E, (\cdot \mid \cdot))$ is an inner product space, it follows from remark (11.16) that we may replace the dual equation by the adjoint equation

$$\dot{z} = -[D_3 f(\lambda_0, t, 0)]^* z.$$

In this case condition (43) reads

$$\int_0^{2\pi} (w(t) \mid D_1 D_3 f(\lambda_0, t, 0)v(t)) \, dt \neq 0. \tag{45}$$

(b)   Relation (43) can be simplified significantly if the linearizations $D_3 f(\lambda_0, t, 0)$ and $D_1 D_3 f(\lambda_0, t, 0)$ are independent of the time, that is to say, if we have:

$$f(\lambda, t, x) = Ax + (\lambda - \lambda_0)Bx + g(\lambda, t, x),$$

where $A, B \in \mathcal{L}(E)$ and $g \in C^k(\Lambda \times \mathbb{R} \times X, E)$, as well as

$$g(\cdot, \cdot, 0) = 0, \quad D_3 g(\lambda_0, \cdot, 0) = 0 \quad \text{and} \quad D_1 D_3 g(\lambda_0, \cdot, 0) = 0.$$

Then the equation, linearized at $(\lambda_0, 0)$, simply becomes $\dot{y} = Ay$. This equation has a nontrivial 2π-periodic solution if and only if 1 is a Floquet multiplier, that is, if

$$1 \in \sigma(e^{2\pi A}).$$

Based on the spectral mapping theorem, this relation is satisfied if and only if

$$i\mathbb{Z} \cap \sigma(A) \neq \emptyset. \tag{46}$$

If $z = x + iy \in E_{\mathbb{C}}$ is an eigenvector of $A_{\mathbb{C}}$ corresponding to the eigenvalue $\lambda \in \mathbb{C}$ with $Im(\lambda) \neq 0$, then $\bar{z} := x - iy \in E_{\mathbb{C}}$ is an eigenvector of $A_{\mathbb{C}}$ corresponding to the eigenvalue $\bar{\lambda}$. And since $Im(\lambda) \neq 0$, x and y are linearly independent in E (cf. section 13, case 2, (b)). Hence, if for some $k \in \mathbb{Z}^*$ we have $ik \in \sigma(A)$, then $-ik$ also belongs to $\sigma(A)$. Consequently, in this case, 1 is an eigenvalue of $e^{2\pi A}$ with geometric multiplicity ≥ 2. We must therefore assume that $0 \in \sigma(A)$ and $i\mathbb{Z}^* \cap \sigma(A) = \emptyset$ in order to apply proposition (26.16).

If $\xi \in \ker(A)$, it immediately follows from the power series of $e^{2\pi A}$ that $e^{2\pi A}\xi = \xi$. Therefore – under the assumption that $i\mathbb{Z}^* \cap \sigma(A) = \emptyset$ – the stationary points $\xi \in \ker(A)$ are the only 2π-periodic solutions of $\dot{x} = Ax$. With this we deduce the following criterion from proposition (26.16):

Let $g \in C^k(\Lambda \times \mathbb{R} \times X, E)$, $k \geq 2$, be 2π-periodic in t and such that $g(\cdot, \cdot, 0) = 0$, $D_3 g(\lambda_0, \cdot, 0) = 0$ and $D_1 D_3 g(\lambda_0, \cdot, 0) = 0$. Moreover, assume that $A, B \in \mathcal{L}(E)$ and $\dim \ker(A) = 1$. Then, if

$$f(\lambda, t, x) = Ax + (\lambda - \lambda_0)Bx + g(\lambda, t, x)$$

and

$$\sigma(A) \cap i\mathbb{Z} = \{0\},$$

condition (43) of proposition (26.16) is equivalent to

$$\langle e', Be \rangle \neq 0, \tag{47}$$

where $\mathbb{R}e = \ker(A)$ and $\mathbb{R}e' = \ker(A')$.

(c)   *If $(E, (\cdot \mid \cdot))$ is an inner product space and if A is symmetric $(A = A^*)$, then (47) is equivalent to*

$$(Be \mid e) \neq 0,$$

where $\mathbb{R}e = \ker(A)$.   □

Hopf Bifurcation

While the preceding investigations pertain to the case when $\dim(\Lambda) = \dim(N) = 1$, we now consider the following case.

Case 2: $\dim(\Lambda) = \dim(N) =: n \geq 2$.   For these investigations we need the following lemma.

(26.18) Lemma. *Let Y and Z be arbitrary Banach spaces and let $\mathrm{Isom}(Y, Z)$ denote the set of all continuous ("toplinear") isomorphisms from Y onto Z. Then $\mathrm{Isom}(Y, Z)$ is open in $\mathcal{L}(Y, Z)$.*

Proof. If there is no toplinear isomorphism between Y and Z, then $\mathrm{Isom}(Y, Z) = \emptyset$. If there exists some $T \in \mathrm{Isom}(Y, Z)$, then the mapping $S \mapsto T^{-1} \circ S$ is a homeomorphism from $\mathcal{L}(Y, Z)$ onto $\mathcal{L}(Y)$ (with inverse $A \mapsto T \circ A$). One easily verifies that this function maps $\mathrm{Isom}(Y, Z)$ onto $\mathcal{GL}(Y)$. Since, by remark (25.6), $\mathcal{GL}(Y)$ is open in $\mathcal{L}(Y)$, the assertion follows. $\qquad\square$

The following theorem represents a generalization of theorem (26.13).

(26.19) Theorem. *Let $k \geq 2$ and assume that $g \in C^k(\Lambda \times X, E)$ is such that $g(\cdot, 0) = 0$. Moreover, let $A := D_2 g(0, 0) \in \mathcal{L}(E)$ and $N := \ker(A)$, as well as $B := D_1 D_2 g(0, 0) \in \mathcal{L}^2(F \times E, E)$ and $B_e := B[\cdot, e]$ for all $e \in E$. Assume that*

$$\dim(F) = \dim(N)$$

and

$$\mathrm{im}(B_e) \oplus \mathrm{im}(A) = E, \quad \forall e \in N \setminus \{0\}. \tag{48}$$

If N_c is an arbitrary complementary subspace for N and if $\mathbb{B}_N$ denotes the unit ball in N, then there exist some $\epsilon > 0$ and a function

$$(\lambda(\cdot), y(\cdot)) \in C(\epsilon \mathbb{B}_N, \Lambda \times \mathcal{L}(N, N_c))$$

with the following properties:

(i)   *In a neighborhood of $(0, 0) \in \Lambda \times E$, $g^{-1}(0)$ consists exactly of the "trivial solution" $\Lambda \times \{0\}$ and the set*

$$M := \{(\lambda(\xi), \xi + y(\xi)\xi) \mid \xi \in \epsilon \mathbb{B}_N\}.$$

(ii) $(\lambda(\cdot), y(\cdot))$ *is $(k-1)$-times differentiable in $\epsilon \mathbb{B}_N \setminus \{0\}$ and $(\lambda(0), y(0)) = (0, 0)$.*

(iii)   *For every $e \in \mathbb{S}_N := \partial \mathbb{B}_N$ we have*

$$[t \mapsto (\lambda(te), y(te))] \in C^k((-\epsilon, \epsilon), \Lambda \times \mathcal{L}(N, N_c)).$$

Proof. We employ the notation of theorem (26.11) and choose $\epsilon > 0$ such that $\epsilon \overline{\mathbb{B}}_N \subseteq U$. Here we may choose, without loss of generality, a Hilbert norm on N (i.e., we may identify N with $\mathbb{R}^n$). Then $\mathbb{S}_N$ is an $(n-1)$-dimensional C^∞-manifold and consequently

$$Y := \Lambda_0 \times (-\epsilon, \epsilon) \times \mathbb{S}_N$$

is a $2n$-dimensional C^∞-manifold. We define

$$a \in C^{k-1}(Y, R_c)$$

by

$$a(\lambda, s, e) := h_0(\lambda, se)e.$$

By theorem (26.11) we then have

$$a(0, 0, \cdot) = 0$$

and

$$D_1 a(0, 0, e) = P_c D_1 D_2 g(0, 0)[\cdot, e] = P_c B_e \tag{49}$$

for every $e \in \mathbb{S}_N$.

Since the bifurcation equation has the form $h_0(\lambda, \xi)\xi = 0$, it follows that (λ, ξ) is a solution of the bifurcation equation for $0 < |\xi| < \epsilon$ if and only if

$$\xi = se \quad \text{for some} \quad 0 < s < \epsilon, \quad e \in \mathbb{S}_N \quad \text{and} \quad a(\lambda, s, e) = 0.$$

Since $\dim(F) = \dim(N) = \dim(R_c)$, condition (48) is equivalent to

$$P_c B_e \in \mathrm{Isom}(F, R_c), \quad \forall e \in N \setminus \{0\}.$$

Since $D_1 a \in C^{k-2}(Y, \mathcal{L}(F, R_c))$, it therefore follows from lemma (26.18) and (49) that

$$(D_1 a)^{-1}(\mathrm{Isom}(F, R_c))$$

is an open neighborhood of $\{0\} \times \{0\} \times \mathbb{S}_N$ in Y. By reducing Λ_0 and ϵ sufficiently, we may assume, without loss of generality, that

$$Y = (D_1 a)^{-1}(\mathrm{Isom}\ (F, R_c)).$$

Therefore 0 is a regular value of a, which implies (e.g., Hirsch [1]) that $a^{-1}(0)$ is an n-dimensional C^{k-1}-submanifold of Y. Because $a(0, 0, e) = 0$, we have

$$\{0\} \times \{0\} \times \mathbb{S}_N \subseteq a^{-1}(0).$$

The compactness of $\{0\} \times \{0\} \times \mathbb{S}_N$ and the fact that for each $e \in \mathbb{S}_N$, $D_1 a(0, 0, e)$ is an isomorphism imply that locally the equation $a(\lambda, s, e) = 0$ can be solved uniquely for λ. By possibly decreasing ϵ, we may assume, without loss of generality, that $a^{-1}(0)$ can be represented as the graph

$$a^{-1}(0) = \{(\tilde{\lambda}(s, e), s, e) \mid (s, e) \in (-\epsilon, \epsilon) \times \mathbb{S}_N\}$$

of some function

$$\tilde{\lambda} \in C^{k-1}((-\epsilon, \epsilon) \times \mathbb{S}_N, \Lambda).$$

We now consider the C^∞-function

$$\varphi : (-\epsilon, \epsilon) \times \mathbb{S}_N \to N, \quad (s, e) \mapsto se.$$

Then $\varphi|(0, \epsilon) \times \mathbb{S}_N$ is a C^∞-diffeomorphism from $(0, \epsilon) \times \mathbb{S}_N$ onto $\epsilon \mathbb{B}_N \setminus \{0\}$ with inverse $\psi : \epsilon \mathbb{B}_N \setminus \{0\} \to (0, \epsilon) \times \mathbb{S}_N$ given by $\psi(\xi) = (|\xi|, \xi/|\xi|)$. We now set

$$\lambda(\xi) := \begin{cases} \tilde{\lambda} \circ \psi(\xi), & \text{if } \xi \in \epsilon \mathbb{B}_N \setminus \{0\} \\ 0, & \text{if } \xi = 0. \end{cases}$$

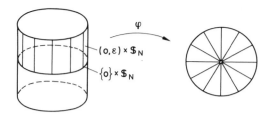

The unique solvability of the equation $a(\lambda, s, e) = 0$ for λ and the fact that $a(0, 0, e) = 0$ imply that $\tilde{\lambda}(0, e) = 0$ for all $e \in \mathbb{S}_N$. From this and from the mean value theorem, we obtain that

$$\tilde{\lambda}(s, e) = s\tilde{\lambda}_0(s, e), \quad \forall (s, e) \in (-\epsilon, \epsilon) \times \mathbb{S}_N,$$

for some appropriate function $\tilde{\lambda}_0 \in C^{k-2}((-\epsilon, \epsilon) \times \mathbb{S}_N, F)$. Consequently, λ has the form

$$\lambda(\xi) = |\xi|\tilde{\lambda}_0 \circ \psi(\xi) \quad \text{for all } \xi \in \epsilon\mathbb{B}_N \setminus \{0\},$$

from which we deduce that

$$\lambda \in C(\epsilon\mathbb{B}_N, \Lambda) \cap C^{k-1}(\epsilon\mathbb{B}_N \setminus \{0\}, \Lambda).$$

Since we clearly have

$$a(\lambda, s, e) = -a(\lambda, -s, -e) \quad \text{for all } (\lambda, s, e) \in Y,$$

the relation $\tilde{\lambda}(-s, -e) = \tilde{\lambda}(s, e)$ now follows from the uniqueness of $\tilde{\lambda}$. From this we obtain for every $e \in \mathbb{S}_N$ and $-\epsilon < t < 0$ the relation

$$\lambda(te) = \lambda(-t(-e)) = \tilde{\lambda}(-t, -e) = \tilde{\lambda}(t, e),$$

which, in turn, implies that $[t \mapsto \lambda(te)] \in C^{k-1}((-\epsilon, \epsilon), \Lambda)$. Setting $y(\xi) := \eta_0(\lambda(\xi), \xi)$ and using theorem (26.11), we obtain the assertion. $\qquad \square$

(26.20) Remarks. (a) Since $E = N \oplus N_c \cong N \times N_c$, one can interpret M as the graph of the function

$$\epsilon\mathbb{B}_N \to F \times N_c, \quad \xi \mapsto (\lambda(\xi), y(\xi)\xi).$$

Hence, in particular, $M \setminus \{(0, 0)\}$ is an n-dimensional C^{k-1}-manifold, where $n := \dim(N)$. The set M can be looked at as the union of the C^{k-1}-curves

$$(-\epsilon, \epsilon) \mapsto (\lambda(te), t(e + y(te)e)), \quad e \in \mathbb{S}_N,$$

through $(0, 0)$. The tangents to these curves at $(0, 0)$ have the form $(\mathbb{R}\mu, \mathbb{R}e)$, where $(\mu, e) \in F \times \mathbb{S}_N$. They are, in particular, transversal to the trivial solution $\Lambda \times \{0\}$.

(b) In the proof of theorem (26.19) we have performed a *blow-up of a singularity*. The function $h(\lambda, \xi) = h_0(\lambda, \xi)\xi$ has a singularity at $(0, 0)$, in the sense that $Dh(0, 0) = 0$. We have "blown-up" this singularity by "blowing-up" the point $\xi = 0 \in N$ into the sphere $\{0\} \times \mathbb{S}_N$ such that $a(\lambda, \xi) = h(\lambda, \xi)/|\xi|$ possesses only regular points on $\{0\} \times \mathbb{S}_N$.

(c) Qualitatively, the bifurcation diagram, again, has the same form as in the case of theorem (26.13). The projection of M into E is a C^{k-1}-manifold if $x \neq 0$, and it is a union of C^{k-1}-curves through 0, which are tangential to $\mathbb{R}e$, $e \in \mathbb{S}_N$, there.

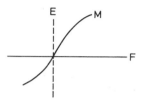

(d) From assumption (48) we deduce that $n \in 2\mathbb{N}^*$. To see this, we note that $P_c B_e \in$ Isom(F, R_c) for all $e \in \mathbb{S}_N$ implies that $\det(P_c B_e) \neq 0$ for all $e \in \mathbb{S}_N$. Now the assertion follows from $\det(P_c B_e) = \det(-P_c B_e) = (-1)^n \det(P_c B_e)$ and the connectedness of $\mathbb{S}_N$.

$\square$

We will now apply the general theorem above to the *Hopf bifurcation*, that is, to the bifurcation problem of a nonconstant periodic solution from a stationary point of an autonomous differential equation. We therefore let $\Lambda \subseteq \mathbb{R}$ and assume that

$$f \in C^k(\Lambda \times X, E),$$

for some $k \geq 2$, satisfies $f(\cdot, 0) = 0$. Then for every $\lambda \in \Lambda$, $0 \in E$ is a critical point of the autonomous equation

$$\dot{x} = f(\lambda, x). \tag{50}$$

We also know that

$$1 \in \sigma(e^{T D_2 f(\lambda_0, 0)}) \tag{51}$$

is a necessary condition for $(\lambda_0, 0)$ to be a bifurcation point for T-periodic solutions of equation (50). However, no particular period is now singled out anymore. Even if $(\lambda_0, 0)$ is a bifurcation point for periodic solutions, one can, in general, only expect the periods of these solutions to lie near T whenever λ is near λ_0. We therefore introduce this unknown period as an additional parameter $\tau \in \mathbb{R}$. Since u is a τ-periodic solution ($\tau > 0$) of the equation $\dot{x} = f(\lambda, x)$ if and only if $v(t) := u(t\tau/2\pi)$ is a 2π-periodic solution of $\dot{y} = \frac{\tau}{2\pi} f(\lambda, y)$, we consider the new equation

$$\dot{y} = \frac{\tau}{2\pi} f(\lambda, y), \quad \lambda \in \Lambda, \ \tau \in (0, \infty), \tag{52}$$

and we seek 2π-periodic solutions near $(T, \lambda_0, 0) \in \mathbb{R} \times \Lambda \times X$.

For a sufficiently small $\delta > 0$, we set $\Sigma := (-\delta, \delta)^2$ and

$$\tilde{f}(\sigma, x) := \frac{\tau + T}{2\pi} f(\lambda + \lambda_0, x), \quad \sigma := (\tau, \lambda),$$

for all $(\sigma, x) \in \Sigma \times X$. We then search for nontrivial 2π-periodic solutions of the parameter-dependent equation

$$\dot{x} = \tilde{f}(\sigma, x) \tag{53}$$

near $(0, 0)$, where $\sigma \in \Sigma \subseteq \mathbb{R}^2$. With the notation of example (26.15 b) now referring to equation (53), we set

$$g(\sigma, \xi) := \xi - u(2\pi, 0, \xi, \sigma)$$

and so for $B := D_1 D_2 g(0, 0) \in \mathcal{L}^2(\mathbb{R}^2 \times E, E)$ we obtain

$$\langle e', B[\cdot, e] \rangle = -\int_0^{2\pi} \langle w(t), D_1 D_2 \tilde{f}(0, 0)[\cdot, v(t)] \rangle \, dt, \tag{54}$$

where v denotes the 2π-periodic solution of the equation

$$\dot{y} = D_2 \tilde{f}(0, 0) y \quad \text{with} \quad v(0) = e, \tag{55}$$

and w is the 2π-periodic solution of the dual equation

$$\dot{z} = -[D_2 \tilde{f}(0, 0)]' z \quad \text{with} \quad w(0) = e' \tag{56}$$

(cf. the discussion preceding theorem (26.16)).

Equation (55) has a nontrivial 2π-periodic solution if and only if

$$1 \in \sigma \left(e^{2\pi D_2 \tilde{f}(0,0)} \right).$$

It follows from the spectral mapping theorem and from the definition of $\tilde{f}$ that this relation is satisfied if and only if we have

$$\frac{2\pi i}{T} \mathbb{Z} \cap \sigma(D_2 f(\lambda_0, 0)) \neq \emptyset.$$

We now make the following *assumptions*:

$$\frac{2\pi i}{T} \mathbb{Z} \cap \sigma(D_2 f(\lambda_0, 0)) = \{\pm 2\pi i / T\} \tag{57}$$

and $2\pi i / T$ *is an eigenvalue of* $D_2 f(\lambda_0, 0)$ *of geometric multiplicity 1.*

To simplify, we set

$$L := D_2 \tilde{f}(0, 0) = (T / 2\pi) D_2 f(\lambda_0, 0).$$

For

$$A := D_2 g(0, 0) = id_E - e^{2\pi L} \in \mathcal{L}(E)$$

and

$$N := \ker(A) = \ker(1 - e^{2\pi L})$$

we have

$$N = [\ker(i - L_{\mathbb{C}}) \oplus \ker(-i - L_{\mathbb{C}})] \cap E,$$

hence, in particular,

$$\dim N = 2,$$

and $e \in N$ if and only if

$$v(t) := t \cdot e := \exp(tL)e, \quad t \in \mathbb{R},$$

is a 2π-periodic solution of (55). [In fact, if $L_{\mathbb{C}}z = iz$, it then follows from the power series representation that $e^{2\pi L_{\mathbb{C}}}z = e^{2\pi i}z = z$. Since $L_{\mathbb{C}}\bar{z} = \overline{L_{\mathbb{C}}z} = -i\bar{z}$, we similarly obtain $e^{2\pi L_{\mathbb{C}}}\bar{z} = e^{-2\pi i}\bar{z} = \bar{z}$. For $\xi := (z + \bar{z})/2$ and $\eta := (z - \bar{z})/2i$ we therefore have $e^{2\pi L}\xi = e^{2\pi L_{\mathbb{C}}}\xi = \xi$ and $e^{2\pi L}\eta = e^{2\pi L_{\mathbb{C}}}\eta = \eta$, which proves that $N \supseteq [\ker(i - L_{\mathbb{C}}) \oplus \ker(-i - L_{\mathbb{C}})] \cap E$. If it were true that $\dim N > 2$, we could find a 2π-periodic solution of $\dot{y} = Ly$ which is not a linear combination of $e^{tL}\xi$ and $e^{tL}\eta$. But from assumptions (57) and from (the proof of) theorem (12.10) it easily follows that this is impossible. We therefore have $N = [\ker(i - L_{\mathbb{C}}) \oplus \ker(-i - L_{\mathbb{C}})] \cap E.$] Therefore $2\pi \cdot e = e$ and consequently $t \cdot e = t \cdot (2\pi \cdot e) = 2\pi \cdot (t \cdot e)$ for every $e \in N$ and $t \in \mathbb{R}$. This shows that N is invariant with respect to the linear flow $\exp(tL)$.

Let $e \in N \setminus \{0\}$ be arbitrary and suppose $\{e_1', e_2'\}$ is an arbitrary basis for $\ker(A')$. Then we have (cf. (36)) $P_c B e = P_c B[\cdot, e] \in \text{Isom}(\mathbb{R}^2, R_c)$ if and only if

$$\begin{vmatrix} \langle e_1', B[u_1, e] \rangle & \langle e_1', B[u_2, e] \rangle \\ \langle e_2', B[u_1, e] \rangle & \langle e_2', B[u_2, e] \rangle \end{vmatrix} \neq 0, \tag{58}$$

where $\{u_1, u_2\}$ is an arbitrary basis for $\mathbb{R}^2$.

We denote the dual linear flow by $t \cdot e'$, where $e' \in \ker(A')$, i.e.,

$$t \cdot e' = \exp(-tL')e', \quad \forall e' \in \ker(A').$$

If we take into account the fact that the integral of a periodic function over an interval of length equal to the period is independent of the position of this interval, we obtain from (54) that

$$\begin{aligned} \langle e', B[\cdot, e] \rangle &= -\int_0^{2\pi} \langle (t - s) \cdot (s \cdot e'), D_1 D_2 \tilde{f}(0, 0)[\cdot, (t - s) \cdot (s \cdot e)] \rangle \, dt \\ &= -\int_{-s}^{2\pi - s} \langle t \cdot (s \cdot e'), D_1 D_2 \tilde{f}(0, 0)[\cdot, t \cdot (s \cdot e)] \rangle \, dt \\ &= -\int_0^{2\pi} \langle t \cdot (s \cdot e'), D_1 D_2 \tilde{f}(0, 0)[\cdot, t \cdot (s \cdot e)] \rangle \, dt \\ &= \langle s \cdot e', B[\cdot, s \cdot e] \rangle \end{aligned} \tag{59}$$

for all $e \in N$, $e' \in \ker(A')$ and $s \in \mathbb{R}$.

Because $\ker(A')$ is invariant with respect to the dual linear flow, we deduce that for each basis $\{e_1', e_2'\}$ of $\ker(A')$ and for every $s \in \mathbb{R} \setminus \{0\}$, $\{s \cdot e_1', s \cdot e_2'\}$ is also a basis for $\ker(A')$. It therefore follows from (59) that $P_c B_e \in \mathrm{Isom}(\mathbb{R}^2, R_c)$ implies

$$P_c B_{t\cdot e} \in \mathrm{Isom}(\mathbb{R}^2, R_c), \quad \forall t \in \mathbb{R}. \tag{60}$$

By corollary (24.22), we have that Ω, the interior of the periodic orbit $\gamma(e) := \{t \cdot e \mid t \in \mathbb{R}\} \subseteq N$, contains a critical point. Since 0 is the only critical point of the linear flow $\exp(tL)$, we have $0 \in \Omega$. Thus for every $\xi \in N \setminus \{0\}$ there exists

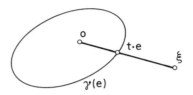

some $\alpha > 0$ and $t \cdot e \in \gamma(e)$ such that $\xi = \alpha(t \cdot e)$. Hence from (60) we obtain that

$$P_c B_\xi = \alpha P_c B_{t\cdot e} \in \mathrm{Isom}(\mathbb{R}^2, R_c), \quad \forall \xi \in N \setminus \{0\},$$

if there exists some $e \in N \setminus \{0\}$ such that $P_c B_e \in \mathrm{Isom}(\mathbb{R}^2, R_c)$. In other words: *If there exists some $e \in N \setminus \{0\}$ such that*

$$\mathrm{im}(B_e) \oplus \mathrm{im}(A) = E,$$

then this relation holds for all $e \in N \setminus \{0\}$.

From the definition of $\tilde{f}$ we obtain

$$D_1 D_2 \tilde{f}(0,0)[\hat{\sigma}, \cdot] = \frac{\hat{\tau}}{T} L + \hat{\lambda} \frac{T}{2\pi} D_1 D_2 f(\lambda_0, 0)$$

for all $\hat{\sigma} := (\hat{\tau}, \hat{\lambda}) \in \mathbb{R}^2$. Since L and e^{tL} commute and since $e^{-tL'} = \left[\left(e^{tL}\right)^{-1}\right]'$, it follows that

$$\langle t \cdot e', L(t \cdot e) \rangle = \langle t \cdot e', t \cdot (Le) \rangle = \langle e', Le \rangle \tag{61}$$

for all $t \in \mathbb{R}$, hence

$$\langle e', B[\hat{\sigma}, e] \rangle = -\int_0^{2\pi} \langle t \cdot e', D_1 D_2 \tilde{f}(0,0)[\hat{\sigma}, t \cdot e] \rangle \, dt$$

$$= -\hat{\tau} \frac{2\pi}{T} \langle e', Le \rangle - \hat{\lambda} \frac{T}{2\pi} \int_0^{2\pi} \langle t \cdot e', D_1 D_2 f(\lambda_0, 0)(t \cdot e) \rangle \, dt.$$

If now we choose $\{u_1, u_2\}$ to be the standard basis for $\mathbb{R}^2$, we then obtain for the determinant in (58) the value

$$\begin{vmatrix} \langle e_1', Le \rangle & \int_0^{2\pi} \langle t \cdot e_1', D_1 D_2 f(\lambda_0, 0)(t \cdot e) \rangle \, dt \\ \langle e_2', Le \rangle & \int_0^{2\pi} \langle t \cdot e_2', D_1 D_2 f(\lambda_0, 0)(t \cdot e) \rangle \, dt \end{vmatrix}. \tag{62}$$

After these preliminaries we can prove the following fundamental theorem about Hopf bifurcation.

(26.21) Theorem. *Let $\Lambda \subseteq \mathbb{R}$, $k \geq 2$ and assume that $f \in C^k(\Lambda \times X, E)$ is such that $f(\cdot, 0) = 0$. Suppose that for some $\lambda_0 \in \Lambda$ and $T_0 > 0$, $L := (T_0/2\pi)D_2 f(\lambda_0, 0)$ satisfies*

$$i\mathbb{Z} \cap \sigma(L) = \{\pm i\} \quad \text{and } i \text{ is a simple eigenvalue of } L. \tag{63}$$

Moreover, let $z := \xi + i\eta \in \ker(L_\mathbb{C} - i)$ and $w := u + iv \in \ker(L'_\mathbb{C} - i)$ and assume that the determinant

$$\begin{vmatrix} \langle u, \eta \rangle & \int_0^{2\pi} \langle t \cdot u, D_1 D_2 f(\lambda_0, 0)(t \cdot \xi) \rangle \, dt \\ \langle v, \eta \rangle & \int_0^{2\pi} \langle t \cdot v, D_1 D_2 f(\lambda_0, 0)(t \cdot \xi) \rangle \, dt \end{vmatrix} \tag{64}$$

is nonzero, where we have set $t \cdot \xi := e^{tL}\xi$ and $t \cdot u := e^{-tL'}u$, as well as $t \cdot v := e^{-tL'}v$. Then the equation

$$\dot{x} = f(\lambda, x)$$

possesses a unique one-parameter family of noncritical periodic orbits $\{\gamma(s) \mid 0 < s < \epsilon\}$ in a neighborhood of $(\lambda_0, 0)$.

More precisely, the following holds: There exists some $\epsilon > 0$ and a function

$$(y(\cdot), T(\cdot), \lambda(\cdot)) \in C^{k-1}((-\epsilon, \epsilon), X \times \mathbb{R} \times \Lambda)$$

satisfying

$$(y(0), T(0), \lambda(0)) = (0, T_0, \lambda_0)$$

and such that $\gamma(s) := \gamma(x(s))$, $0 < |s| < \epsilon$, is a noncritical $T(s)$-periodic orbit of $\dot{x} = f(\lambda(s), x)$ through $x(s) \in E$, where

$$x(s) := s(\xi + y(s)), \quad -\epsilon < s < \epsilon.$$

If $0 < s_1 < s_2 < \epsilon$, then $\gamma(s_1) \neq \gamma(s_2)$, and there exists a neighborhood of $(\lambda_0, 0, T_0) \in \Lambda \times X \times \mathbb{R}$ such that every noncritical periodic orbit contained in it belongs to the family $\{\gamma(s) \mid 0 < s < \epsilon\}$.

Proof. Since $\ker(L_\mathbb{C} - i) = [\mathrm{im}(L'_\mathbb{C} - i)]^\perp$ (cf. the remarks preceding lemma (26.12), which, of course, also hold in case $\mathbb{K} = \mathbb{C}$) and because $L'_\mathbb{C} = (L')_\mathbb{C}$, i is a simple eigenvalue of L'.

One easily verifies that $\{u, v\}$ is a basis for $[\ker(L'_\mathbb{C} - i) \oplus \ker(L'_\mathbb{C} + i)] \cap E' = \ker(A')$. Since, in addition, the equation $L\xi = -\eta$ follows from $L_\mathbb{C}(\xi + i\eta) = i(\xi + i\eta) = -\eta + i\xi$, we see that the determinant in (64) agrees with (62) (where $e := \xi$ and $\{e'_1, e'_2\} = \{u, v\}$) modulo the factor -1. Hence it follows from (62) and the considerations preceding the theorem that the conditions of theorem (26.19) are satisfied for the function g. Consequently, in a sufficiently small neighborhood of $(0, 0) \in \Sigma \times X$, the set of zeros of g is given by $M \cup (\Sigma \times \{0\})$, where for

$x \neq 0$, M is a two-dimensional C^{k-1}-manifold and $M \cap (\Sigma \times \{0\}) = \{(0,0)\}$. Furthermore, M has the global parametrization

$$\{(\hat{\sigma}(\alpha), \alpha + \hat{y}(\alpha)\alpha) \mid \alpha \in \epsilon \mathbb{B}_N\}, \tag{65}$$

where

$$(\hat{\sigma}, \hat{y}) \in C^{k-1}(\epsilon \mathbb{B} \setminus \{0\}, \Sigma \times \mathcal{L}(N, N_c)) \cap C(\epsilon \mathbb{B}, \Sigma \times \mathcal{L}(N, N_c))$$

and

$$(\hat{\sigma}(0), \hat{y}(0)) = (0, 0)$$

for some $\epsilon > 0$ sufficiently small and some arbitrary complementary subspace N_c of $N := \ker(A)$. For each $(\sigma, x) \in M$, $u(\cdot, 0, x, \sigma)$ is a 2π-periodic solution of $\dot{x} = \tilde{f}(\sigma, x)$ through x, and this solution is nonconstant if and only if $(\sigma, x) \neq (0, 0)$. If (σ, x) is taken to be sufficiently close to $(0, 0)$, then the whole orbit

$$\Gamma(\sigma, x) := \{(\sigma, u(t, 0, x, \sigma)) \mid 0 \leq t \leq 2\pi\}$$

lies in M and every point of M lies on exactly one of these orbits.

Let $Q : E \to N$ denote the projection on N along N_c. Since σ is constant along every orbit $\Gamma(\sigma, x)$ and since (65) is a global parametrization of M, the projection $Q \circ pr_E(\Gamma(\sigma, x))$ of $\Gamma(\sigma, x)$ into N is a C^{k-1}-Jordan curve $\tilde{\gamma}(\alpha)$ through the point $\alpha = Qx \in N$ (if $x \neq 0$). Now $h(\alpha) := Q\tilde{f}(\hat{\sigma}(\alpha), \alpha + \hat{y}(\alpha)\alpha)$ defines a C^{k-1}-vector field in a deleted neighborhood of $0 \in N$, which is tangent to the orbits $\tilde{\gamma}(\alpha)$ and has the unique critical point $0 \in N$. If $\Omega(\alpha)$ denotes the interior

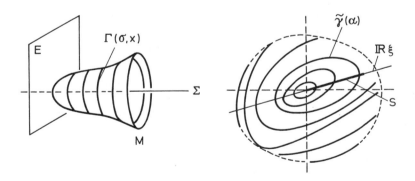

of $\tilde{\gamma}(\alpha)$, it follows from corollary (24.22) (and considering corollary (21.6 viii)) that $0 \in \Omega(\alpha)$. Hence all curves $\tilde{\gamma}(\alpha)$, $\alpha \neq 0$, "go" around the origin of N, and every point of some appropriate neighborhood U of $0 \in N$ lies on exactly one of these curves.

Let $\varphi(t) := h(t\xi)$ for $|t| < \epsilon$. Then $\varphi \in C^{k-1}((-\epsilon, \epsilon), N)$,

$$\dot{\varphi}(0) = QD_2\tilde{f}(0, 0)\xi = QL\xi = -\eta,$$

and $\{\xi, \eta\}$ is a basis for N. This implies that in some deleted neighborhood of 0, the vector field h is transversal to the line $\mathbb{R}\xi$. Therefore there exists some $\epsilon > 0$ so that $S := \{s\xi \mid 0 < s < \epsilon\}$ is a local section for the flow induced by h and such that every orbit $\tilde{\gamma}(\alpha)$, which is sufficiently close to $0 \in N$, intersects S in a unique point (cf. the discussion in section 24). We now set

$$y(s) := \hat{y}(s\xi)\xi, \quad \sigma(s) := \hat{\sigma}(s\xi), \tag{66}$$

as well as

$$(T(s), \lambda(s)) := \sigma(s) + (T_0, \lambda_0) \tag{67}$$

for all $|s| < \epsilon$. Then the functions $y(\cdot)$, $T(\cdot)$ and $\lambda(\cdot)$ have the properties as claimed.

$\square$

(26.22) Remarks. (a) The proof above shows that $y(s) \in N_c$ for all $s \in (-\epsilon, \epsilon)$, where N_c is a complementary subspace of

$$N = [\ker(L_{\mathbb{C}} - i) \oplus \ker(L_{\mathbb{C}} + i)] \cap E$$

in E. Here $y(\cdot)$, of course, depends on the choice of N_c.

(b) For a fixed N_c, the functions $y(\cdot)$, $\lambda(\cdot)$ and $T(\cdot)$ depend on the choice of $\xi \in N \setminus \{0\}$. The exact dependence can be seen from (66) and (67).

(c) Although, in the proof of theorem (26.21), we only made use of the fact that i is a geometrically simple eigenvalue of L, condition (64) implies that i must be an algebraically simple eigenvalue of L.

In order to see this, we set $C := L_{\mathbb{C}} - i$ and assume that 0 is an eigenvalue of C of multiplicity $n > 1$. Then $E_{\mathbb{C}}$ has a decomposition $E_{\mathbb{C}} = Z_0 \oplus Z_1$ which reduces C, $C = C_0 \oplus C_1$, such that $\sigma(C_0) = \{0\}$ and $C_1 \in \mathcal{GL}(Z_1)$. There exists a basis $\{z_1, \ldots, z_n\}$ for Z_0 such that $C_0 z_j = z_{j-1}$, $j = 1, \ldots, n$, and $z_0 := 0$, from which it follows, in particular, that $\ker(C_0) = \ker(C) \subseteq \operatorname{im}(C)$ (cf. section 12). Since $\ker(C') = [\operatorname{im}(C)]^{\perp}$, we have that $\langle w, a \rangle = 0$ for all $a \in \operatorname{im}(C)$. Since, finally, $L_{\mathbb{C}} z = iz$ implies $L_{\mathbb{C}} \bar{z} = -i\bar{z}$, and therefore $\bar{z} \in \ker(L_{\mathbb{C}} + i) = \ker(C + 2i) \subseteq Z_1 \subseteq \operatorname{im}(C)$, it follows that $\eta = (z - \bar{z})/2i \in \operatorname{im}(C)$ with $z := z_1$, from which we obtain $\langle w, \eta \rangle = \langle u, \eta \rangle + i\langle v, \eta \rangle = 0$. Consequently, in this case the first column of the determinant (64) is the zero column, which implies the assertion.

(d) With the notation of the proof above, we have $\dot{\sigma}(0) = 0$. If, in fact, we had $\dot{\sigma}(0) \neq 0$, then $\sigma(s) = s\dot{\sigma}(0) + o(s)$ would imply the existence of some $\epsilon_0 > 0$ so that $\sigma(-s) \neq \sigma(s)$ for all $0 < s < \epsilon_0$. But then $\hat{\sigma}$ would not be constant along the orbits $\tilde{\gamma}(\alpha)$ near the origin (because the line $\mathbb{R}\xi$ intersects every such orbit twice – on both sides of zero). $\square$

The Eigenvalue Condition

The following considerations serve to find a geometric interpretation of condition (64).

(26.23) Lemma. *If i is a simple eigenvalue of L, then for every eigenvector*

$$z := \xi + i\eta \in \ker(L_{\mathbb{C}} - i)$$

there exists a functional

$$w := u + iv \in \ker(L'_{\mathbb{C}} - i)$$

such that

$$\langle u, \xi \rangle = -\langle v, \eta \rangle = 1 \quad and \quad \langle u, \eta \rangle = \langle v, \xi \rangle = 0.$$

With this choice of z and w, the determinant in (64) has the value

$$\int_0^{2\pi} \langle t \cdot u, D_1 D_2 f(\lambda_0, 0)(t \cdot \xi) \rangle \, dt. \tag{68}$$

Proof. Since i, and therefore also $-i$, is a simple eigenvalue of $L_{\mathbb{C}}$, the direct sum decomposition

$$E_{\mathbb{C}} = \ker(L_{\mathbb{C}} - i) \oplus \ker(L_{\mathbb{C}} + i) \oplus H,$$

where

$$H := \bigoplus_{\substack{\mu \in \sigma(L) \\ \mu \neq \pm i}} \ker[(L - \mu)^{m(\mu)}],$$

reduces the operator $L_{\mathbb{C}}$. If $M := H \cap E$, it follows that the direct sum decomposition $E = N \oplus M$ reduces the operator L. We choose a basis $\{e_1, \ldots, e_m\}$ for E such that $e_1 := \xi$ and $e_2 := \eta$, where $z := \xi + i\eta$ is an arbitrary eigenvector of $L_{\mathbb{C}}$ corresponding to the eigenvalue i. We let $\{e'_1, \ldots, e'_m\}$ denote the corresponding dual basis. Then we have $Le_1 = -e_2$, $Le_2 = e_1$ and $L(M) \subseteq M$. Hence for every $x = \alpha e_1 + \beta e_2 + \zeta \in N \oplus M$ we have

$$\langle L'e'_j, x \rangle = \langle e'_j, Lx \rangle = \alpha \langle e'_j, Le_1 \rangle + \beta \langle e'_j, Le_2 \rangle + \langle e'_j, L\zeta \rangle$$
$$= -\alpha \langle e'_j, e_2 \rangle + \beta \langle e'_j, e_1 \rangle + \langle e'_j, L\zeta \rangle.$$

From this we deduce

$$\langle L'e'_1, x \rangle = \beta = \langle e'_2, x \rangle$$

and

$$\langle L'e'_2, x \rangle = -\alpha = \langle -e'_1, x \rangle.$$

We thus obtain

$$L'e'_1 = e'_2 \quad and \quad L'e'_2 = -e'_1,$$

which implies

$$L'_{\mathbb{C}}(e'_1 - ie'_2) = i(e'_1 - ie'_2).$$

Setting $u := e'_1$ and $v := -e'_2$, the assertion follows.   □

The following "perturbation theorem" improves proposition (25.12) in the case of a simple eigenvalue.

(26.24) Proposition. *Let Y denote a finite dimensional complex Banach space and let $A(\cdot) \in C^k(\Lambda, \mathcal{L}(Y))$ for some $k \geq 1$. For some $\lambda_0 \in \Lambda$ let $\mu_0 \in \mathbb{C}$ be a simple eigenvalue of $A(\lambda_0)$ with corresponding eigenvector y_0. Then there exist neighborhoods, U of μ_0 in $\mathbb{C}$ and V of λ_0 in Λ, as well as functions $\mu(\cdot) \in C^k(V, U)$ and $y(\cdot) \in C^k(V, Y \setminus \{0\})$ with the following properties*:

(i) $(\mu(\lambda_0), y(\lambda_0)) = (\mu_0, y_0)$.
(ii) *For each $\lambda \in V$ we have*

$$A(\lambda)y(\lambda) = \mu(\lambda)y(\lambda).$$

(iii) $\mu(\lambda)$ *is the only eigenvalue of $A(\lambda)$ in U and $\mu(\lambda)$ is simple.*

Proof. We identify Y with $\mathbb{R}^{2\dim Y}$ and $\mathbb{C}$ with $\mathbb{R}^2$, and we define

$$T \in C^k(\Lambda \times [Y \times \mathbb{C}], Y \times \mathbb{C})$$

by

$$T(\lambda, (y, \mu)) := (A(\lambda)y - \mu y, |y|^2 - 1).$$

We then have $T(\lambda_0, (y_0, \mu_0)) = 0$ and

$$D_2 T(\lambda_0, (y_0, \mu_0))(\hat{y}, \hat{\mu}) = ([A(\lambda_0) - \mu_0]\hat{y} - \hat{\mu}y_0, 2(y_0 \mid \hat{y}))$$

for all $(\hat{y}, \hat{\mu}) \in Y \times \mathbb{C}$. If $(\hat{y}, \hat{\mu}) \in \ker D_2 T(\lambda_0, (y_0, \mu_0))$, then

$$\hat{y} \perp y_0, \quad [A(\lambda_0) - \mu_0]\hat{y} = \hat{\mu}y_0 \in \ker[A(\lambda_0) - \mu_0]. \qquad (69)$$

Since μ_0 is a simple eigenvalue of $A(\lambda_0)$, $\ker[A(\lambda_0) - \mu_0]$ reduces the operator $A(\lambda_0) - \mu_0$. It thus follows from (69) that $(\hat{y}, \hat{\mu}) = (0, 0)$, that is, $\ker D_2 T(\lambda_0, (y_0, \mu_0)) = \{0\}$. Therefore $D_2 T(\lambda_0, (y_0, \mu_0)) \in \mathcal{GL}(Y \times \mathbb{C})$ and now the implicit function theorem implies the existence of neighborhoods, U of μ_0 in $\mathbb{C}$, V of λ_0 in Λ and W of y_0 in Y, as well as a C^k-function $(y(\cdot), \mu(\cdot)) : V \to W \times U$ such that

$$(\lambda, (y, \mu)) \in V \times [W \times U], \quad T(\lambda, (y, \mu)) = 0 \quad \Leftrightarrow \quad \lambda \in V, \quad (y, \mu) = (y(\lambda), \mu(\lambda)).$$

This shows that for every $\lambda \in V$, $\mu(\lambda)$ is an eigenvalue of $A(\lambda)$ with corresponding eigenvector $y(\lambda)$ and that there are no other eigenvalues of $A(\lambda)$ in U. Hence, by proposition (25.12), $\mu(\lambda)$ is a simple eigenvalue of $A(\lambda)$. $\qquad \square$

After these preliminaries we can give a geometric interpretation of assumption (64) of theorem (26.21) about the Hopf bifurcation.

(26.25) Theorem. *Let $\Lambda \subseteq \mathbb{R}$, $k \geq 2$ and $f \in C^k(\Lambda \times X, E)$ be such that $f(\cdot, 0) = 0$. Assume that for some $\lambda_0 \in \Lambda$ and some $T_0 > 0$ we have*:

$$\frac{2\pi i}{T_0} \mathbb{Z} \cap \sigma(D_2 f(\lambda_0, 0)) = \{\pm 2\pi i/T_0\} \tag{70}$$

and $2\pi i/T_0$ is a simple eigenvalue of $D_2 f(\lambda_0, 0)$.

In addition, let $\mu(\lambda)$ denote the eigenvalue of $D_2 f(\lambda, 0) \in \mathcal{L}(E)$, which is well-defined in some neighborhood V of λ_0, such that $\mu(\lambda_0) = 2\pi i/T_0$. If

$$\frac{d}{d\lambda}[Re\,\mu(\lambda_0)] \neq 0, \tag{71}$$

then in some neighborhood of $(\lambda_0, 0)$, the equation $\dot{x} = f(\lambda, x)$ has a unique one-parameter family of noncritical periodic orbits $\{\gamma(s) \mid 0 < s < \epsilon\}$ which converge to the critical point 0 as $s \to 0$. More precisely, the assertions of theorem (26.21) hold.

Proof. We set $Y := E_{\mathbb{C}}$ and $A(\lambda) := [D_2 f(\lambda, 0)]_{\mathbb{C}}$. We then have

$$A(\cdot) \in C^{k-1}(\Lambda, \mathcal{L}(Y))$$

and $A(\lambda_0)$ has the simple eigenvalue $\mu_0 := 2\pi i/T_0$. By lemma (26.23), we can choose

$$z_0 := \xi + i\eta \in \ker(A(\lambda_0) - \mu_0)$$

and

$$w := u + iv \in \ker([A(\lambda_0)]' - \mu_0)$$

such that

$$\langle u, \xi \rangle = -\langle v, \eta \rangle = 1 \quad \text{and} \quad \langle u, \eta \rangle = \langle v, \xi \rangle = 0. \tag{72}$$

For every $t \in \mathbb{R}$ we have

$$t \cdot z_0 := e^{tL_{\mathbb{C}}} z_0 = t \cdot \xi + i(t \cdot \eta) \in \ker(A(\lambda_0) - \mu_0)$$

and

$$t \cdot w := e^{-tL'_{\mathbb{C}}} w = t \cdot u + i(t \cdot v) \in \ker([A(\lambda_0)]' - \mu_0),$$

as well as

$$\langle t \cdot w, t \cdot z_0 \rangle = \langle w, z_0 \rangle = \langle u, \xi \rangle - \langle v, \eta \rangle = 2.$$

It follows from proposition (26.24) that $\mu(\cdot) \in C^{k-1}(V, \mathbb{C})$, $\mu(\lambda_0) = \mu_0$ and that for every $t \in \mathbb{R}$ there exists a function $z_t(\cdot) \in C^{k-1}(V, Y)$ such that $z_t(\lambda_0) = t \cdot z_0$ and

$$A(\lambda)z_t(\lambda) = \mu(\lambda)z_t(\lambda), \quad \forall \lambda \in V.$$

Differentiating this equation at the point λ_0, we obtain

$$DA(\lambda_0)(t \cdot z_0) = D\mu(\lambda_0)(t \cdot z_0) + [\mu_0 - A(\lambda_0)]Dz_t(\lambda_0).$$

By applying the functional $t \cdot w$ to this equation, we obtain, based on the consideration above, the relation

$$\langle t \cdot w, DA(\lambda_0)(t \cdot z_0) \rangle = 2D\mu(\lambda_0),$$

hence

$$2Re\, D\mu(\lambda_0) = \langle t \cdot u, D_1 D_2 f(\lambda_0, 0)(t \cdot \xi) \rangle$$
$$- \langle t \cdot v, D_1 D_2 f(\lambda_0, 0)(t \cdot \eta) \rangle$$

for all $t \in \mathbb{R}$. By integrating this equation we get

$$4\pi Re\, D\mu(\lambda_0) = \int_0^{2\pi} \langle t \cdot u, D_1 D_2 f(\lambda_0, 0)(t \cdot \xi) \rangle \, dt \tag{73}$$
$$- \int_0^{2\pi} \langle t \cdot v, D_1 D_2 f(\lambda_0, 0)(t \cdot \eta) \rangle \, dt.$$

It follows from (59) that

$$\int_0^{2\pi} \langle t \cdot v, D_1 D_2 f(\lambda_0, 0)(t \cdot \eta) \rangle \, dt \tag{74}$$
$$= \int_0^{2\pi} \langle t \cdot (s \cdot v), D_1 D_2 f(\lambda_0, 0)(t \cdot (s \cdot \eta)) \rangle \, dt$$

for all $s \in \mathbb{R}$. We will now show that

$$(\pi/2) \cdot \eta = \xi \quad \text{and} \quad (\pi/2) \cdot v = -u. \tag{75}$$

It then follows from (73) and (74) that

$$2\pi Re\, D\mu(\lambda_0, 0) = \int_0^{2\pi} \langle t \cdot u, D_1 D_2 f(\lambda_0, 0)(t \cdot \xi) \rangle \, dt. \tag{76}$$

Therefore, based on the second part of lemma (26.23), the condition (64) of theorem (26.21) is satisfied and hence the assertion is proved.

Since E has the direct sum decomposition $E = N \oplus M$, which reduces L, and since E' can be identified with $N' \oplus M'$ in a natural way, it suffices, in order to prove (75), to consider $L|N$.

We identify N with $\mathbb{R}^2$ by means of the basis $\{\xi, \eta\}$. Then $L\xi = -\eta$ and $L\eta = \xi$ imply that $L|N$ is represented by the matrix

$$\begin{bmatrix} 0 & 1 \\ -1 & 0 \end{bmatrix}.$$

Consequently, the linear flow $e^{tL}|N = e^{tL|N}$ represents a clockwise rotation about the origin (cf. section 13, case 2: (b)). If $t = \pi/2$, we therefore have $t \cdot \eta = \xi$ and $t \cdot \xi = -\eta$. It follows from $\langle t \cdot z, t \cdot \zeta \rangle = \langle z, \zeta \rangle$ for all $z \in E'$ and $\zeta \in E$ that

$$\langle t \cdot v, \xi \rangle = \langle t \cdot v, t \cdot \eta \rangle = \langle v, \eta \rangle = -1$$

and

$$\langle t \cdot v, \eta \rangle = -\langle t \cdot v, t \cdot \xi \rangle = -\langle v, \xi \rangle = 0,$$

which implies $t \cdot v = -u$. Therefore (75) holds and the theorem is proved.   □

Periodic Solutions of Hamiltonian Systems

As a simple application of the Hopf bifurcation theorem we will prove the *Liapunov center theorem* about the existence of small periodic solutions of autonomous Hamiltonian systems. In the following theorem we will identify $\mathbb{R}^{2n}$ with its dual space in the usual way, that is, by means of the Euclidean inner product (cf. the remark (11.16 b)). If $H : U \to \mathbb{R}^{2n}$, then $H' := DH$ denotes the gradient and $H'' := D^2H$ the Hessian of H. Furthermore, $J \in \mathcal{L}(\mathbb{R}^n \times \mathbb{R}^n)$ denotes the symplectic normal from.

(26.26) Theorem. *Let $U \subseteq \mathbb{R}^{2n}$ be open and $H \in C^k(U, \mathbb{R})$, $k \geq 3$, and assume that $z_0 \in U$ is a critical point of H. Moreover, assume that for some $\omega > 0$, $i\omega$ is a simple eigenvalue of $JH''(z_0)$ and*

$$\sigma(JH''(z_0)) \cap \omega i\mathbb{Z} = \{\pm i\omega\}.$$

Then, in some neighborhood of z_0, the Hamiltonian system

$$\dot{z} = JH'(z)$$

possesses a unique one-parameter family of noncritical periodic orbits $\{\gamma(s) \mid 0 < s < \epsilon\}$. More precisely, we have the following: There exist C^{k-2}-functions

$$z(\cdot) : (-\epsilon, \epsilon) \to U, \quad T(\cdot) : (-\epsilon, \epsilon) \to \mathbb{R}_+$$

so that

$$z(0) = z_0 \quad and \quad T(0) = 2\pi/\omega$$

and such that for $0 < |s| < \epsilon$, $\gamma(s) := \gamma(z(s))$ is a noncritical $T(s)$-periodic orbit through $z(s) \in U$. If $0 < s_1 < s_2$, then $\gamma(s_1) \neq \gamma(s_2)$ and there exists a neighborhood of z_0 such that every periodic orbit in it belongs to the family $\{\gamma(s) \mid 0 \leq s < \epsilon\}$.

Proof. By means of a translation we can shift z_0 to the origin and thus, without loss of generality, we may assume that $z_0 = 0$. We then define $f \in C^{k-1}(\mathbb{R} \times U, \mathbb{R}^{2n})$ by

$$f(\lambda, z) := JH'(z) + \lambda H'(z).$$

Then $f(\cdot, 0) = 0$ and

$$D_2 f(0, 0) = JH''(0), \quad \text{as well as} \quad D_1 D_2 f(0, 0) = H''(0).$$

Hence condition (63) of theorem (26.21) is satisfied, where $T_0 := 2\pi/\omega$ and $\lambda_0 := 0$. It follows from

$$(H'(z) \mid f(\lambda, z)) = (H'(z) \mid JH'(z)) + \lambda|H'(z)|^2 = \lambda|H'(z)|^2$$

that for every fixed $\lambda \in \mathbb{R}$, H is a Liapunov function for $\dot{z} = f(\lambda, z)$, where $\dot{H}(z) = \lambda|H'(z)|^2$. Since, by assumption, $0 \notin \sigma(JH''(0))$, and therefore also $0 \notin \sigma(H''(0))$, 0 is a nondegenerate critical point. It then follows from the inverse function theorem that $H'(z) \neq 0$ for all nonzero z in some neighborhood of $z = 0$. Hence for all $\lambda \neq 0$ and all z in some neighborhood of the origin we have $\dot{H}(z) \neq 0$, which implies that for all $\lambda \neq 0$, the vector field $f(\lambda, \cdot)$ has no noncritical periodic orbits near $z = 0$. The assertion therefore follows from theorem (26.21) if we can show that condition (64) is satisfied (in this case we have $\lambda(s) = 0$ for all $s \in (-\epsilon, \epsilon)$).

We now set $E := \mathbb{R}^{2n}$ and $L := JB$, where $B := (1/\omega)H''(0)$. Since $i \in \sigma(L)$, there exists a basis $\{\xi, \eta\}$ for $N := \ker(A)$, where $A := id_E - e^{2\pi L}$, such that

$$L\xi = -\eta \quad \text{and} \quad L\eta = \xi. \tag{77}$$

From $J^* = J^{-1}$, $B^* = B$ and $z = e^{2\pi L}z$ we obtain

$$Jz = Je^{2\pi L}J^{-1}Jz = e^{2\pi JLJ^{-1}}Jz = e^{-2\pi L^*}Jz.$$

Therefore $Jz \in \ker(A^*)$ for all $z \in \ker(A)$. Since $J \in \mathcal{GL}(E)$, it follows that $J[\ker(A)] = \ker(A^*) = [\mathrm{im}(A)]^{\perp}$. Because i is a simple eigenvalue of L, we know that $E = \ker(A) \oplus \mathrm{im}(A)$ is correct. Consequently, for every $z \in N \setminus \{0\}$ there exists some $x \in N$ such that $(Jz \mid x) \neq 0$ (for otherwise we would have $Jz \in [\mathrm{im}(A)]^{\perp} \cap [\ker(A)]^{\perp} = [\mathrm{im}(A)]^{\perp} \cap \mathrm{im}(A) = \{0\}$). Since $(J\xi \mid \xi) = (J\eta \mid \eta) = 0$, we conclude that $(J\xi \mid \eta) \neq 0$ and $(J\eta \mid \xi) \neq 0$. Because $JLJ^{-1} = L^*$, we obtain from (77) that $L^*J\xi = JLJ^{-1}J\xi = -J\eta$ and, similarly, that $L^*J\eta = J\xi$. If we set $u := -J\eta$ and $v := J\xi$, we then have $u + iv \in \ker([D_2 f(0,0)]^*_{\mathbb{C}} - \omega i)$. And since $D_1 D_2 f(0,0) = \omega B$, we finally obtain for the determinant (64)

$$-\omega(J\xi \mid \eta) \int_0^{2\pi} (e^{-tL^*}u \mid Be^{tL}\xi)\, dt. \tag{78}$$

It follows from $B = J^*L$ that $Be^{tL}\xi = J^*e^{tL}L\xi = Je^{tL}\eta = -Je^{tL}J^{-1}u = -e^{tJLJ^{-1}}u = -e^{-tL^*}u$, from which we obtain the value

$$\omega(J\xi \mid \eta) \int_0^{2\pi} |e^{-tL^*}u|^2\, dt \neq 0$$

for (78). This implies the assertion. $\square$

(26.27) Remarks. (a) The exact statement of the Hopf bifurcation theorem (26.21) implies that the periodic orbits of $\dot{z} = JH'(z)$ near z_0 form a two-dimensional C^{k-2}-manifold, which can be represented as the graph of a function on $N := \ker(id_{\mathbb{R}^{2n}} - e^{(2\pi/\omega)H''(z_0)})$ (and is "foliated" by the orbits).

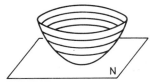

(b) If we write the differential equation

$$-\ddot{x} = f(x), \quad f \in C^2(\mathbb{R}, \mathbb{R}),$$

as a first order system, we obtain a Hamiltonian system in $\mathbb{R}^2$ with Hamiltonian

$$H(x, \dot{x}) = \dot{x}^2/2 + \int_0^x f(\xi)\, d\xi$$

(cf. section 3). Therefore $H'(x, \dot{x}) = (f(x), \dot{x})$, and $z_0 := (x, \dot{x})$ is a critical point of H if and only if $z_0 = (x_0, 0)$, where $f(x_0) = 0$. Since

$$JH''(z_0) = \begin{bmatrix} 0 & 1 \\ -f'(x_0) & 0 \end{bmatrix},$$

the eigenvalues of $JH''(z_0)$ are given by $\pm\sqrt{-f'(x_0)}$. If $f'(x_0) > 0$, i.e., if $U(x) := \int_0^x f(\xi)\, d\xi$ has a strict local minimum at x_0, we have $\sigma(JH''(z_0)) = \{\pm i\omega\}$, where $\omega := \sqrt{f'(x_0)} > 0$. In this case, the Liapunov center theorem asserts the existence of a one-parameter family of noncritical periodic orbits near x_0. This assertion agrees with the elementary qualitative discussion based on the conservation of energy, as was carried out in example (3.4 b).   □

Problems

1. Let E be a finite dimensional real Banach space, $X \subseteq E$ and let $\Lambda \subseteq \mathbb{R}$ be open. Assume that $\Phi \in C^2(\Lambda \times X, E)$ is such that $\Phi(\cdot, 0) = 0$ and $D_2\Phi(\lambda, 0) = I - \lambda K$ for some $K \in \mathcal{L}(E)$. Prove that, if $\lambda_0 \in \Lambda$ and $1/\lambda_0$ is an eigenvalue of K of odd multiplicity, then $(\lambda_0, 0)$ is a bifurcation point of the equation $\Phi(\lambda, x) = 0$.

2. Give a qualitative sketch of the set of all zeros of the function

$$\mathbb{R} \times \mathbb{R}^2 \to \mathbb{R}^2, \quad (\lambda, (x, y)) \mapsto ((1 + \lambda)x - y + x^2, (1 + \lambda)y - x + y^2).$$

(Remark: This is a simple example of *secondary bifurcation*, that is, a situation when the bifurcating solution itself branches.)

3. Under the assumptions of theorem (26.11), the bifurcation equation has the form $h(\lambda, \xi) = h_0(\lambda, \xi)\xi$, where $h_0(\lambda, \xi) = h_{01}(\lambda)\lambda + h_{02}(\lambda, \xi)\xi$. If $P_c D_1 D_2 g(0, 0)|F \times N$ is zero, show that we have

$$h_0(\lambda, \xi) = h_{11}(\lambda)[\lambda]^2 + h_{02}(\lambda, \xi)\xi.$$

Furthermore, show that $h_{11}(0)$ can be calculated from g. Generalize this result, including the case when $h_{11}(0)$ vanishes, etc.

4. Apply problem 3 in case $F = \mathbb{R}$, $\dim(N) = 1$, to discuss the bifurcation of equation $g(\lambda, x) = 0$ if the "transversality condition" $B[\ker(A)] \not\subseteq \operatorname{im}(A)$ of theorem (26.13) is not satisfied.

5. Show that under the assumptions of the Hopf bifurcation theorem no two-sided bifurcation can occur.

6. Show that the system

$$\dot{u} = \lambda u + v + 6u^2$$
$$\dot{v} = -u + \lambda v + vw$$
$$\dot{w} = -u + (\lambda^2 - 1)v - w + u^2$$

has a Hopf bifurcation at $\lambda_0 = 0$ of period $T_0 = 2\pi$.

27. Stability of Bifurcating Solutions

In examples (26.8) we have seen that, in general, the stability properties of the trivial solution change at a bifurcation point. It is the goal of this section to prove, in simple cases, some general results about the stability of bifurcating solutions.

We begin with a simple technical result of a general nature.

(27.1) Lemma. *Let X and Y be Banach spaces. Then the inverse map*

$$f : \operatorname{Isom}(X, Y) \to \mathcal{L}(Y, X), \qquad T \mapsto T^{-1},$$

is infinitely often differentiable and

$$Df(T)S = -T^{-1}ST^{-1}, \qquad \forall S \in \mathcal{L}(X, Y). \tag{1}$$

Proof. It follows from lemma (26.18) that $\operatorname{Isom}(X, Y)$ is open in $\mathcal{L}(X, Y)$. So let $T \in \operatorname{Isom}(X, Y)$ be fixed, and let $S \in \mathcal{L}(X, Y)$ be such that $T + S \in \operatorname{Isom}(X, Y)$. Then

$$\begin{aligned} f(T + S) - f(T) &= (T + S)^{-1} - T^{-1} = (T + S)^{-1}(T - (T + S))T^{-1} \\ &= -(T + S)^{-1}ST^{-1} \end{aligned} \tag{2}$$

and hence

$$f(T + S) - f(T) = -T^{-1}(I + ST^{-1})^{-1}ST^{-1}. \tag{3}$$

Since $\mathcal{GL}(Y)$ is open and since by remark (25.6) the inverse map is continuous on $\mathcal{GL}(Y)$, it follows from (2) that $f(T + S) - f(T) \to 0$ as $S \to 0$, that is, f is continuous. From (3) we now obtain

$$f(T + S) - f(T) + T^{-1}ST^{-1} = [T^{-1} - (T + S)^{-1}ST^{-1}]ST^{-1} = o(\|S\|)$$

as $S \to 0$ in $\mathcal{L}(X, Y)$. Consequently f is differentiable and $Df(T)$ has the form given in (1).

Now set

$$g(T_1, T_2)(S) := -T_1 S T_2, \quad \forall T_1, T_2 \in \mathcal{L}(Y, X), \ S \in \mathcal{L}(Y, X).$$

Then g is a bilinear mapping, i.e.,

$$g \in \mathcal{L}^2(\mathcal{L}(Y, X), \mathcal{L}(\mathcal{L}(X, Y), \mathcal{L}(Y, X))),$$

and hence it is infinitely often continuously differentiable. In addition, we have

$$Df(T) = g(T^{-1}, T^{-1}) = g(f(T), f(T)).$$

Consequently, Df is a C^k-function whenever f is a C^k-function. From this the assertion follows by induction. □

Bifurcation from Simple Eigenvalues

In what follows, we again let E denote a finite dimensional real Banach space. Moreover, let X be open in E and assume that Λ is an open interval in $\mathbb{R}$ so that $(0, 0) \in \Lambda \times X$. Let $g \in C^k(\Lambda \times X, E)$, $k \geq 2$, be such that $g(0, 0) = 0$ and assume that 0 *is a simple eigenvalue of* $A := D_2 g(0, 0) \in \mathcal{L}(E)$ and

$$\ker(A) = \mathbb{R}e$$

for some $e \in E \setminus \{0\}$. Finally, suppose

$$Be \notin \operatorname{im}(A),$$

where $B := D_1 D_2 g(0, 0)$.

Under these assumptions it follows from theorem (26.13) that the set of all zeros of g near (0,0) consists exactly of the λ-axis $\Lambda \times \{0\}$ and a C^{k-1}-curve Γ,

$$(\lambda(\cdot), x(\cdot)) : (-\epsilon, \epsilon) \to \Lambda \times X$$

such that $(\lambda(0), x(0)) = (0, 0)$ and

$$x(s) = s(e + y(s)), \quad \forall s \in (-\epsilon, \epsilon). \tag{4}$$

Here we have $y(0) = 0$ and $y(\cdot) \in C^{k-1}((-\epsilon, \epsilon), N_c)$, where N_c denotes a complementary subspace of $N := \ker(A)$ in E. In particular, we have

$$g(\lambda(s), x(s)) = 0, \quad \forall s \in (-\epsilon, \epsilon).$$

Since 0 is a simple eigenvalue of A, it follows from proposition (26.24) – after possibly reducing ϵ – that there exists a C^{k-1}-function

$$(\kappa(\cdot), u(\cdot)) : (-\epsilon, \epsilon) \to \mathbb{R} \times E$$

satisfying $(\kappa(0), u(0)) = (0, e)$, $u(s) \neq 0$ for all $s \in (-\epsilon, \epsilon)$ and

$$D_2 g(\lambda(s), x(s)) u(s) = \kappa(s) u(s). \tag{5}$$

Indeed, by applying proposition (26.24) to the complexification of $A(s) := D_2 g(\lambda(s), x(s))$ we obtain the existence of a mapping $(\kappa(\cdot), u(\cdot)) \in C^{k-1}((-\epsilon, \epsilon), \mathbb{C} \times E_\mathbb{C})$ such that $(\kappa(0), u(0)) = (0, e)$, $u(s) \neq 0$ and $[A(s)]_\mathbb{C} u(s) = \kappa(s) u(s)$ for all $s \in (-\epsilon, \epsilon)$. By conjugation, it follows from this that $[A(s)]_\mathbb{C} \overline{u(s)} = \overline{\kappa(s)} \, \overline{u(s)}$. From the uniqueness assertion, contained in proposition (26.24 iii), we now obtain $\overline{u(s)} = u(s)$ and $\overline{\kappa(s)} = \kappa(s)$ and therefore (5).

After these preliminaries we can now prove the following theorem.

(27.2) Theorem. *Let the assumptions above be satisfied, that is, the assumptions of theorem (26.13), and suppose that 0 is a simple eigenvalue of $D_2 g(0, 0)$. Furthermore, let $(\lambda(\cdot), x(\cdot)) \in C^{k-1}((-\epsilon, \epsilon), \mathbb{R} \times E)$ be the "curve" Γ of nontrivial solutions of $g(\lambda, x) = 0$ through $(0, 0)$ and let $\kappa(s) \in \sigma(D_2 g(\lambda(s), x(s)))$ denote the unique "continuation of the eigenvalue 0 of $D_2 g(0, 0)$ along the curve Γ." Then there exists – after possibly reducing ϵ – some function $\alpha \in C^{k-2}((-\epsilon, \epsilon), \mathbb{R})$ such that*

$$\kappa(s) = \alpha(s) s \dot{\lambda}(s), \quad \forall s \in (-\epsilon, \epsilon),$$

and

$$\alpha(0) = -\langle e', D_1 D_2 g(0, 0) e \rangle \neq 0,$$

where $e' \in \ker([D_2 g(0, 0)]')$ is determined by $\langle e', e \rangle = 1$.

Proof. Let $Q := \langle e', \cdot \rangle e \in \mathcal{L}(E)$. Then $\ker(Q) = \operatorname{im}(A) =: N_c$ and $E = N \oplus N_c$, and Q is the projection on N along N_c. Moreover, let $u(s) = \beta(s) e + v(s)$, where $v(s) := (id_E - Q) u(s)$ and $\beta(s) = \langle e', u(s) \rangle$. Since $\beta(0) = 1$, we may assume, without loss of generality, that $\beta(s) \neq 0$ for all $s \in (-\epsilon, \epsilon)$. Then, if we divide equation (5) by $\beta(s)$, we may assume, without loss of generality, that

$$u(s) = e + v(s), \quad \text{where} \quad v(s) \in N_c, \quad \forall s \in (-\epsilon, \epsilon). \tag{6}$$

From (4) we obtain

$$\dot{x}(s) = e + z(s), \quad \text{where} \quad z(s) \in N_c, \quad \forall s \in (-\epsilon, \epsilon), \tag{7}$$

and differentiating the identity $g(\lambda(s), x(s)) = 0$ gives us

$$D_1 g(\lambda(s), x(s)) \dot{\lambda}(s) + D_2 g(\lambda(s), x(s)) \dot{x}(s) = 0, \quad \forall s \in (-\epsilon, \epsilon). \tag{8}$$

We now set

$$B(s) := \int_0^1 D_1 D_2 g(\lambda(s), \tau x(s)) d\tau.$$

Then $B(\cdot) \in C^{k-2}((-\epsilon, \epsilon), \mathcal{L}(E))$, and from $g(\cdot, 0) = 0$ and the mean value theorem it follows that

$$D_1 g(\lambda(s), x(s)) = B(s) x(s) = s B(s)[e + y(s)], \quad \forall s \in (-\epsilon, \epsilon).$$

This together with (8) implies the relation

$$s\dot{\lambda}(s)B(s)\dot{x}(s) = -D_2 g(\lambda(s), x(s))\dot{x}(s). \tag{9}$$

We now define

$$T \in C^{k-2}((-\epsilon, \epsilon) \times [\mathbb{R} \times N_c], E)$$

by

$$T(s, (\alpha, a)) := B(s)\dot{x}(s) + D_2 g(\lambda(s), x(s))a + \alpha(\dot{x}(s) - s\dot{\lambda}(s)a)$$

and determine $(\alpha_0, a_0) \in \mathbb{R} \times N_c$ such that $Be = -\alpha_0 e - Aa_0$. Because $A|N_c \in \mathcal{GL}(N_c)$, (α_0, a_0) is uniquely determined and

$$T(0, (\alpha_0, a_0)) = 0. \tag{10}$$

Since, moreover, we have

$$D_2 T(0, (\alpha_0, a_0))(\hat{\alpha}, \hat{a}) = \hat{\alpha}e + A\hat{a}, \quad (\hat{\alpha}, \hat{a}) \in \mathbb{R} \times N_c, \tag{11}$$

it follows that $D_2 T(0, (\alpha_0, a_0))$ is an isomorphism from $\mathbb{R} \times N_c$ onto $E = N \oplus N_c$. Thus from the implicit function theorem we deduce – after possibly reducing ϵ – the existence of a unique function

$$(\alpha, a) \in C^{k-2}((-\epsilon, \epsilon), \mathbb{R} \times N_c)$$

such that $(\alpha(0), a(0)) = (\alpha_0, a_0)$ and

$$B(s)\dot{x}(s) + D_2 g(\lambda(s), x(s))a(s) + \alpha(s)(\dot{x}(s) - s\dot{\lambda}(s)a(s)) = 0, \quad \forall s \in (-\epsilon, \epsilon).$$

We now multiply this equation by $s\dot{\lambda}(s)$ and obtain, by considering (9),

$$D_2 g(\lambda(s), x(s))[\dot{x}(s) - s\dot{\lambda}(s)a(s)] = s\dot{\lambda}(s)\alpha(s)[\dot{x}(s) - s\dot{\lambda}(s)a(s)]$$

for all $-\epsilon < s < \epsilon$. From proposition (26.24) and from the uniqueness of the functions considered, we now deduce the relation

$$\kappa(s) = s\dot{\lambda}(s)\alpha(s), \quad \forall s \in (-\epsilon, \epsilon).$$

To calculate $\alpha_0 = \alpha(0)$, we apply e' to the equation $Be + \alpha_0 e + Aa_0 = 0$. Then we have

$$\langle e', Be \rangle + \alpha_0 + \langle e', Aa_0 \rangle = 0,$$

and, since $a_0 \in N_c$ and $Aa_0 \in N_c = \ker(e')$, we obtain $\alpha_0 = -\langle e', Be \rangle$. This proves the theorem.   □

(27.3) Remarks.   (a)   In the theorem above we have denoted the unique continuation of the eigenvalue 0 of $D_2 g(0, 0)$ along the curve of nontrivial solutions by $\kappa(\cdot)$. Similarly, based on proposition (26.24), one can "continue" the eigenvalue 0 of $D_2 g(0, 0)$ "along the trivial solution," that is, after possibly reducing Λ, there exists a unique function

$$(\mu(\cdot), v(\cdot)) \in C^{k-1}(\Lambda, \mathbb{R} \times E)$$

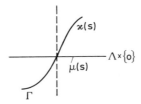

such that $(\mu(0), v(0)) = (0, e)$, $v(\lambda) \neq 0$ and

$$D_2g(\lambda, 0)v(\lambda) = \mu(\lambda)v(\lambda), \quad \forall \lambda \in \Lambda. \tag{12}$$

Differentiating this equation at $\lambda = 0$, we obtain

$$Be + A\dot{v}(0) = \dot{\mu}(0)e$$

and, by applying e', we get

$$\langle e', Be \rangle = \dot{\mu}(0).$$

We therefore have the following:

Under the assumptions of theorem (27.2) we have

$$\kappa(s) = \alpha(s)s\dot{\lambda}(s), \tag{13}$$

where $\alpha \in C^{k-2}((-\epsilon, \epsilon), \mathbb{R})$ and

$$\alpha(0) = -\dot{\mu}(0) \neq 0, \tag{14}$$

and $\mu(\cdot)$ denotes the "continuation of the eigenvalue 0 of $D_2g(0,0)$ along the trivial solution," as defined by (12). *The geometric meaning of condition* (14) *is that as λ traverses the interval $(-\epsilon, \epsilon)$, a simple eigenvalue of $D_2g(\lambda, 0)$ passes through zero with nonvanishing velocity.*

(b)   The importance of the relation $\kappa(s) = \alpha(s)s\dot{\lambda}(s)$ comes from the fact that in certain cases the sign of $\kappa(s)$ can be deduced from the geometric position of the curve of nontrivial solutions Γ. As we will see in the examples that follow, this kind of information is important

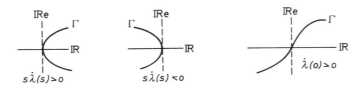

for stability questions in connection with differential equations. If $s\dot{\lambda}(s) > 0$ for $s \neq 0$, we have *supercritical bifurcation*. If $s\dot{\lambda}(s) < 0$ for $s \neq 0$, then we have *subcritical bifurcation*. While if $\dot{\lambda}(s)$ does not change its sign – the case of *transcritical bifurcation* – then the bifurcation curve "crosses" the $\mathbb{R}e$-axis at the origin.

If we now assume that $\dot{\mu}(0) > 0$, that is to say, if we assume that the "critical" eigenvalue of $D_2 g(\lambda, 0)$ passes through the origin from left to right with positive velocity, then we see that in the case of supercritical bifurcation we have $\kappa(s) < 0$, in the case of subcritical bifurcation we have $\kappa(s) > 0$, while in the case of transcritical bifurcation we have $\kappa(s) > 0$ on one "branch" of Γ and $\kappa(s) < 0$ on the other.   □

(27.4) Examples. (a) *The stability of critical points bifurcating from simple eigenvalues.*

Let $X \subseteq E$ be open and suppose $\Lambda \subseteq \mathbb{R}$ is an open interval such that $(0,0) \in \Lambda \times X$. Assume that $f \in C^2(\Lambda \times X, E)$ satisfies $f(\cdot, 0) = 0$. We then consider the autonomous parameter-dependent differential equation

$$\dot{x} = f(\lambda, x).$$

We assume that 0 is a simple eigenvalue of $D_2 f(0,0) \in \mathcal{L}(E)$ whose unique C^1-continuation $\mu(\lambda)$ satisfies $\dot{\mu}(0) > 0$, i.e., there exists a simple eigenvalue $\mu(\lambda)$ of $D_2 g(\lambda, 0)$ which passes through the origin (from left to right) with positive velocity at $\lambda = 0$. Based on remark (27.3 a), the assumptions of theorem (26.13) are then satisfied. Hence the set of zeros of f near $(0,0)$ consists exactly of the "trivial solution" $\Lambda \times \{0\}$ and a C^1-curve Γ which crosses $\Lambda \times \{0\}$ transversely at $(0,0)$.

In addition, we now assume that

$$Re[\sigma(D_2 f(0,0)) \setminus \{0\}] < 0$$

that is, all nonzero eigenvalues of $D_2 f(0,0)$ lie in the open left half plane of $\mathbb{C}$. It then follows from the upper semicontinuity of the spectrum (proposition (25.5)) that for all $\lambda \in \Lambda$ sufficiently close to 0 and for all sufficiently small $|s|$, the inclusions

$$Re[\sigma(D_2 f(\lambda, 0)) \setminus \{\mu(\lambda)\}] < 0$$

and

$$Re[\sigma(D_2 f(\lambda(s), x(s)) \setminus \{\kappa(s)\}] < 0$$

hold, where $(\lambda(\cdot), x(\cdot))$ is the parametrization of Γ from theorem (26.13) and $\kappa(\cdot)$ denotes the continuation of the eigenvalue 0 of $D_2 g(0,0)$ along the bifurcation branch Γ.

The assumptions above, in particular, imply that if $\lambda < 0$, then $x = 0$ is an asymptotically (Liapunov) stable critical point of the differential equation $\dot{x} = f(\lambda, x)$, while if $\lambda > 0$, then $x = 0$ is unstable (cf. theorem (15.6)). In the case of supercritical bifurcation ($s\dot{\lambda}(s) > 0$),

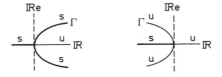

it follows from (27.3 b) and theorem (27.2), as well as theorem (15.6) that near the origin Γ consists completely of asymptotically (Liapunov) stable critical points of the differential equation $\dot{x} = f(\lambda(s), x)$. If we have the case of subcritical bifurcation ($s\dot{\lambda}(s) < 0$), then, in a neighborhood of the origin, Γ consists entirely of unstable critical points, while in the case of transcritical bifurcation, Γ consists of a stable and an unstable branch.

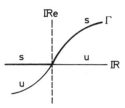

This manifestation is also known as *exchange of stability*. For simple, concrete examples illustrating these phenomena, we refer to examples (26.8 b (β)) and (26.8 b (δ)) (where in the latter we have the case $\dot{\mu}(0) < 0$.

(b)   *The stability of periodic bifurcating solutions in the nonautonomous case.*

Let $\Lambda \subseteq \mathbb{R}$ and $X \subseteq E$ be open and assume that $f \in C^2(\Lambda \times \mathbb{R} \times X, E)$ is such that $f(\cdot, \cdot, 0) = 0$. Moreover, assume that f is 2π-periodic in $t \in \mathbb{R}$ and suppose that for some $\lambda_0 \in \Lambda, 1$ is a simple Floquet multiplier of the linearized equation

$$\dot{y} = D_3 f(\lambda_0, t, 0)y, \tag{15}$$

and that all other Floquet multipliers of this equation lie in the interior of the unit disc in $\mathbb{C}$. In addition, let $v \in C^1(\mathbb{R}, E)$ be a 2π-periodic solution of (15) and let $w \in C^1(\mathbb{R}, E')$ denote a 2π-periodic solution of the dual equation

$$\dot{z} = -[D_3 f(\lambda_0, t, 0)]' z$$

such that $\langle w(0), v(0) \rangle = 1$. Finally, we set

$$\alpha(0) := \int_0^{2\pi} \langle w(t), D_1 D_3 f(\lambda_0, t, 0)v(t) \rangle dt \neq 0.$$

Then, by proposition (26.16), there exists a C^1-curve Γ through $(\lambda_0, 0)$ in $\Lambda \times BC^1(\mathbb{R}, E)$,

$$(-\epsilon, \epsilon) \to \Lambda \times BC^1(\mathbb{R}, E), \quad s \mapsto (\lambda(s), u(s)),$$

satisfying $u(s) \neq 0$ for all $s \neq 0$ and such that in some neighborhood of $(\lambda_0, 0)$ in $\Lambda \times BC(\mathbb{R}, E)$, the set of all 2π-periodic solutions of the equation $\dot{x} = f(\lambda, t, x)$ consists exactly of Γ and the trivial solutions $\Lambda \times \{0\}$.

Assume now that $\alpha(0) < 0$. It then follows from formula (42) of section 26 and remark (27.3 a) (also see the proof of proposition (26.16)) that for all λ in some left-neighborhood of λ_0, all Floquet multipliers of

$$\dot{y} = D_3 f(\lambda, t, 0)y \tag{16}$$

lie in the interior of the unit disc, while for all λ in some right-neighborhood of λ_0, at least one Floquet multiplier of (16) has a modulus greater than 1. It therefore follows from theorem (23.1) that the zero solution of the 2π-periodic equation $\dot{x} = f(\lambda, t, x)$ is asymptotically Liapunov stable for all $\lambda_0 - \epsilon < \lambda < \lambda_0$, while for all $\lambda_0 < \lambda < \lambda_0 + \epsilon$, the zero solution is unstable (where $\epsilon_0 > 0$ is an appropriate positive number).

If we now have $s\dot\lambda(s) > 0$ *for all* $0 < |s| < \epsilon$ (the case of supercritical bifurcation), it follows (as in (a)) that all Floquet multipliers of the equation

$$\dot{y} = D_3 f(\lambda(s), t, u(s))y$$

lie in the interior of the unit disc in $\mathbb{C}$. Therefore every point on Γ near $(\lambda_0, 0)$ is an asymptotically Liapunov stable 2π-periodic solution of $\dot{x} = f(\lambda, x)$, where $\lambda = \lambda(s)$. If, on the other hand, $s\dot\lambda(s) < 0$ for all $0 < |s| < \epsilon$ (subcritical bifurcation), then it follows from theorems (27.2) and (23.1) that all 2π-periodic bifurcating solutions on Γ near $(\lambda_0, 0)$ are unstable.

We leave it to the reader to formulate and prove the corresponding statements for the case of transcritical bifurcation and for the case $\alpha(0) > 0$. $\square$

Concrete Stability Criteria

In order to make explicit statements about the stability properties of bifurcating solutions in concrete cases, we must know whether we have supercritical, subcritical or transcritical bifurcation. In what follows we will give some simple criteria to resolve this question, where we begin with the simplest case, the transcritical bifurcation.

(27.5) Proposition. *Let $\Lambda \subseteq \mathbb{R}$ and $X \subseteq E$ be open and assume that $g \in C^3(\Lambda \times X, E)$ satisfies $g(\cdot, 0) = 0$. Moreover, let $A := D_2 g(0,0)$ and $B := D_1 D_2 g(0,0)$, as well as $\ker(A) = \mathbb{R}e$ and $\ker(A') = \mathbb{R}e'$, where $\langle e', e \rangle = 1$. Then if $Be \notin \mathrm{im}(A)$ and*

$$\langle e', D_2^2 g(0,0)[e]^2 \rangle \neq 0,$$

$(0,0)$ is a transcritical bifurcation point of the equation $g(\lambda, x) = 0$.

Proof. Since the assumptions of theorem (26.13) are satisfied, the set of all solutions of the equation $g(\lambda, x) = 0$ near $(0,0)$ consists exactly of the λ-axis $\Lambda \times \{0\}$ and a C^1-curve Γ,

$$(\lambda(\cdot), x(\cdot)) : (-\epsilon, \epsilon) \to \Lambda \times X,$$

which crosses the λ-axis transversely at $(0,0)$. Based on the proof of theorem (26.13), we have

$$\varphi(\lambda(s), s) = 0, \quad \forall s \in (-\epsilon, \epsilon), \tag{17}$$

where φ is defined by

$$\varphi(\lambda, s) := \langle e', h_0(\lambda, se)e \rangle. \tag{18}$$

By differentiating (17), we obtain

$$D_1\varphi(\lambda(s), s)\dot{\lambda}(s) + D_2\varphi(\lambda(s), s) = 0 \qquad (19)$$

and so, since $D_1\varphi(0,0) = \langle e', Be \rangle \neq 0$ (cf. formula (38) of section 26), we have

$$\dot{\lambda}(0) = -D_2\varphi(0,0)/\langle e', Be \rangle. \qquad (20)$$

From (18) we obtain

$$D_2\varphi(0,0) = \langle e', D_2 h_0(0,0)[e]^2 \rangle,$$

and therefore, based on formula (19) of theorem (26.11), we get

$$D_2\varphi(0,0) = \langle e', D_2^2 g(0,0)[e]^2 \rangle/2$$

(cf. formula (36) of section 26). Therefore $\dot{\lambda}(0) \neq 0$, which implies that $\dot{\lambda}(s)$ does not change its sign near $s = 0$. Now the assertion follows from remark (27.3 b).□

(27.6) Remark. The proof above shows that

$$\dot{\lambda}(0) = -\langle e', D_2^2 g(0,0)[e]^2 \rangle/2\langle e', Be \rangle. \qquad (21)$$

From this the "direction" of the bifurcation curve Γ near (0,0) can be determined.   □

In the following proposition we will give sufficient conditions for super- and subcritical bifurcation.

(27.7) Proposition. *Let* $\Lambda \subseteq \mathbb{R}$ *and* $X \subseteq E$ *be open and assume that* $g \in C^3(\Lambda \times X, E)$ *satisfies* $g(\cdot, 0) = 0$. *Furthermore, let* $A := D_2 g(0,0)$ *and* $B := D_1 D_2 g(0,0)$, *as well as* $\ker(A) = \mathbb{R}e$ *and* $\ker(A') = \mathbb{R}e'$ *be such that* $\langle e', e \rangle = 1$. *Also assume that* $Be \notin \mathrm{im}(A)$ *and*

$$D_2^2 g(0,0)[e]^2 = 0. \qquad (22)$$

If

$$\frac{\langle e', D_2^3 g(0,0)[e]^3 \rangle}{\langle e', Be \rangle} < 0 \qquad [resp. > 0], \qquad (23)$$

then (0,0) *is a supercritical* [*resp. subcritical*] *bifurcation point of the equation* $g(\lambda, s) = 0$.

Proof. With the notation of the proof of proposition (27.5) and using (21) and (22), we find that $\dot{\lambda}(0) = 0$. By differentiating (19) at $s = 0$, we thus obtain

$$\ddot{\lambda}(0) = -D_2^2\varphi(0,0)/\langle e', Be \rangle. \qquad (24)$$

From the definition of φ in (18) we deduce

$$D_2^2\varphi(0,0) = \langle e', D_2^2 h_0(0,0)[e]^3 \rangle.$$

Since by theorem (26.11) we have

$$h_0(\lambda, \xi) = h_{01}(\lambda)\lambda + h_{02}(\lambda, \xi)\xi, \quad \xi \in \ker(A) =: N, \qquad (25)$$

we obtain

$$D_2^2\varphi(0,0) = 2\langle e', D_2 h_{02}(0,0)[e]^3\rangle. \tag{26}$$

(Formally differentiating (25) would require that $h_{02} \in C^2$ and hence (cf. theorem (26.11)) $g \in C^4$. However, under the assumption that $g \in C^3$, one easily verifies, by applying the definition of the derivative directly, that (25) is twice differentiable with respect to ξ at the point $(0,0)$ and that the second derivative is given by (26).) From formula (22) of section 26 we obtain

$$D_2 h_{02}(0,0) = \int_0^1 (1-t)t\, dt D_2^3 h(0,0) = (1/6)D_2^3 h(0,0),$$

hence

$$D_2^2\varphi(0,0) = \langle e', D_2^3 h(0,0)[e]^3\rangle/3. \tag{27}$$

To calculate the right-hand side of this equation we differentiate equation (26) of section 26 at the point $(\lambda, \xi) = (0,0)$. Since $D_2\eta(0,0) = 0$ and $\mathrm{im}(D_2 g(0,0)) \subseteq \ker(P_c)$, we find that

$$D_2^3 h(0,0)[e]^3 = P_c D_2^3 g(0,0)[e]^3 + 3P_c D_2^2 g(0,0)[D_2^2\eta(0,0)[e]^2, e]. \tag{28}$$

If we consider (22), then, by differentiating identity (11) of section 26, we obtain

$$PD_2 g(0,0)D_2^2\eta(0,0)[e]^2 = 0.$$

Since $D_2^2\eta(0,0)[e]^2 \in N_c$ and $\ker[D_2 g(0,0)|N_c] = \{0\}$, it follows that $D_2^2\eta(0,0)[e]^2 = 0$. Hence we finally deduce from (28) that

$$D_2^3 h(0,0)[e]^3 = P_c D_2^3 g(0,0)[e]^3. \tag{29}$$

Because $P_c = \langle e', \cdot\rangle e$ and $\langle e', e\rangle = 1$, we finally have, using (29), (27) and (24), that

$$\ddot{\lambda}(0) = -\langle e', D_2^3 g(0,0)[e]^3\rangle/3\langle e', Be\rangle. \tag{30}$$

Now $\dot{\lambda}(0) = 0$ implies that $\dot{\lambda}(s) = s\ddot{\lambda}(0) + o(s)$, and therefore

$$s\dot{\lambda}(s) = s^2\ddot{\lambda}(0) + o(s^2) \quad \text{as} \quad s \to 0.$$

From this, together with (30), the assertion follows. $\qquad\square$

(27.8) Remarks.   (a) If, instead of (22), we merely assume that

$$P_c D_2^2 g(0,0)[e]^2 = 0,$$

then, by differentiating identity (11) of section 26, we obtain

$$PD_2^2 g(0,0)[e]^2 + PD_2 g(0,0)D_2^2\eta(0,0)[e]^2 = 0.$$

From this we deduce the expression

$$D_2^2\eta(0,0)[e]^2 = -[D_2 g(0,0)|N_c]^{-1}PD_2^2 g(0,0)[e]^2,$$

which is basically known. With this expression and using (28), (27) and (24), we obtain the relation

$$\ddot{\lambda}(0) = -\frac{\langle e', D_2^3 g(0,0)[e]^3\rangle + 3\langle e', D_2^2 g(0,0)[D_2^2 \eta(0,0)[e]^2, e]\rangle}{3\langle e', Be\rangle}$$

from which, in certain cases, $\ddot{\lambda}(0)$, and thereby the direction of the bifurcation curve, can be explicitly calculated.

(b)   If we introduce a basis for E with e as the first basis vector, and thereby identify E with $\mathbb{R}^m$, then

$$D_2^k g(0,0)[e]^k = \frac{\partial^k g}{(\partial x^1)^k}(0,0) \quad \text{for all } k \in \mathbb{N}.$$

If 0 is a simple eigenvalue of $A = D_2 g(0,0)$ and if we identify $E' = (\mathbb{R}^m)'$ with $\mathbb{R}^m$ by means of the standard basis (cf. remark (11.16 b)), then e' agrees exactly with the first basis vector $e_1 = (1,0,\dots,0)$. Therefore

$$\langle e', D_2^k g(0,0)[e]^k\rangle = \frac{\partial^k g^1}{(\partial x^1)^k}(0,0), \quad \forall k \in \mathbb{N},$$

where g^1 denotes the first component of $g = (g^1, \dots, g^m)$.

(c)   With the aid of the previous remark we find that in the case of example (26.8 b) (where $g(\lambda, x) := f(\lambda, x^1, x^2) = (\lambda a x^1 - a(x^1)^3, bx^2)$) we have

$$D_2^2 g(0,0)[e]^2 = 0$$

and

$$\frac{\langle e', D_2^3 g(0,0)[e]^3\rangle}{\langle e', Be\rangle} = -6.$$

Therefore proposition (27.7) insures supercritical bifurcation, which agrees with example (26.8 b). $\square$

Hopf Bifurcation

It is the goal of the following considerations to prove an analogue of theorem (27.2) for the case of Hopf bifurcation.

So let $\Lambda \subseteq \mathbb{R}$ and $X \subseteq E$ be open and assume that $f \in C^k(\Lambda \times X, E)$, $k \geq 2$, is such that $f(\cdot, 0) = 0$. We consider the parameter-dependent autonomous differential equation

$$\dot{x} = f(\lambda, x).$$

As in the last section, we assume that there exist some $\lambda_0 \in \Lambda$ and $T_0 > 0$ such that $2\pi i/T_0$ is a simple eigenvalue of $D_2 f(\lambda_0, 0)$ and that $k2\pi i/T_0 \notin \sigma(D_2 f(\lambda_0, 0))$ for all $k \in \mathbb{Z} \setminus \{\pm 1\}$. Then i is a simple eigenvalue of

$$L := (T_0/2\pi)D_2 f(\lambda_0, 0)$$

and we choose an arbitrary eigenvector

$$\xi + i\eta \in \ker(L_{\mathbb{C}} - i).$$

It follows from lemma (26.23) that we can find an eigenfunctional

$$\xi' + i\eta' \in \ker(L'_{\mathbb{C}} - i)$$

such that

$$\langle \xi', \xi \rangle = -\langle \eta', \eta \rangle = 1 \quad \text{and} \quad \langle \xi', \eta \rangle = \langle \eta', \xi \rangle = 0. \tag{31}$$

For all σ in some appropriate neighborhood $\Sigma \subseteq \mathbb{R}^2$ of the origin, we again set

$$\sigma := (\tau, \lambda) \quad \text{and} \quad \tilde{f}(\sigma, x) := \frac{\tau + T_0}{2\pi} f(\lambda + \lambda_0, x).$$

Then v is a $(T_0 + \tau)$-periodic solution of the equation $\dot{x} = f(\lambda + \lambda_0, x)$ if and only if

$$u(t) := v(t(T_0 + \tau)/2\pi) \tag{32}$$

is a 2π-periodic solution of the equation

$$\dot{x} = \tilde{f}(\sigma, x).$$

We again let $u(\cdot, 0, x, \sigma)$ denote the global solution of the equation $\dot{y} = \tilde{f}(\sigma, y)$ satisfying the initial condition $y(0) = x$, and we set

$$g(\sigma, x) := x - u(2\pi, 0, x, \sigma). \tag{33}$$

Finally, we assume that

$$\int_0^{2\pi} \langle t \cdot \xi', D_1 D_2 f(\lambda_0, 0)(t \cdot \xi) \rangle \, dt \neq 0,$$

where we have again used the abbreviations

$$t \cdot \xi := e^{tL} \xi \quad \text{and} \quad t \cdot \xi' := e^{-tL'} \xi'.$$

Based on theorem (26.21) and lemma (26.23), we know that for some $\epsilon > 0$ there exists a C^{k-1}-function

$$(y(\cdot), T(\cdot), \lambda(\cdot)) : (-\epsilon, \epsilon) \to X \times \mathbb{R}_+ \times \Lambda$$

satisfying

$$(y(0), T(0), \lambda(0)) = (0, T_0, \lambda_0)$$

and such that with

$$x(s) := s(\xi + y(s)), \quad -\epsilon < s < \epsilon, \tag{34}$$

$\gamma(s) := \gamma(x(s))$, $0 < |s| < \epsilon$, is a noncritical $T(s)$-periodic orbit of $\dot{x} = f(\lambda(s), x)$ through $x(s) \in E$. By remark (26.22 a), we know that

$$y(s) \in N_c, \quad \forall\, s \in (-\epsilon, \epsilon),$$

where N_c denotes a (fixed) complementary subspace of $N := \ker[D_2 g(0,0)]$ in E.

For every $s \in (-\epsilon, \epsilon)$, let

$$u_s(t) := u(t, 0, x(s), \sigma(s)), \quad t \in \mathbb{R}.$$

Then u_s is a 2π-periodic solution of the equation $\dot{y} = \tilde{f}(\sigma(s), y)$. Hence we have

$$g(\sigma(s), u_s(t)) = 0, \quad \forall\, t \in \mathbb{R}.$$

By differentiating this identity at $t = 0$, we obtain

$$D_2 g(\sigma(s), x(s))\dot{u}_s(0) = 0, \tag{35}$$

where we have

$$\dot{u}_s(0) = \tilde{f}(\sigma(s), x(s)).$$

Since $\tilde{f}(\cdot, 0) = 0$, it follows from the mean value theorem that

$$\tilde{f}(\sigma(s), x(s)) = \int_0^1 D_2 \tilde{f}(\sigma(s), \alpha x(s))\, d\alpha\, x(s) \tag{36}$$

for all $s \in (-\epsilon, \epsilon)$. Set

$$u(s) := s^{-1} \tilde{f}(\sigma(s), x(s)), \quad -\epsilon < s < \epsilon.$$

It then follows from (34) and (36) that $u(\cdot) \in C^{k-1}((-\epsilon, \epsilon), E)$ and

$$u(0) = D_2 \tilde{f}(0,0)\xi = -\eta$$

(cf. the proof of lemma (26.33)). We thus deduce from (35) the relation

$$D_2 g(\sigma(s), x(s))u(s) = 0, \quad \forall\, s \in (-\epsilon, \epsilon). \tag{37}$$

After these preliminaries we can prove an analogue of relation (5).

(27.9) Lemma. *There exist C^{k-1}-functions*

$$\kappa(\cdot) : (-\epsilon, \epsilon) \to \mathbb{R}, \quad v(\cdot) : (-\epsilon, \epsilon) \to E$$

such that

$$D_2 g(\sigma(s), x(s))v(s) = \kappa(s)v(s) + \langle \eta', v(s)\rangle u(s), \quad -\epsilon < s < \epsilon, \tag{38}$$

and

$$\kappa(0) = 0, \quad \text{as well as} \quad v(0) = \xi.$$

Proof. Let

$$A(s) := D_2 g(\sigma(s), x(s)) + \langle \eta', \cdot\rangle u(s), \quad -\epsilon < s < \epsilon.$$

Then $A \in C^{k-1}((-\epsilon, \epsilon), \mathcal{L}(E))$ and

$$A(0) = D_2 g(0, 0) - \langle \eta', \cdot \rangle \eta. \tag{39}$$

Since i is a simple eigenvalue of L, 0 is an eigenvalue of $D_2 g(0,0) = id_E - e^{2\pi L}$ of multiplicity 2. Because $\{\xi, \eta\}$ is a basis for $N := \ker[D_2 g(0,0)]$ and since there exists some N_c (namely $N_c := \operatorname{im}[D_2 g(0,0)]$) such that the direct sum $E = N \oplus N_c$ reduces the operator $D_2 g(0,0)$, it follows from (39) that $E = N \oplus N_c$ also reduces the operator $A(0)$. From (31) and (39) we obtain

$$A(0)\xi = 0 \quad \text{and} \quad A(0)\eta = \eta.$$

Hence $\sigma(A(0)|N) = \{0, 1\}$, which shows that 0 is a simple eigenvalue of $A(0)$ with corresponding eigenvector ξ. Now the assertion follows from proposition (26.24) – after ϵ has been reduced appropriately – since, as in the proof of (5), the uniqueness statement of proposition (26.24) implies that $\kappa(s)$ and $v(s)$ are real-valued. $\qquad \square$

The following lemma represents the analogue of theorem (27.2).

(27.10) Lemma. *There exists some* $\alpha(\cdot) \in C^{k-2}((-\epsilon, \epsilon), \mathbb{R})$ *such that*

$$\kappa(s) = \alpha(s) s \dot{\lambda}(s) \quad \text{and} \quad \alpha(0) = \frac{T_0}{2\pi} \int_0^{2\pi} \langle t \cdot \xi', D_1 D_2 f(\lambda_0, 0)(t \cdot \xi) \rangle \, dt.$$

Proof. Differentiating the identity $g(\sigma(s), x(s)) = 0$, we obtain

$$D_1 g(\sigma(s), x(s))\dot{\sigma}(s) + D_2 g(\sigma(s), x(s))\dot{x}(s) = 0, \quad \forall s \in (-\epsilon, \epsilon). \tag{40}$$

Setting

$$B^1(s) := \frac{\partial}{\partial \tau} g(\sigma(s), x(s)) \quad \text{and} \quad B^2(s) := \frac{\partial}{\partial \lambda} g(\sigma(s), x(s)),$$

we have

$$D_1 g(\sigma(s), x(s))\dot{\sigma}(s) = B^1(s)\dot{\tau}(s) + B^2(s)\dot{\lambda}(s).$$

If we let $v(\cdot, 0, x, \lambda)$ denote the global solution of the IVP $\dot{y} = f(\lambda, y)$, $y(0) = x$, it follows from (32) that

$$u(2\pi, 0, x(s), (\tau(s), \lambda(s))) = v(T_0 + \tau(s), 0, x(s), \lambda(s)).$$

Together with (33), we deduce from this that

$$B^1(s) = -\frac{\partial}{\partial \tau} u(2\pi, 0, x(s), \sigma(s)) = -\frac{\partial}{\partial t} v(T_0 + \tau(s), 0, x(s), \lambda(s))$$

$$= -f(\lambda_0 + \lambda(s), x(s)) = -\frac{2\pi}{T_0 + \tau(s)} \tilde{f}(\sigma(s), x(s))$$

$$= -\gamma(s) s u(s),$$

where $\gamma(s) := 2\pi/(T_0 + \tau(s))$, and therefore

$$B^1(s)\dot{\tau}(s) = -\gamma(s)u(s)s\dot{\tau}(s), \quad \forall s \in (-\epsilon, \epsilon). \tag{41}$$

With $C(s) := \int_0^1 \frac{\partial}{\partial \lambda} D_2 g(\sigma(s), \alpha x(s))d\alpha$ it follows from the mean value theorem that

$$B^2(s) = \frac{\partial}{\partial \lambda} g(\sigma(s), x(s)) = C(s)x(s),$$

hence

$$B^2(s)\dot{\lambda}(s) = C(s)[\xi + y(s)]s\dot{\lambda}(s), \quad \forall s \in (-\epsilon, \epsilon). \tag{42}$$

We have

$$C(0)\xi = \frac{\partial}{\partial \lambda} D_2 g(0, 0)\xi = B[e_2, \xi], \tag{43}$$

where $B := D_1 D_2 g(0, 0)$ and $e_2 := (0, 1) \in \mathbb{R}^2$ is a standard basis vector. Taking (41) and (42) into account, (40) has the form

$$D_2 g(\sigma(s), x(s))\dot{x}(s) - \gamma(s)u(s)\dot{\tau}(s)s + C(s)[\xi + y(s)]s\dot{\lambda}(s) = 0. \tag{44}$$

Let now $P : E \to N_c := \mathrm{im}[D_2 g(0, 0)]$ denote the projection along N and let $v(\cdot)$ be the function in lemma (27.9). Then

$$v(s) = \langle \xi', v(s)\rangle \xi - \langle \eta', v(s)\rangle \eta + Pv(s).$$

Since $\langle \xi', v(0)\rangle = \langle \xi', \xi\rangle = 1$, we may assume, without loss of generality, that $\langle \xi', v(s)\rangle \neq 0$ for all $s \in (-\epsilon, \epsilon)$. Therefore, if we replace $v(s)$ by $v(s)/\langle \xi', v(s)\rangle$, we may assume, without loss of generality, that

$$v(s) = \xi + w(s), \quad \forall s \in (-\epsilon, \epsilon),$$

where $w(s) := -\langle \eta', v(s)\rangle \eta + Pv(s)$. In particular, we have $w(0) = 0$.

We now subtract equation (44) from equation (38) and obtain

$$\begin{aligned} D_2 g(\sigma(s), x(s))[v(s) - \dot{x}(s)] + \gamma(s)u(s)\dot{\tau}(s)s - C(s)[\xi + y(s)]s\dot{\lambda}(s) \\ = \kappa(s)v(s) + \langle \eta', v(s)\rangle u(s), \end{aligned} \tag{45}$$

were we have $v(s) - \dot{x}(s) = w(s) - y(s) - s\dot{y}(s)$, hence, in particular, $v(0) - \dot{x}(0) = 0$.

We will now first show that there exists a function

$$(\rho, \delta, \eta_c) \in C^{k-2}((-\epsilon, \epsilon), \mathbb{R} \times \mathbb{R} \times N_c)$$

satisfying $\rho(0) = 1$ and

$$\rho(s)v(s) - \dot{x}(s) = \delta(s)u(s) + \eta_c(s), \quad \forall s \in (-\epsilon, \epsilon). \tag{46}$$

In order to see this, we define

$$T \in C^{k-2}((-\epsilon, \epsilon) \times \mathbb{R}^2, N)$$

by

$$T(s, (\rho, \delta)) := (id - P)[\rho v(s) - \dot{x}(s) - \delta u(s)], \quad \forall s \in (-\epsilon, \epsilon).$$

We then have $T(0, (1, 0)) = 0$ and

$$D_2 T(0, (1, 0))(\hat{\rho}, \hat{\delta}) = \hat{\rho}\xi + \hat{\delta}\eta, \quad (\hat{\rho}, \hat{\delta}) \in \mathbb{R}^2.$$

Therefore $T(0, (1, 0)) \in \mathrm{Isom}(\mathbb{R}^2, N)$ and so the implicit function theorem implies – after possibly reducing ϵ – the existence of a unique function $(\rho, \delta) \in C^{k-2}((-\epsilon, \epsilon), \mathbb{R}^2)$ such that $(\rho(0), \delta(0)) = (1, 0)$ and $T(s, (\rho(s), \delta(s))) = 0$ for all $s \in (-\epsilon, \epsilon)$. Now $\eta_c(s) := \rho(s)v(s) - \dot{x}(s) - \delta(s)u(s)$ has the required properties.

It is clear that we can replace the function v by ρv in (45). Since $u(s) \in \ker[D_2 g(\sigma(s), x(s))]$, we thus obtain from (45) and (46) the relation

$$D_2 g(\sigma(s), x(s))\eta_c(s) + \gamma(s)u(s)\dot{\tau}(s)s - C(s)[\xi + y(s)]s\dot{\lambda}(s)$$
$$= \kappa(s)v(s) + \langle \eta', v(s)\rangle u(s)$$

for all $s \in (-\epsilon, \epsilon)$, where we again have $\eta_c(0) = 0$.

We now try the "ansatz"

$$\eta_c(s) = a(s)s\dot{\lambda}(s), \quad \kappa(s) = \alpha(s)s\dot{\lambda}(s) \tag{47}$$

and

$$\gamma(s)s\dot{\tau}(s) - \langle \eta', v(s)\rangle = \beta(s)s\dot{\lambda}(s), \tag{48}$$

for some $a(\cdot) : (-\epsilon, \epsilon) \to N_c$ and $\alpha(\cdot), \beta(\cdot) : (-\epsilon, \epsilon) \to \mathbb{R}$. From this and (45) we obtain

$$s\dot{\lambda}(s)[D_2 g(\sigma(s), x(s))a(s) - \alpha(s)v(s) + \beta(s)u(s) - C(s)(\xi + y(s))] = 0.$$

We define a mapping

$$S(\cdot) \in C^{k-1}((-\epsilon, \epsilon), \mathcal{L}(\mathbb{R}^2 \times N_c, E))$$

by

$$S(s)(\alpha, \beta, a) := D_2 g(\sigma(s), x(s))a - \alpha v(s) + \beta u(s).$$

Since

$$S(0)(\alpha, \beta, a) = D_2 g(0, 0)a - \alpha\xi - \beta\eta,$$

$S(0)$ is an isomorphism from $\mathbb{R}^2 \times N_c$ onto E. So, based on lemma (26.18), – and by possibly reducing ϵ – we may assume, without loss of generality, that $S(\cdot) \in C^{k-1}((-\epsilon, \epsilon), \mathrm{Isom}(\mathbb{R}^2 \times N_c, E))$.

We consider the equation

$$D_2 g(\sigma(s), x(s))a(s) - \alpha(s)v(s) + \beta(s)u(s) = C(s)[\xi + y(s)], \tag{49}$$

which, because of (43), reduces to

$$D_2 g(0, 0)a(0) - \alpha(0)\xi - \beta(0)\eta = B[e_2, \xi] = B_\xi e_2 \tag{50}$$

if $s = 0$.

Applying ξ' to (50), we obtain

$$\alpha(0) = -\langle \xi', B[e_2, \xi] \rangle, \tag{51}$$

and applying η' to (50) gives us

$$\beta(0) = \langle \eta', B[e_2, \eta] \rangle. \tag{52}$$

Therefore equation (50) has the unique solution $(\alpha(0), \beta(0), a(0))$, where $\alpha(0)$, $\beta(0)$ and $a(0)$ are given by (51), (52) and $[D_2 g(0,0)|N_c]^{-1}[B_\xi e_2 + \alpha(0)\xi + \beta(0)\eta]$, respectively. Since $S(s)$ is an isomorphism, it follows from lemma (27.1) that

$$s \mapsto (\alpha(s), \beta(s), a(s)) := [S(s)]^{-1} C(s)[\xi + y(s)]$$

defines a C^{k-2}-function which satisfies (51), (52) and (49). Therefore the assumptions (47) and (48) are justified.

Since by formula (54) of section 26 we have

$$\langle \xi', B[e_2, \xi] \rangle = -\int_0^{2\pi} \langle t \cdot \xi', \frac{\partial}{\partial \lambda} D_2 \tilde{f}(0,0)(t \cdot \xi) \rangle dt$$

$$= -\frac{T_0}{2\pi} \int_0^{2\pi} \langle t \cdot \xi', D_1 D_2 f(\lambda_0, 0)(t \cdot \xi) \rangle dt,$$

the assertion now follows from (51).   □

After these preliminaries we can prove the following theorem about the *stability in the case of Hopf bifurcation*.

(27.11) Theorem. *Let $\Lambda \subseteq \mathbb{R}$ and $X \subseteq E$ be open and assume that $f \in C^2(\Lambda \times X, E)$ is such that $f(\cdot, 0) = 0$. Assume that for some $\lambda_0 \in \Lambda$ and some $\omega_0 > 0$ we have:*

$$i\omega_0 \mathbb{Z} \cap \sigma(D_2 f(\lambda_0, 0)) = \{\pm i\omega_0\}$$

and $i\omega_0$ is a simple eigenvalue of $D_2 f(\lambda_0, 0)$.

Furthermore, let $\mu(\lambda) \in \sigma(D_2 f(\lambda, 0))$ denote the unique continuation·of the eigenvalue $i\omega_0$ of $D_2 f(0, 0)$ along the trivial solution $\Lambda \times \{0\}$ and assume that

$$\frac{d}{d\lambda}[Re\, \mu(\lambda_0)] > 0. \tag{53}$$

Then there exist some $\epsilon > 0$ and a C^1-function

$$(\lambda(\cdot), x(\cdot), T(\cdot)) : (-\epsilon, \epsilon) \to \Lambda \times X \times \mathbb{R}_+$$

such that $(\lambda(0), x(0), T(0)) = (\lambda_0, 0, 2\pi/\omega_0)$ and with the following properties:

(i)   *For every $s \in (0, \epsilon)$, $\gamma(s) := \gamma(x(s))$ is a noncritical $T(s)$-periodic orbit of the equation $\dot{x} = f(\lambda(s), x)$ through $x(s)$.*

(ii) *$\gamma(s) \neq \gamma(t)$ for all $0 < s < t < \epsilon$.*

(iii) *There exists a neighborhood of $(\lambda_0, 0, 2\pi/\omega_0) \in \Lambda \times X \times \mathbb{R}$ which contains no other noncritical periodic orbits of $\dot{x} = f(\lambda, x)$.*

If

$$s\dot\lambda(s) > 0 \quad for \quad s > 0 \quad (\text{``supercritical bifurcation''}) \tag{54}$$

and

$$Re[\sigma(D_2 f(\lambda_0, 0)) \setminus \{\pm i\omega_0\}] < 0, \tag{55}$$

every orbit $\gamma(s)$ is stable.
    If, however, we have

$$s\dot\lambda(s) < 0 \quad for \quad s > 0 \quad (\text{``subcritical bifurcation''}), \tag{56}$$

then every orbit $\gamma(s)$, $0 < s < \epsilon$, is unstable.

Proof. The existence of periodic orbits $\gamma(s)$ with the indicated properties is the assertion of theorem (26.25).
    Since $u(\cdot, 0, x, \sigma)$ is the global solution of the IVP $\dot y = \tilde f(\sigma, y)$, $y(0) = x$, the eigenvalues of $D_3 u(2\pi, 0, x(s), \sigma(s))$ are precisely the Floquet multipliers of the orbit $\gamma(s)$. Thus $\mu \in \mathbb{C}$ is a Floquet multiplier of $\gamma(s)$ if and only if $1 - \mu$ is an eigenvalue of $D_2 g(\sigma(s), x(s))$.
    It follows from (37) that

$$D_2 g(\sigma(s), x(s))u(s) = 0, \quad \forall s \in (-\epsilon, \epsilon), \tag{57}$$

which reflects the fact that 1 is always a Floquet multiplier of $\gamma(s)$. From formula (76) of section 26, lemma (27.10) and from (53) we deduce that

$$\frac{d}{d\lambda}[Re\,\mu(\lambda_0)] = \frac{1}{2\pi} \int_0^{2\pi} \langle t \cdot \xi', D_1 D_2 f(\lambda_0, 0)(t \cdot \xi)\rangle \, dt$$
$$= \alpha(0)/T_0 > 0.$$

It therefore follows from lemma (27.10) – after possibly reducing ϵ – that

$$\text{sgn}\,\kappa(s) = \text{sgn}\,s\dot\lambda(s), \quad \forall s \in (0, \epsilon). \tag{58}$$

By assumptions (54) and (56) we have that

$$w(s) := v(s) + u(s)\langle\eta', v(s)\rangle/\kappa(s)$$

is well-defined for $0 < s < \epsilon$ and, using (57) and (38), we have

$$D_2 g(\sigma(s), x(s))w(s) = \kappa(s)w(s), \quad \forall s \in (0, \epsilon).$$

Therefore $\gamma(s)$ possesses the Floquet multiplier $1 - \kappa(s)$ for all $0 < s < \epsilon$.
    Let U and V be disjoint neighborhoods of $\sigma(e^{2\pi L}) \setminus \{1\}$ and $\{1\}$, respectively (where $L := (1/\omega_0)D_2 f(\lambda_0, 0)$). Since $e^{2\pi L} = D_3 u(2\pi, 0, 0, \sigma(0))$ and based on proposition (25.12), we may assume, without loss of generality, that 1 and $1 - \kappa(s)$ are the only Floquet multipliers of $\gamma(s)$ in V. Since 1 is an eigenvalue of $e^{2\pi L}$ of multiplicity 2, it follows from proposition (25.12) that 1 is a simple Floquet multiplier of $\gamma(s)$ for all $0 < s < \epsilon$ if $\kappa(s) \neq 0$ holds. It follows from (58)

that this condition is always satisfied under our assumptions, and therefore for all $0 < s < \epsilon$, $\gamma(s)$ is a hyperbolic periodic orbit. Since in the case of subcritical bifurcation (58) implies the inequality $1 - \kappa(s) > 1$, it follows in this case, based on theorem (23.8), that $\gamma(s)$ is unstable. If, on the other hand, we have the case of supercritical bifurcation, then $1 - \kappa(s) < 1$. By (55), as well as the spectral mapping theorem and the upper semicontinuity of the spectrum, we may assume, without loss of generality, that all Floquet multipliers of $\gamma(s)$, distinct from 1 and $1 - \kappa(s)$, lie in the interior of the unit disc in the complex plane. Thus, based on theorem (23.8), $\gamma(s)$ is stable for all $s \in (0, \epsilon)$.   □

(27.12) Remarks.   (a) The condition

$$\frac{d}{d\lambda}[Re\,\mu(\lambda_0)] > 0$$

means that the pair of conjugate complex eigenvalues $\mu(\lambda)$ and $\overline{\mu(\lambda)}$ "crosses" the imaginary axis at $+i\omega_0$ and $-i\omega_0$, respectively, with positive velocity as λ goes from $\lambda_0 - \epsilon_0$ to $\lambda_0 + \epsilon_0$. The case of the reversed inequality can of course be reduced to the case above by changing f to $-f$.

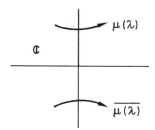

(b)   Let $\xi + i\eta$ be an eigenvector of $[D_2 f(\lambda_0, 0)]_{\mathbb{C}}$ corresponding to the eigenvalue $i\omega_0$ and let $\xi' + i\eta'$ be an eigenvector corresponding to the eigenvalue $i\omega_0$ of $[D_2 f(\lambda_0, 0)]'_{\mathbb{C}}$ and such that

$$\langle \xi', \xi \rangle = -\langle \eta', \eta \rangle = 1 \quad \text{and} \quad \langle \xi', \eta \rangle = \langle \eta', \xi \rangle = 0.$$

Then

$$\frac{d}{d\lambda}[Re\,\mu(\lambda_0)] = \frac{1}{2\pi} \int_0^{2\pi} \langle t \cdot \xi', D_1 D_2 f(\lambda_0, 0)(t \cdot \xi) \rangle \, dt.$$

(c)   Based on the upper semicontinuity of the spectrum, the conditions

$$\frac{d}{d\lambda}[Re\,\mu(\lambda_0)] > 0$$

and

$$Re[\sigma(D_2 f(\lambda_0, 0)) \setminus \{\pm i\omega_0\}] < 0$$

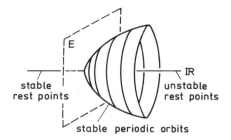

stable
rest points

unstable
rest points

stable periodic orbits

imply that if $\lambda_0 - \epsilon_0 < \lambda < \lambda_0$, then all eigenvalues of $D_2 f(\lambda, 0)$ lie in the open, left complex half plane. Hence for $\lambda_0 - \epsilon_0 < \lambda < \lambda_0$, 0 is an asymptotically stable critical point of the vector field $f(\lambda, \cdot)$. If $\lambda_0 < \lambda < \lambda_0 + \epsilon_0$, then at least one eigenvalue of $D_2 f(\lambda, 0)$ lies in the open, right complex half plane. Therefore the trivial solution of $\dot{x} = f(\lambda, x)$ is unstable if $\lambda_0 < \lambda < \lambda_0 + \epsilon_0$. So in the case of supercritical bifurcation an *exchange of stability* occurs, that is to say, as λ_0 is passed, the stability of the trivial solution goes over to the nontrivial periodic solutions bifurcating from the trivial solution. On the other hand, in the case of subcritical bifurcation, we have a complete *"loss of stability."* $\square$

Concrete Stability Formulae

For practical applications of theorem (27.11) we again need criteria which make it possible to determine whether we have subcritical or supercritical bifurcation. We now turn to this problem and again we make use of the notation introduced above.

(27.13) Lemma. *Let the assumptions of theorem (27.11) be satisfied with $f \in C^3(\Lambda \times X, E)$ and set*

$$\zeta := -[D_2 g(0,0)|N_c]^{-1} P D_2^2 g(0,0)[\xi]^2 \tag{59}$$

and

$$\gamma := \langle \xi', D_2^3 g(0,0)[\xi]^3 \rangle + 3\langle \xi', D_2^2 g(0,0)[\zeta, \xi] \rangle.$$

Then if $\gamma \neq 0$, we have

$$\operatorname{sgn} s\dot{\lambda}(s) = \operatorname{sgn} \gamma$$

for all $s \in (0, \epsilon)$.

Proof. We know from the proofs of theorems (26.19) and (26.21) that

$$h_0(\sigma(s), s\xi)\xi = 0, \quad \forall\, s \in (-\epsilon, \epsilon),$$

where h_0 has the meaning as given in theorem (26.11). By differentiating this identity we find that

$$D_1 h_0(\sigma(s), s\xi)[\dot{\sigma}(s), \xi] + D_2 h_0(\sigma(s), s\xi)[\xi]^2 = 0.$$

Since by remark (26.22 d) we have $\dot{\sigma}(0) = 0$, it follows by differentiating once more at $s = 0$ that

$$D_1 h_0(0, 0)[\ddot{\sigma}(0), \xi] + D_2^2 h_0(0, 0)[\xi]^3 = 0.$$

As in the proof of proposition (27.7) and in remark (27.8), we obtain

$$D_2^2 h_0(0, 0)[\xi]^3 = (1/3) P_c D_2^3 g(0, 0)[\xi]^3 + P_c D_2^2 g(0, 0)[\zeta, \xi],$$

where ζ is given by (59).

From theorem (26.11) we know that

$$D_1 h_0(0, 0)[\ddot{\sigma}(0), \xi] = P_c D_1 D_2 g(0, 0)[\ddot{\sigma}(0), \xi] = P_c B_\xi \ddot{\sigma}(0).$$

We thus obtain – by considering formula (54) of section 26 –

$$\begin{aligned}
\gamma &= 3\langle \xi', D_2^2 h_0(0, 0)[\xi]^3 \rangle = -3\langle \xi', D_1 h(0, 0)[\ddot{\sigma}(0), \xi] \rangle \\
&= -3\langle \xi', B[\ddot{\sigma}(0), \xi] \rangle \\
&= 3 \int_0^{2\pi} \langle t \cdot \xi', D_1 D_2 \tilde{f}(0, 0)[\ddot{\sigma}(0), t \cdot \xi] \rangle \, dt \\
&= \frac{3}{T_0} \int_0^{2\pi} \langle t \cdot \xi', L(t \cdot \xi) \rangle \, dt \, \ddot{\tau}(0) \\
&\quad + \frac{3 T_0}{2\pi} \int_0^{2\pi} \langle t \cdot \xi', D_1 D_2 f(\lambda_0, 0)(t \cdot \xi) \rangle \, dt \, \ddot{\lambda}(0).
\end{aligned}$$

Since

$$\langle t \cdot \xi', L(t \cdot \xi) \rangle = \langle t \cdot \xi', t \cdot (L\xi) \rangle = -\langle \xi', \eta \rangle = 0,$$

it follows from remark (27.12 b) that

$$\gamma = 3 T_0 \frac{d}{d\lambda} [Re \, \mu(\lambda_0)] \ddot{\lambda}(0),$$

hence

$$\operatorname{sgn} \ddot{\lambda}(0) = \operatorname{sgn} \gamma.$$

Since $\dot{\lambda}(0) = 0$, we have

$$s\dot{\lambda}(s) = s^2 \ddot{\lambda}(0) + o(s^2) \quad \text{as} \quad s \to 0,$$

from which the assertion follows.   □

We are therefore left with the problem of calculating the expression γ. In order to simplify, we make the following *assumption*

$$P D_2^2 g(0, 0)[\xi]^2 = 0. \tag{60}$$

Then $\zeta = 0$ and consequently

$$\gamma = \langle \xi', D_2^3 g(0,0)[\xi]^3 \rangle.$$

Assumption (60) *is satisfied, in particular, if* $\dim(E) = 2$. Because

$$g(0,\xi) = \xi - u(2\pi, 0, \xi, 0)$$

we find that

$$D_2 g(0,\xi) = id_E - D_3 u(2\pi, 0, \xi, 0),$$
$$D_2^2 g(0,\xi) = -D_3^2 u(2\pi, 0, \xi, 0), \tag{61}$$
$$D_2^3 g(0,0) = -D_3^3 u(2\pi, 0, 0, 0).$$

Since $u(\cdot, 0, \xi, 0)$ is the global solution of the IVP

$$\dot{x} = \tilde{f}(0, x), \quad x(0) = \xi,$$

if follows from the differentiability theorem (9.5) that for $u(t) := u(t, 0, 0, 0)$ we have

$$D_3 \dot{u}\xi = D_2 \tilde{f}(0,0) D_3 u\xi = LD_3 u\xi,$$
$$D_3^2 \dot{u}[\xi]^2 = LD_3^2 u[\xi]^2 + D_2^2 \tilde{f}(0,0)[D_3 u\xi]^2 \tag{62}$$

and

$$D_3^3 \dot{u}[\xi]^3 = LD_3^3 u[\xi]^3 + 3D_2^2 \tilde{f}(0,0)[D_3^2 u[\xi]^2, D_3 u\xi]$$
$$+ D_2^3 \tilde{f}(0,0)[D_3 u\xi]^3, \tag{63}$$

as well as $D_3 u(0)\xi = \xi$, $D_3^2 u(0)[\xi]^2 = 0$ and $D_3^3 u(0)[\xi]^3 = 0$. Hence we obtain from (61), (63) and the variation of constants formula that

$$D_2^3 g(0,0)[\xi]^3 = -\int_0^{2\pi} e^{(2\pi-\tau)L} D_2^3 \tilde{f}(0,0)[D_3 u(\tau)\xi]^3 \, d\tau$$
$$-3\int_0^{2\pi} e^{(2\pi-\tau)L} D_2^2 \tilde{f}(0,0)[D_3^2 u(\tau)[\xi]^2, D_3 u(\tau)\xi] \, d\tau. \tag{64}$$

It follows from (62) that

$$D_3 u(t)\xi = e^{tL}\xi \tag{65}$$

and

$$D_3^2 u(t)[\xi]^2 = \int_0^t e^{(t-\tau)L} D_2^2 \tilde{f}(0,0)[e^{\tau L}\xi]^2 \, d\tau \tag{66}$$

for all $t \in \mathbb{R}$. Because $e^{2\pi L'}\xi' = \xi'$, we thus obtain from (64) that

$$\langle \xi', D_2^3 g(0,0)[\xi]^3 \rangle = -\int_0^{2\pi} \langle e^{-\tau L'}\xi', D_2^3 \tilde{f}(0,0)[e^{\tau L}\xi]^3 \rangle d\tau$$
$$-3\int_0^{2\pi} \langle e^{-\tau L'}\xi', D_2^2 \tilde{f}(0,0)[D_3^2 u(\tau)[\xi]^2, e^{\tau L}\xi] \rangle d\tau. \tag{67}$$

Since the direct sum decomposition $E = N \oplus N_c$, where $N = \ker[D_2 g(0,0)] = \ker[id_E - e^{2\pi L}]$ and $N_c = im[D_2 g(0,0)]$, reduces the operator $D_2 g(0,0)$, we may assume, without loss of generality, that $E = N$. We now identify E with $\mathbb{R}^2$ by means of the basis $\{\xi, \eta\}$ and $(\mathbb{R}^2)'$ with $\mathbb{R}^2$ with respect to the standard basis (cf. remark (11.16 b)). Then L is represented by the matrix

$$\begin{bmatrix} 0 & 1 \\ -1 & 0 \end{bmatrix}.$$

It thus follows from section 13 (cf. section 13, case 4) that e^{tL} has the matrix representation

$$\begin{bmatrix} \cos t & \sin t \\ -\sin t & \cos t \end{bmatrix}. \tag{68}$$

Using the abbreviations $c(t) := \cos t$ and $s(t) := \sin t$, we consequently obtain

$$e^{\tau L}\xi = (c(\tau), -s(\tau)) = e^{-\tau L'}\xi'. \tag{69}$$

Finally, if we set

$$\varphi(z) := f(\lambda_0, z) \quad \text{for all} \quad z = (x, y) \in \mathbb{R}^2 \, (= N = E),$$

then for the first integral in (67) we obtain the expression

$$\frac{1}{\omega_0} \int_0^{2\pi} [D_1^3 \varphi^1 c^4 - 3D_1^2 D_2 \varphi^1 c^3 s + 3D_1 D_2^2 \varphi^1 c^2 s^2 - D_2^3 \varphi^1 c s^3$$
$$- D_1^3 \varphi^2 c^3 s + 3D_1^2 D_2 \varphi^2 c^2 s^2 - 3D_1 D_2^2 \varphi^2 c s^3 + D_2^3 \varphi^2 s^4] \, dt,$$

where we have set $\varphi = (\varphi^1, \varphi^2)$. Since the integrals of the odd functions $c^3 s$ and $c s^3$ vanish, and since we have

$$\int_0^{2\pi} c^4(t)dt = \int_0^{2\pi} s^4(t) \, dt = 3\pi/4,$$

as well as

$$\int_0^{2\pi} c^2(t)s^2(t) \, dt = \pi/4,$$

it follows that the first integral in (67) has the value

$$\frac{3\pi}{4\omega_0} [D_1^3 \varphi^1 + D_1 D_2^2 \varphi^1 + D_1^2 D_2 \varphi^2 + D_2^3 \varphi^2]. \tag{70}$$

In order to calculate the second integral in (67), we must first determine the function (66). To this end, we note that the vector $\omega_0 D_2^2 \tilde{f}(0,0)[e^{\tau L}\xi]^2$ is given by

$$(\varphi_{11}^1 c^2 - 2\varphi_{12}^1 cs + \varphi_{22}^1 s^2, \, \varphi_{11}^2 c^2 - 2\varphi_{12}^2 cs + \varphi_{22}^2 s^2),$$

where we have used the abbreviation

$$\varphi_{jk}^i := D_j D_k \varphi^i.$$

Because

$$e^{-\tau L} = \begin{bmatrix} c(\tau) & -s(\tau) \\ s(\tau) & c(\tau) \end{bmatrix},$$

we obtain for the components of the vector $\omega_0 e^{-\tau L} D_2^2 \tilde{f}(0,0)[e^{\tau L}\xi]^2$ the formulae

$$\varphi_{11}^1 c^3 - 2\varphi_{12}^1 c^2 s + \varphi_{22}^1 s^2 c - \varphi_{11}^2 sc^2 + 2\varphi_{12}^2 s^2 c - \varphi_{22}^2 s^3 \tag{71}$$

and

$$\varphi_{11}^1 c^2 s - 2\varphi_{12}^1 cs^2 + \varphi_{22}^1 s^3 + \varphi_{11}^2 c^3 - 2\varphi_{12}^2 c^2 s + \varphi_{22}^2 cs^2. \tag{72}$$

Because

$$\int_0^t c^2 s \, d\tau = -(c^3(t) - 1)/3 \quad \text{and} \quad \int_0^t cs^2 \, d\tau = s^3(t)/3, \tag{73}$$

as well as

$$\int_0^t c^3 \, d\tau = c^2 s + (2/3)s^3 \quad \text{and} \quad \int_0^t s^3 \, d\tau = -cs^2 - 2(c^3 - 1)/3, \tag{74}$$

we get by integrating (71) and (72), respectively,

$$\begin{aligned} &\varphi_{11}^1[c^2 s + (2/3)s^3] + (2/3)\varphi_{12}^1(c^3 - 1) + \varphi_{22}^1 s^3/3 \\ &+ \varphi_{11}^2(c^3 - 1)/3 + (2/3)\varphi_{12}^2 s^3 + \varphi_{22}^2[cs^2 + (2/3)(c^3 - 1)], \end{aligned} \tag{75}$$

and

$$\begin{aligned} &- \varphi_{11}^1(c^3 - 1)/3 - (2/3)\varphi_{12}^1 s^3 - \varphi_{22}^1[cs^2 + (2/3)(c^3 - 1)] \\ &+ \varphi_{11}^2[c^2 s + (2/3)s^3] + (2/3)\varphi_{12}^2(c^3 - 1) + \varphi_{22}^2 s^3/3. \end{aligned} \tag{76}$$

If we multiply the vector with components given by (75) and (76) from the left by the matrix (68), we obtain the vector

$$(a, b) := \omega_0 D_3^2 u(t)[\xi]^2 = \omega_0 e^{tL} \int_0^t e^{-\tau L} D_2^2 \tilde{f}(0,0)[e^{\tau L}\xi]^2 \, d\tau,$$

hence

$$\begin{aligned} a =& \varphi_{11}^1[c^3 s + (2/3)cs^3 - s(c^3 - 1)/3] + (2/3)\varphi_{12}^1[c(c^3 - 1) - s^4] \\ &+ \varphi_{22}^1[cs^3/3 - cs^2 - (2/3)s(c^3 - 1)] \\ &+ \varphi_{11}^2[c(c^3 - 1)/3 + c^2 s^2 + (2/3)s^4] \\ &+ (2/3)\varphi_{12}^2[cs^3 + s(c^3 - 1)] \\ &+ \varphi_{22}^2[c^2 s^2 + (2/3)c(c^3 - 1) + s^4/3] \end{aligned}$$

and

$$b = - \varphi_{11}^1[c^2 s^2 + (2/3)s^4 + c(c^3 - 1)/3]$$
$$- (2/3)\varphi_{12}^1[s(c^3 - 1) + cs^3]$$
$$- \varphi_{22}^1[s^4/3 + c^2 s^2 + (2/3)c(c^3 - 1)]$$
$$+ \varphi_{11}^2[-s(c^3 - 1)/3 + c^3 s + (2/3)cs^3]$$
$$+ (2/3)\varphi_{12}^2[-s^4 + c(c^3 - 1)]$$
$$+ \varphi_{22}^2[-cs^3 - (2/3)s(c^3 - 1) + cs^3/3].$$

Using the abbreviations from above, the second integral in (67) becomes

$$\frac{1}{\omega_0^2} \int_0^{2\pi} [\varphi_{11}^1 ac^2 + \varphi_{12}^1(-acs + bc^2) - \varphi_{22}^1 bcs \tag{77}$$
$$- \varphi_{11}^2 acs + \varphi_{12}^2(as^2 - bcs) + \varphi_{22}^2 bs^2] \, d\tau.$$

In order to evaluate this expression, we note that the integral of an odd periodic function over an interval of length equal to the period is zero. Considering this fact and inserting the expressions for a and b in (77), we obtain

$$\frac{1}{\omega_0^2} \int_0^{2\pi} \{\varphi_{11}^1 \, \varphi_{12}^1[c^3(c^3 - 1)/3 - 2c^2 s^4 - 2c^4 s^2 + cs^2(s^3 - 1)/3]$$
$$+ \varphi_{11}^1 \, \varphi_{11}^2[c^3(c^3 - 1)/3 + s^2 c(c^3 - 1)/3]$$
$$+ \varphi_{11}^1 \, \varphi_{22}^2[(2/3)c^3(c^3 - 1) + c^2 s^4/3 - (2/3)s^6 - s^2 c(c^3 - 1)/3]$$
$$+ \varphi_{12}^1 \, \varphi_{22}^1[c^2 s^4 - c^4 s^2 + (4/3)s^2 c(c^3 - 1) - (2/3)c^3(c^3 - 1)]$$
$$+ \varphi_{12}^1 \, \varphi_{12}^2[(2/3)c^3(c^3 - 1) - (2/3)s^6]$$
$$+ \varphi_{22}^1 \, \varphi_{11}^2[s^2 c(c^3 - 1) - c^4 s^2]$$
$$+ \varphi_{22}^1 \, \varphi_{22}^2[-c^2 s^4/3 - s^6/3]$$
$$+ \varphi_{11}^2 \, \varphi_{12}^2[-c^2 s^4/3 + (2/3)s^6 - s^2 c^4]$$
$$+ \varphi_{12}^2 \, \varphi_{22}^2[(5/3)c^2 s^4 + 2s^2 c(c^3 - 1) - s^6/3]\} \, d\tau.$$

It follows from (74) that $\int_0^{2\pi} c^3 = 0$, and, using integration by parts, one easily verifies that

$$\int_0^{2\pi} c^6 = \int_0^{2\pi} s^6 = 5 \int_0^{2\pi} c^4 s^2 = 5 \int_0^{2\pi} s^4 c^2 = 5\frac{\pi}{8}.$$

Thus for the expression above we obtain the value

$$\frac{\pi}{4\omega_0^2}[-\varphi_{11}^1 \varphi_{12}^1 + \varphi_{11}^1 \varphi_{11}^2 - \varphi_{12}^1 \varphi_{22}^2 - \varphi_{22}^1 \varphi_{22}^2 + \varphi_{11}^2 \varphi_{12}^2 + \varphi_{12}^2 \varphi_{22}^2]. \tag{78}$$

Substituting (70) and (78) into (67), we finally get

$$\langle \xi', D_2^3 g(0,0)[\xi]^3 \rangle = -\frac{3\pi}{4\omega_0} \{ D_1^3 \varphi^1 + D_1 D_2^2 \varphi^1 + D_1^2 D_2 \varphi^2 + D_2^3 \varphi^2$$

$$+ \frac{1}{\omega_0}[-\varphi_{11}^1 \varphi_{12}^1 + \varphi_{11}^1 \varphi_{11}^2 - \varphi_{12}^1 \varphi_{22}^1 - \varphi_{22}^1 \varphi_{22}^2 \qquad (79)$$

$$+ \varphi_{11}^2 \varphi_{12}^2 + \varphi_{12}^2 \varphi_{22}^2] \}.$$

After these preparations we are ready to prove a stability criterion for Hopf bifurcation.

(27.14) Theorem. *Let the assumptions of theorem (27.11) be satisfied with $f \in C^3(\Lambda \times X, E)$. Moreover, let $E = N \oplus N_c$ be a direct sum decomposition of E which reduces $D_2 f(\lambda_0, 0)$, where*

$$N = [\ker\{[D_2 f(\lambda_0,0)]_{\mathbb{C}} - i\omega_0\} \oplus \ker\{[D_2 f(\lambda_0,0)]_{\mathbb{C}} + i\omega_0\}] \cap E.$$

Introduce a basis for N such that the matrix representation of $D_2 f(\lambda_0, 0)|N$ has the form

$$\begin{bmatrix} 0 & \omega_0 \\ -\omega_0 & 0 \end{bmatrix}.$$

In addition, set

$$\varphi(x, y, z) := f(\lambda_0, (x, y, z)), \quad ((x, y), z) \in N \times N_c,$$

and let

$$\varphi = (\varphi^1, \varphi^2, \varphi^3),$$

where $(\varphi^1, \varphi^2) \in N$ and $\varphi^3 \in N_c$, and, of course, (x, y) resp. (φ^1, φ^2) denote the coordinates with respect to the above basis for N. Finally, let

$$\varphi_1 := D_1 \varphi(0), \quad \varphi_2 := D_2 \varphi(0), \quad \varphi_{11} := D_1^2 \varphi(0), \quad etc.$$

and

$$\delta := \omega_0[\varphi_{111}^1 + \varphi_{122}^1 + \varphi_{112}^2 + \varphi_{222}^2] - \varphi_{11}^1 \varphi_{12}^1 + \varphi_{11}^1 \varphi_{11}^2$$
$$- \varphi_{12}^1 \varphi_{22}^1 - \varphi_{22}^1 \varphi_{22}^2 + \varphi_{11}^2 \varphi_{12}^2 + \varphi_{12}^2 \varphi_{22}^2,$$

and assume that

$$\varphi_{11}^3 = 0. \qquad (80)$$

If now $\delta < 0$, we have supercritical bifurcation, and the periodic orbits branching off of the trivial solution are asymptotically stable if

$$Re[\sigma(D_2 f(\lambda_0, 0)) \setminus \{\pm i\omega_0\}] < 0.$$

If, on the other hand, $\delta > 0$, we have subcritical bifurcation, and the periodic orbits branching off of the trivial solution are unstable.

Proof. Because

$$\varphi_{11}^3 = P D_2^2 g(0,0)[\xi]^2,$$

it follows from (80) that $\zeta = 0$, hence, together with (78), we have that

$$\delta = -\frac{4\omega_0^2}{3\pi}\langle\xi', D_2^3 g(0,0)[\xi]^3\rangle = -\frac{4\omega_0^2}{3\pi}\gamma. \tag{81}$$

Therefore the assertion follows from theorem (27.11) and lemma (27.13).   □

(27.15) Remarks.   (a) If the simplifying assumption

$$\varphi_{11}^3 = PD_2^2 g(0,0)[\xi]^2 = 0$$

is not satisfied, we must replace the value δ by $\delta + \tilde{\delta}$ in the preceding theorem, where

$$\tilde{\delta} := -\frac{4\omega_0^2}{\pi}\langle\xi', D_2^2 g(0,0)[\zeta,\xi]\rangle \tag{82}$$

and

$$\zeta := -[D_2 g(0,0)|N_c]^{-1} PD_2^2 g(0,0)[\xi]^2$$

(cf. lemma (27.13) and (81)). Because $LP = PL$, we obtain from (61) and (66) that

$$PD_2^2 g(0,0)[\xi]^2 = -\int_0^{2\pi} e^{(2\pi-\tau)L} PD_2^2 \tilde{f}(0,0)[e^{\tau L}\xi]^2 \, d\tau.$$

Because

$$[D_2 g(0,0)|N_c]^{-1} e^{2\pi L} P = [(id - e^{2\pi L})|N_c]^{-1} e^{2\pi L} P$$
$$= [e^{-2\pi L|N_c} - id_{N_c}]^{-1} P,$$

we therefore have

$$\zeta = \omega_0^{-1}[e^{-2\pi L|N_c} - id_{N_c}]^{-1} \int_0^{2\pi} e^{-\tau L} PD_2^2 f(\lambda_0, 0)[e^{\tau L}\xi]^2 \, d\tau$$
$$= \omega_0^{-1}[e^{-2\pi L|N_c} - id_{N_c}]^{-1} \int_0^{2\pi} e^{-\tau L|N_c}[\varphi_{11}^3 c^2(\tau)$$
$$- 2\varphi_{12}^3 s(\tau)c(\tau) + \varphi_{22}^3 s^2(\tau)] \, d\tau,$$

where we again have set $c(\tau) := \cos(\tau)$ and $s(\tau) := \sin(\tau)$.

If $L|N_c$ is known, that is, if the spectrum of L is explicitly known, then ζ can be explicitly calculated from (83) – and therefore $\tilde{\delta}$ from (82). If δ is replaced by $\delta + \tilde{\delta}$ in theorem (27.14), one obtains (in principle) an explicitly verifiable stability criterion without the additional assumption that $\varphi_{11}^3 = 0$.

(b) The calculation of $\tilde{\delta}$, as indicated in (a), becomes fairly easy if L is semisimple, i.e., if L is diagonalizable over $\mathbb{C}$. We then can find a direct sum decomposition of N_c, $N_c = N_1 \oplus \ldots \oplus N_k$, which reduces $L|N_c$, $L|N_c = L_1 \oplus \ldots \oplus L_k$, such that N_j is either one or two-dimensional. In the first case, L_j represents multiplication by a real number (a real eigenvalue of L). In the second case, N_j has a basis such that the matrix representation of L_j with respect to this basis has the form

$$\begin{bmatrix} \alpha & \beta \\ -\beta & \alpha \end{bmatrix}.$$

This is the case if and only if $L_\mathbb{C}$ has an eigenvalue of the form $\alpha + i\beta$ with $\beta \neq 0$.   □

(27.16) Examples.   (a) We consider the special Liénard equation

$$\ddot{u} + (u^2 - \lambda)\dot{u} + u = 0$$

(cf. example (24.25)). If $x := (y, z)$ and

$$f(\lambda, x) := (z, (\lambda - y^2)z - y),$$

this equation is equivalent to

$$\dot{x} = f(\lambda, x) \quad \text{in} \quad \mathbb{R}^2.$$

With the usual identifications, we have $f(\cdot, 0) = 0$ and

$$D_2 f(\lambda, 0) = \begin{bmatrix} 0 & 1 \\ -1 & \lambda \end{bmatrix}.$$

Therefore the eigenvalues of $D_2 f(\lambda, 0)$ are $(\lambda/2) \pm \sqrt{(\lambda/2)^2 - 1}$. Consequently i is a simple eigenvalue of $D_2 f(0, 0)$ and for $\mu(\lambda) := (\lambda/2) + i\sqrt{1 - (\lambda/2)^2}$ we have

$$\frac{d}{d\lambda}[Re\,\mu(0)] = 1/2.$$

Hence the assumptions of theorem (27.11) are satisfied and therefore we have a Hopf bifurcation with period $T_0 = 2\pi$ at the point $\lambda_0 = 0$. Using the notation of theorem (27.14), we have $\delta = \varphi_{112}^2 = -2$. We therefore have supercritical bifurcation and the nontrivial periodic orbits, branching off of 0, are asymptotically stable.

(b)   We consider the system

$$\dot{u} = u + v + w^2$$
$$\dot{v} = (\lambda - 2)u + (\lambda - 1)v - u^3 - u^2 v$$
$$\dot{w} = -w + \lambda u^3 - w^3$$

in $\mathbb{R}^3$, or, with the obvious identifications,

$$\dot{x} = f(\lambda, x).$$

Then $f(\lambda, 0) = 0$ and

$$D_2 f(\lambda, 0) = \begin{bmatrix} 1 & 1 & 0 \\ \lambda - 2 & \lambda - 1 & 0 \\ 0 & 0 & -1 \end{bmatrix}.$$

Therefore the eigenvalues of $D_2 f(\lambda, 0)$ are -1 and $(\lambda/2) \pm i\sqrt{1 - (\lambda/2)^2}$. Consequently i is a simple eigenvalue of $D_2 f(0, 0)$ and for $\mu(\lambda) := (\lambda/2) + i\sqrt{1 - (\lambda/2)^2}$ we have $d(Re\,\mu(0))/d\lambda = 1/2$. Hence the assumptions of theorem (27.11) are satisfied and therefore we have a Hopf bifurcation with period $T_0 = 2\pi$ at the point $\lambda_0 = 0$.
    Because

$$D_2 f(0, 0) = \begin{bmatrix} 1 & 1 & 0 \\ -2 & -1 & 0 \\ 0 & 0 & -1 \end{bmatrix},$$

the direct sum decomposition $\mathbb{R}^3 = \mathbb{R}^2 \oplus \mathbb{R}$ reduces the operator $D_2 f(0,0)$ and, if $L :=$ $D_2 f(0,0)|\mathbb{R}^2$, we have

$$L = \begin{bmatrix} 1 & 1 \\ -2 & -1 \end{bmatrix}.$$

In order to apply theorem (27.14), we must introduce a basis for $\mathbb{R}^2$ so that the corresponding matrix representing L has the form

$$\begin{bmatrix} 0 & 1 \\ -1 & 0 \end{bmatrix}.$$

Based on our general considerations above, we will therefore determine a complex eigenvector $z = \xi + i\eta$ corresponding to the eigenvalue i of $L_{\mathbb{C}}$, i.e., we solve the equation $z^1 + z^2 = iz^1$ (the second equation $-2z^1 - z^2 = iz^2$ is automatically satisfied). A nontrivial solution is given by $z = (1, i - 1)$. Hence $\xi = (1, -1)$ and $\eta = (0, 1)$ form a basis for $\mathbb{R}^2$ and the corresponding matrix representation of L has the desired form. Since $\xi = e_1 - e_2$ and $\eta = e_2$, that is, since $e_1 = \xi + \eta$ and $e_2 = \eta$, we have

$$f(0, (u, v, w)) = f(0, ue_1 + ve_2 + we_3) = f(0, u\xi + (u + v)\eta + we_3).$$

From this it follows that

$$\begin{aligned} \varphi(\alpha, \beta, \gamma) &:= f(0, \alpha\xi + \beta\eta + \gamma e_3) = f(0, (\alpha, \beta - \alpha, \gamma)) \\ &= (\beta + \gamma^2)e_1 + (-\alpha - \beta - \alpha^2\beta)e_2 - (\gamma + \gamma^3)e_3 \\ &= (\beta + \gamma^2)\xi + (-\alpha - \alpha^2\beta + \gamma^2)\eta - (\gamma + \gamma^3)e_3. \end{aligned}$$

Therefore the coordinates of φ with respect to the new basis (ξ, η, e_3) of $\mathbb{R}^3$ are given by

$$\varphi(\alpha, \beta, \gamma) = (\beta + \gamma^2, -\alpha - \beta\alpha^2 + \gamma^2, -\gamma - \gamma^3).$$

Consequently, we have $\varphi_{11}^3 = 0$ and

$$\varphi_{111}^1 = \varphi_{122}^1 = \varphi_{222}^2 = 0, \quad \varphi_{112}^2 = -2,$$

as well as

$$\varphi_{11}^1 = \varphi_{12}^1 = \varphi_{22}^1 = \varphi_{11}^2 = \varphi_{12}^2 = 0.$$

From this we obtain $\delta = \varphi_{112}^2 = -2$. Based on theorem (27.14), we have supercritical bifurcation, and the noncritical periodic orbits branching out of the stationary point are asymptotically stable. □

Problems

1. Show that for the *Van der Pol equation*

$$\ddot{u} + \lambda(u^2 - 1)\dot{u} + u = 0$$

we have a Hopf bifurcation at $\lambda_0 = 0$. What is $\lambda(s)$, $0 < s < \epsilon$, in this case?

2. Derive from example (27.16 a), *without calculating*, that for the special Liénard equation

$$\ddot{u} + (\lambda - u^2)\dot{u} + u = 0$$

we have a Hopf bifurcation with period 2π at $\lambda_0 = 0$, and that the noncritical orbits, branching out of the trivial solution, are unstable.

3. Discuss the problem of Hopf bifurcation and the stability of the bifurcating solutions for the system

$$\dot{u} = \lambda u + v + au^2 + bv^2$$
$$\dot{v} = -u + \lambda v + cu^2 + dv^2,$$

where $a, b, c, d \in \mathbb{R}$.

4. Derive a relation, similar to lemmas (27.9) and (27.13), from which one can obtain information about the behavior of the period $T(s) = T_0 + \tau(s)$ near $s = 0$.

5. Calculate the value of $\tilde{\delta}$ in remark (27.15 a) for the case $\dim(E) = 3$.

Notes

The following comments are concerned with a few, selected theorems which ordinarily are not found in text books on ordinary differential equations, either because they are not part of the standard material, or because they are more recent. In all those cases when no references are given, we are dealing with classical results and they can be found in most books on ordinary differential equations listed in the references.

Section 3: The Legendre transformation plays an important role in convex analysis. There, within the framework of duality theory, it is given the appropriate abstract formulation which is the starting point for numerous applications. For an account of this theory, next to some applications, we refer to the book by Ekeland-Temam [1]. Recently, this abstract Legendre transformation has been applied successfully to the study of periodic solutions of Hamiltonian systems (cf. Clarke [1], Clarke-Ekeland [1]).

Section 4: Functional analytic treatments of partial differential equations, along the lines of ordinary differential equations, are presented, for instance, in the texts of Barbu [1], Gajewski-Gröger-Zacharias [1], Henry [1], Krein [1], Lakshmikantham-Leela [1] and Tanabe [1].

Section 10: The elementary results on (semi)flows found in this text are essentially taken from Bhatia-Hajek [1], Bhatia-Szegö [1] and Sell [1] (see also Saperstone [1]). It is well-known (cf. Carlson [1]) that a semiflow can be reparametrized in such a way that it becomes a global semiflow. Since in general situations (e.g., for partial differential equations) it is more appropriate and more natural to work with the given semiflow, we develop the theory of local semiflows in this book to the point where it is useful for the geometric understanding of the phenomena considered.

Section 13: For the topological classification of linear flows we follow, for the most part, essentially Arnold [2] and Irwin [1].

Section 16: The definition of invariance for semiflows differs from the one given in Bhatia-Hajek [1]. (It corresponds to the "weak invariance" given there.) However, our definition seems to be better suited for our needs and seems to be used more often in the recent literature. The proofs of (16.5) and the approximation lemma (16.7) are taken from Deimling [1].

The subtangent condition plays an important role also within parabolic systems and general evolution equations in infinite dimensional spaces. (See, for instance, Amann [2], Mawhin [1]).

Section 18:   With regard to the general statements about Liapunov functions we essentially follow Walker [1]. Lemma (18.12) goes back to Mazur (cf. Dunford-Schwartz [1, Chapter V.9]). Theorem (18.15) appears to be new. It is a mathematically precise formulation of the intuitive principle which says that a set is asymptotically stable if at every boundary point the vector field properly points inward.

Section 19:   In this section we essentially follow Irwin [1]. For more general results about traveling waves we refer to Fife [1].

Section 20:   Representation formulae, as given in proposition (20.4), are studied at length in Daleckii-Krein [1].

Section 21:   The idea of the proof of the fundamental lemma (21.3) seems to go back to Milnor [1]. The simple proof of Borsuk's theorem is due to Gromes [1]. In the remaining parts of this section we follow Amann [1].

Section 22:   The concept of guiding functions was introduced by Krasnosel'skii [1]. Most results of this section go back to him, while most proofs given are new in some technical details. A consequence of this is that our proofs of lemmas (22.4) and (22.8) can easily be modified so that they also remain true for the Leray-Schauder degree in the infinite dimensional case. This leads to important, general results (cf. Amann [3]).

A different approach to obtain existence theorems for periodic solutions of differential equations is to transform the problem into a fixed point problem in some appropriate function space. This method is presented in detail in volume 2 of Rouche-Mawhin [1] and in Mawhin [1]. Essentially, it is restricted to the nonautonomous case, as is the method set forth here.

By means of abstract variational methods from nonlinear functional analysis it is possible to obtain existence and multiplicity results for periodic solutions of autonomous and nonautonomous Hamiltonian systems. For some of the developments in this area we refer to Amann-Zehnder [1] and Rabinowitz [1]. Their treatment goes beyond the scope of this text, since they make use of more penetrating methods from nonlinear functional analysis.

Section 23:   The proofs of the theorems about the "asymptotic period" and the "asymptotic phase" essentially follow the presentation in Hirsch-Smale [1]. The main idea of the proof of the Poincaré Recurrence Theorem was taken from Siegel-Moser [1].

Section 24:   The proof given for theorem (24.15) comes from Krasnosel'skii-Perow-Powolozki-Sabrejko [1]. Different proofs can be found in Hartman [1] and Klingenberg [1]. The connection between degree and winding number is a classical result of global analysis (and is essentially due to H. Hopf) (cf. Siegberg [1]).

Section 25:   The continuation methods treated here are the simplest local results that follow, more or less, directly from the implicit function theorem. The question of the global behavior of the set of periodic solutions is, of course, of great (current) interest. This question is closely related to the bifurcation problem treated in the next section and it requires penetrating topological and analytical methods, which would go beyond the scope of this book. For several interesting investigations in this direction we refer to the paper by Mallet-Paret-Yorke [1], as well as to the very general Morse-type theory of Conley [1].

Section 26: Bifurcation problems play an important role in many areas of ordinary differential equations (see, e.g., Helleman [1]), partial differential equations (e.g., Bardos-Lasry-Schatzman [1]), nonlinear functional analysis (e.g., Amann-Bazley-Kirchgässner [1]). The Liapunov-Schmidt reduction employed here is one of the standard tools in this area. However, it seems that until now nobody has explicitly observed that the bifurcation equation has the simple form $h_0(\lambda, \xi)\xi = 0$, which is the basis for the almost trivial proof of theorem (26.13) (see, however, Hale [1]). It is clear that theorem (26.11) – and thereby also the subsequent theorems – remain true in the infinite dimensional case if one assumes that $D_2 g(0, 0)$ is a Fredholm operator.

The theorem about the bifurcation from simple eigenvalues was first proved by Crandall-Rabinowitz [3].

Theorem (26.19) and the resulting simple proof of the Hopf bifurcation theorem are new. The standard proof techniques of the Hopf bifurcation theorems make use of either the center manifold theorem (e.g., Marsden-McCracken [1], Carr [1]), functional analytic methods in infinite dimensional Banach spaces (e.g., Crandall-Rabinowitz [1,2]) or power series (e.g., Hopf [2], Ioos-Joseph [1]). It should be clear to specialists how the proof given here can be carried over to the case of parabolic evolution equations. For a similar application of the technique of "blowing up" a singularity we refer to Marsden [1]. I owe remark (26.20 d) to A. Vanderbauwehde.

The perturbation theorem (26.24) is also valid in the infinite dimensional case if λ_0 is an isolated, simple eigenvalue of $A(\lambda_0)$. In this case the Dunford-Taylor functional calculus must be employed (cf. Kato [1]). The simple proof given in this English edition I owe to A. Stahel.

For pointers to the numerous investigations of bifurcation problems in "non-generic" situations we refer to Ioos-Joseph [1], as well as for investigations about the branching off of "invariant tori."

Finally, it should be remarked that in a series of interesting papers (e.g., Alexander-Yorke [1], Chow & Mallet-Paret [1], Chow & Mallet-Paret & Yorke [1], Ize [1]) the "global" bifurcation of periodic orbits is studied. For this deeper topological methods are employed.

The proof given here for the Liapunov Center Theorem follows Schmidt [1]. An important assumption of the theorem is the "nonresonance condition" $\sigma(JH''(z_0)) \cap \omega i \mathbb{Z} = \{\pm i\omega\}$. Weinstein [1,2] has shown that even without this assumption the Hamiltonian vector field JH' has at least n periodic orbits near z_0

if $H''(z_0)$ is positive definite (see also Moser [1]). An interesting global version of this result was proved by Ekeland-Lasry [1]. The proofs of these results require deeper methods of nonlinear functional analysis and algebraic topology.

Section 27:   Theorem (27.2) goes back to Crandall-Rabinowitz [4]. With the obvious modifications, the proof given here is also valid in the infinite dimensional case. The proofs of lemmas (27.9) and (27.10) – and therefore the stability theorem (27.11) – are modifications of the corresponding results in Crandall-Rabinowitz [1]. Similar statements can be found in Ioos-Joseph [1]. The construction of η_c in the proof of lemma (27.10) is due to W.-J. Beyn and M. Stiefenhofer. They pointed out a gap in the original proof in the German edition.

The stability criterion of theorem (27.14) agrees with the formula given in Marsden-McCracken [1, p. 126], which is valid in the two-dimensional case. However, our derivation is considerably simpler and shorter (cf. the footnote on p. 125 in Marsden-McCracken [1]). In the book cited, an algorithm is given to derive a corresponding stability criterion for the case $n \geq 3$. Yet, it seems – at least in the form given – to be restricted to the case $n = 3$ (cf. the formula for Δ on p. 135 of the book cited). Our "algorithm," given in remark (27.15), is valid for any dimension.

Bibliography

Abraham, R., and J.E. Marsden:
[1] Foundations of Mechanics. Benjamin, Reading, Mass., 1978.

Abraham, R., and J. Robbin:
[1] Transversal Mappings and Flows. Benjamin, New York, 1967.

Alexander, J., and J.A. Yorke:
[1] Global bifurcation of periodic orbits. Amer. J. Math., 100 (1978), 263-292.

Amann, H.:
[1] Lectures on Some Fixed Point Theorems. IMPA, Rio de Janeiro, 1974.
[2] Invariant sets and existence theorems for semi-linear parabolic and elliptic systems. J. Math. Anal. Appl. 65 (1978), 432–467.
[3] A note on degree theory for gradient mappings. Proc. Amer. Math. Soc. 85 (1982), 591-595.

Amann, H., N. Bazley, K. Kirchgässner, eds.:
[1] Applications of Nonlinear Analysis in the Physical Sciences. Pitman, Boston, 1981.

Amann, H., and S. Weiss:
[1] On the uniqueness of the topological degree. Math. Z. 130 (1973), 39-54.

Amann, H., and E. Zehnder:
[1] Periodic solutions of asymptotically linear Hamiltonian systems. Manuscripta Math. 32 (1980), 149-189.

Arnold, V.I.:
[1] Mathematical Methods of Classical Mechanics. Springer Verlag, New York, 1978.
[2] Ordinary Differential Equations. MIT-Press, Cambridge, Mass., 1973.

Arnold, V.I., and A. Avez:
[1] Problèmes ergodiques de la méchanique classique. Gauthier-Villars, Paris, 1967. (English edition: Ergodic Problems of Classical Mechanics. W. A. Benjamin, New York, 1968.)

Barbu, V.:
[1] Nonlinear Semigroups and Differential Equations in Banach Spaces. Nordhooff, Leyden, 1976.

Bardos, C., J.M. Lasry, M. Schatzman, eds.:
[1] Bifurcation and Nonlinear Eigenvalue Problems. Lecture Notes in Math. # 782, Springer Verlag, Berlin, 1980.

Behnke, H., and F. Sommer:
[1] Theorie der analytischen Funktionen einer komplexen Veränderlichen. Springer Verlag, Berlin, 1965.

Berger, M.S.:
[1] Nonlinearity and Functional Analysis. Academic Press, New York, 1977.

Bhatia, N.P., and O. Hajek:
[1] Local Semi-Dynamical Systems. Lecture Notes in Math. # 90, Springer Verlag, Berlin 1969.

Bhatia, N.P., and G.P. Szegö:
[1] Stability Theory of Dynamical Systems. Springer Verlag, Berlin, 1970.
[2] Dynamical Systems: Stability Theory and Applications. Lecture Notes in Math. # 35, Springer Verlag, Berlin, 1967.

Braun, M.:
[1] Differential Equations and their Applications, Third Edition. Springer Verlag, New York, 1984. (German edition: Differentialgleichungen und ihre Anwendungen. Springer Verlag, Berlin, 1979.)

Bröcker, T.:
[1] Analysis in mehreren Variablen. Teubner, Stuttgart, 1980.

Carlson, D.H.:
[1] A generalization of Vinograd's theorem for dynamical systems. J. Diff. Eqns. 11 (1972), 193-201.

Carr, J.:
[1] Applications of Center Manifold Theory. Springer Verlag, New York, 1981.

Chow, S.-N., and J. Mallet-Paret:
[1] Fuller's Index and Global Hopf Bifurcation. J. Diff. Eqns. 29 (1978), 66-85.

Chow, S.-N., J. Mallet-Paret, J.A. Yorke:
[1] Global Hopf Bifurcation from a Multiple Eigenvalue. Nonlinear Analysis, T.M. & A. 2 (1978), 753-763.

Clarke, F.:
[1] Periodic Solutions to Hamiltonian Inclusions. J. Diff. Eqns. 40 (1981), 1-6.

Clarke, F., and I. Ekeland:
[1] Hamiltonian trajectories having prescribed minimal period. Comm. Pure Appl. Math. 33 (1980), 103-116.

Coddington, E.A., and N.A. Levinson:
[1] Theory of Ordinary Differential Equations. McGraw-Hill, New York, 1955.

Conley, C.:
[1] Isolated Invariant Sets and the Morse Index. CBMS Regional Conference Series in Math. # 38, Amer. Math. Soc., Providence, 1976.

Conway, J.B.:
[1] Functions of One Complex Variable. Springer Verlag, New York, 1973.

Crandall, M.G., and P.H. Rabinowitz:
[1] The Hopf Bifurcation Theorem. MRC Technical Summary Report # 1604, Univ. Wisconsin, Madison, 1976.

[2] Mathematical Theory of Bifurcation. In C. Bardos and D. Bessis, eds.: Bifurcation Phenomena in Math. Phys. and Related Topics. D. Reidel Pub. Comp., Dordrecht, 1980.

[3] Bifurcation from simple eigenvalues. J. Funct. Anal. 8 (1971), 321-340.

[4] Bifurcation, Perturbation of Simple Eigenvalues, and Linearized Stability. Arch. Rat. Mech. Anal., 52 (1973), 161-180.

Cronin, J.:
[1] Differential Equations. M. Dekker Inc., New York, 1980.

Daleckii, Ju.L., and M.G. Krein:
[1] Stability of Solutions of Differential Equations in Banach Spaces. AMS Transl. Math. Monographs 43, Providence, R.I., 1974.

D'Heedene, R.N.:
[1] A third order autonomous differential equation with almost periodic solutions. J. Math. Anal. Appl. 3 (1961), 344-350.

Deimling, K.:
[1] Ordinary Differential Equations in Banach Spaces. Lecture Notes in Math. # 596, Springer Verlag, Berlin 1977.

[2] Nichtlineare Gleichungen und Abbildungsgrade. Springer Verlag, Berlin, 1974.

Dieudonné, J.:
[1] Foundations of Modern Analysis. Academic Press, New York, 1969.

Dunford, N., and J.T. Schwartz:
[1] Linear Operators. Part I. Interscience, New York, 1957.

Dugundji, J.:
[1] Topology. Allyn Bacon, Boston, 1966.

Eisenack, G., and C. Fenske:
[1] Fixpunkttheorie. B.I. Mannheim, 1978.

Ekeland, I., and J.M. Lasry:
[1] On the number of periodic trajectories for a Hamiltonian flow on a convex energy surface. Annals of Math. 112 (1980), 283-319.

Ekeland, I., and R. Temam:
[1] Convex Analysis and Variational Problems. North Holland, Amsterdam, 1976.

Fife, P.:
[1] Mathematical Aspects of Reacting and Diffusing Systems. Lecture Notes in Biomathematics # 28, Springer Verlag, Berlin, 1979.

Fleming, W.H.:
[1] Functions of Several Variables. Springer Verlag, New York, 1977.

Gajewski, H., K. Gröger and K. Zacharias:
[1] Nichtlineare Operatorgleichungen und Operatordifferentialgleichungen. Akademie-Verlag, Berlin, 1974.

Greub, W.H., S. Halperin, and R.J. Vanstone:
[1] Curvature, Connections and Cohomology. Academic Press, New York, 1972.

Gromes, W.:
[1] Ein einfacher Beweis des Satzes von Borsuk. Math Z. 178 (1981), 399-400.

Guillemin, V., and A. Pollack:
[1] Differential Topology. Prentice-Hall, Englewood Cliffs, N.J., 1974.

Hahn, W.:
[1] Theorie und Anwendung der direkten Methode von Ljapunov. Springer Verlag, Berlin, 1967.

Hale, J.:
[1] Ordinary Differential Equations. Wiley, New York, 1969.
[2] Theory of Functional Differential Equations. Springer Verlag, New York, 1977.
[3] Generic bifurcation with applications. In R.J. Knops (ed.): Nonlinear analysis and mechanics: Heriot-Watt Symp., vol. I, 59-157, Pitman, London, 1977.

Hartman, Ph.:
[1] Ordinary Differential Equations. J. Wiley & Sons, New York, 1964.

Hastings, S.P., and J.D. Murray:
[1] The existence of oscillatory solutions in the Field-Noyes model for the Belousov-Zhabotinskii reaction. SIAM J. Appl. Math. 28 (1975), 678-688.

Hastings, S.P., J.J. Tyson, and D. Webster:
[1] Existence of periodic solutions for negative feedback cellular control systems. J. Diff. Eqns. 25 (1977), 39-64.

Helleman, H.G. (ed.):
[1] Nonlinear Dynamics. Annals New York Acad. Sci., 357 (1980), New York Acad. Sci., New York, N.Y.

Henry, D.:
[1] Geometric Theory of Semilinear Parabolic Equations. Lecture Notes in Math. # 840, Springer Verlag, Berlin, 1981.

Hewitt, E., and K. Stromberg:
[1] Real and Abstract Analysis. Springer Verlag, Berlin, 1965.

Hirsch, M.:
[1] Differential Topology. Springer Verlag, New York, 1976.

Hirsch, M.W., and S. Smale:
[1] Differential Equations, Dynamical Systems, and Linear Algebra. Academic Press, New York, 1964.

Holmann, H., and H. Rummler:
[1] Alternierende Differentialformen. B.I. Mannheim, 1972.

Hopf, E.:
[1] Ergodentheorie. Ergebnisse der Math., Springer Verlag, Berlin, 1937.
[2] Abzweigung einer periodischen Lösung von einer stationären Lösung eines Differentialsystems. Ber. Math.-Phys. Kl. Sächs. Akad. Wiss. Leipzig, 94 (1942), 1-22. (An engl. transl. can be found in Marsden & McCracken [1].)

Ioos, G., and D.D. Joseph:
[1] Elementary Stability and Bifurcation Theory. Springer Verlag, New York, 1980.

Irwin, M.C.:
[1] Smooth Dynamical Systems. Academic Press, London, 1980.

Ize, J.:
[1] Bifurcation theory for Fredholm operators. Memoirs Amer. Math. Soc., 7, No. 174, Providence, 1976.

Jacobs, K.:
[1] Neuere Methoden und Ergebnisse der Ergodentheorie. Ergebnisse der Math., Springer Verlag, Berlin, 1960.

Kamke, E.:
[1] Differentialgleichungen: Lösungsmethoden und Lösungen I. Teubner, Stuttgart, 1977.

Kato, T.:
[1] Perturbation Theory for Linear Operators. Springer Verlag, New York, 1966.

Kirchgraber, U., and E. Stiefel:
[1] Methoden der analytischen Störungsrechnung und ihre Anwendungen. Teubner, Stuttgart, 1978.

Klingenberg, W.:
[1] Eine Vorlesung über Differentialgeometrie. Springer Verlag, Berlin, 1973. (English edition: A Course in Differential Geometry. Springer Verlag, New York, 1978.)

Kneser, H.:
[1] Funktionentheorie. Vandenhoeck & Rupprecht, Göttingen, 1958.

Knobloch, H.W., and F. Kappel:
[1] Gewöhnliche Differentialgleichungen. Teubner, Stuttgart, 1974.

Krasnosel'skii, M.A.:
[1] Translation Along Trajectories of Differential Equations. Transl. Math. Monographs AMS, Providence, R.I., 1968.

Krasnosel'skii, M.A., A.I. Perow, A.I. Powolozki, P.P. Sabrejko:
[1] Vektorfelder in der Ebene. Akademie-Verlag, Berlin, 1966.

Krein, S.G.:
[1] Linear Differential Equations in Banach Space. Amer. Math. Soc., Providence, 1972.

Lakshmikantam, V., and S. Leela:
[1] Nonlinear Differential Equations in Abstract Spaces. Pergamon Press, Oxford, 1981.

Lang, S.:
[1] Real Analysis. Addison-Wesley, Reading, Mass., 1969.

LaSalle, J.P.:
[1] The Stability of Dynamical Systems. CBMS Regional Series Appl. Math., Soc. Ind. Appl. Math., Philadelphia, 1976.

Li, B.:
[1] Periodic Orbits of Autonomous Ordinary Differential Equations: Theory and Applications. Nonlinear Anal., T.M. & A. 5 (1981), 931-958.

Lloyd, N.G.:
[1] Degree Theory. Cambridge Univ. Press, Cambridge, 1978.

Mallet-Paret, J., and J.A. Yorke:
[1] Snakes: oriented families of periodic orbits, their sources, sinks, and continuation. J. Diff. Eqns. 43 (1982), 419-450.

Marsden, J.E.:
[1] Qualitative methods in bifurcation theory. Bull. Amer. Math. Soc., 84 (1978), 1125-1148.

Marsden, J.E., and M. McCracken:
[1] The Hopf Bifurcation and Its Applications. Springer Verlag, New York, 1976.

Martin, Jr., R.H.:
[1] Nonlinear Operators & Differential Equations in Banach Spaces. J. Wiley & Sons, New York, 1976.

Mawhin, J.:
[1] Topological Degree Methods in Nonlinear Boundary Value Problems. CBMS Regional Conference Series Math. # 40. Amer. Math. Soc., Providence, 1977.

Milnor, J.W.:
[1] Topology from the Differentiable Viewpoint. Univ. Press of Virginia, Charlottesville, 1965.

Moser, J.:
[1] Stable and Random Motions in Dynamical Systems. Princeton Univ. Press, Princeton, N.J., 1973.
[2] Periodic orbits near an equilibrium and a theorem of Alan Weinstein. Comm. Pure Appl. Math., 29 (1976), 727-747.

Murray, J.D.:
[1] Lectures on Nonlinear-Differential-Equation Models in Biology. Clarendon Press, Oxford, 1977.

Nemytskii, V.V., and V.V. Stepanov:
[1] Qualitative Theory of Differential Equations. Princeton Univ. Press, Princeton, N.J., 1960.

Palmer, K.J.:
[1] A generalization of Hartman's linearization theorem. J. Math. Anal. Appl. 41 (1973), 753-758.
[2] Linearization near an integral manifold. J. Math. Anal. Appl. 51 (1975), 243-255.
[3] Qualitative behavior of a system of ODE near an equilibrium point – A generalization of the Hartman-Grobman theorem. Sonderforschungsbereich 72, Univ. Bonn, Reprint no. 372, Bonn, 1980.

Rabinowitz, P.H.:
[1] Periodic Solutions of Hamiltonian Systems. Comm. Pure Appl. Math. 31 (1978), 157-184.

Reiffen, H.J., and H.W. Trapp:
[1] Einführung in die Analysis I-III. B.I. Mannheim, 1973.

Reissig, R., G. Sansone, R. Conti:
[1] Non-linear Differential Equations of Higher Order. Noordhoff, Leyden, 1974.

Rouche, N., P. Habets, M. Laloy:
[1] Stability Theory by Liapunov's Direct Method. Springer Verlag, New York, 1977.

Rouche, N., and J. Mawhin:
[1] Equations Différentielles Ordinaires. Masson, Paris, 1973. (English edition: Ordinary Differential Equations. Pitman, Boston, 1980.)

Rudin, W.:
[1] Real and Complex Analysis. McGraw-Hill, New York, 1970.

Saperstone, S.H.:
[1] Semidynamical Systems in Infinite Dimensional Spaces. Springer Verlag, New York, 1981.

Sansone, G., and R. Conti:
[1] Non-linear Differential Equations. Pergamon Press, Oxford, 1964.

Schäfer, H.H.:
[1] Topological Vector Spaces. Springer Verlag, New York, 1971.

Schmidt, D.S.:
[1] Hopf's bifurcation theorem and the center theorem of Liapunov. In J.E. Marsden-M. McCracken: "The Hopf Bifurcation and its Applications," Springer Verlag, New York, 1976, pp. 95-103.

Schubert, H.:
[1] Topologie. Teubner, Stuttgart, 1964. (English edition: Topology. Allyn and Bacon, Boston, 1968.

Sell, G.R.:
[1] Topological Dynamics and Ordinary Differential Equations. Math. Studies # 33, Van Nostrand, New York, 1971.

Siegberg, H.W.:
[1] Some historical remarks concerning degree theory. Amer. Math. Monthly 88 (1981), 125-139.

Siegel, C.L., and J.K. Moser:
[1] Lectures on Celestial Mechanics. Springer Verlag, Berlin, 1971.

Spivak, M.:
[1] A Comprehensive Introduction to Differential Geometry, vol. I. Publish or Perish Inc., Boston, 1970.
[2] Calculus on Manifolds. W. A. Benjamin, Inc., New York, 1965.

Tanabe, H.:
[1] Equations of Evolution. Pitman, London, 1979.

Triebel, H.:
[1] Höhere Analysis. VEB Deutscher Verlag der Wissenschaften, Berlin, 1972.

Tyson, J.J.:
[1] The Belousov-Zhabotinskii Reaction. Lecture Notes in Biomath. # 10, Springer Verlag, Berlin, 1976.

Walker, J.A.:
[1] Dynamical Systems and Evolution Equations, Theory and Applications. Plenum Press, New York, 1980.

Walter, R.:
[1] Differentialgeometrie. B.I. Mannheim, 1978.
[2] Einführung in die lineare Algebra. Vieweg, Braunschweig, 1982.

Walter, W.:
[1] Gewöhnliche Differentialgleichungen. Heidelberger Taschenbücher, Springer Verlag, Berlin, 1972.

Weinstein, A.:
[1] Normal modes for nonlinear Hamiltonian systems. Inv. Math. 20 (1973), 47-57.
[2] Bifurcations and Hamilton's principle. Math. Zeitschr. 159 (1978), 235-248.

Yosida, K.:
[1] Functional Analysis. Springer Verlag, Berlin, 1965.

Zeidler, E.:
[1] Vorlesungen über nichtlineare Funktionsanalysis I – Fixpunktsätze. Teubner, Leipzig, 1976. (English edition: Nonlinear Functional Analysis and its Applications I – Fixed-Point Theorems. Springer Verlag, New York, 1986.)

Index